YO-AFY-113

MECHANISTIC PRINCIPLES OF ENZYME ACTIVITY

Edited by

Joel F. Liebman
Arthur Greenberg

VCH

Joel F. Liebman
Department of Chemistry
University of Maryland Baltimore County
Baltimore, Maryland 21228

Arthur Greenberg
Division of Chemistry
New Jersey Institute of Technology
Newark, NJ 07102

Library of Congress Cataloging-in-Publication Data

Mechanistic principles of enzyme activity / edited by Joel F. Liebman,
  Arthur Greenberg.

    p.   cm. — (Molecular structure and energetics ; v. 9)
  Includes bibliographies and index.
  ISBN 0-89573-706-X
  1. Enzymes. 2. Enzyme kinetics. 3. Molecular association.
I. Liebman, Joel F. II. Greenberg, Arthur. III. Series.
QD461.M629 1986 vol. 9
[QP601]
541.2′2 s—dc19
[574.19′25]

QP
601
.M4
1988

88-19229
CIP

Printed in the United States of America.

ISBN-0-89573-706-X   VCH Publishers
ISBN-3-527-269-07-X   VCH Verlagsgesellschaft

Published jointly by:
VCH Publishers, Inc.
220 East 23rd Street
Suite 909
New York, NY 10010

VCH Verlagsgesellschaft mbH
P.O. Box 10 11 16
D-6940 Weinheim
Federal Republic of Germany

MOLECULAR STRUCTURE AND ENERGETICS

Series Editors

Joel F. Liebman
University of Maryland Baltimore County

Arthur Greenberg
New Jersey Institute of Technology

Advisory Board

Other Volumes in the Series

Chemical Bonding Models
Physical Measurements
Studies of Organic Molecules
Biophysical Aspects
Advances in Boron and the Boranes
Modern Models of Bonding and Delocalization
Structure and Reactivity
Fluorine-Containing Molecules
Environmental Influences and Recognition in Enzyme Chemistry
From Atoms to Polymers: Isoelectronic Analogies

Contributors

Steven A. Benner, Swiss Federal Institute of Technology, Universitat Strasse 16, CH-8092 Zurich, Switzerland

Thomas C. Bruice, Department of Chemistry, University of California, Santa Barbara, California 93106

David W. Christianson, University of Pennsylvania, Philadelphia, Pennsylvania 19104

Donald J. Creighton, University of Maryland Baltimore County, Baltimore, Maryland 21228

Anthony W. Czarnik, Department of Chemistry, The Ohio State University, Columbus, Ohio 43210

Dabney White Dixon, Department of Chemistry and Laboratory for Microbial and Biological Sciences, Georgia State University, Atlanta, Georgia 30303

David K. Lavallee, Department of Chemistry, Hunter College of the City University of New York, New York, New York 10021

William N. Lipscomb, Department of Chemistry, Harvard University, Cambridge, Massachusetts 02138

Tayebeh Pourmotabbed, University of Maryland Baltimore County, Baltimore, Maryland 21228

Richard L. Schowen, Department of Chemistry, University of Kansas, Lawrence, Kansas 66045

Series Foreword

Molecular structure and energetics are two of the most ubiquitous, fundamental and, therefore, important concepts in chemistry. The concept of molecular structure arises as soon as even two atoms are said to be bound together since one naturally thinks of the binding in terms of bond length and interatomic separation. The addition of a third atom introduces the concept of bond angles. These concepts of bond length and bond angle remain useful in describing molecular phenomena in more complex species, whether it be the degree of pyramidality of a nitrogen in a hydrazine, the twisting of an olefin, the planarity of a benzene ring, or the orientation of a bioactive substance when binding to an enzyme. The concept of energetics arises as soon as one considers nuclei and electrons and their assemblages, atoms and molecules. Indeed, knowledge of some of the simplest processes, e.g., the loss of an electron or the gain of a proton, has proven useful for the understanding of atomic and molecular hydrogen, of amino acids in solution, and of the activation of aromatic hydrocarbons on airborne particulates.

Molecular structure and energetics have been studied by a variety of methods ranging from rigorous theory to precise experiment, from intuitive models to casual observation. Some theorists and experimentalists will talk about bond distances measured to an accuracy of 0.001 Å, bond angles to 0.1°, and energies to 0.1 kcal/mol and will emphasize the necessity of such precision for their understanding. Yet other theorists and experimentalists will make equally active and valid use of such seemingly ill-defined sources of information as relative yields of products, vapor pressures, and toxicity. The various chapters in this book series use as their theme "Molecular Structure and Energetics," and it has been the individual authors' choice as to the mix of theory and of experiment, of rigor and of intuition that they have wished to combine.

As editors, we have asked the authors to explain not only "what" they know but "how" they know it and explicitly encouraged a thorough blending of data and of concepts in each chapter. Many of the authors have told us that writing their chapters have provided them with a useful and enjoyable (re)education. The chapters have had much the same effect on us and we trust readers will share our enthusiasm. Each chapter stands autonomously as a combined review and tutorial of a major research area. Yet clearly there are interrelations between them and to emphasize this coherence we have tried to have a single theme in each volume. Indeed the first four volumes of this series were written in parallel, and so for these there is an even higher degree of unity. It is this underlying unity of molecular structure and energetics with all of chemistry that marks the series and our efforts.

Another underlying unity we wish to emphasize is that of the emotions and of the intellect. We thus enthusiastically thank Alan Marchand for the opportunity to write a volume for his book series, which grew first to multiple volumes, and then became the current, autonomous series for which this essay is the foreword. We also wish to emphasize the support, the counsel, the tolerance and the encouragement we have long received from our respective parents, Murray and Lucille, Murray and Bella; spouses, Deborah and Susan; parents-in-law, Jo and Van, Wilbert and Rena; and children, David and Rachel. Indeed, it is this latter unity, that of the intellect and of emotions, that provides the motivation for the dedication for this series:

"To Life, to Love, and to Learning."

Joel F. Liebman
Baltimore, Maryland

Arthur Greenberg
Newark, New Jersey

Introduction

This volume is the first of two books treating the activities and properties of enzymes from the perspective of structure and energetics. It establishes mechanistic principles and describes many of the means for deducing them. The second volume, *Environmental Influences and Recognition in Enzyme Chemistry,* considers the interactions between parts of an enzyme and the enzyme and other proteins and substrates, as well as with the solvent.

The first chapter, by D. Christianson and W. Lipscomb, describes their X-ray crystallographic studies of zinc protease inhibitor complexes. The evidence suggests the possibility of a "stop-action sequence" picture of the enzymatic mechanism, and additionally supports the view that zinc acts mainly as a water activator rather than directly acting as an electrophile at the carbonyl oxygen. The following chapter by S. Benner summarizes the principles of stereoelectronic effects which have been observed in many guises over the most recent decades, but have only recently been interrelated. After estimating the magnitudes of these effects, the author relates their importance to enzymatic specificity and activity. Chapter 3 by A. Czarnik, describes the well-known accelerating effects of intramolecularity which compose part of the catalytic effect induced by the formation of enzyme-substrate complexes. The most subtle issue is the disentangling of proximity (entropy effects, effective concentration, orbital alignment) and strain in the acceleration of rates.

The fourth chapter, by R. Schowen, examines a fundamental topic, protolytic general catalysis, with particular emphasis on charge-relay catalysis in serine proteases. The approach is very wide-ranging, including descriptions of reaction surfaces and spectroscopic and crystallographic results, as well as biomimetic and physical organic mechanistic probes. D. W. Dixon, in the next chapter, explores some fundamental aspects of electron transfer. These include spatial (structural and conformational) and energetic requirements. She discusses the methods of electron transfer measurement. These considerations are applied toward understanding electron transfer in cytochromes c and b_5. The sixth chapter, by T. C. Bruice, concerns itself with mechanistic studies of the organometallic chemistry of iron protoporphyrin-IX mixed function oxidases. The approach includes the preparation and isolation of model compounds for postulated oxygen-transferring intermediates. The following chapter, by D. Lavallee, considers the very formation of these and other metalloporphyrins. Key questions include the stage at which the metal atom is inserted during biosynthesis and the fundamental mechanistic factors responsible for the sequence. A second chapter from T. C. Bruice, again applies mechanistic organic chemistry to the study of flavin and porphyrin mixed function oxidases enzymes. The two chapters by Bruice emphasize the behavior of synthetic intermediates. The final chapter in this enzymes volume, by D. Creighton and T. Pourmotabbed, concerns glutathione-dependent aldehyde oxidation reactions. The relevant enzymes are important as eliminators of undesirable glycolysis by-products and even xenobiotics. They have the interesting ability to handle multiple equilibrium forms of aldehydes including diastereotopic thiohemiacetals and even (presumably) *cis* and *trans*-enediol intermediates.

Contents

Series Foreword   v

Introduction   vii

1. Structural Aspects of Zinc Protease Mechanisms   1
David W. Christianson and William N. Lipscomb

    1. Introduction   1
    2. Catalytic Role of Zinc   2
    3. Difference Fourier Method   5
    4. Structure-Function Studies of Carboxypeptidase A   11
    5. Further Elucidation of Zinc Protease Mechanism   21
    Acknowledgments   23
    References   23

2. Stereoelectronic Analysis of Enzymatic Reactions   27
Steven A. Benner

    1. Introduction   27
    2. Stereoelectronic Principles   28
    3. Large Stereoelectronic Effects in Enzymatic Reactions   41
    4. Medium-Sized Stereoelectronic Effects   47
    5. Small Stereoelectronic Effects   56
    6. Conclusions   70
    Acknowledgments   71
    References   71

3. Intramolecularity: Proximity and Strain   75
Anthony W. Czarnik

    1. Introduction   75
    2. Definitions   77
    3. Rate Accelerations Due to Covalently Enforced Proximity   79
    4. Rate Accelerations Due to Noncovalently Enforced Proximity   95
    5. Rate Accelerations Due to Covalently Enforced Strain   102
    6. Rate Accelerations Due to Noncovalently Enforced Strain   104
    7. Conclusion   112
    Acknowledgments   112
    References   113

4. Structural and Energetic Aspects of Protolytic Catalysis by Enzymes: Charge-Relay Catalysis in the Function of Serine Proteases   **119**
Richard L. Schowen

    1. Introduction   119
    2. Crystallographic and NMR-Spectroscopic Studies   136
    3. Chemical Models for Charge-Relay Catalysis   141
    4. Theoretical Models for Charge-Relay Catalysis   146
    5. Proton-Inventory Studies   156
    6. Summary and Prospects   162
    7. Appendix: Some Points of Detail   163
  References   164

5. Electron Transfer in Cytochromes C and B_5   **169**
Dabney White Dixon

    1. Introduction   169
    2. Structure, Theory, and Experimental Methods   171
    3. Factors Controlling Electron Transfer   184
    4. Intramolecular Electron Transfer   196
    5. Intermolecular Electron Transfer   204
    6. Miscellaneous Topics   215
  Acknowledgments   217
  References   217

6. Chemical Studies Related to Iron Protoporphyrin-IX Mixed Function Oxidases   **227**
Thomas C. Bruice

    1. The Enzymes   228
    2. Chemical Preparation of Higher Valent Iron-Oxo Porphyrin Species   230
    3. Electrochemical Generation of Higher Valent Iron-Oxo Porphyrin Species   233
    4. Higher-Valent Manganese-Oxo Species   234
    5. Mechanisms of Oxygen Transfer from Percarboxylic Acids and Alkyl Hydroperoxides to Metal(III) Porphyrins   236
    6. A Model for the Catalase Reaction: The Mechanism for the Formation of Oxygen on Reaction of an Iron(II) Porphyrin with Hydrogen Peroxide   245
    7. The Mechanism of Reaction of Hydrogen Peroxide with Manganese(III) Porphyrin   248

8. The Rebound Mechanism for Oxygen Insertion into Carbon-
   Hydrogen Bonds   249
9. Mechanisms of Dealkylation Reactions   253
10. Reaction of *N,N*-Dimethylaniline *N*-Oxides with Metal(III)
    Porphyrins   255
11. The Mechanism of the Epoxidation of Alkenes   262
Acknowledgments   273
References   273

7. **Porphyrin Metalation Reactions in Biochemistry**   **279**
   David K. Lavallee

   1. Introduction: Biological Porphyrin Metalation   279
   2. Metalation of Porphyrins *in vitro*   280
   3. Formation of Metal Complexes of Macrocycles *in vivo*   286
   4. Ferrochelatase Inhibition and Chlorophyll Biosynthesis   307
   5. Pathological Conditions Associated with Ferrochelatase   309
   6. Conclusion   311
   Acknowledgments   311
   References   311

8. **Chemical Studies and the Mechanism of Flavin Mixed Function**
   Oxidase Enzymes   **315**
   Thomas C. Bruice

   1. Chemical Studies Related to the Flavoenzyme Mixed Function
      Oxidases: General Aspects   316
   2. Historical Aspects and Postulations of Mechanisms   319
   3. The Practical Synthesis of Models of the 4a,5-Hydroperoxy-
      flavins and the Postulated 6-Amino-5-oxo-3*H*,5*H*-uracil Inter-
      mediate   321
   4. The Reaction of 1,5-Dihydroflavins with Molecular Oxygen   326
   5. Monooxygen Transfer from 4a,5-Dihydro-4a-Hydroperoxy-5-
      Alkyllumiflavins   331
   6. Comparison of the Reactions of Flavoenzyme Mixed-Function
      Oxidase Enzymes and 5-Ethyl-4a,5-Dihydro-4a-Hydroperoxy-
      lumiflavin Which Result in Heteroatom Oxygenation   338
   7. Reactions Involving Nucleophilic Additions to Carbonyl
      Functions   339
   8. Hydroxylation of Electron-Rich Aromatic Rings   342
   Acknowledgments   350
   References   350

9. **Glutathione-Dependent Aldehyde Oxidation Reactions**   **353**
   Donald J. Creighton and Tayebeh Pourmotabbed

   1. Introduction   353
   2. Distribution of Glutathione, Methylglyoxal and Formaldehyde
      in Cells   356
   3. The Glyoxalase Enzyme System   358
   4. Kinetic Model for the Conversion of Methylglyoxal to D-lactate in Erythrocytes   369
   5. Optimization of Efficiency in the Glyoxalase Pathway   374
   6. The Formaldehyde Dehydrogenase/S-Formylglutathione Hydrolase Enzyme System   378
   7. Future Directions   382
  Acknowledgments   382
  References   382

Addendum   **387**

Index   **389**

CHAPTER 1

Structural Aspects of Zinc Protease Mechanisms

David W. Christianson

University of Pennsylvania, Philadelphia, Pennsylvania

William N. Lipscomb

Harvard University, Cambridge, Massachusetts

CONTENTS

1. Introduction 1
2. Catalytic Role of Zinc 2
3. Difference Fourier Method 5
4. Structure–Function Studies of Carboxypeptidase A 11
5. Further Elucidation of Zinc Protease Mechanisms 21
Acknowledgments 23
References 23

1. INTRODUCTION

Enzymes which require a divalent zinc ion for the catalytic hydrolysis of peptide substrates comprise a class of enzymes known informally as "zinc proteases." Although only a handful of these enzymes have yielded themselves to successful X-ray crystallographic investigation, it is becoming clear that their active sites have evolutionarily converged upon a common three-dimensional arrangement of catalytic residues. This is not an unexpected consequence: assuming that an enzyme preferentially binds the transition

1

state of a chemical reaction[1] rather than simply reactant or product molecules, enzymes which catalyze similar reactions via identical mechanisms will prefer the binding of an identical transition-state structure. The binding of this common transition-state structure is accommodated by similar geometric locations of enzyme residues complementary to the transition state in each related enzyme. For the zinc proteases, this transition-state structure must involve the catalytically required zinc ion.

There is currently a wealth of information regarding various zinc proteases. Unfortunately, X-ray crystallographic investigations of only four of these enzymes are at various stages of completion, thus limiting structural comparisons. Nevertheless, X-ray structures of the zinc exoprotease carboxypeptidase A (CPA)[2] and its complexes with various substrate and transition-state analogues[3-9] are available from recent high-resolution crystallographic studies. This enzyme, isolated[10] in pure form in 1937 and subsequently shown to require zinc for activity,[11] is regarded as the prototypical zinc protease. The X-ray structure of CPA was reported[12] in 1968 and was subsequently refined at 1.54 Å resolution.[2]

The structure of the zinc endoprotease thermolysin (TLN) has been determined and refined at 1.6 Å resolution,[13] and its complexes with many interesting inhibitors have been reported.[14-18] One of these, the TLN–phosphoramidon complex,[15] resembles a structure isosteric with an actual proteolytic transition state. Additionally, two refined TLN-inhibitor complexes were recently reported involving identically bound phosphonamidate and phosphonate transition-state analogs.[19] Nevertheless, all of the TLN-inhibitor structures display interesting interactions from which mechanistic inferences can be based. In addition to CPA and TLN, the structures of two other zinc proteases have been reported, although not to the fine detail exemplified by CPA and TLN. Carboxypeptidase B,[20] 49% homologous with CPA,[21] is nearly identical in structure to CPA. The structure of an interesting dipeptidyl hydrolase, D-alanyl-D-alanyl carboxypeptidase, sometimes referred to as "peptidase G," has been reported.[22] Although the high-resolution structures of these zinc proteases have not been reported, nor have their interactions with strongly bound inhibitors been investigated, general structural features of their active sites, which they share with CPA and TLN, can be used to infer similarities in hydrolytic mechanisms.

2. CATALYTIC ROLE OF ZINC

Assuming that these four zinc proteases have converged in structure to complement the binding of one chemically identical transition state along the reaction coordinate of peptide hydrolysis, and that the transition-state structure requires the participation of the active-site zinc ion, we now elaborate further upon a common role of the zinc ion. It is certainly involved as a classical electrophilic catalyst; ie, it provides electrostatic stabilization for negatively charged intermediates formed during the course of a hydrolytic reaction. By extension, the zinc ion will likewise stabilize the fractional negative charges formed in the transition states flanking such intermediates.[23] However, an electrophilic role for zinc which involves the initial coordination of the scissile carbonyl of the substrate, an interaction which would polarize the $C=O$ double bond and make it more susceptible to nucleophilic attack, has been traditionally ascribed to the active-site zinc ion of CPA since its three-dimensional structure was elucidated.[12] A similar role for the active site metal ions of other zinc proteases was also considered in comparative studies.[24–26]

This ion-dipole interaction was further substantiated by studies of metal ion catalysis observed in the hydrolysis of model compounds.[27,28] However, in these model compounds, the ion-dipole interaction is facilitated by— indeed, is perhaps only a consequence of—a chelate interaction involving the labile carbonyl and some other moiety on the model compound. Fife and Przystas recently found that such a chelate effect is required for the hydrolysis of N-acylimidazoles in solution.[28] The enzymatic hydrolysis of typical peptide substrates cannot proceed by a zinc–water mechanism if such a chelate interaction occurs to the active-site metal ion of zinc proteases. Moreover, this chelate interaction requires an unblocked amino group in a terminal P_1 residue of a peptide. Therefore, the scissile carbonyl of more complex peptide substrates may not have as great a propensity toward zinc coordination as do unblocked dipeptides. A chelate interaction is observed in the binding of the poor hydrolytic substrate, glycyl-L-tyrosine, to CPA, where both the scissile carbonyl and amino terminus are coordinated to the active site zinc ion.[9] This interaction, quite probably nonproductive, had been used to justify the importance of a zinc–carbonyl interaction. Results from the binding of hydroxamic acid inhibitors to TLN were also interpretable in a similar light.[16] See Figure 1-1 for a comparison of these similar binding modes.

What, then, is the catalytic role of the zinc ion? The importance of a formal zinc–carbonyl interaction may decrease in light of structural and model compound studies. The model compound results of Fife and Przystas do indicate that zinc promotes a water molecule as a potent nucleophile in both proteolytic and esterolytic reactions. Additionally, work by Groves and Olson[29] has shown that zinc-coordinated water can have a pK_a as low as 7 (ie, be hydroxide-like or nucleophilic at optimum pH). A similar role can be envisioned in the zinc-containing proteases. Moreover, in an enzyme active site the zinc–water system is complemented by other enzyme residues which can perform general acid/base functions. Such residues may enhance, or even further promote, the nucleophilicity of zinc-bound water. For instance, the catalytic elements defining the S_1'/S_1 catalytic region in all the zinc proteases of known three-dimensional structure can be summarized as follows: base, water, zinc, electrophile (Figure 1-2). Previously, it was thought that only the base, water, and zinc ion were catalytically important; thus, only these elements were considered in prior active-site compari-

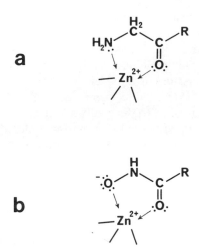

Figure 1-1. (a) Mode of binding of glycyl-L-tyrosine to the active site zinc ion of CPA (R, remainder of molecule). Both the scissile peptide bond and the amino terminus (as the free base) coordinate to zinc as a chelate. This binding mode, although quite probably nonproductive, has been invoked in support of the catalytic importance of a zinc–carbonyl interaction. (b) Mode of binding of hydroxamic acid derivatives to the active site zinc ion of TLN (R, remainder of molecule). Both the carbonyl and the ionized hydroxamate oxygen coordinate to zinc in bidentate fashion. This binding mode has also been used to justify the catalytic importance of a zinc–carbonyl interaction in substrate hydrolysis. Note the overall similarity of this binding mode to that displayed in (a), particularly with regard to the chelate interaction with the metal ion.

sons.[24–26] The electrophile is an additional feature, and its catalytic importance is suggested in view of results of crystallographic studies of CPA[3–8] and TLN,[14] and model building studies of TLN.[30] Residues serving as the base could be ionized Glu or perhaps Asp; those serving as the electrophile could be protonated Arg, His, or perhaps Lys. These residues, including the zinc–water couple, are probably involved in the transition state of the proteolytic reaction, assuming a promoted-water pathway (see Figure 1-3). The key feature of this scheme is the involvement of the additional electrophile in the

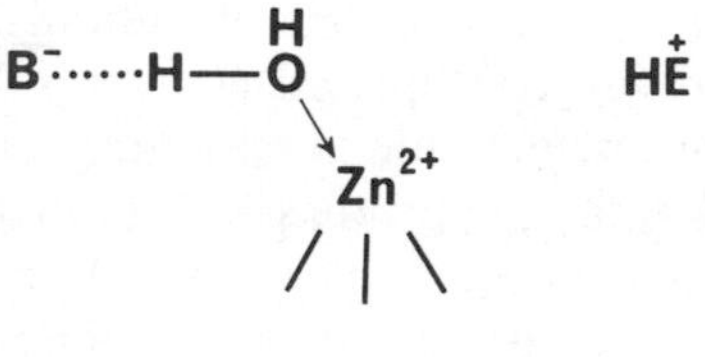

$$B^{-}\cdots\cdots H\!-\!\overset{\text{H}}{\underset{\displaystyle Zn^{2+}}{O}}\qquad\qquad H\overset{+}{E}$$

Figure 1-2. The "catalytic tetrad" which is found in the native zinc proteases of reasonably well determined three-dimensional structure. The base (B) appears as either Glu or Asp; the electrophile (E) can be Arg, His, or Lys. Although there appears to be a vacant coordination site on the "electrophile" side of the zinc ion, high resolution X-ray work on CPA and TLN does not reveal a bound water molecule at this site in the native enzymes. However, this site is often observed occupied upon the binding of transition-state analogues to each enzyme.

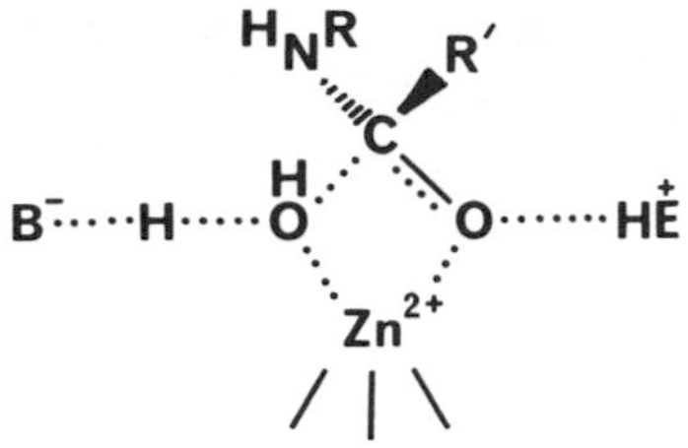

Figure 1-3. The transition state of a proteolytic reaction at the active site of the generalized zinc protease depicted in Figure 1-2. Note that the zinc ion is involved as a classical electrophilic catalyst, ie, it is helping to stabilize the negative charge of the transition state (and of the energetically nearby tetrahedral intermediate). Since the zinc ion is obligatory for catalysis by the zinc proteases, this scheme accurately depicts one catalytic role of the active-site zinc ion. However, the zinc ion need not be the sole agent responsible for the polarization of the scissile carbonyl in a binding mode preceeding this transition state. Likewise, it need not be the sole agent responsible for the nucleophilic promotion of its bound water molecule—this task may be shared substantially between the metal ion and the base.

active site of the zinc protease. A possible role for this electrophile would involve a hydrogen bond to the scissile carbonyl of the substrate, allowing for the attack of a zinc/base-promoted water molecule, with concurrent or subsequent coordination of the developing oxyanion to the zinc ion. The initial coordination of the scissile carbonyl to zinc would raise the pK_a of the zinc-bound water molecule and, perhaps, make it less nucleophilic enough to suppress its attack at the stubborn peptide carbonyl. Therefore, it is an attractive possibility that the zinc ion may be only partially involved, if at all, in the polarization of the substrate carbonyl. Instead, the role of zinc may be simply to promote the attack of a water molecule, with the assistance of a nearby general base. Subsequently the zinc can help to stabilize the negative charge carried by the tetrahedral intermediate and its flanking transition states.

3. DIFFERENCE FOURIER METHOD

When studying the three-dimensional aspects of enzyme mechanism, it is, of course, necessary to have a reliable X-ray structure available for analysis. Of tantamount importance are the structures of the enzyme complexed with ligands such as inhibitors, pseudosubstrates, or actual substrates trapped at low temperature. If there is a reasonably well-determined structure of a native enzyme, the structure of its complexes with ligand molecules can be determined in many cases by using the difference Fourier method. There is, however, one condition: crystals of the enzyme–ligand complex must be isomorphous with those of the native enzyme. Hence, this method cannot be used for enzymes which undergo large conformational changes upon the binding of ligands, such as allosteric enzymes. Furthermore, the active site of the enzyme must be accessible to ligand molecules which diffuse through the solvent-filled interstices of the crystal (protein crystals typically contain about 50% solvent). Alternatively, the enzyme and ligand can be cocrystallized, but the crystals may not be isomorphous with those of the native enzyme.

The difference Fourier method was first introduced to the study of protein–ligand interactions in 1964, when Stryer and colleagues studied the binding of azide ion to myoglobin.[31] This method allows one to obtain the difference of the electron density functions of two isomorphous structures. For example, a difference electron density map involving an enzyme–ligand (EL) complex and a native enzyme (E) will yield density corresponding to the ligand molecule, as well as any enzyme residues which may have moved to accommodate the binding of the ligand. All other electron density corresponding to the native enzyme usually cancels out. A one-dimensional example is illustrated in Figure 1-4. The ligand, as well as any significant conformational changes of enzyme residues, is built into the three-dimensional difference electron density map; the resulting EL model can undergo subsequent least-squares refinement[32] to be brought into better agreement with the observed data. Indeed, one method of obtaining structure factor derivatives for use in calculating least-squares atomic shifts involves the estimation of gradients from a simple difference electron density map. Alternatively, the gradients can be calculated analytically. The use of gradients estimated from difference maps, however, can save almost an order of magnitude of the computer time required for protein refinement.

The reliable use of difference Fourier methods requires a native enzyme structure with very good phases. The greater the uncertainty in the native phases, the less clear and continuous the difference density (ie, that which corresponds to the ligand) will be. Also, Fourier series termination errors will affect the difference density. Because the Fourier synthesis is a truncated series, regardless of whether observed or calculated structure factors are used as coefficients, errors in the electron density function will manifest themselves by preventing the convergence of the function in a small but finite interval about the actual value of the function. However, the greater the amount of data (ie, the higher the resolution), the greater the number of terms in the synthesis and, therefore, the better the convergence. It is also important to use the highest resolution data possible in order to counter the effects of series termination. These effects can be especially pronounced around the sites of electron-rich metal ions in metalloenzymes.

In the difference electron density function, however, series termination errors can cancel out, but only in places where there is no difference density! Normally, these errors will manifest themselves around the sites of atoms which are not phased on in the synthesis (ie, the difference density). For example, in a difference electron density map calculated with Fourier coefficients $(|F_{EL}| - |F_E|) \exp(i\phi_E)$, the difference density will be subject to the

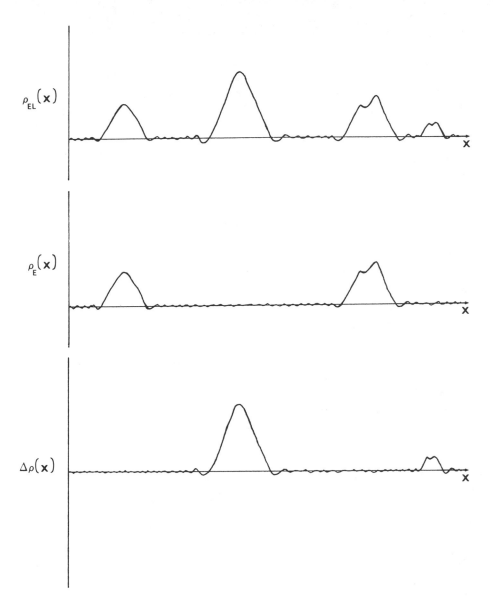

Figure 1-4. A one-dimensional depiction of a difference Fourier map. The density of the enzyme–ligand (EL) complex (top) minus the density of the native enzyme (E) (middle) results in the density of the ligand molecule, plus any conformational changes of particular enzyme residues (bottom). Note the subsidiary minima and maxima surrounding the more pronounced peaks. This phenomenon is due to the fact that a Fourier synthesis composed of a finite number of terms will not converge within a small interval around the origin. Known as series termination errors, these effects will be more pronounced if fewer terms are included in the Fourier synthesis. For the crystallographer, this means that the large minima adjacent to major peaks in difference Fourier maps could affect the observed electron density of the ligand, and especially so if only low-resolution data are available. Such effects could be particularly acute around the sites of the electron-rich metal ions of the zinc proteases. The preferred way to minimize such detrimental effects is to collect the highest resolution data possible, for this which will make the subsidiary peaks of termination error less pronounced.

effects of series termination. This can result in the blur, or even the net cancellation, of electron density corresponding to some atoms of the ligand. Such effects only serve to confound the fitting of the molecular model to the difference density, as well as the subsequent refinement of the enzyme–ligand complex.

In the calculation of difference electron density maps involving enzyme-ligand complexes, it is necessary to evaluate the validity of the approximation that the phase angles for the EL complex and native enzyme are approximately equal. In the case of a small (~20 nonhydrogen atoms) ligand bound to a relatively large (~2000 nonhydrogen atoms) enzyme, the overall contribution of ligand scattering to the Fourier synthesis is negligible and the approximation is generally very good. The following discussion is intended to illustrate the mathematical validity of the difference Fourier method.

The electron density equation for the native enzyme (designated by subscript "E") is given by the following function:

$$\rho_E(\mathbf{r}) = \frac{1}{V} \sum_{\mathbf{h}} |F_E| \exp(i\phi_E) \exp[-2\pi i(\mathbf{h} \cdot \mathbf{r})] \tag{1-1}$$

where $\rho(\mathbf{r})$ is the electron density at the Cartesian coordinate defined by the vector $\mathbf{r}$, V is the volume of the unit cell, $|F|$ is the structure factor amplitude, ϕ is the phase angle, and $\mathbf{h}$ contains the crystallographic indices hkl. Note that there is a unique structure factor amplitude and corresponding phase for each reflection $\mathbf{h}$; however, a subscript $\mathbf{h}$ has been omitted from amplitude and phase expressions in the electron density equation for the sake of clarity.

The electron density function for the enzyme–ligand complex (designated by subscript EL) can be likewise represented:

$$\rho_{EL}(\mathbf{r}) = \frac{1}{V} \sum_{\mathbf{h}} |F_{EL}| \exp(i\phi_{EL}) \exp[-2\pi i(\mathbf{h} \cdot \mathbf{r})] \tag{1-2}$$

If the contribution from the ligand, F_L, to the Fourier synthesis is small, then $\phi_{EL} \cong \phi_E$ (see Figure 1-5), and an approximate electron density function for the EL complex can be constructed as follows (assuming that the native and complexed enzyme are isomorphous):

$$\rho_{EL}(\mathbf{r}) \cong \frac{1}{V} \sum_{\mathbf{h}} |F_{EL}| \exp(i\phi_E) \exp[-2\pi i(\mathbf{h} \cdot \mathbf{r})] \tag{1-3}$$

Given this simplifying approximation, the difference density ($\Delta\rho(\mathbf{r})$) between the EL complex and the native enzyme can be expressed by

$$\Delta\rho(\mathbf{r}) =$$

$$\rho_{EL}(\mathbf{r}) - \rho_E(\mathbf{r}) = \frac{1}{V} \sum_{\mathbf{h}} (|F_{EL}| - |F_E|) \exp(i\phi_E) \exp[-2\pi i(\mathbf{h} \cdot \mathbf{r})] \tag{1-4}$$

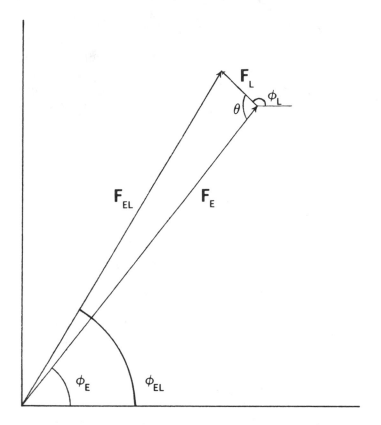

Figure 1-5. An Argand diagram. The abcissa is the real component, the ordinate imaginary, of the same (ie, the same crystallographic index *hkl*) structure factor before (E) and after (EL) a ligand has bound to the enzyme. The structure factor corresponding to only the ligand atom (L) shows the relationship between those of the native enzyme and the enzyme–ligand complex. Note that the approximation which legitimizes the difference Fourier method, that $\phi_{EL} = \phi_E$, holds true, at least when $|F_E|$ is not too small and when the difference $|F_{EL}| - |F_E|$ is not large. It is interesting to note that this relationship begins to break down for smaller values of $|F_E|$ and $|F_{EL}|$. It also breaks down when the ligand atom is too large, ie, when $|F_L|$ becomes significantly large so that the approximation $\phi_{EL} = \phi_E$ becomes invalid.

In the derivation of Equation 1-4, it is assumed that $\phi_{EL} \cong \phi_E$. The true electron density corresponding to the ligand (designated by subscript L) is actually represented by the following equation:

$$\rho_L(\mathbf{r}) = \frac{1}{V} \sum_{\mathbf{h}} |F_L| \exp(i\phi_L) \exp[-2\pi i(\mathbf{h} \cdot \mathbf{r})] \qquad (1\text{-}5)$$

This function, however, is experimentally insoluble; the crystallographer must resort to Equation 1-4 as an approximation to Equation 1-5. How valid is this approximation that $\Delta\rho(\mathbf{r}) \cong \rho_L(\mathbf{r})$? Consider the following argument in reference to Figure 1-5.

Equation [1-6] is true by the law of cosines:

$$|F_{EL}|^2 = |F_L|^2 + |F_E|^2 - 2|F_L||F_E| \cos(\theta) \tag{1-6}$$

Now, θ can be expressed as $\phi_E - \phi_L + \pi$; since $\cos(\theta) = -\cos(-\theta + \pi)$, Equation 1-6 can be rewritten and rearranged as follows:

$$|F_{EL}|^2 - |F_E|^2 = |F_L|^2 + 2|F_L||F_E| \cos(\phi_L - \phi_E) \tag{1-7}$$

which, upon factoring, yields

$$(|F_{EL}| - |F_E|)(|F_{EL}| + |F_E|) = |F_L|^2 + 2|F_L||F_E| \cos(\phi_L - \phi_E) \tag{1-8}$$

or, more conveniently written,

$$(|F_{EL}| - |F_E|) = \frac{|F_L|^2}{(|F_{EL}| + |F_E|)} + \frac{2|F_L||F_E| \cos(\phi_L - \phi_E)}{(|F_{EL}| + |F_E|)} \tag{1-9}$$

Since $\cos(\phi_L - \phi_E) = \frac{1}{2}\exp[i(\phi_L - \phi_E)] - \frac{1}{2}\exp[-i(\phi_L - \phi_E)]$, Equation 1-10 is obtained upon simplification and multiplication by $\exp(i\phi_E)$ throughout:

$$(|F_{EL}| - |F_E|) \exp(i\phi_E) = \frac{|F_L|^2}{(|F_{EL}| + |F_E|)} \exp(i\phi_E)$$

$$+ \frac{|F_L||F_E|}{(|F_{EL}| + |F_E|)} \exp(i\phi_L)$$

$$+ \frac{|F_L||F_E|}{(|F_{EL}| + |F_E|)} \exp(2i\phi_E - i\phi_L) \tag{1-10}$$

The left-hand side of this equation represents the magnitude and phase of the difference Fourier coefficient $|\Delta F| \exp(\phi_E)$ used in the $\Delta\rho(\mathbf{r})$ calculation of Equation 1-4, which is an approximation to Equation 1-5. The right-hand side of Equation 1-10 shows the actual constitution of the difference Fourier coefficients $|\Delta F| \exp(i\phi_E)$ used to approximate $|F_L| \exp(i\phi_L)$. An analysis of the right-hand side of Equation 1-10 shows the validity of the approximation of Equation 1-4 to Equation 1-5 as follows.

Since $|F_L|$ is very small in comparison with $|F_E|$ and $|F_{EL}|$ (see Figure 1-5), the first term makes no substantial contribution to the Fourier synthesis.

The second term provides an image of the ligand (ie, the difference density). Since its contribution is derived as one-half of $|F_L| \exp(i\phi_L)$ (if $|F_E| \cong |F_{EL}|$), the peaks corresponding to the ligand in the difference electron density map will appear at about half height, assuming that the ligand is bound in 100% occupancy to the enzyme, and that 100% of the reflections with intensities greater than 2σ were measurable. Luzzati[33] has shown that the peak heights of atoms not included in phasing (ie, the calculation of ϕ_E) depend on the number of atoms excluded. These peak heights approach one-half as the number of unphased atoms approaches zero.

The last term results in noise. However, since there is no correlation between $-\phi_L$ and $2\phi_{EL}$, it makes no substantial contribution to the Fourier synthesis.

Hence, the following relation results from the simplification of Equation 1-10:

$$(|F_{EL}| - |F_E|) \exp(i\phi_E) \cong \frac{1}{2} |F_L| \exp(i\phi_L) \qquad (1\text{-}11)$$

given that F_E is much greater than F_L and therefore F_E approximately equals F_{EL}. This means that $\Delta\rho(\mathbf{r})$ approximately equals $\frac{1}{2} \rho_L(\mathbf{r})$, where $\Delta\rho(\mathbf{r})$ and $\rho_L(\mathbf{r})$ are defined in Equations 1-4 and 1-5, respectively. The factor of one-half in scaling between these two functions is considered as follows.

In practice, it is often convenient to use difference maps calculated with Fourier coefficients $2|F_{EL}| - |F_E|$, ie, $2(|F_{EL}| - |F_E|) + |F_E|$, and phases calculated from the native enzyme, for this will make the scales of electron densities between enzyme and ligand atoms more nearly equal. If, in the interaction with the ligand, active-site enzyme residues move appreciably, structure factors and phases must be calculated as part of the F_E component in which the atoms for these particular residues have been omitted. The resultant difference Fourier between the observed structure factors and those calculated as described will result in unbiased electron density for the particular residues in question—the resultant density will not be dependent in any way upon the residue's original location in the native enzyme. The density for the ligand is, of course, unbiased as well, since it is not included in the phase calculation. This omission procedure cannot be performed for too many residues and/or ligands at a time, however, since the greater the number of atoms in the unphased component of a difference map calculation, the less correct is the assumption that $\phi_{EL} \cong \phi_E$ (Figure 1-5).*

4. STRUCTURE–FUNCTION STUDIES OF CARBOXYPEPTIDASE A

In order to evaluate reasonable mechanistic pathways followed by zinc proteolytic enzymes containing the "tetrad" of Figure 1-2, detailed structural studies of these enzymes, including difference Fourier studies with appropriate ligand molecules, are necessary. Although mechanistic discussion in this review presumes a promoted-water hydrolytic pathway, there are still those who defend nucleophilic hydrolytic pathways (at least for CPA) which

---

* For an excellent overview of the difference Fourier method, the reader is referred to Chapter 14 of *Protein Crystallography* by T. L. Blundell and L. N. Johnson (Academic, New York: 1976), after which this section is inspired.

would involve a metastable anhydride intermediate.[34,35] This mechanistic alternative will be considered toward the end of this section. Since more structural studies have been performed on CPA with substrate and transition state analogues, the ensuing discussion will relate the results of these studies and outline the implications for the greater class of zinc proteases. Following the notation of Figure 1-2, the active site base of CPA is Glu-270, the electrophile, Arg-127. Additionally, enzyme residues Tyr-248, Arg-145, and Asn-144 are involved in binding the P_1' residue of the substrate—specifically, in binding its terminal carboxylate moiety. These interactions account for the observed specificity of the enzyme toward the hydrolysis of C-terminal amino acids from peptide or ester substrates. The interaction of CPA with inhibitors possessing a carbonyl moiety (such as a ketone or aldehyde) isosteric with the scissile carbonyl of actual substrates has been the subject of recent X-ray crystallographic investigation.[3-5,7] These inhibitors are remarkable due to the fact that their carbonyls can display catalytically pertinent reactivity when bound to the enzyme. Just as an enzyme can facilitate the addition of a water molecule or enzyme-bound nucleophile to the scissile carbonyl of substrates, it may do the same for the nonlabile carbonyl of ketones or aldehydes. If this occurs, however, the resulting tetrahedral "intermediate"—indeed, transition-state analogue—cannot collapse to give products. The comparison of different binding modes exhibited by a group of inhibitors which can react in varying degrees (in such dead-end fashion) might be used to tentatively "map out" a reasonable mechanistic pathway. This approach is similar in spirit to the reaction coordinate method inspired by Bürgi and Dunitz in the study of small-molecule reactivity.[36]

The three ketones studied by use of the difference Fourier method were 2-benzyl-3-p-methoxybenzoylpropanoic acid (BMP);[4,37,38] 2-benzyl-4-oxo-5,5,5-trifluoropentanoic acid (TFP);[5,39] and 5-benzamido-2-benzyl-4-oxopentanoic acid (BOP).[7,40] Additionally, the aldehyde 2-benzyl-3-formylpropanoic acid (BFP)[3,41] was investigated. The structural relationships between these inhibitors and an actual peptide substrate can be compared in Figure 1-6. The aldehyde BFP and the α-trifluoroketone TFP are exceptional in that they possess extremely electrophilic carbonyl carbons. Such electrophilicity would certainly favor the addition of a nucleophile—either a water molecule or a nucleophilic enzyme residue—when bound to CPA. Depending upon the species which adds, the structural result might provide support for either the promoted-water or the anhydride catalytic pathway. The other two ketones, BMP and BOP, are not remarkable in terms of the reactivity expected at the carbonyl moiety, and at first glance they might be expected to bind to the enzyme as the free carbonyl. The actual results, however, obtained from X-ray crystallographic studies of these inhibitors are surprising and even unexpected with regard to the supposedly "unreactive" ketones. All of the inhibitors share the same interactions deep within the S_1' hydrophobic pocket of the enzyme: The benzyl group of each inhibitor is nestled within this "specificity pocket," and the terminal carboxylate

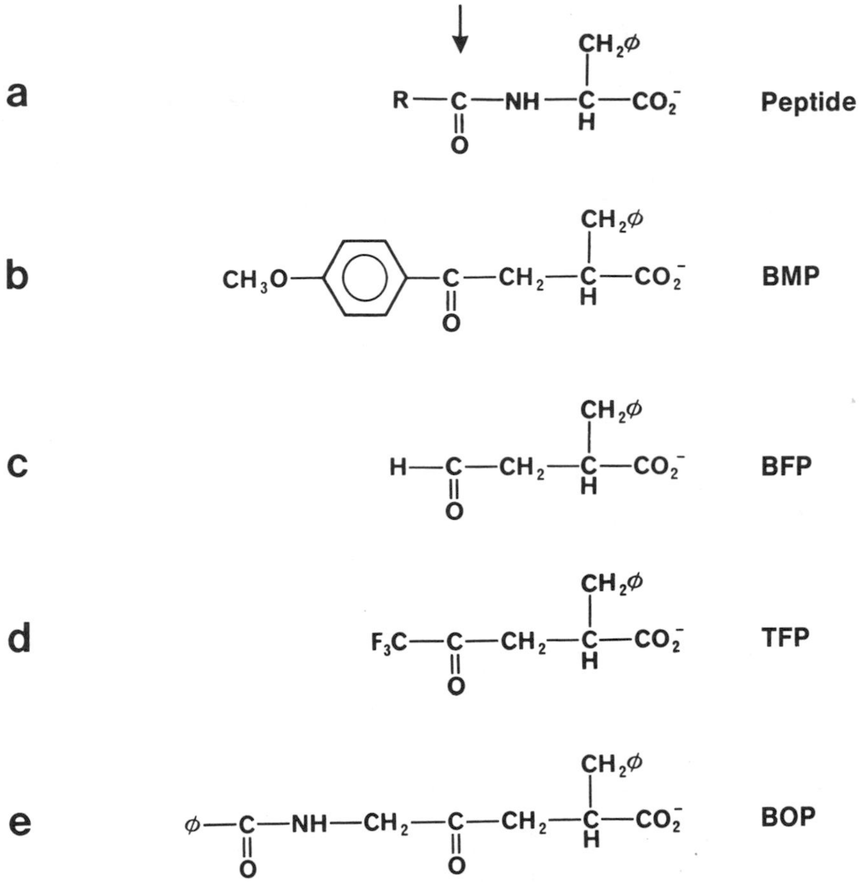

Figure 1-6. (a) Typical C-terminal portion of a peptide substrate for CPA (R, remainder of peptide). An arrow denotes the scissile amide linkage. The enzyme exhibits preferred specificity toward substrates possessing a free terminal carboxylate as well as a bulky aromatic C-terminal side chain such as phenylalanine. Note throughout this figure that ϕ denotes a phenyl group. (b) The ketonic substrate analog 2-benzyl-3-*p*-methoxybenzoylpropanoic acid (BMP). This compound binds to CPA as the intact ketone, with the carbonyl hydrogen bonded to the positively charged guanidinium moiety of Arg-127. (c) The aldehyde substrate analog 2-benzyl-3-formylpropanoic acid (BFP). This inhibitor binds to CPA as the *gem*-diol hydrate at the carbonyl carbon. Given the notable electrophilicity of the aldehyde carbonyl, this was not an unexpected consequence. (d) The ketonic substrate analog 2-benzyl-4-oxo-5,5,5-trifluoropentanoic acid (TFP). Because of an extremely electrophilic carbonyl carbon (due to the α-trifluoro moiety) this compound binds to the enzyme as the hydrate species. (e) The ketonic substrate analog 5-benzamido-2-benzyl-4-oxopentanoic acid (BOP). Because the ketone carbonyl is not remarkable in terms of its expected reactivity, this inhibitor was expected to bind to the enzyme with an intact carbonyl. Surprisingly, it was observed to bind as the carbonyl hydrate adduct, just like BFP and TFP. In this case, as well as the others, the enzyme may well have facilitated the addition of a water molecule to a carbonyl moiety. If this is the case, the enzyme may have provided a model for the first step of substrate hydrolysis, ie, facilitating the addition of a water molecule to the scissile carbonyl moieties of peptides and esters.

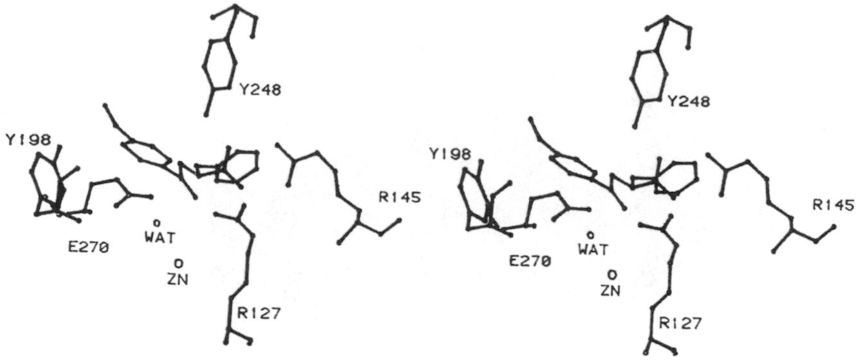

Figure 1-7. A stereoview of the CPA–BMP complex. Pertinent active-site residues are indicated by their standard one-letter abbreviations (eg, E, glutamate; R, arginine; Y, tyrosine) and sequence number. The active-site zinc ion and its bound water molecule are indicated. Note that the carbonyl of BMP is hydrogen bonded to the positively charged guanidinium moiety of Arg-127. BMP does not perturb the zinc–ligand environment; the zinc–water interaction of the native enzyme is preserved in this binding mode. The water–carbonyl carbon distance is 3.1 Å; The water molecule is oriented nicely toward the π^* orbital of the carbonyl. The observed structure may provide a model for the Michaelis complex, or precatalytic complex, of an actual hydrolytic reaction. The product of such a promoted attack of water would be a tetrahedral intermediate with two geminal oxygens much like the hydrated carbonyl adducts, except that the actual tetrahedral intermediate would bear a negative charge. The scissile carbonyl need not be substantially coordinated to zinc in order for this first catalytic step to occur.

is engaged in a double salt link with the guanidinium moiety of Arg-145. Additionally, Tyr-248 is in the "down" conformation in order to donate a hydrogen bond in anti orientation to the inhibitor carboxylate. However, all similarities end with these interactions.

BMP binds to CPA as the free ketone as expected. However, the ketonic oxygen does not coordinate to zinc, as might have been predicted if a zinc–carbonyl interaction were dominant in an actual hydrolytic mechanism. Instead, the ketone carbonyl accepts a bifurcated hydrogen bond from the positively charged guanidinium moiety of Arg-127 (Figure 1-7). The zinc-bound water of the native enzyme remains on zinc, and this water appears to be "poised" for attack of the π^* orbital of the ketone carbonyl carbon (only 3.1 Å away). If this structure does indeed represent the precatalytic complex of a hydrolytic reaction, it may provide support for a zinc/base-promoted water hydrolytic mechanism and not for any mechanism involving a covalent link to Glu-270 or a water-excluding zinc–carbonyl interaction.

The aldehyde BFP and the α-trifluoroketone TFP bind to CPA as hydrate adducts (ie, as the *gem*-diol moiety at the former carbonyl carbon). For illustrative purposes, a stereoview of the CPA–TFP complex is presented in Figure 1-8. Quite surprisingly, the ketone BOP binds in hydrated fashion as well, even though it has an "unreactive" carbonyl. Although not yet proven chemically, the enzyme may have facilitated the addition of a water mole-

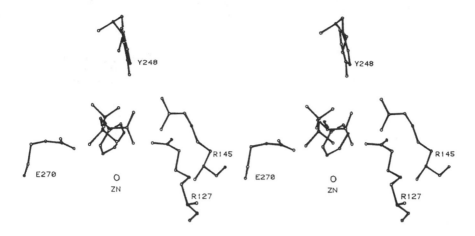

Figure 1-8. A stereoview of the CPA–TFP complex. Pertinent active-site residues are indicated by their standard one-letter abbreviations and sequence number. Note the *gem*-diol moiety in the region above the zinc ion; in each of the hydrated inhibitors studied so far, different zinc–oxygen distances for the hydrate have been observed (see Table 1-2); ie, there is not one particularly "correct" orientation of a tetrahedral hydrate, or perhaps of an actual tetrahedral intermediate of a hydrolytic reaction. In fact, different orientations may imply different models of structures along the reaction coordinate of the hydrolysis reaction. Nevertheless, the tetrahedral structure depicted illustrates the structural result of the direct attack of water at a scissile carbonyl.

cule across the carbonyl of BOP. Alternatively, the enzyme may have selected these hydrated forms preferentially from the solutions in which crystals were soaked, not only for BOP but also for BFP and TFP. Nevertheless, these binding phenomena provide an excellent structural example of how an enzyme favors the binding of structures resembling a transition state rather than those resembling reactants or products. The *gem*-diol moiety is reminiscent of a tetrahedral intermediate which would result from the direct attack of a zinc/base-promoted water molecule directly at the scissile carbonyl of an actual substrate (see Figure 1-9). The observed hydrate structures do not provide a structural model for an intermediate along the anhydride pathway. The collapse of the tetrahedral intermediate represented by the hydrate structures would require proton donation from Glu-270 to the leaving amino group. Although this particular proton donor would imply cis addition of water across the peptide bond rather than the more favorable trans addition, the stereochemically required inversion of pyramidal nitrogen is quite rapid (on the order of 10^8/s for ammonia) and, therefore, well within the turnover rate for typical substrate hydrolysis (k_1 for Bz-Gly-Gly-Phe is 1200/min[42]). The keto/hydrate binding behavior of these inhibitors is summarized in Table 1-1.

It is notable that there may not be one particular orientation of a tetrahedral intermediate, as represented by the three *gem*-diol hydrate adducts, as well as the tetrahedral phosphonic acid cleaved from the phosphonamidate

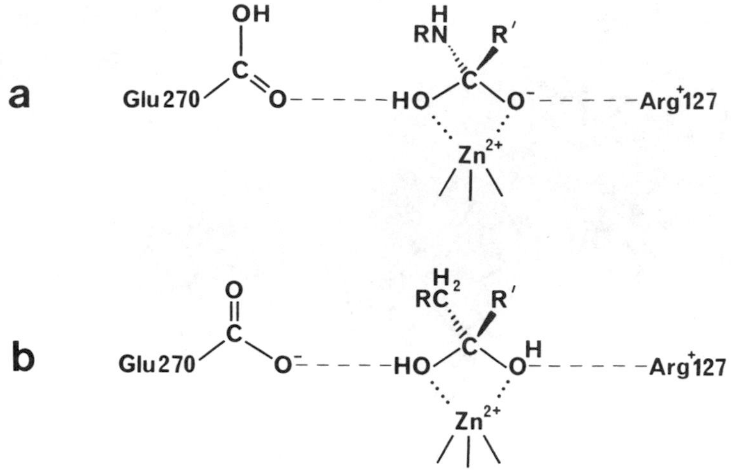

Figure 1-9. (a) An actual tetrahedral intermediate of a proteolytic reaction. It may or may not be coordinated in bidentate fashion to the zinc ion as illustrated. As the actual chemical intermediate, it might be asymmetrically bound due to the negative charge of the oxyanion. It is interesting to consider that the transition state, which would bear a more symmetrically disposed charge across the two geminal oxygens, may be the species which favors true bidentate coordination to the metal ion. (b) A generalized depiction of the tetrahedral *gem*-diol hydrates as observed bound to the active site of CPA. Note how this binding mode resembles the structure in (a). There is actually great flexibility observed in the binding of these hydrates to the zinc ion; see Table 1-2 for a summary of pertinent distance information.

inhibitor   *N*-[[[(benzyloxycarbonyl)amino]methyl]hydroxyphosphinyl]-L-phenylalanine (ZGP′, Figure 1-10).[6,43] The phosphonic acid hydrolyzed from ZGP′ was observed to bridge the zinc ion and Arg-127, and the complex may mimic an early stage of the promoted-water reaction coordinate. The interactions of tetrahedral zinc-bound inhibitors are summarized in Table 1-2. Assuming that a promoted-water hydrolytic mechanism is favored, a considerable variation of contacts between the enzyme and a tetrahedral intermediate as represented by these inhibitors is implied. Enzyme residues and the active-site zinc ion are observed to move in order to accommodate specific

TABLE 1-1.   Summary of Carbonyl-Containing
Inhibitor Binding to Carboxypeptidase A

Inhibitor[a]	Predominant form in aqueous solution	Species bound to enzyme
BMP	Ketone	Ketone
BFP	Aldehyde	Hydrate
TFP	Hydrate	Hydrate
BOP	Ketone	Hydrate

[a] See Figure 1-6.

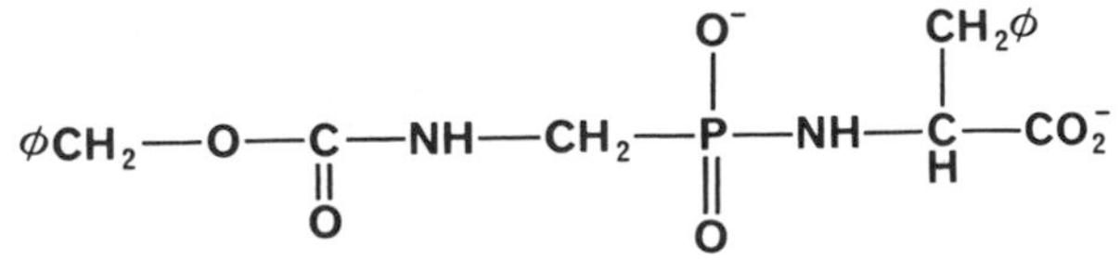

$$\phi CH_2-O-\underset{\underset{O}{\|}}{C}-NH-CH_2-\underset{\underset{O}{\|}}{\overset{\overset{O^-}{|}}{P}}-NH-\underset{\underset{H}{|}}{\overset{\overset{CH_2\phi}{|}}{C}}-CO_2^-$$

Figure 1-10. The CPA inhibitor *N*-[[[(benzyloxycarbonyl)amino]methyl]hydroxyphosphinyl]-L-phenylalanine (ZGP'). Although it is depicted here as the intact phosphonamidate, it was observed bound to CPA as the hydrolyzed species, ie, the corresponding phosphonic acid plus phenylalanine. As such, it formally represents a products complex; however, with the tetrahedral phosphonate bound to zinc, the complex still resembles an enzyme transition-state analog complex.

binding modes, each of which may represent some structure along the reaction coordinate of an actual hydrolytic reaction. The observed rotational flexibility of the tetrahedral moiety implies that there is flexibility in the attack of the scissile carbonyl group by zinc-bound water, stabilization of the tetrahedral intermediate (and its flanking transition states), and dispersal of products through progressive reorientation away from Arg-127 toward Glu-270. The tetrahedral intermediate might require certain discrete orientations and contacts depending upon its position along the reaction coordinate (ie, whether it is forming or collapsing). Additionally, an oxyanion would probably be drawn closer to zinc than would the hydroxyl moieties of the observed hydrate structures. Interactions within the S_1' hydrophobic pocket, and in other subsites on the enzyme known to be of kinetic significance,[44,45] surely influence the orientation and stability of the tetrahedral intermediate.

These difference Fourier studies of CPA transition-state analogs largely imply, but do not prove, that the enzyme favors the promoted-water hydro-

TABLE 1-2. Observed Orientations of Tetrahedral Transition State Analogues

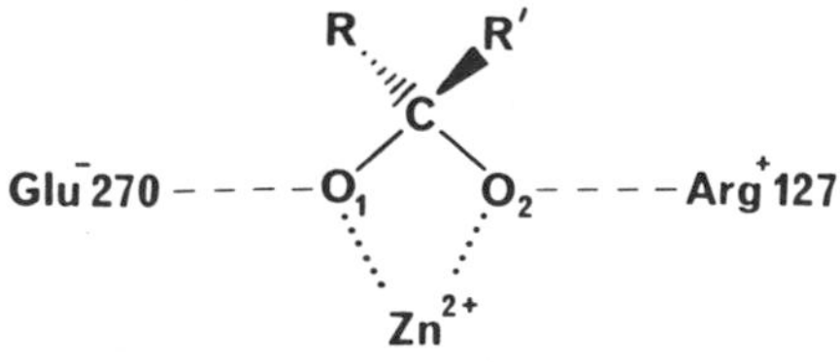

$$\text{Glu}^-270 ----O_1 \qquad O_2 ---- \text{Arg}^+127$$

with R, R' on C above, and Zn^{2+} below.

| Interaction | Distance ± 0.2 Å | | | |
	CPA–ZGP'[a]	CPA–BFP	CPA–TFP	CPA–BOP
Glu-270-01	3.4	2.4	2.6	2.6
01-Zn	2.2	2.7	3.4	2.9
02-Zn	3.3	2.5	2.6	2.5
Arg-127-02	2.7	3.6	3.2	3.2

[a] See Figure 1-10; only the *hydrolyzed* phosphonamidate was observed bound to the enzyme.

lytic mechanism. In fact, chemical data from other laboratories suggest that some ester substrates might proceed through a hydrolytic mechanism involving a mixed anhydride intermediate.[34,35,46,47] The observation, however, in some of these low temperature studies—that an intermediate accumulates that has two nonprotein zinc ligands—is consistent with the tetrahedral intermediate of the promoted-water pathway in which the two geminal oxygens are stabilized by the zinc ion. Alternatively, if one or more of these particular ester substrates do indeed follow the anhydride pathway, they may be "forced" to do so in order to avoid severe steric interactions with the enzyme. The bulky chromophore in conjugation with the scissile carbonyls of these substrates may prevent the substrate from achieving the proper transition-state structure for the promoted attack of water. The CPA–BMP complex may illustrate such an effect: the p-methoxybenzoyl group of BMP probably prevents it from binding to CPA as the hydrate in observable occupancy, whereas the better substrate analog BOP can bind to CPA as the hydrate. Indeed, if BMP is superimposed on the BOP coordinates, severe interactions occur between the p-methoxybenzoyl group of BMP and enzyme residue Phe-279. These interactions are unavoidable: no single-bond rotations of a hydrated BMP model can achieve an acceptable "catalytic" conformation. Perhaps, then, the anhydride pathway may be favored over the promoted-water pathway for those particular substrates which have certain steric limitations.

Bearing these considerations in mind, as well as those raised in the second section of this review, a reasonable promoted-water hydrolytic mechanism can be postulated for CPA which is consistent with the current structural and chemical data. The first step involves the hydrogen bonding of the substrate carbonyl to Arg-127 (Figure 1-11). The positively charged guanidinium moiety of this residue could polarize the scissile carbonyl, making it more susceptible to nucleophilic attack by a water molecule promoted by zinc and Glu-270. Formerly, the role of carbonyl polarization was ascribed solely to the zinc ion. However, such a step would increase the pK_a of zinc-bound water, making it less nucleophilic. By letting this water molecule attack the scissile carbonyl polarized by Arg-127 (and perhaps only partially, if at all, by zinc), the potent nucleophilicity of zinc-bound water would be preserved and even enhanced by its strong hydrogen bond with the γ-carboxylate of Glu-270. Riordan[48] has shown that one unidentified arginyl residue is required for the hydrolysis of peptides, but not for esters. If this arginyl residue is, in fact, Arg-127, we can consider differences between proteolysis and esterolysis in light of the electrophilic roles considered for zinc and Arg-127. The more reactive carbonyl groups of esters may coordinate to zinc and still be good hydrolytic substrates for the resultantly less nucleophilic zinc-bound water.

As the tetrahedral intermediate is formed, or immediately subsequent to its formation, the developing oxyanion could shift its association from Arg-127 to zinc (Figure 1-12). This shift would involve a clockwise "rotation" of

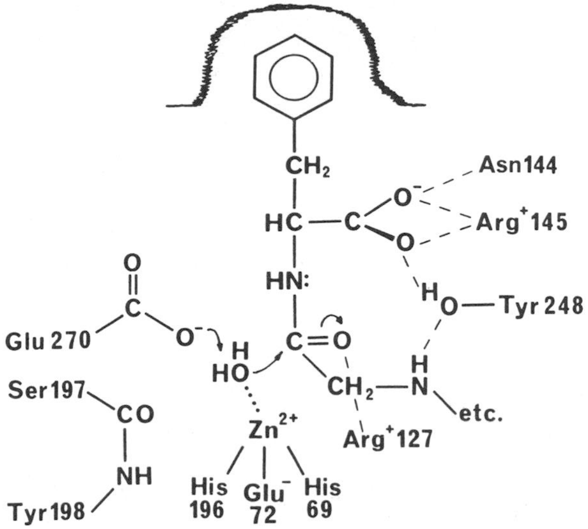

Figure 1-11. A typical peptide substrate, with terminal phenylalanine, is shown bound to the active site of CPA. The P$_1'$ benzyl group resides in the hydrophobic pocket, or "specificity pocket," of the enzyme, and Arg-145 provides a double salt link with the terminal carboxylate of the substrate. Both Tyr-248 and Asn-144 provide hydrogen bonds in anti orientation to this terminal carboxylate. These interactions account for the enzyme's great specificity toward substrates possessing C-terminal amino acids with bulky aromatic side chains. Tyr-248 also accepts a hydrogen bond from the amide NH of the penultimate peptide bond and, thus, may provide specificity toward substrates possessing such a penultimate peptide bond. Note the cis peptide bond between Ser-197 and Tyr-198. Importantly, the scissile carbonyl is shown to be polarized by Arg-127 and perhaps only partly, if at all, by the zinc ion. A Glu-270/zinc-promoted water molecule is shown to attack the polarized carbonyl of the substrate.

the tetrahedral center depicted in Table 1-2. The oxyanion contact with Arg-127 can be retained for additional stabilization—this interaction is observed in the binding of some of the tetrahedral hydrates already discussed. However, the positively charged zinc ion would probably favor a close coordination interaction with the negative oxyanion. Also facilitated by this rotation would be the transfer of a proton from the former zinc-bound water molecule to Glu-270. The resulting hydroxyl of the tetrahedral intermediate could retain a hydrogen bond with the carboxylic acid carbonyl of the now-protonated Glu-270. In addition to a hydrogen bond with the backbone carbonyl of Ser-197, this hydroxyl could also retain some coordination to zinc for additional stabilization. Hence, such a tetrahedral intermediate could be stabilized through a pentacoordinate zinc ion (counting the three enzyme residues Glu-72, His-69, and His-196 as single ligands). The collapse of the tetrahedral intermediate would involve a required proton transfer from the newly protonated Glu-270 to the leaving amino group.[14] After an intervening proton transfer, probably mediated by Glu-270, the product complex illus-

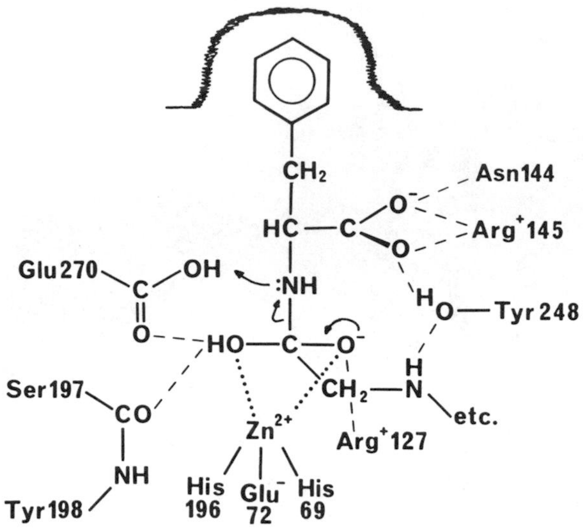

$$
\begin{array}{c}
\text{CH}_2 \\
\text{HC—C} \overset{O}{\underset{O}{\big<}} \cdots \text{Asn144} \\
\text{>Arg}^+\text{145}
\end{array}
$$

Glu270—C—OH ⤵ :NH ... H O—Tyr 248

Ser197—CO ... HO—C—O⁻ CH₂—N H etc.

Zn²⁺   Arg⁺127

Tyr198—NH His 196 Glu⁻ 72 His 69

Figure 1-12. The tetrahedral intermediate resulting from the previous step (Figure 1-11). Subsequent to, or concurrent with, its formation, the developing oxyanion moves to zinc for additional electrostatic stabilization. However, its contact with Arg-127 can be retained for an energetically favorable hydrogen bond. The hydroxyl of the tetrahedral intermediate can remain coordinated to zinc, although the negatively charged oxyanion might be drawn closer to the positively charged zinc ion. Additionally, the hydroxyl moiety can donate a bifurcated hydrogen bond to the backbone carbonyl oxygen of Ser-197 as well as the carbonyl oxygen of the now-protonated Glu-270. If the tetrahedral intermediate has its two geminal oxygens coordinated to zinc, the postulated mechanism would involve a pentacoordinate zinc ion as a metastable intermediate. This intermediate can collapse to form products, with required proton donation by Glu-270 to the leaving amino group as depicted.

trated in Figure 1-13 would result. The proximity of the carboxylate of Glu-270 to the product carboxylate causes an unfavorable electrostatic interaction which probably facilitates product release, even though the zinc ion probably masks much of the charge of the product carboxylate. This carboxylate has been shown to accept a hydrogen bond from Tyr-248 in the complex of CPA with the potato inhibitor,[49] but in a more recent study of a novel enzyme–substrate–product complex[50] structurally similar to that of Figure 1-13, the carboxylate is observed to bridge the zinc ion and Arg-127; no contact is made with Tyr-248.

Remember that the mechanistic sequence depicted in Figures 1-9–1-11 is generalizable, at the very least, to other zinc proteases of known three-dimensional structure possessing the tetrad of Figure 1-2. For CPA, Glu-270 is the "base" and Arg-127 is the "electrophile." All other residues shown simply provide specificity in substrate binding. This mechanistic sequence is likely generalizable to many other zinc proteases which have not yet yielded three-dimensional structures, including the pharmaceutical targets angioten-

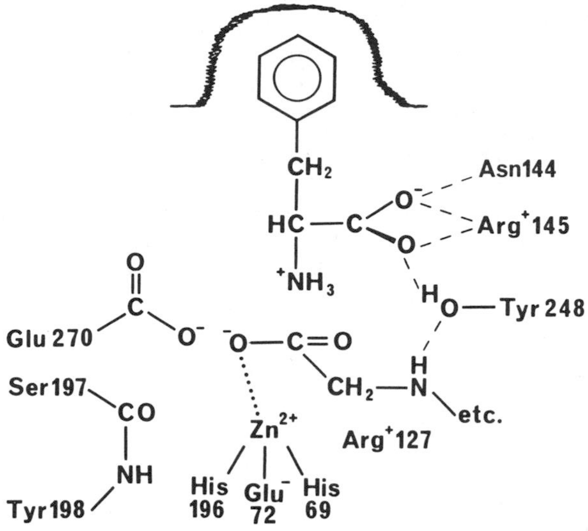

Figure 1-13. Hydrolysis products are shown just after the final step depicted in Figure 1-12, with an intervening proton transfer between the product carboxylate and ammonium group (this proton transfer is probably mediated by Glu-270). The product carboxylate can have one oxygen on the zinc ion, and the other can be hydrogen bonded to Tyr-248 or Arg-127. The electrostatic interaction between the negatively charged carboxylates of the product and Glu-270 probably facilitates a product release mechanism.

sin-converting enzyme, collagenase, and enkephalinase, to name but a few. Since the design of enzyme inhibitors for such enzymes is of significant clinical importance, new metalloprotease inhibitors might now be designed in order to exploit various aspects of the proposed mechanism. In particular, inhibitors might be designed to target interaction with the "electrophile," probably Arg, His, or Lys, and thus provide a new route toward the rational design of drug molecules targeted for zinc proteases in vivo.

5. FURTHER ELUCIDATION OF ZINC PROTEASE MECHANISMS

There are still many questions regarding the structure and function of zinc proteases; by no means are the puzzles anywhere near completion. However, the current decade has brought with it a variety of new techniques which are beginning to shed new light on the general understanding of enzymatic catalysis. These techniques have been (or are currently being) applied to various problems in the understanding of the catalytic mechanisms of zinc proteases.

Site-directed mutagenesis (SDM) has provided a method by which individual amino acids of an enzyme can be replaced by others. Hence, if a particular amino acid is suspected to be of catalytic importance in the function of a particular enzyme, it can be replaced with a nonfunctional amino acid and the resultant mutant enzyme checked for activity. For instance, Tyr-248 of CPA was long suspected to function as a general acid[51] or as a nucleophile/zinc ligand.[52,53] However, Gardell and colleagues[54] were able to modify the CPA gene from the rat by substituting a phenylalanine residue at position 248. The activities of both wild-type and mutant enzymes were assayed for peptidase and esterase activity; the mutant enzyme displayed near-normal behavior in both cases. More recently,[55] substitutions have been made for Glu-270 which have radically altered enzyme activity. The general conclusion from these studies is that Tyr-248 is not obligatory for catalysis, whereas Glu-270 is. However, at first glance one cannot discount an alternate reaction mechanism followed by the mutant enzyme in the case of the Tyr-248 substitution. The study must be accompanied by detailed chemical and kinetic analysis similar to the host of studies already performed on the native enzyme. Furthermore, such studies only pinpoint residues which are necessary for catalysis—SDM methods will not reveal what specific function the catalytic residues perform. This problem is left for the enzymologists and crystallographers.

Low-temperature crystallography can be used, in conjunction with low-temperature enzymological methods, to elucidate the structure and function of catalytically important enzyme residues caught before, during, or just after the act of catalysis. Low-temperature enzymological methods have been applied to CPA,[34,46,47] but the results were interpreted to imply the isolation of the putative anhydride intermediate. In light of recent crystallographic studies of the enzyme, it is possible that the tetrahedral intermediate of the promoted-water mechanism, with both geminal oxygens coordinated to zinc (and possibly one to Arg-127), would also rationalize the observed spectroscopic data. Alternatively, enhanced product binding in viscous cryosolvent systems, or even simultaneous substrate-product binding, may occur. Low-temperature crystallographic studies of CPA have since commenced,[9] and further studies in progress will help to deduce the actual nature of the spectroscopically observed intermediates of ester hydrolysis. These studies may also show that differences between peptide and ester hydrolysis do not lie in vastly different binding modes, as suggested by Vallee and colleagues,[42,52,53] but instead lie in the association of the scissile substrate carbonyl with either zinc or Arg-127 and different rate-determining steps involving these two electrophiles.

Another excellent method for evaluating the stability of such binding modes is that of molecular dynamics (MD).[56] In many ways, the results of MD studies are complementary to the results of X-ray crystallography. MD methods can verify the thermodynamically favorable binding modes observed crystallographically for tightly bound inhibitors as minimum energy structures. MD methods can, in fact, go one step further. Crystallographers can have considerable problems trying to isolate an unreacted substrate bound to an enzyme, or even an enzyme-bound reaction intermediate. However, MD methods can predict just how an unreacted substrate might bind to an enzyme and thereby complement the results of the crystallographer. Of course, there is always the question of whether the minimum energy binding mode is the productive, or catalytically active, structure. Therefore, the observation of particular minimum energy binding modes via X-ray or MD methods must be made in conjunction with the results of other chemical studies. MD methods have been applied to apo-CPA,[57] and their application toward the study of TLN-inhibitor complexes has also been recently reported.[58] All of these methods, in conjunction with the more traditional chemical studies, will help to elucidate further the catalytic mechanisms of not only CPA but the whole family of zinc proteases. Furthermore, such studies may even highlight features as yet unconsidered in the function of these enzymes.

ACKNOWLEDGMENTS

We wish to thank the National Institutes of Health for Grant GM 06920 which supported much of the described research. Additionally, W. R. Grace & Co. is acknowledged for their support, and D.W.C. thanks AT&T Bell Laboratories for a doctoral fellowship during his study at Harvard University.

REFERENCES

1. Pauling, L. *Nature (London)* **1948**, *161*, 707–709
2. Rees, D. C. ; Lewis, M.; Lipscomb, W. N. *J. Mol. Biol.* **1983**, *168*, 367–387.
3. Christianson, D. W.; Lipscomb, W. N. *Proc. Natl. Acad. Sci. U.S.A.* **1985**, *82*, 6840–6844.
4. Christianson, D. W.; Kuo, L. C.; Lipscomb, W. N. *J. Am. Chem. Soc.* **1985**, *107*, 8281–8283.
5. Christianson, D. W.; Lipscomb, W. N. *J. Am. Chem. Soc.* **1986**, *108*, 4998–5003.
6. Christianson, D. W.; Lipscomb, W. N. *J. Am. Chem. Soc.* **1986**, *108*, 545–546.
7. Christianson, D. W.; David, P. R.; Lipscomb, W. N. *Proc. Natl. Acad. Sci. U.S.A.* **1987**, *84*, 1512–1515.
8. Christianson, D. W.; Lipscomb, W. N. in *Zinc Enzymes*, Bertini, I.; Luchinat, C.; Maret, W.; Zeppezauer, M., Eds. "Mechanistic Inferences from the Binding of Ligands to Carboxypeptidase A." Birkhauser: Boston, **1986**, pp. 121–132.
9. Christianson, D. W.; Lipscomb, W. N. *Proc. Natl. Acad. Sci. U.S.A.* **1986**, *83*, 7658–7572.
10. Anson, M. L. *J. Gen. Physiol.* **1937**, *20*, 777–780.
11. Vallee, B. L.; Neurath, H. *J. Am. Chem. Soc.* **1954**, *76*, 5006–5007.
12. Lipscomb, W. N.; Hartsuck, J. A.; Reeke, G. N.; Quiocho, F. A.; Bethge, P. H.; Ludwig, M. L.; Steitz, T. A.; Muirhead, H.; Coppola, J. C. *Brookhaven Symp. Biol.* **1968**, *21*, 24–90.

13. Holmes, M. A.; Matthews, B. W. *J. Mol. Biol.* **1982,** *160,* 623–639.
14. Monzingo, A. F.; Matthews, B. W. *Biochemistry* **1984,** *23,* 5724–5729.
15. Weaver, L. H.; Kester, W. R.; Matthews, B. W. *J. Mol. Biol.* **1977,** *114,* 119–132.
16. Holmes, M. A.; Matthews, B. W. *Biochemistry* **1981,** *20,* 6912–6920.
17. Kester, W. R.; Matthews, B. W. *Biochemistry* **1977,** *16,* 2506–2516.
18. Bolognesi, M. C.; Matthews, B. W. *J. Biol. Chem.* **1979,** *254,* 634–639.
19. Tronrud, D. E.; Holden, H. M.; Matthews, B. W. *Science* **1987,** *235,* 571–574.
20. Schmid, M. F.; Herriott, J. R. *J. Mol. Biol.* **1976,** *103,* 175–190.
21. Titani, K.; Ericsson, L. H.; Walsh, K. A.; Neurath, H. *Proc. Natl. Acad. Sci. U.S.A.* **1975,** *72,* 1666–1670.
22. Dideberg, O.; Charlier, P.; Dive, G.; Joris, B.; Frere, J. M.; Ghuysen, J. M. *Nature (London)* **1982,** *299,* 469–470.
23. Hammond, G. S. *J. Am. Chem. Soc.* **1955,** *77,* 334–338.
24. Lipscomb, W. N. *Annu. Rev. Biochem.* **1983,** *52,* 17–34.
25. Argos, P.; Garavito, R. M.; Eventoff, W.; Rossmann, M. G. *J. Mol. Biol.* **1978,** *126,* 141–158.
26. Kester, W. R.; Matthews, B. W. *J. Biol. Chem.* **1977,** *252,* 7704–7710.
27. Woolley, P. *Nature (London)* **1975,** *258,* 677–682.
28. Fife, T. H.; Przystas, T. J. *J. Am. Chem. Soc.* **1986,** *108,* 4631–4636.
29. Groves, J. T.; Olson, J. R. *Inorg. Chem.* **1985,** *24,* 2715–2717.
30. Hangauer, D. G.; Monzingo, A. F.; Matthews, B. W. *Biochemistry* **1984,** *23,* 5730–5741.
31. Stryer, L.; Kendrew, J. C.; Watson, H. C. *J. Mol. Biol.* **1964,** *8,* 96–104.
32. Hendrickson, W. A.; Konnert, J. In "Biomolecular Structure, Function, and Evolution"; Srinivasan, R., Ed.; Pergamon Press: London, 1981, pp. 43–47.
33. Luzzati, V. *Acta Crystallogr.* **1953,** *6,* 142–192.
34. Makinen, M. W.; Kuo, L. C.; Dymowski, J. J.; Jaffer, S. *J. Biol. Chem.* **1979,** *254,* 356–366.
35. Sander, M. E.; Witzel, H. *Biochem. Biophys. Res. Commun.* **1985,** *132,* 681–687.
36. Bürgi, H. B.; Dunitz, J. D. *Acc. Chem. Res.* **1983,** *16,* 153–161.
37. Suigimoto, T.; Kaiser, E. T. *J. Am. Chem. Soc.* **1978,** *100,* 7750–7751.
38. Suigimoto, T.; Kaiser, E. T. *J. Am. Chem. Soc.* **1979,** *101,* 3946–3951.
39. Gelb, M. H.; Svaren, J. P.; Abeles, R. H. *Biochemistry* **1985,** *24,* 1813–1817.
40. Grobelny, D.; Goli, U. B.; Galardy, R. E. *Biochemistry* **1985,** *24,* 7612–7617.
41. Galardy, R. E.; Kortylewicz, Z. P. *Biochemistry* **1984,** *23,* 2083–2087.
42. Auld, D. S.; Holmquist, B. *Biochemistry* **1974,** *13,* 4355–4361.
43. Jacobsen, N. E.; Bartlett, P. A. *J. Am. Chem. Soc.* **1981,** *103,* 654–657.
44. Abramowitz, N.; Schecter, I.; Berger, A. *Biochem. Biophys. Res. Commun.* **1967,** *29,* 862–867.
45. Kuo, L. C.; Fukuyama, J. M.; Makinen, M. W. *J. Mol. Biol.* **1983,** *163,* 63–105.
46. Kuo, L. C.; Makinen, M. W. *J. Biol. Chem.* **1982,** *257,* 24–27.
47. Kuo, L. C.; Makinen, M. W. *J. Am. Chem. Soc.* **1985,** *107,* 5255–5261.
48. Riordan, J. F. *Biochemistry* **1973,** *12,* 3915–3923.
49. Rees, D. C.; Lipscomb. W. N. *J. Mol. Biol.* **1982,** *160,* 475–498.
50. Christianson, D. W.; Lipscomb, W. N. *J. Am. Chem. Soc.* **1987,** *109,* 5536–5538.
51. Lipscomb, W. N. *Acc. Chem. Res.* **1982,** *15,* 232–238.
52. Vallee, B. L.; Galdes, A.; Auld, D. S.; Riordan, J. F. In "Metal Ions in Biology", Vol 5; Spiro, T. G., Ed.; Wiley: New York, 1983, pp. 25–75.
53. Vallee, B. L.; Galdes, A. *Adv. Enzymol.* **1984,** *56,* 283–430.
54. Gardell, S. J.; Craik, C. S.; Hilvert, D.; Urdea, M. S.; Rutter, W. J. *Nature (London)* **1985,** *317,* 551–555.
55. Gardell, S. J. Abstracts, Genetic Physico-Chemical Approaches for Analysis of Biological Catalysts. University of Florence: Florence, Italy, June 16–20, 1986.
56. Karplus, M.; McCammon, J. A. *Ann. Rev. Biochem.* **1983,** *53,* 263–300.

57. Makinen, M. W. In "Zinc Enzymes"; Bertini I.; Luchinat, C.; Maret, W.; and Zeppezauer, M., Eds.; Birkhauser: Boston, 1986, pp. 215–224.
58. Bash, P. A.; Singh, U. C.; Brown, F. K.; Langridge, R.; Kollman, P. A. *Science* **1987,** *235,* 574–576.

CHAPTER 2

Stereoelectronic Analysis of Enzymatic Reactions

Steven A. Benner

**Swiss Federal Institute of Technology,
Zurich, Switzerland**

CONTENTS

1. Introduction.. 27
2. Stereoelectronic Principles 28
3. Large Stereoelectronic Effects in Enzymatic Reactions 41
4. Medium-Sized Stereoelectronic Effects 47
5. Small Stereoelectronic Effects 56
6. Conclusions... 70
Acknowledgments....................................... 71
References 71

1. INTRODUCTION

Stereoelectronic theory concerns those aspects of chemical structure and reactivity that can be explained in terms of the arrangement of lone pairs of electrons, bonds and localized orbitals in three-dimensional space.[1,2] To the organic chemist, stereoelectronic theory is especially well known as an explanation for the relative energies of different anomers of sugars and related compounds, the reactivity of acetals and orthoesters, and the mechanistic details of enolizations, decarboxylations, and hydrolysis reactions. Less familiar to the organic chemist, of course, are problems in enzymatic reaction mechanisms and the application of stereoelectronic theory to solve these problems.

Biological chemists, in contrast, are largely unaware of stereoelectronic theory. Even those who are familiar with its broad outlines are skeptical that a theory dealing with orbitals, their relative energies, and their geometric arrangement will be of any help in addressing problems presented by complex biological catalysts. Further, biological chemists often find theory daunting, especially that phrased in the language of physical organic chemistry.

This survey attempts to bridge this gap. We shall briefly discuss stereoelectronic theory, focusing on a few areas in organic chemistry where the theory has been applied and estimating the energetic importance of different stereoelectronic phenomena. Then we shall apply stereoelectronic theory to the analysis of several specific enzymatic reactions. We shall present stereoelectronic hypotheses, several of them new, to illustrate how the theory can help explain certain aspects of enzymatic behavior and provide working hypotheses that make predictions about enzymatic behavior. Finally, we shall discuss briefly the connection between chemical reactivity and natural selection, the process whereby enzymatic catalytic sites have evolved to reflect underlying chemical fundamentals.

At the outset, we should warn the reader that stereoelectronic theory is, in fact, a collection of different theories, postulates, and hypotheses, many of which are only incompletely supported by experimental data. Therefore, application of stereoelectronic theory in any field is not necessarily secure.

For the organic chemists, many of whom have embraced stereoelectronic theory, this review seeks to point out some of the more problematical aspects of the theory. There remains a need for more research on organic models to better establish the origin of "stereoelectronic" effects, the range of chemical phenomena to which stereoelectronic analyses are applicable, and the magnitude of stereoelectronic effects.

For the biological chemist, this review aims to illustrate how stereoelectronic theory can be used to discuss the mechanistic details of organic reactions in the active sites of enzymes. Particular stress will be given to those cases where stereoelectronic analysis makes predictions about the behavior of individual enzymes. Many of these predictions will undoubtedly be tested by experiment in the near future.

2. STEREOELECTRONIC PRINCIPLES

A. Stereoelectronic Effects in Organic Chemistry

Organic chemistry is dominated by "structural theory," which attempts to explain the reactivity of organic molecules in terms of the arrangement of their constituent atoms in three-dimensional space. While this "traditional" approach has been successful in explaining a wide range of behaviors in organic molecules, it has proven to be only partly satisfactory. Important elements of organic reactivity remain unexplained unless structural theory includes an analysis of the spatial arrangement of orbitals, bonds, and unshared pairs of electrons as well as atoms in a molecule. Analyses of molecular behavior that focus on these aspects of molecular structure are gathered here together under the rubric *stereoelectronic theory*.

For example, the barrier to interconversion of cis and trans isomers in olefins is generally explained by noting that a double bond is formed when there is overlap between p orbitals on adjacent atoms (Figure 2-1). This overlap is absent when the double bond is "twisted," and the molecule in twisted conformation therefore has higher energy than in a planar conformation. The requirement for a planar olefin system, embodied in "Bredt's rule" as taught in introductory organic chemistry, is stereoelectronic in the most general sense of the term. There is energetic stabilization associated with the overlap of two orbitals that together contain a sum of two electrons (Figure 2-2). Thus, molecules are expected to adopt conformations where such overlap is possible.

Many elements of stereoelectronic theory are simply restatements of the general principle that it is energetically unfavorable to twist double bonds. However, in recent years, stereoelectronic theory has come to focus on the orientation of nonbonding "lone pairs" of electrons relative to neighboring atoms and bonds. Here, the two orbitals are generally an orbital containing a lone pair of electrons (eg, an sp^3 orbital on an oxygen) and a vacant orbital,

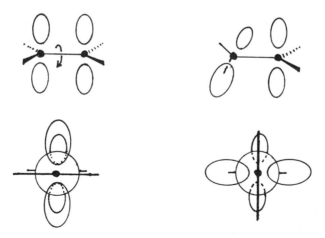

Figure 2-1. Overlap between p orbitals on adjacent atoms in a planar system (left), and in a system twisted 180° (right), looking at the π system from the side (top drawings) and looking down the axis of the carbon–carbon bond (bottom drawings).

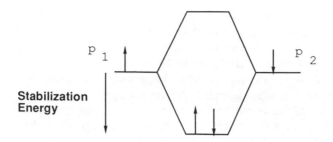

Figure 2-2. In classical bonding theory, when two p orbitals have correct overlap, they mix to form a π (bonding) orbital and a π^* (antibonding) orbital. The energy of a system with two electrons is lowered if this overlap can occur, as the two electrons can occupy the π orbital with lower energy.

generally an antibonding orbital associated with a σ bond (Figure 2-3). Stereoelectronic theory argues that partial donation of electron density from the filled orbital into the vacant antibonding orbital is energetically favorable, and that molecules will adopt conformations which, by providing the proper orbital overlap, will allow this donation to take place.

There is hardly an area of structural analysis of organic compounds containing heteroatoms that has been untouched by stereoelectronic analysis.

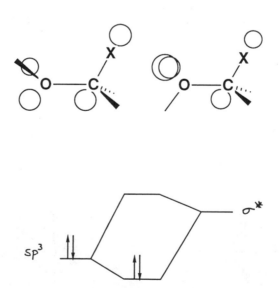

Figure 2-3. A filled sp³ orbital (eg, a lone pair of electrons on oxygen) next to a σ^* orbital that is relatively low in energy (eg, one associated with a carbon–halogen bond). The system can achieve lower energy, at least in principle, if it adopts a conformation where the sp³ and σ^* orbitals overlap (left, correct overlap). This overlap occurs when the C—X bond is *antiperiplanar* to a lone pair of electrons in an sp³ orbital. No overlap is shown on the right.

Two excellent monographs reviewing this area have been written.[1,2] The reader is referred to them for an extensive discussion of the development and application of stereoelectronic theory by two of its most important exponents.

Stereoelectronic theory is used to interpret both the structure and the reactivity of organic molecules in general. However, two groups of compounds played an important role in developing the structural applications of the theory. The first group comprises oxacyclohexane rings bearing an electronegative substituent at the 2 position; an example is 2-chloro-1,3-dioxane (Figure 2-4). These molecules often prefer conformations where the chloro substituent is axial.[3]

A second group, related structurally to the first but more biologically relevant, are sugar derivatives, whose conformational properties have been studied extensively. These studies have shown that electronegative substituents in the anomeric position (carbon atom 1) often prefer to be axial. Lemieux and co-workers coined the term "the anomeric effect" to describe the second phenomenon, and the term has come to be used widely to describe the first type of phenomenon as well.[4]

In organic chemistry, a difference between fact and expectation is called an "effect." For steric reasons, cyclohexanes with equatorial substituents are energetically more stable than those with axial substituents. However, oxacyclohexanes with an electron withdrawing group attached to the carbon adjacent to the ring oxygen in an *axial* position are generally preferred. Thus, the "structural anomeric effect" is defined as the difference between two energy differences: (1) the difference in energy between axial and equatorial anomers and (2) the difference in energy between two epimers of an analogously substituted cyclohexane.[5]

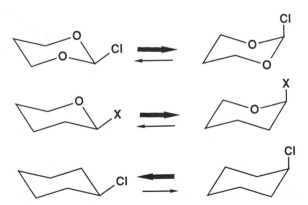

Figure 2-4. The chloro substituent prefers to be axial in the dioxacyclohexane system (top). This contrasts with the normal conformational behavior of cyclohexanes (bottom), where the equatorial conformer is more stable by 3–10 kcal/mol, depending on the substituent. Sugars and other oxacyclohexane species (middle) also prefer the axial substituent more than would be expected based on substituted cyclohexanes as models.

There are two competing explanations for the structural anomeric effect (Figure 2-5). The first, preferred by many organic chemists, argues that the anomeric effect arises from an interaction between a lone pair of electrons centered on oxygen and the adjacent vacant antibonding orbital (σ^*) orbital associated with the carbon–heteroatom bond. In the axial conformer of

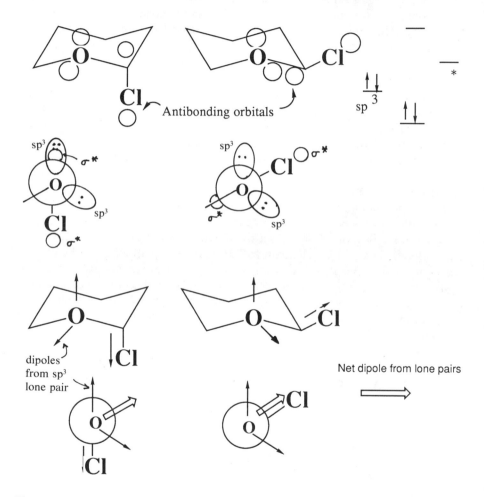

Figure 2-5. Two explanations for the "anomeric effect." At the top is a view of 2-chloro-oxacyclohexane from the side; the bottom shows projection down the oxygen–carbon bond. The "electronic" interpretation (top) notes that when the chloro substituent is in the equatorial position, the antibonding (σ^*) orbital associated with the carbon–chlorine bond does not overlap well with the adjacent sp³ orbital. In the axial conformer, the sp³ orbital is antiperiplanar to the C—Cl bond, overlap occurs, and the molecule is therefore more stable. In the dipolar interpretation for the anomeric effect (bottom), it is noted that in the equatorial conformer (right), the dipole associated with the C—Cl bond is aligned with the net dipole arising from the two lone pairs. This is energetically less favorable than the situation in the axial conformer (left), where the dipoles are not aligned.

2-chlorooxacyclohexane, the lone pair of electrons on the ring oxygen is antiperiplanar to the σ bond joining the carbon and the chlorine and is correctly oriented to allow donation of electrons from the lone pair into the antibonding orbital of this bond. This is expected to be energetically stabilizing. In the equatorial conformer, the orbital bearing the lone pair is nearly orthogonal to the antibonding orbital; the two orbitals do not overlap, and no attendant stabilization can be realized.

In this view, the anomeric effect occurs because the axial conformer is stabilized with respect to the equatorial conformer by a favorable electronic interaction.

The alternative interpretation centers on dipolar interactions between the lone pair of electrons on oxygen and the electron withdrawing substituent. The lone pair creates a dipole, with its negative end pointing away from the oxygen in the direction of the lone pair. Likewise, a carbon–heteroatom bond is polarized with the negative end of the dipole pointing towards the heteroatom. Orienting these two dipoles parallel to each other is presumed to be undesirable. For example, in the equatorial conformer of 2-chlorooxacyclohexane (but not in the axial conformer), the two dipoles are oriented parallel. Thus, in this view, the equatorial conformer is destabilized with respect to the axial conformer by an unfavorable dipole interaction.

Crystallographic data have been especially powerful in persuading many organic chemists to adopt the first explanation in preference to the latter. Consider, for example, the structure of 2,3-*cis*-dichloro-1,4-dioxacyclohexane (Figure 2-6). The carbon–chlorine bond joining the axial chlorine substituent to the ring is significantly longer than the carbon–chlorine bond joining the equatorial chlorine substituent to the ring.[3] Further, the bond joining the carbon attached to the axial chlorine and the adjacent oxygen is considerably shorter than the bond joining the carbon bearing the equatorial chlorine and its adjacent oxygen.

The electronic interpretation of the anomeric effect implies a donation of electrons from a lone pair on oxygen into the antibonding orbital of the bond

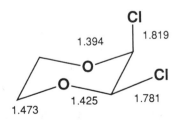

Figure 2-6. In 2,3-dichloro-1,4-dioxacyclohexane, the axial C—Cl bond is longer than the equatorial C—Cl bond, and the O—C bond to the carbon bearing the axial chlorine is shorter than the O—C bond to the carbon bearing the equatorial chlorine. This is consistent with the electronic interpretation of the anomeric effect, but does not "prove" that this is the sole, or even principal source of the effect.

to the axial substituent. This means that there is more than a "single bond" between the oxygen and the anomeric carbon in the cases where the orbitals and bonds are correctly disposed. Further, donation of electron density into the antibonding orbital of the axial carbon–chlorine bond implies that there is less than a "single bond" to the axial chlorine. Thus, stereoelectronic analysis, and the electronic interpretation of the anomeric effect in particular, accounts for the subtle differences in bond lengths observed in this molecule.

However, it is generally inadvisable to presume that the behavior of organic molecules arises from a single interaction. This is true as well in stereoelectronic theory. There are now data that argue persuasively that dipolar interactions also have a role in explaining the structure of these compounds, along with steric and other factors. Indeed, much of the data suggests that orbital overlap may not be the most important origin of the anomeric effect in at least some compounds.

For example, the magnitude of the anomeric effect in sugar derivatives is dependent on the nature of the solvent[6,7] in a direction that strongly suggests that dipolar interactions are at least partly responsible for the anomeric effect. Only in less polar solvents do derivatives of sugars show a preference for the axial anomer.[4,8,9,10] The direction and magnitude of the structural dependence on the dielectric constant of the solvent are most clearly consistent with the explanation for the anomeric effect that considers dipolar interactions. In more polar solvents, the anomeric effect is weaker. In less polar solvents, the anomeric effect is stronger.

In some compounds, the electronic explanation for the observed conformational behavior is clearly inadequate. Particularly dramatic is the example provided by the studies of Lemieux[11] and Paulsen and their co-workers[12] on a triacetoxyxylopyranose derivative with an imidazoyl substituent at the anomeric position (Figure 2-7). In the unprotonated form, an equilibrium mixture contains 35% of a conformer with the imidazoyl group equatorial and three acetoxy groups axial. This demonstrates a preference for the imidazoyl group to be equatorial, unexpected based on considerations of orbital overlap alone.

However, when the imidazoyl group is protonated, *only* the equatorial isomer is observed. Naive considerations suggest that the relevant antibonding orbital in the protonated molecules has lower energy than in the unprotonated molecule. Therefore, one naively would expect that the axial conformer would be more preferred in the protonated form than in the unprotonated form.

This is an example of the "reverse anomeric effect."[13] The reverse anomeric effect, defined for observations that contradict expectations based on the anomeric effect, is here the observation that positively charged groups at an anomeric center prefer to be equatorial. The reverse anomeric effect contradicts explanations for the anomeric effect that are based on simple ideas of orbital overlap and remains a major problem in electronic interpretations of structure and reactivity in glycosides and related compounds.

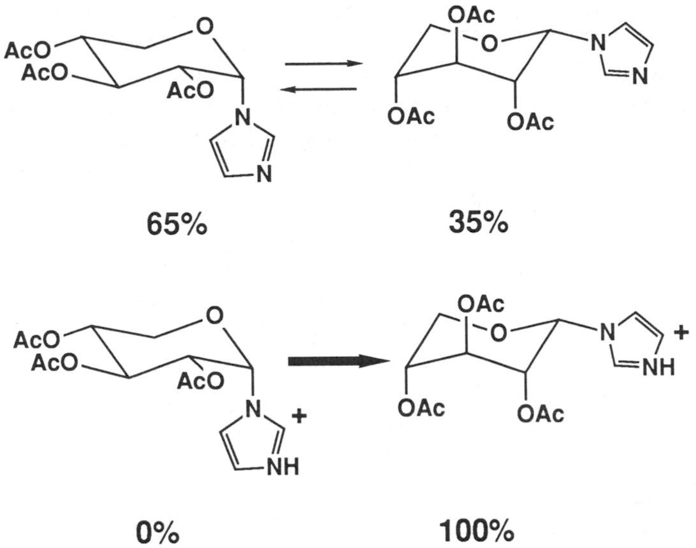

65%   **35%**

0%   **100%**

Figure 2-7. The fact that the protonated imidazole prefers to be equatorial more than the unprotonated imidazole is unexpected by those who believe in the electronic interpretation of the anomeric effect. The relevant σ^* orbital species should be lower in energy, and the attendent energetic stabilization greater, in the protonated species than in the unprotonated species. These data are consistent with the dipolar explanation for the anomeric effect.

Similar analyses may be used to explain the conformations of molecules containing atoms other than oxygen and having rings of different size or no rings at all. For example (Figure 2-8), dimethoxymethane adopts a conformation in the gas phase where both methyl groups are gauche to a carbon–oxygen bond.[14] In this conformation, each carbon–oxygen bond is antiperiplanar to a lone pair of electrons. This conformational preference can be explained in terms of a "generalized anomeric effect" where lone pairs of electrons antiperiplanar to carbon–oxygen bonds are able to form stabilizing interactions with the antibonding orbitals of those bonds. Alternatively, the

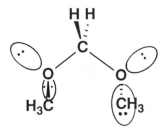

Figure 2-8. The conformation of dimethoxymethane in the gas phase has a lone pair of electrons antiperiplanar to both carbon–oxygen bonds.

effect can be explained in terms of dipolar interactions as discussed previously.

Comparable arguments are often used rather successfully to describe the structure of molecules with heteroatoms other than oxygen. Again, it is important to note that, while these analyses generally appear successful, isolated problems remain and suggest that other interactions are partly responsible for observed molecular behavior. For example, in cyclic phosphate esters with chloro or methoxy attached to the phosphorus, the axial conformation is favored over the equatorial by 1.4–1.5 kcal/mol.[15] This is consistent with expectations based on the antiperiplanar lone pair hypothesis. However, when the substituent is a methyl group, the axial conformer is also preferred by 1 kcal/mol. Further, when the substituent is an unprotonated methylamino group, the equatorial isomer is slightly preferred. These results are not readily anticipated by a view that argues that the anomeric effect arises from a stabilizing interaction between lone pairs of electrons and vacant antibonding orbitals.

Thus, the anomeric effect describes a systematic deviation from the expected behavior of organic molecules that deserves an explanation. The anomeric effect is undoubtedly a result of a combination of several microscopic factors. In applying stereoelectronic analysis to biochemistry, it is important to keep this fact in mind. As we shall see later, the potentially multiple origins of the anomeric effect make analyses more complicated, but not necessarily less rewarding.

Stereoelectronic analyses can also be used to explain the reactivity of organic molecules, and one frequently sees references to a "kinetic anomeric effect."[16] Here, the orientation of lone pairs of electrons in three-dimensional space is presumed to influence the relative energies of different transition states, often for competing reactions.

One of the early arguments that invoked orbital overlap to explain a difference in rates for two competing reactions was made by Corey and Sneen. In keto steroids, axial protons α to the ketone were found to exchange with solvent via enolization more rapidly than equatorial protons (Figure 2-9).[17] This is understandable in terms of a simple stereoelectronic argument. The bond to the axial hydrogen has rather good overlap with the π system of the adjacent carbonyl group; the overlap between the equatorial bond and the π system is rather poor. As the overlap with the π system is largely responsible for the stabilization of the incipient carbanion in the enolization process, the axial hydrogen is expected to be the easier to remove.

The kinetic anomeric effect was developed by Deslongchamps and co-workers,[2] who used the theory to explain the rates of removal of hydrogen atoms attached to carbons also bonded to heteroatoms and the rates of hydrolysis of orthoesters. In both cases, the rates were proposed to be dependent on the orientation of lone pairs of electrons relative to bonds that were breaking. Again, their arguments were formally similar, if more elaborate, to those outlined previously. To assist an atom as it departs with a pair

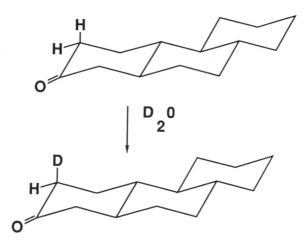

Figure 2-9. In certain keto steroids, base-catalyzed exchange of protons α to the carbonyl group results in washing in of label from solvent in the axial position much faster than in the equatorial position.

of electrons from a reaction center, atoms bonded to the reaction center should bear lone pairs of electrons arranged antiperiplanar to the departing group. In this position, electron density from the lone pairs was proposed to enter the antibonding orbital adjacent to it, weakening the bond and making it more susceptible to reaction.

Kirby,[1] Perrin,[18] and many others have further explored the effect of the relative orientation of lone pairs of electrons antiperiplanar to σ bonds having low-lying antibonding orbitals. Kirby has noted that such bonds are longer with antiperiplanar lone pairs of electrons than without and has correlated bond length with kinetic reactivity. Cieplak has further extended these ideas to include interactions between antibonding orbitals and lone pairs of electrons on atoms not directly bonded to the reacting center.[19]

Other reactions have been similarly examined using stereoelectronic theory. For example, stereoelectronic analysis argues that S_N2 reactions at saturated centers are subject to stringent geometric requirements.[20] If a lone pair on the attacking nucleophile must overlap with the antibonding orbital from the bond to the leaving group, the attacking nucleophile must approach 180° from the departing group. This has, in fact, been shown to be (at least approximately) the case.[21]

Analyses based on a "kinetic anomeric effect" can be rather complicated. "Anomeric effects" both in ground states and in transition states must be considered. Multiple conformations of ground states can exist, each with different relative orientations of orbitals and lone pairs of electrons. Deciding which of several stereoelectronically distinct conformers is undergoing reaction is challenging, a fact embodied in the Curtin–Hammett principle.

The casual reader of the stereoelectronic literature can spot some examples where the arguments perhaps lack full rigor. First, it is important to note that kinetic effects are concerned with the relative rates of reactions, and experiments that examine the product ratios to obtain evidence for a kinetic anomeric effect must examine reactions that are under kinetic control.

Further, stereoelectronic interactions that are presumed to stabilize a ground state are often the same ones that are presumed to activate a molecule for reaction. The clearest examples are from reactions at the anomeric center of sugars, where the same stereoelectronic arguments are occasionally used to predict both that an axial anomer will be more stable than the equatorial anomer *and* that the axial anomer will be more reactive than the equatorial anomer. For such arguments to be true, the stereoelectronic interaction must be stronger in the transition state than it is in the ground state.

Further, molecules that display conformational flexibility often have conformations with different stereoelectronic interactions. If these different conformers are in rapid equilibrium compared to the rate of the reaction, the Curtin–Hammett principle requires that the reaction pathways from either the major or the minor conformer be kinetically indistinguishable. The apparent energy of activation in either case corresponds to the energy difference between the lowest ground state and the lowest transition state. This is true regardless of whether or not the lowest transition state is reached directly from the lowest ground state.

Finally, even if one conformer is identified as the reacting species, it is incorrect to assume that the orbital overlap in the transition state arising from that conformer will be the same as in that reacting ground state. If the transition state for a reaction is extremely late (''productlike''), it will have largely lost in its structure the information that tells it which of perhaps several possible ground states it came from. Thus, a late transition state having favorable stereoelectronic interactions may in principle be approached from a ground state that lacks those favorable interactions.

In short, stereoelectronic theory presents the full range of intellectual challenges and pitfalls of any theory in physical organic chemistry. Sinnott and Bennett have recently provided an excellent review of some of the pitfalls that can accompany a casual stereoelectronic analysis.[22]

B. Stereoelectronic Analysis and Enzymatic Reactions

The same principles that govern the reactivity of molecules in solution are presumed to govern the reactivity of molecules in the active sites of enzymes. Thus, considerations of interactions between orbitals and electrons in three-dimensional space that are valid in solution are likely to be valid in the active sites of enzymes as well.

However, for reactions in the active sites of enzymes, the discussion must change its form. In the active site, the energies of different conformations of substrate molecules are dominated by interactions between the substrate and amino acid residues in the protein. Further, the reactivity of molecules in the active site is dominated by the interactions of amino acid residues with the transition states for possible reaction paths. With enzyme–substrate interactions on the order of 10–20 kcal/mol, these interactions cannot be neglected.

The amino acid residues present in the active site are a product of biological evolution. Therefore, there are two ways to interpret structural and behavior features of a specific protein. The first focuses on function, where natural selection has presumably produced proteins with behaviors that best assist the survival of the host organism. The second focuses on the history of the enzymatic lineage, where random mutations were randomly fixed in ancestral proteins during the divergent evolution of the protein. Some features of the active site may result of historical accident and have no particular purpose.[23,24]

The relative importance of accident and selection in the evolution of modern proteins is not clear.[25] However, it is clear that natural selection is the only mechanism for obtaining functional behavior in the active site. Natural selection acts at the level of the organism; in principle, enzymes will obey stereoelectronic principles only if organisms containing enzymes that do not obey stereoelectronic principles are less fit to survive and reproduce.

Therefore, the central question becomes this: Is the fitness of an organism less if it contains enzymes that do not conform to stereoelectronic principles? The question is esoteric to chemists and biologists alike. It concerns two fields, evolutionary biology and physical organic chemistry, and there are few scientists in either field who are familiar with (let alone are willing to tolerate) the other. However, this is the question that must be addressed to understand the importance of stereoelectronics in enzymatic catalysis.

Those interested in the subject presently hold two contradicting points of view. On one hand, rate enhancements of 12 orders of magnitude (17 kcal/mol) are not uncommon in enzymatic reactions. Stereoelectronic effects are often only 1–2 kcal/mol in magnitude. It might be argued that the latter are negligible with respect to the former and that natural selection is unlikely to be influenced by them.

On the other hand, if enzymes have evolved to be optimal catalysts, and if stereoelectronic factors favor one mode of reaction over another, one might argue that the enzyme will have evolved an active site that exploits the stereoelectronic principle rather than one that fights it. Indeed, if natural selection refines the active site quite highly, it may be that "perfect" enzymes will follow stereoelectronic rules that would be lost in the "noise" in normal chemical reactions.

Thus, the organic chemist interested in stereoelectronic effects is concerned "only" with characterizing the effects and determining their magnitudes. The biological chemist must also ascertain whether effects of this nature and of this magnitude have any impact on the fitness of an organism struggling to survive and reproduce.

A word must be said here about how little is known about the impact of enzymatic behaviors in general on natural selection.[25] We have addressed this issue at length elsewhere.[26] We cannot rule out a priori the hypothesis that natural selection produces enzymes that conform to subtle stereoelectronic rules. However, such ideas are often the subject of extensive controversy. It appears as if nature makes chemical distinctions in enzymes worth approximately a kilocalorie per mole in energy. This implies that, at least in some cases, energetic distinctions at this level in a single enzyme have an impact on fitness.[24,26] However, ascribing a functional purpose to any subtle chemical trait of a biological macromolecule is likely to be disputed from some quarter. And, as we shall see later, some of the stereochemical behaviors are subtle indeed.

Fortunately, for stereoelectronic control, one may (at least in principle) make the argument in the reverse direction. One could deduce the nature and magnitude of a stereoelectronic effect by studying model reactions. Based on these studies, one might then, as a hypothesis, *assume* that such stereoelectronic hypotheses have governed the evolution of enzymatic active sites. Finally, one could *test* this hypothesis by examining the behavior of individual enzymes. If a sufficient number of enzymes conforms to this hypothesis, one might *conclude* that stereoelectronic effects are sufficiently large to have an impact on the survival of the host organism.

Even this approach has some problems. Stereoelectronic predictions are often "either–or" propositions likely to be confirmed 50% of the time by random events. Thus, a statistically significant number of enzymes catalyzing analogous reactions must be examined in order to draw statistically significant conclusions.

Further, the several enzymes examined should not be closely related. Closely homologous enzymes are expected to behave similarly even if there is no adaptive reason for similar behavior, just as human siblings are expected to look alike regardless of whether the specific appearance confers survival advantage. Thus, two enzymes behaving according to one stereoelectronic proposition confirm this proposition most strongly if the two enzymes are not homologous. Unfortunately, proving "nonhomology" for a pair of enzymes is extremely difficult.

C. The Magnitude of Stereoelectronic Effects

The magnitude of stereoelectronic effects is quite variable. At one extreme, the twisting of an olefin costs 40–50 kcal/mol. At the other extreme, the magnitude of the anomeric effect used to explain the preference of electronegative substituents at the anomeric center of glycosides is only 1–2 kcal/mol. In between are stereoelectronic effects with intermediary energetic implications.

In our discussion, the following pattern will emerge. Large stereoelectronic effects (40–50 kcal/mol) can generally be rigorously ascribed to orbital

overlap effects. Enzymatic reactions nearly uniformly conform to stereoelectronic rules of this energetic magnitude. Stereoelectronic rules with
intermediate energetic consequences (~10–20 kcal/mol), such as a preference for anti addition to an olefin, also seem to be obeyed by enzymes.
However, small stereoelectronic effects (1–2 kcal/mol) may be best explained not as outcomes of orbital interactions intrinsic to a molecular structure, but rather as a combination of orbital, dipolar, or solvent interactions
(vide supra). As solvation and dipolar interactions are quite different in the
active sites of enzymes, it is quite possible that these do not dominate the
reactivity of organic molecules in the active site. Predictions of enzymatic
behavior based on such small effects are only infrequently followed by enzymes. In some cases (eg, in vinylogous elimination reactions) stereoelectronic expectations seem to be uniformly *violated*.

3. LARGE STEREOELECTRONIC EFFECTS IN ENZYMATIC REACTIONS

A. Enolizations

One of the first rules taught in introductory courses in organic chemistry is
that substituents attached to a carbon–carbon double bond must lie in a
plane. Bredt's rule forbids double bonds at "bridgeheads" in bicyclic compounds where the double bond would be twisted. In carbon–carbon double
bonds, the barrier to rotation is approximately 45 kcal/mol. This is a large
number even when compared with the energies corresponding to the rate
enhancements brought about by enzymes. Not surprisingly, when "anti-
Bredt" compounds have been prepared by ingenious organic chemists,
they are uniformly quite reactive. It seems unlikely that enzymes would
have evolved to violate Bredt's rule, especially in a way nonproductive to
catalysis.

The hypothesis that axial hydrogens are removed in the enolization of
ketones that are part of six-membered rings (the "Corey–Sneen rule," vide
supra)[17] may be interpreted as arising from the requirement that the double
bond of an enol(ate) must be planar (Figure 2-9). Of course, the simplest
argument (referring only to the geometries of the ground state and the product enol) is not entirely rigorous. For example, if the transition state for the
enolization is very late (productlike) and largely planar, the departing proton
is essentially gone in the transition state. Such a transition state could (in
principle) be reached by the abstraction of either the axial or equatorial
hydrogen; by the time the transition state is achieved, the information is lost
as to whether an axial or an equatorial hydrogen was removed. Alternatively, one might say that if the transition state is formed with substantial
atomic reorganization (including essentially complete removal of the pro-

ton), either the axial or equatorial proton could leave via a transition state with satisfactory overlap.

Here, the caveat is perhaps moot. Enolizations do not seem to occur by transition states that are very late. Further, axial hydrogens are generally the ones that are abstracted, even outside of the active site. This then is our *first* *stereoelectronic* *"rule."* How well can it be applied to enzymology?

Enzymatic enolizations are common. Often, there is insufficient information regarding the conformation in the active site to know whether the enolization occurs according to the Corey–Sneen rule. However, in the enzyme triose-phosphate isomerase, the "principle" seems to apply. The enzyme catalyzes two enolization reactions. The stereochemistry of proton removal is known, as is the crystal structure of the enzyme. The crystal structure suggests that the enolizations proceed as expected based on the stereoelectronic rationalizations (Figure 2-10).[27]

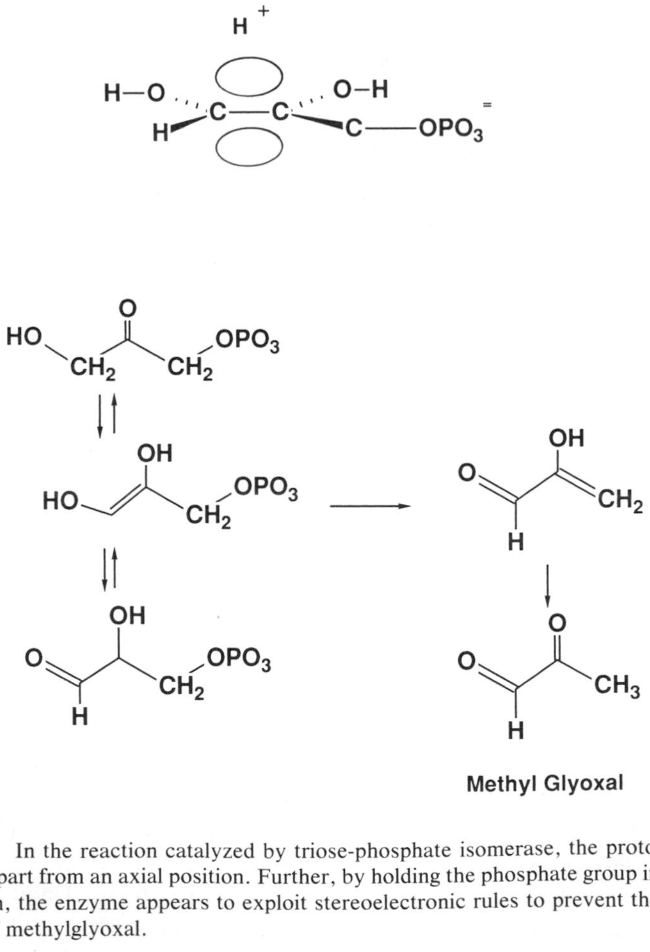

Methyl Glyoxal

Figure 2-10. In the reaction catalyzed by triose-phosphate isomerase, the proton appears to enter and depart from an axial position. Further, by holding the phosphate group in the plane of the π system, the enzyme appears to exploit stereoelectronic rules to prevent the undesirable formation of methylglyoxal.

Further, the enzyme appears to use stereoelectronic control to prevent an undesired side reaction. The intermediate enediol can readily lose phosphate to give, after tautomerization, methylglyoxal. This reaction path is often seen in solution reactions of triose phosphates.[28] Stereoelectronically, the fragmentation requires that the C—O bond holding phosphate at the 3-carbon of the enediol intermediate must be perpendicular to the plane of the π system (Figure 2-10).

Elimination of phosphate to form methylglyoxal is an undesirable side reaction in triose-phosphate isomerase, and one might expect the active site to have evolved the use of stereoelectronic principles to prevent the reaction. Indeed, in the bound triose phosphate, the phosphorus–oxygen bond is such that it lies in the plane of the adjacent π system. It appears as if stereoelectronic principles have been "exploited" in the active site to prevent an undesired side reaction.[27]

Finally, it has been called to the attention of this author[128] that a similar situation exists in aspartate aminotransferase, an enzyme dependent on pyridoxal phosphate. Again, elimination of the phosphate, in this case from the cofactor, from an intermediate in the reaction sequence is undesirable. Again, orientation of the potentially scissile bond in the plane of the pyridine ring makes this undesired reaction stereoelectronically disfavored.

B. Decarboxylations of β-Keto Acids

In enzymatic decarboxylations of β-keto Acids, an enol, or an analogously unsaturated species such as an enamine, is generally a presumed intermediate. Two mechanisms for catalyzing the decarboxylation of β-keto acids have been explored. In one class, exemplified by the enzyme acetoacetate decarboxylase from *Clostridium acetobutylicum*, the active site contains a lysine residue. The amino group of the lysine forms a Schiff's base with the keto group of acetoacetate, and the decarboxylation proceeds via an intermediate enamine (Figure 2-11).[29]

The enamine is expected to be planar, and stereoelectronic principles predict that the reactive conformer is one where the bond broken as carbon

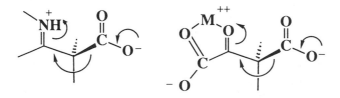

Figure 2-11.   Planar representations of the decarboxylation of β-keto acids via the Schiff's base and metal complex mechanisms. In the stereoelectronically allowed transition states, the breaking carbon–carbon bond is perpendicular to the carbonyl system.

dioxide is released should be "axial." In other words, the bond should be held in the active site so that there is maximal overlap with the adjacent π^* orbital.[30] Studies on "model" compounds, including bridgehead carboxylates,[31] argue (albeit with the same caveat as discussed previously) that decarboxylations occur only when the leaving carboxyl group is orthogonal to the plane of the π system. We may then formulate a *second stereoelectronic "rule"* and examine enzymatic decarboxylations where the principle might be tested.

In the enzyme acetoacetate decarboxylase, only one of the two enantiomers of the conformationally constrained substrate analog cyclohexanone-2-carboxylate is decarboxylated.[32] This experimental result provides some information about the spatial disposition of functional groups in the active site. In this case, decarboxylation is presumed to occur from the axial conformer bound in the active site. However, in the absence of crystallographic information, no conclusion can be drawn regarding stereoelectronic control in this system.

However, in the pathway for the oxidative removal of methyl groups from C-4 of steroids, stereoelectronic preferences appear to be violated. In the conversion of lanosterol to cholesterol, the two methyl groups at position 4 of the A ring are sequentially oxidized to carboxylic acids and then decarboxylated (Figure 2-12). The decarboxylation presumably occurs via an enol or another stereochemically similar intermediate formed from lanosterone.

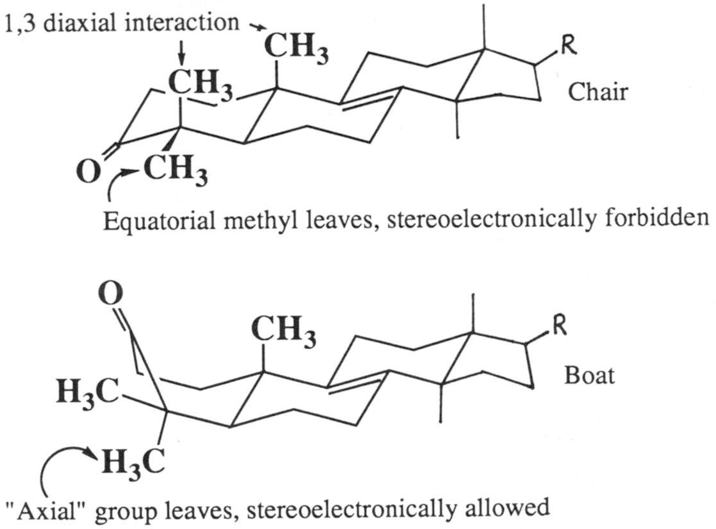

Figure 2-12. In the enzymatic oxidative demethylation of lanosterol, the group that is formally *equatorial* is oxidized and lost in each step. However, the boat conformer of lanosterone is expected to be low in energy because of an unfavorable 1,3 diaxial interaction. In the boat conformer, a substituent that would be equatorial in the chair is now axial and can leave (after oxidation to a carboxyl group) via a stereoelectronically allowed transition state.

Again, stereoelectronic predictions are that the carboxyl group being removed must occupy an axial (or β) position for much the same reason as the exchange of axial protons α to the carbonyl group is preferred. In the axial position, the breaking carbon–carbon bond can overlap with the π system of the adjacent carbonyl. The enol that is thus created is planar. In the equatorial (or α) position, the overlap between the breaking bond and the adjacent π system is small.

However, the first methyl group of lanosterone that is removed is the α-methyl group. This group is in the *equatorial* position in a standard chair conformation. Decarboxylation is with retention of configuration, leaving the second methyl group, β, in what is formally an "axial" position. However, before being decarboxylated, the second methyl group is epimerized into the 4α position. It is then lost in turn by oxidative decarboxylation from what is again formally an *equatorial* position.[32–35]

As stereoelectronic principles argue that *axial* and not equatorial groups should be lost, this pathway appears to violate stereoelectronic preferences twice. Two decarboxylations occur where the leaving electrophile (in this case, carbon dioxide) departs from an "equatorial" position. Departure of the carboxyl group formally leaves behind a carbanion intermediate. An electron pair in an equatorial position is not appropriately oriented in space to have proper overlap with the adjacent, stabilizing, carbonyl π system.

While this enzyme may indeed violate stereoelectronic rules, a conformational ambiguity common in stereoelectronic analysis suggests an alternative conclusion. Six-membered rings are conformationally flexible, and the A ring of lanosterone is especially so. In the conventional chair conformation, the methyl substituents at positions 4 and 10 have an unfavorable 1,3 diaxial interaction. Further, in the 3-keto species that is undergoing reaction, the boat does not have an unfavorable 1,4 interaction that would normally increase the energy of the boat conformer. Thus, in lanosterone, the boat conformer is expected to be unusually stable, and the chair conformer unusually unstable (Figure 2-12).

If we assume that the decarboxylation proceeds from a boat conformer, the stereoelectronic principles are not violated. In the boat conformation (with the carbonyl carbon at the "flagpole"), the α-methyl group at the 4 position (equatorial in the chair conformer) is now axial with an orientation relative to the adjacent carbonyl group that permits proper overlap.

Decarboxylation from a boat is consistent with data regarding the bromination of lanosterone, which might be viewed as a "model" system.[36] Here, the 2α-bromo steroid is the primary product. Again, these results are consistent with two interpretations: attack of an enol according to stereoelectronic principles to give the boat conformer of the product, or attack of the enol in violation of stereoelectronic rules to give a chair conformer.

More complicated is the oxidative decarboxylation of 4-methyl groups in steroids catalyzed by enzymes found in algae.[37] Here, the α-methyl group is again lost from an equatorial position, but the decarboxylation occurs with inversion of configuration. Thus, the carboxylate departs from the bottom of the molecule and the proton adds from the top. For stereoelectronic rules to hold, the decarboxylation must occur when the ring is in a boat conformation, while protonation of the intermediate enol is allowed only if the product is a chair.

While this proposition seems ad hoc, there is a kernel of chemical reasonableness to it. Were the carbon dioxide to leave from the β position in a chair conformer (consistent with stereoelectronic rules), the carboxylate in an axial position would have a severe 1,3 diaxial interaction with the methyl group in the 10 position. This is the same bad interaction that presumably forces lanosterone to adopt a boat conformation. Thus, it is reasonable that the boat conformation of the decarboxylating species is energetically accessible. Further, the attack of a small proton on the enol is not as likely to be constrained by steric factors. Addition of a proton from the top of the enol to produce a chair, which is now the more stable conformer, is reasonable.

C. Pyridoxal Enzymes

Reactions involving pyridoxal cofactors similarly involve the moving of double bonds. In most cases, a bond of the substrate covalently bonding to the pyridoxal cofactor is broken. Again, in the decarboxylation of amino acids, the racemization of amino acids, or the elimination of substituents from the β or γ positions of amino acids, stereoelectronic principles argue that the group being removed must be perpendicular to the plane of the Schiff's base in the active site (Figure 2-13). This expectation was first suggested by Dunathan,[38] who noted that this arrangement would produce a double bond that is not twisted. This, then, is a *third stereoelectronic prediction* that needs to be tested by examining enzymatic reactions.

In different enzymes employing pyridoxal as a cofactor, different bonds of the substrate are broken. Stereoelectronic arguments can be used to predict

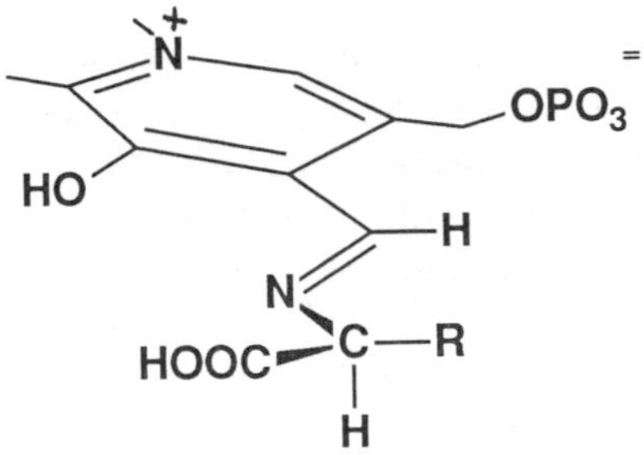

Figure 2-13. Adduct between pyridoxal phosphate and an amino acid in a conformation where the removal of the α proton of the amino acid is stereoelectronically allowed.

that the bond that is broken in each case will be oriented in the active site perpendicular to the plane of the pyridoxal ring. This prediction is testable by crystallographic analysis. Unfortunately, only one pyridoxal-dependent enzyme has a known crystal structure, aspartate aminotransferase. Precise geometric information cannot be obtained from the structure. However, a detailed analysis by Kirsch et al. suggests strongly that the structural information is at least not inconsistent with stereoelectronic principles.[39]

D. Conclusions

One might conclude from this discussion that enzymes have evolved not to violate those stereoelectronic preferences that have large (>30 kcal/mol) energetic consequences. However, these rules are in one view the least interesting, as these are the preferences that we least expect to be violated. We now proceed to examine stereoelectronic rules that have smaller energetic consequences.

4. MEDIUM-SIZED STEREOELECTRONIC EFFECTS

A. Biosynthesis of Olefins

In the departure of a nucleophile and an electrophile from vicinal atoms on a carbon skeleton to form an olefin, both "syn" and "anti" transition state geometries are possible (Figure 2-14). Intermediary geometries form twisted olefins which, following the previous discussion, presumably are not allowed. By microscopic reversibility, both syn and anti stereochemical modes are also available for addition reactions to olefins.

Stereoelectronic considerations suggest that the anti orientation is preferred in concerted elimination reactions. By microscopic reversibility, anti addition is therefore preferred in the addition reaction. In the transition state for the anti elimination, the electrons from the bond to the departing electrophile are oriented such that they can donate into the antibonding orbital of the bond to the departing nucleophile.

Figure 2-14. Anti (left) and syn (right) eliminations to form olefins. In each case, the microscopic reverse reaction must follow the same stereochemical path.

In contrast, stereoelectronic principles make no explicit statement regarding the preferred stereochemical course of a stepwise addition/elimination reaction. Stereoelectronics would be equally consistent with either the syn or the anti mode.

The stereoelectronic preference for anti eliminations in a concerted reaction is a preference, not a requirement. It is frequently but incorrectly assumed by biological chemists that only the anti geometry is consistent with a concerted elimination. This is not likely to be correct. Indeed, many examples of syn elimination are known that do not obviously proceed via intermediates.[40] Again, issues regarding solvation and the nature of the leaving group are of concern. The magnitude of any "intrinsic" preference for anti elimination is not known. It may be larger than 5 kcal/mol; it is certainly much less than the 40 kcal/mol associated with the rotation around a double bond.

Many enzymes catalyze addition/elimination reactions. Hydro-lyases are enzymes that add the elements of water to an olefin. Ammonia lyases add the elements of ammonia to an olefin. Both classes of enzymes are well studied stereochemically. Fumarase and aspartate ammonia-lyase, a representative enzyme from each class, both proceed via an anti transition state geometry[41-43] as expected for a stereoelectronically preferred concerted elimination reaction.

The existence of two enzymes with anti elimination modes might suggest that this stereochemical mode is selected for stereoelectronic reasons, and that anti addition to olefins will occur as a "rule." While this functional conclusion is reasonable, it is obviously not forced by these data, even after recognizing that so far only two enzymes have been considered. There exists a competing historical argument that suggests that fumarase and aspartase are evolutionarily related and that the common anti stereospecificity of the two enzymes is the result of a conserved historical accident. Supporting this historical model are recent data showing that the amino acid sequences of some fumarases and some aspartases are approximately 40% identical.[44]

Nevertheless, many other enzymes catalyze analogous elimination processes with an overall anti stereoselectivity. These include the addition/elimination reaction catalyzed by arginosuccinase,[45] aconitase,[46] oleic acid dehydratase from *Pseudomonas*,[47] adenylosuccinase,[48] malease,[49] and enolase.[50] Two additional ammonia lyases, phenylalanine ammonia-lyase and histidine ammonia-lyase, also remove the fragments of H—NH$_2$ in an anti mechanism.[51,52]

Other reactions that are analogous in the broadest sense also proceed via anti transition states. For example, the decarboxylative elimination of 5-pyrophosphomevalonate (3,5-dihydroxy-3-methylpentanoic acid, 5-pyrophosphate) proceeds with an overall anti stereoselectivity.[53]

Based on these examples, one is tempted to construct the following argument. Stereoelectronic principles suggest that 1,2 elimination reactions are preferred in the anti stereochemical sense. A large number of enzymes produce anti eliminations. In view of the wide range of substrates, it seems unlikely (but not impossible) that all of these enzymes are related and that the common stereochemical course reflects divergent evolution from a common ancestor. Therefore, one might argue that the common stereochemical course reflects convergent evolution of active sites that conform to stereoelectronic principles for functional reasons. Again, this argument requires an assumption that enzymes that catalyzed syn eliminations would be less able to contribute to the survival of the host organism.

However, this line of reasoning is unsatisfactory, as there are many examples of enzymes that catalyze syn addition/elimination reactions. For example, the enzymatic formation of shikimate involves the elimination of the elements of water with a syn stereochemical relationship.[54] So do the reactions catalyzed by cis-cis-muconate cycloisomerase,[55] enoyl-CoA hydratase,[56] dehydroquinate synthase,[57] β-hydroxydecanoylthioester dehydratase,[58] yeast fatty-acid synthetase,[59] methylglutaconyl-CoA hydratase,[60] and dehydroquinate dehydratase.[61] Methacrylate is converted to β-hydroxyisobutyrate, presumably via the intermediacy of methacrylyl-CoA, by a syn addition.[62] 3-Dehydroshikimate is converted to protocatechuate via a syn elimination.[63] Cyclization of carboxymuconic acid to give β-carboxymuconolactone is also syn.[64]

The most obvious way to account for syn elimination in these cases is to assume that the elimination reactions catalyzed by the first class of enzymes is *concerted* while that catalyzed by the second class of enzymes is *stepwise*. For stepwise reactions, one has a choice of several functional hypotheses that argue for syn elimination as the preferred stereochemical course. The simplest is that active sites with a single base are "better" for a stepwise elimination than those with multiple functional groups.[65] In a syn elimination, the base can act both to abstract a proton in the first step and then, in the protonated form, assist the departure of the nucleophilic group in the second step.

Such a set of hypotheses is clearly ad hoc. Further, independent evidence to support these hypotheses requires demonstrating that individual enzymatic reactions are either stepwise or concerted. Traditionally, such demonstrations are among the most difficult in chemistry.

Simple examination of the structure of the substrates lends support to the hypotheses. In many cases where the elimination is syn (and presumably stepwise), the departing proton leaves behind an electron pair that is adjacent to a group that is chemically very stabilizing (eg, a thioester or a ketone) (Figure 2-15). However, in those cases where the elimination is anti (and presumably concerted), the proton that is leaving is adjacent to a group that is *not* chemically well suited to stabilize an adjacent carbanion (eg, a carboxylate). In the first case, a hypothetical "carbanion" intermediate is relatively stable; the hypothesis of a stepwise reaction is reasonable. In the second case, the hypothetical carbanion is not stable; the hypothesis of a concerted reaction is, therefore, again reasonable.

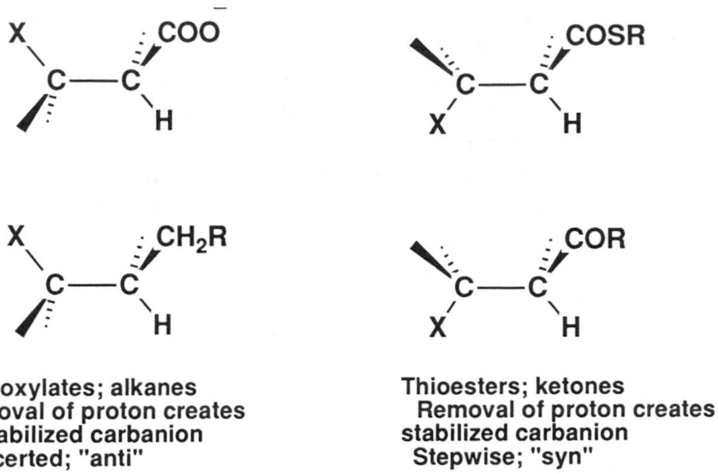

Carboxylates; alkanes
Removal of proton creates
unstabilized carbanion
Concerted; "anti"

Thioesters; ketones
Removal of proton creates
stabilized carbanion
Stepwise; "syn"

Figure 2-15. A stereoelectronic rule predicting the stereospecificity of enzymes that catalyze addition/elimination reactions. Where an unstabilized carbanion would be an intermediate in a stepwise mechanism, the reaction is concerted and the transition state geometry is anti. For stabilized carbanions or cations, stereospecificity is not determined by stereoelectronic principles.

This sort of argument is not without precedent in the enzymological literature. For example, Schwab and Klassen proposed that the stereochemical course of allylic rearrangements likewise is stepwise or concerted, syn or anti, depending on the intrinsic reactivity of the substrate molecule.[66] While there are perhaps some exceptions to this proposal,[67,68] the suggestion remains important as a potential working hypothesis.

One might attempt to apply this hypothesis to three enzymes catalyzing three related reactions whose stereospecificities have been recently determined. These reactions are (1) the cyclization of *cis-cis*-muconate to give muconolactone, a syn addition reaction; (2) the cyclization of *cis-cis*-carboxymuconate to give β-carboxymuconolactone, also a syn addition reaction; and (2) cyclization of *cis-cis*-carboxymuconate to give γ-carboxymuconolactone, an anti addition (Figure 2-16).[69] The stereospecificities of these enzymes were examined to test an intriguing hypothesis of Ornston that sequential enzymes in a pathway are related.[70] The last enzyme was expected to have the same stereospecificity as the first two.

The fact that the third enzyme catalyzed an anti addition/elimination reaction contradicted these expectations. The investigators proposed that perhaps the anti addition to a dicarboxylated double bond proceeded via a transition state of lower energy than a syn addition because the two carboxylate anions are somewhat farther apart in the anti transition state than in the syn transition state. This is not a stereoelectronic hypothesis and is problematic from a theoretical point of view.

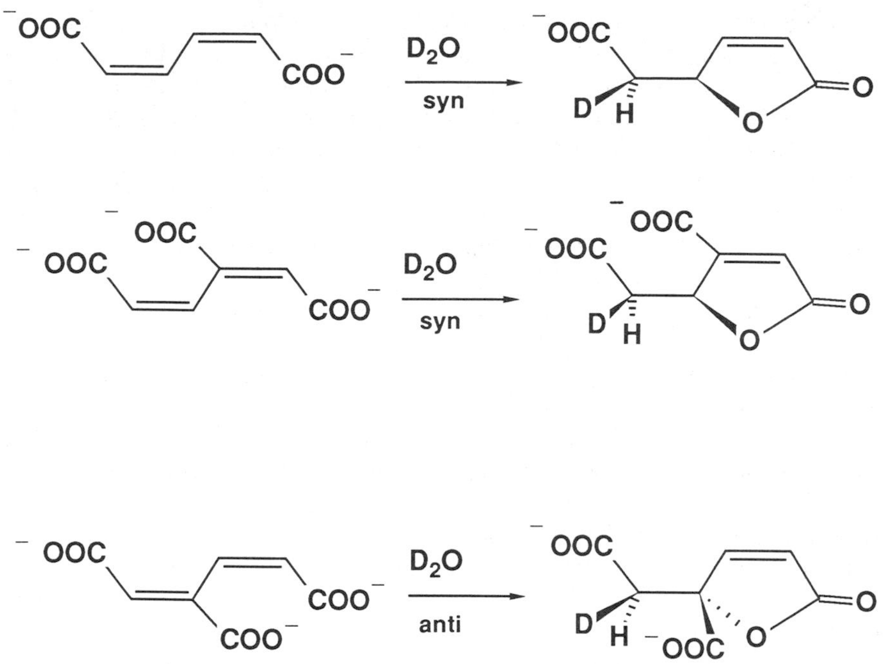

Figure 2-16. In the reactions that proceed via a syn transition state, a hypothetical carbocation intermediate is "less" bad than in those reactions that proceed via an anti transition state.

However, the stereoelectronic hypothesis suggested above would rationalize these data if the first and second reactions are nonconcerted while the third is concerted. Naively, a carbonium ion intermediate in the first two reactions is expected to be more stable (or perhaps, less unstable) than such an intermediate in the third reaction.

Together, these hypotheses form a coherent functional picture describing the stereochemical choices made by enzymes catalyzing the formation of olefins by elimination reactions. The picture is based on the notion of functional adaptation of proteins to reflect stereoelectronic hypotheses. As a stimulus to further research, such hypotheses are invaluable. However, they are not without problems. To illustrate the potential level of confusion, four cases are mentioned below where these hypotheses seem unsatisfactory to explain the behavior of individual enzymes.

First, methylaspartase appears to be paradoxical in light of these hypotheses, as it catalyzes both a syn and an anti addition to mesaconate to form a mixture of L-threo- and L-erythro-β-methylaspartate. The V_{max} for syn elimination is only 1% of that for anti elimination.[71] There is reported to be an exchange of proton with solvent, suggesting a carbanion intermediate. The significance of this result is not clear.

Second, in the biosynthesis of aromatic amino acids, the elimination of water from dehydroquinate to form dehydroshikimate is syn. This is as expected from the previous discussion because the putative carbanion intermediate is adjacent to a ketone and is, therefore, likely to be quite stable (Figure 2-17). However, in the most stable chair conformation of the substrate, the hydrogen that is abstracted is equatorial, not axial as expected by stereoelectronic theory. Indeed, in order for the hydrogen to become axial, the substrate must either adopt a chair conformation with three groups axial or must adopt a boat conformation.[72] The simplest suggestion is based on the assumption that the reaction is stepwise and proceeds via an enamine intermediate between the substrate and a lysine. One must then postulate a conformational change in the enamine to avoid violation of stereoelectronic rules.

Third, it has been suggested some fungi catalyze the conversion of mevalonic acid to 3-methyl-5-hydroxypent-2-enoic acid via a syn elimination.[73] This is potentially an exception to the rule. However, the studies were done by feeding labeled compound to whole cells. Therefore, the experiments do not rule out the intermediacy of a coenzyme A derivative, which is expected to undergo a syn elimination (vide supra).

Finally, recent experiments by Cleland and co-workers have raised new questions about whether enzymes that "ought" to catalyze concerted eliminations in fact do so. For example, phenylalanine ammonia-lyase catalyzes an anti elimination. This stereospecificity might be expected based on the model outlined previously if we assume that a hypothetical carbanion intermediate next to a phenyl ring is "unstabilized." However, Hermes, Cleland, and co-workers recently reported[74] that the elimination of ammonia from dihydrophenylalanine catalyzed by this enzyme follows a stepwise mechanism.

There are three potential problems with any interpretation of this result. First, dihydrophenylalanine is an unnatural substrate; the mechanism may be different with the natural substrate. Second, the putative carbanion of

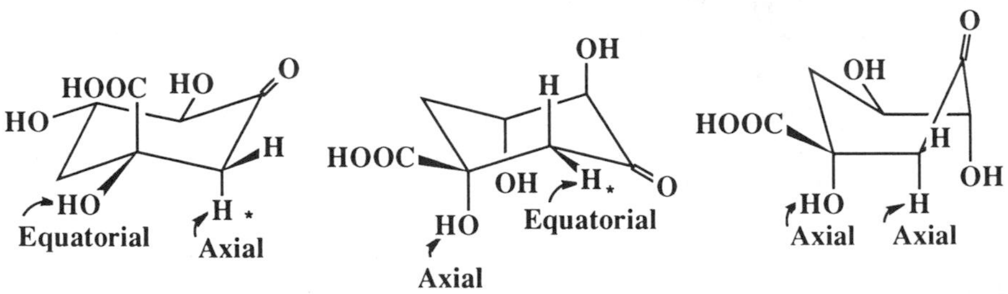

Figure 2-17. The stereoelectronic dilemma in the elimination of water from dehydroquinate. In one chair conformer (left, the most stable chair), the proton that leaves is correctly oriented for a stereoelectronically allowed transition state. In the other chair conformer (middle), the hydroxyl is correctly oriented. In the boat conformer, both are correctly oriented. The boat conformer is often assumed to be the conformer in the active site. Of course, this requires the assumption that the reaction is concerted, which it probably is not.

dihydrophenylalanine is completely unstabilized. Third, the mechanism of the enzyme is obscure in other regards; especially problematic is the role of a dehydroalanine residue in the active site.[75] Nevertheless, the stereospecificity of phenylalanine ammonia-lyase (and perhaps other enzymes in this series as well) may not be interpretable as a confirming instance of the stereoelectronic "rule" proposed previously.

As these examples illustrate, there is still ample opportunity for research in this area. However, we must conclude that the stereoelectronic hypotheses as presented here are plausible. At least in those cases where stereoelectronic preferences correspond to energy differences between transition states on the order of 20 ± 10 kcal/mol, there is as yet no reason to believe that active sites have not evolved to conform to stereoelectronic preferences.

B. Vinylogous Dehydrations

In the context of this discussion of 1,2 elimination reactions, it is appropriate to consider briefly vinylogous analogs of such reactions. The stereochemical aspects of model vinylogous reactions are not well understood. For example, while the S_N2 reaction at carbon proceeds exclusively with inversion of configuration in model systems, the S_N2' seems to display no overwhelming stereochemical preference.[76,77] Simple stereoelectronic analysis suggests that the reaction should occur with syn addition (Figure 2-18).[78] Many efforts have been made to detect such a stereochemical preference, but all have proven to be biased by other factors built into the model or have given ambiguous results. Thus, the stereoelectronic preference for syn or anti transition states must be relatively small. In the absence of hard data, we can only guess that it is less than 5 kcal/mol.

The conversion of 5-enolpyruvylshikimate-3-phosphate to chorismate is a reaction that is a vinylogous elimination of water; here a diene is formed (Figure 2-19). The elimination is anti, not syn as is expected from naive

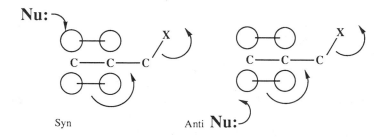

Figure 2-18. The S_N2' reaction involves the attack of a nucleophile on an allylic halide, for example. The double bond moves to the right, and the nucleophile adds in the 3 position. Stereoelectronically "allowed" is at left, stereoelectronically "forbidden" is at right.

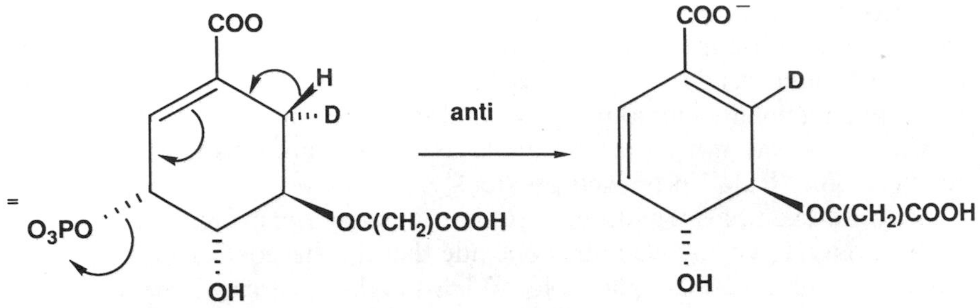

Figure 2-19. The vinylogous elimination in the biosynthesis of chorismate is anti, contrary to stereoelectronic predictions that suggest this reaction should be preferably syn.

stereoelectronic analysis.[79] This has been interpreted as ruling out a concerted mechanism.[80] However, such a strong conclusion is not warranted by model studies that suggest that the stereoelectronic preference is small and may be dominated by steric, polar, or solvent effects (vide supra).

In the absence of definitive data, one is tempted to construct a hypothesis in the reverse direction. The vinylogous elimination is anti in the enzyme. We might *presume* that the enzyme has evolved to be optimally efficient by selecting the preferred stereochemical mode for the elimination. We might then *conclude* that the preferred mode for a vinylogous elimination is anti. Such a hypothesis directs the experimental problem back to the organic chemist, who must design better systems to explore subtle preferences in organic reactivity.

One step in the biosynthesis of hydroxyphenylpyruvate involves an oxidation of an alcohol to a ketone and a vinylogous β-decarboxylation (Figure 2-20). The stereochemistry of the vinylogous oxidation/decarboxylation/ elimination is anti, again violating stereoelectronic expectations for a concerted reaction. One might observe that given the stereochemistry of the substrate, the oxidation/decarboxylation/elimination must be anti to yield the desired product. This observation suggests that the system is not a good one to test stereoelectronic theory; at the very least, the reaction is expected to be stepwise.

There are two problems with this view. First, although the stereochemistry of the substrate requires an anti elimination to yield the desired product, the stereochemistry of the starting material is itself dependent on an earlier enzymatic reaction, where erythrose 4-phosphate is chosen as the starting material for the pathway. Were stereoelectronic effects truly important here, one could argue that the pathway would have begun with threose 4-phosphate. This argument is too complicated to be considered in detail here. However, as prephenate is a precursor to both tyrosine and phenylalanine, one must be produced by an anti elimination and one by a syn elimination.

More serious is evidence that suggests that the reaction catalyzed by this enzyme is *concerted*. Using data from some elegant experiments measuring

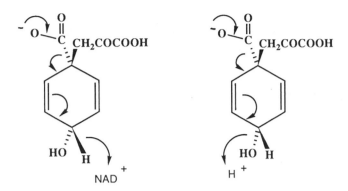

Figure 2-20. The vinylogous oxidative decarboxylation in the biosynthesis of hydroxyphen-ylpyruvate is anti, again contrary to stereoelectronic predictions that suggest this reaction should be preferably syn. Left—Enzymatic, concerted; right—solution, stepwise.

kinetic isotope effects, Hermes, Cleland, and co-workers argued that the oxidation and decarboxylative elimination occur simultaneously.[81]

A concerted, anti vinylogous elimination clearly violates stereoelectronic expectations. Thus, we are forced to believe either that the enzyme violates stereoelectronic preferences or that our conception of stereoelectronic effects in this system is incorrectly formulated. Interestingly, a model reaction for the acid-catalyzed decarboxylative elimination of phenylpyruvate in solution was shown to be stepwise (Figure 2-20).[81] This reaction is syn, the stereoelectronically allowed course for a concerted process.

In light of these results, simple stereoelectronic analysis as applied to vinylogous reactions simply is not satisfactory. It would be interesting to examine more closely an enzyme catalyzing a third reaction, the decarboxylative elimination of prephenate (Figure 2-20) to form phenylpyruvate. The reaction stereochemistry is syn, the stereoelectronically approved geometry for a concerted reaction. There is as yet no evidence to suggest whether this reaction is concerted or stepwise.

Thus, there are three vinylogous eliminations in the pathway for the synthesis of aromatic amino acids. One is syn and two are anti. One of the anti reactions is concerted; in the other two cases, it is not known whether the reactions are concerted or stepwise. If the formation of phenylpyruvate from prephenate is stepwise and the formation of chorismate is concerted, stereoelectronic theory has failed to account for all three reactions. At best, the working hypothesis must be that the preferred stereochemical mode for concerted vinylogous reactions is "anti-stereoelectronic." Needless to say, a theoretical basis for such a hypothesis needs to be developed.

However, the reader should appreciate that a concerted oxidative decarboxylation of prephenate violates the organic chemist's intuitions about orbital overlap and conformational analysis. 1,4-Cyclohexadiene systems might prefer boat conformations. Yet, in the boat conformer with one of the two bonds being broken in the transition state, if the reaction is concerted, it must be equatorial. Thus, the Hermes–Cleland suggestion requires either that a bond that is broken be essentially orthogonal to a π system or that the reactive conformation is not a boat.

5. SMALL STEREOELECTRONIC EFFECTS

A variety of energetically smaller stereoelectronic phenomena are hypothesized to control the conformation or reactivity of organic molecules in solution. Whether similar phenomena control structure and reactivity in the active site is still unknown. Indeed, proposals that they do are among the most hotly disputed in biological chemistry.

A. Oxidoreductases

Nicotinamide cofactors have the functionally significant heterocycle attached as a glycosidic residue on a ribose sugar. The stereospecificity of dehydrogenases dependent on nicotinamide cofactors has been widely studied. The two hydrogens at the 4 position of the reduced nicotinamide ring are diastereotopic (Figure 2-21). Some enzymes transfer the *pro-R* (A) hydrogen, others the *pro-S* (B) hydrogen, and the enzymes are approximately equally distributed between the two stereochemical types.

Stereoelectronic arguments have been applied to the analysis of the stereospecificity of dehydrogenases acting on alcohols dependent on nicotinamide cofactors. One proposal was that the lone pair of electrons on the nitrogen of the reduced nicotinamide ring prefer to be antiperiplanar to the carbon–oxygen bond of the ribose ring.[82,83] This suggestion is simply a statement of an "exo anomeric effect" and is expected to be reflected in a hindered rotation around the carbon–nitrogen glycosidic bond (Figure 2-22).

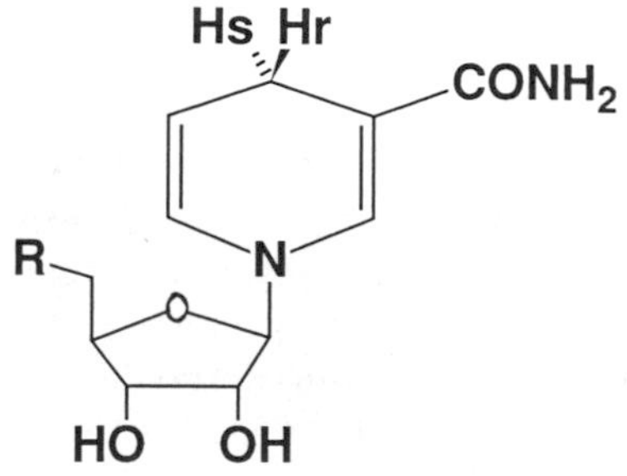

Figure 2-21. The two hydrogens at the 4 position of the reduced nicotinamide pyridine ring are different. Enzymes distinguish between the two hydrogens. About half of the enzymes studied so far transfer H_R, about half transfer H_S.

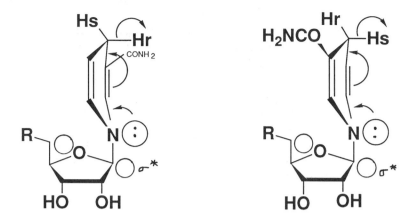

Figure 2-22. In reduced nicotinamide cofactors, an exo anomeric effect is expected because the lone pair of electrons on nitrogen is adjacent to the antibonding orbital (σ^*) associated with the carbon–oxygen bond of the ribose ring. Stereoelectronic theory predicts that the molecule will adopt a conformation where donation of electron density from the lone pair into the antibonding orbital is possible. Two conformations, syn (right, H_S transferred) and anti (left, H_R transferred), exist where this donation is possible.

To facilitate overlap, the nitrogen is expected to be distorted from planarity to increase the amount of overlap between the lone pair of electrons on nitrogen and the adjacent σ^* orbital associated with the carbon–oxygen bond. This nonplanarity is expected to cause the dihydronicotinamide ring to distort into a boat conformation (Figure 2-22).

Distortion into a boat conformation makes the hydrogen at the 4 position of the nicotinamide ring syn to the lone pair on nitrogen axial, and the hydrogen anti to the lone pair equatorial. Based on the simple stereoelectronic principles discussed previously, the axial hydrogen is expected to be kinetically easier to transfer for two reasons. First, it is axial. Second, it is vinylogous to a lone pair of electrons in a syn conformation (Figure 2-22).

The exo anomeric effect is expected to make two conformations of NADH lower in energy, one where the carboxamide of the dihydronicotinamide ring is syn to the ribose ring, the other where the conformation is anti. Thus, the stereoelectronic prediction is that the *pro-R* hydrogen at the 4 position should be transferred from the anti conformer, the *pro-S* hydrogen transferred from the syn conformer.

These arguments are, of course, abstract and suggest only that there should be a stereoelectronic preference in the absence of an enzyme for one conformer to transfer one hydrogen and the other conformer to transfer the other. The exo anomeric effect is expected to be small, on the order of 1–2 kcal/mol. Whether enzymes have evolved active sites that reflect such a small stereoelectronic effect is, of course, the issue.

The crystal structures of several dehydrogenases are known, and Rossmann and co-workers[84] first noted that, in five cases, enzymes that evolved to transfer the *pro-R* hydrogen bind the cofactor in an anti conformation while enzymes that have evolved to transfer the *pro-S* hydrogen bind the cofactor in the syn conformation. These results are consistent with stereoelectronic predictions, and the analysis was the basis of a successful, but controversial, theory explaining the stereospecificity of dehydrogenases dependent on nicotinamide cofactors.[82]

This correlation between cofactor orientation and stereospecificity of hydrogen transfer is consistent with stereoelectronic expectations and is quite interesting. However, it clearly may have a nonstereoelectronic basis. Perhaps relevant to this point is the report that glutathione reductase, an enzyme that transfers the *pro-S* hydrogen, has been reported to bind the cofactor in the anti conformation.[85] Unfortunately, coordinates for the bound cofactor have not been published, and it is difficult to know exactly how seriously this apparent exception violates the kinetic stereoelectronic suggestion. It may be that this enzyme violates stereoelectronic predictions or that one-electron chemistry operating with fewer geometric constraints may play an important role in this enzyme (vide infra).

One potential nonstereoelectronic interpretation of these data is a historical one.[86] One might presume that enzymes that transfer the *pro-R* hydrogen are all homologous and that the binding of cofactor in the anti conformation is a highly conserved, but nonfunctional, structural feature of the active site. The suggestion has been made in many forms and is not without merit. However, the recent arguments that the folding of the peptide backbone makes it impossible for enzymes binding one conformer to evolve from enzymes binding the other is inconsistent with what is generally known about the adaptation of proteins.[87] Further, it appears that at least two pairs of enzymes with opposite stereospecificities are related.[86]

The stereoelectronic argument has been occasionally misunderstood. For example, it has been argued that the α anomers of NAD$^+$ cannot display the same exoanomeric effect as the β anomers.[88] This argument is not correct (Figure 2-23). Further, it has been claimed that the stereoelectronic argument cannot be correct because the order of binding determines the relative position of substrate and cofactor in the active site.[89] While this may be true, it is irrelevant to the central question: Are stereoelectronic effects important in determining the relative orientation of bound species? Presumably, the order of binding is itself an evolutionary variable that could adapt to conform to chemical principles.

Analogous arguments can be applied to enzymatic reduction of flavins and deazaflavins. Here again, the lone pair of electrons on nitrogen can, at least in principle, be oriented syn or anti to a departing "hydride." Reduced flavins adopt butterfly conformations,[90] and deazaflavins have proven to be valuable stereochemical probes of the reaction geometry. However, to the extent that flavin reactions (or, for that matter, reactions with nicotinamide cofactors) involve single electron transfers, stereoelectronic considerations are not likely to be applicable. The geometric constraints on electron trans-

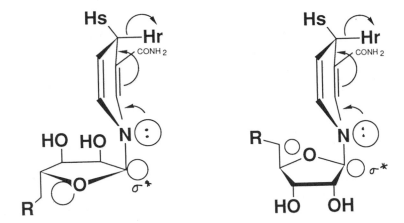

Figure 2-23. Oppenheimer has suggested that the ability of some dehydrogenases to handle the α anomer of nicotinamide is a test of the stereoelectronic interaction shown in Figure 2-22. The argument is that the α and β anomers differ "fundamentally" in the relative orientation of "orbitals of the furanose ring oxygen" and the dihydropyridine ring. While there are clearly geometric differences between the α and β anomers of NADH that might be classified as fundamental, both anomers have analogous exo anomeric interactions.

fer reactions are only incompletely explored;[91] there may, in fact, be none. Thus, the interpretation of stereochemistry in these systems is not likely to be simple. This consideration may ultimately be important in understanding the active site in glutathione reductase.

B. Group Transfer Reactions

Inversion of stereochemistry in the binuclear nucleophilic attack at a saturated carbon (the S_N2 reaction) is one of the best known stereochemical details of organic reactions. In stereoelectronic theory, the preferred transition state is one where the lone pair of electrons on the attacking nucleophile is geometrically able to overlap with the σ* orbital associated with the bond holding the leaving group to the reaction center.

Thus, the incoming nucleophile is expected to be oriented 180° from the leaving group. Experimentally, angles of attack that are approximately 120° are not tolerated.[20,21] Thus, the geometric constraints on the attack in the S_N2 reaction seem to be as rigorous as those for any reaction in organic chemistry. These experimental results suggest that the bending force constant in the transition state is rather high.

In enzymatic reactions, evidence from kinetic isotope effects suggests that the transition state is still "tighter" than for analogous displacement reactions in solution.[92] Indeed, one commonly suggested mechanism for catalysis of the S_N2 reaction proposes a precise alignment of the attacking nucleophile with respect to the reacting center, followed by "compression" of the transition state. Not surprisingly, in all enzymatic reactions studied so far, individual S_N2 displacements occur with inversion of configuration.

In the absence of crystallographic data on methyltransferases, it is difficult to carry the discussion further. However, an interesting controversy has developed regarding stereoelectronic effects in analogous displacement reactions at phosphorus. Here, in contrast to displacements at carbon, pentavalent intermediates are possible. The pentacoordinate intermediates can undergo pseudorotation to give different intermediates, which in principle can decompose to give displacement with net retention of configuration at the reaction center.[93] Further, substituents attached to the reaction center bear several lone pairs of electrons. The orientation of these lone pairs with respect to forming and breaking bonds may be analyzed in stereoelectronic terms.

This discussion was initiated by Gorenstein and co-workers.[94] They first analyzed the conformation of phosphodiesters that is preferred based on stereoelectronic arguments. Stereoelectronic considerations suggest that the preferred conformation will be one where lone pairs on oxygen are antiperiplanar to other phosphorus oxygen bonds. Thus, the bond connecting the oxygen and the ester substituent is gauche to the adjacent phosphorus–oxygen bond (Figure 2-24). This "gauche–gauche" conformer is analogous to the conformation of dimethoxymethane, again presumed to result from stereoelectronic interactions.

Gorenstein and co-workers calculated a series of energies for the various conformations of phosphodiesters.[94] The calculated energy differences between the various conformations are quite small. The gauche–gauche con-

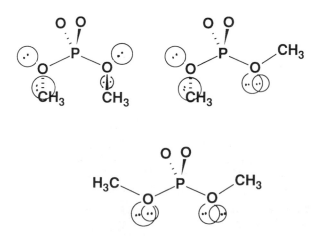

Figure 2-24.   The gauche–gauche (upper left) conformer of dimethylphosphate places two lone pairs antiperiplanar to phosphorus–oxygen bonds and is expected to be the most stable for stereoelectronic reasons. The trans–trans conformer (bottom) has phosphorus–carbon bonds antiperiplanar to both phosphorus–oxygen bonds and is expected to be the least stable. The gauche–trans conformer is at the upper right.

formation, with two stereoelectronically "correct" interactions, was assigned a reference conformational energy of 0.0 kcal/mol. The calculated energy for the gauche–trans conformation was 0.14 kcal/mol (one "correct" interaction), and 0.88 kcal/mol for the trans–trans conformation (no "correct" interactions). Given this rather small energy difference, the preference for the gauche–gauche interaction in phosphate diesters is not expected to be large. Similar conclusions might be drawn from experimental work by Gerlt and co-workers.[95]

In an attempt to extend this structural analysis to a kinetic one, Gorenstein et al. noted that, for the hydrolysis of a phosphodiester bond, the stereoelectronically preferred conformation is one where the breaking phosphorus–oxygen bond is antiperiplanar to a lone pair of electrons on the adjacent ester oxygen. Further, the lone pairs on the oxygen that is departing should not be antiperiplanar to the adjacent phosphorus ester oxygen. This situation is obtained in a gauche–trans conformation, where the oxygen atom bonded to phosphorus with the gauche conformation is "pushing" and the phosphorus–oxygen bond with the trans dihedral angle is breaking.

Finally, Gorenstein and co-workers noted that, in transfer RNA, one bond susceptible to ribonuclease cleavage in fact has this gauche–trans conformation. They argued that this conformation made the bond more reactive and noted that several other bonds on the surface of the tRNA molecule that did not have this reactive conformation were not cleaved by RNase.

While this analysis is theoretically reasonable and valuable as a working hypothesis, crystallographic data on ribonuclease may shortly become available to test it directly. Petsko and co-workers have examined, at low temperatures, cocrystals of ribonuclease and various transition state analogs, substrates, and inhibitors bound in the active site.[96]

C. Reactions at Carbonyl Groups

Stereoelectronic analysis applied to the reaction of nucleophiles with carbonyl groups has been especially important. The most stable transition state geometry should be one where an electron pair on the attacking nucleophile can overlap with the π^* antibonding orbital of the carbonyl group. Thus, there is a "trajectory" for the attack of the nucleophile (Figure 2-25). Pio-

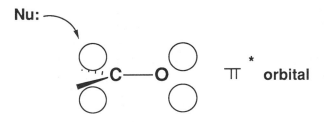

Figure 2-25.   Trajectory for the attack of a nucleophile on a carbonyl system.

neered in the early 1970s by Burgi and Dunitz,[97] the argument was originally based on an analysis of crystal structures of molecules containing both nucleophiles and carbonyl groups.

Based on this idea, Bizzozero, Zweifel, and Dutler have provided a rather thorough stereoelectronic analysis of reactions catalyzed by serine proteases.[98,99] These enzymes catalyze the hydrolysis of peptide bonds, via acyl enzyme species, that are formed by the attack of a serine residue in the active site, presumably via a tetrahedral intermediate.

Bizzozero and Zweifel[98] noted that, by the principle of microscopic reversibility, the stereoelectronic principles controlling the destruction of the tetrahedral intermediate must also control the formation of the tetrahedral intermediate. The stereoelectronically preferred route for the decomposition of the tetrahedral intermediate to form starting amide and free enzyme is one where a lone pair of electrons on nitrogen is antiperiplanar to the departing serine nucleophile. Thus, by microscopic reversibility, attack of serine on the amide carbonyl to form a tetrahedral intermediate must *create* a nitrogen with a lone pair of electrons antiperiplanar to the serine. In this geometry, given the crystal structure of chymotrypsin, the lone pair is pointing in the direction of the solvent.

For the tetrahedral intermediate to become an acyl enzyme intermediate, the amide nitrogen must invert for the nitrogen lone pair to accept a proton. In most mechanisms, His-57 is assigned the role of proton source for this step. However, to accept a proton from His-57, the nitrogen must suffer inversion in order to redirect the lone pair toward the histidine proton. This inversion also stabilizes the tetrahedral intermediate with respect to decomposition to give starting material. With the lone pair pointing toward His-57, it is no longer antiperiplanar to the bond joining the oxygen of Ser-195 to the tetrahedral intermediate.

This analysis was applied to explain the inability of chymotrypsin to hydrolyze peptide bonds with proline and to analyze in detail the crystal structures of trypsin with various inhibitors bound in the active site. In the latter case, the possibility that the nitrogen does not invert, but rather is protonated directly from solvent, was analyzed in detail.

Application of the principle of microscopic reversibility to the opposite direction of a reaction where stereoelectronic effects are well understood in only one direction is a useful intellectual approach. It is rigorous, it often provides new insights, and it is frequently overlooked.

For example, fused β-lactams are also presumed to hydrolyze via tetrahedral intermediates formed with a serine residue in the active site of a β-lactamase. The five- and four-membered rings of the tetrahedral intermediate almost certainly must be "cis" to avoid a highly strained species. Thus, the lone pair of electrons on nitrogen in this tetrahedral intermediate must be cis to the hydrogen at the carbon shared by the two rings. Thus, to be stereoelectronically allowed, the serine hydroxyl must attack from the "endo" side of the β-lactam (Figure 2-26).

Endo attack of a fused β-lactam is hindered compared to "exo" attack, as

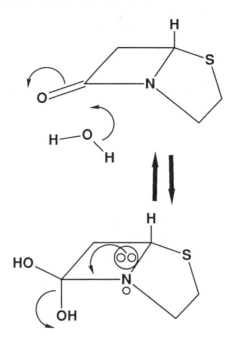

Figure 2-26.   If the stereoelectronically allowed decomposition of the tetrahedral intermediate is one where the lone pair of electrons on nitrogen can assist the departure of the nucleophile, then the stereoelectronically allowed direction for nucleophilic attack on the lactam is from the endo side (the bottom in this drawing) of the lactam. This is the opposite of the attack from the top that would be predicted on steric grounds.

the fused five-membered ring lies astride the allowed trajectory for the attack. Of course, fused β-lactams are rather stable in aqueous solution considering the ring strain that normally should encourage attack. This may, in part, be explained by the fact that the sterically preferred mode of attack (exo) is disallowed stereoelectronically, while the stereoelectronically allowed mode of attack (endo) is hindered sterically.[129]

The crystal structure of β-lactamase is currently under investigation.[100] When these studies are complete, the enzymatic structure will make a statement, albeit for a single case, as to whether this enzyme preferred to construct an active site that conforms to steric constraints or to stereoelectronic constraints. However, this discussion would not be complete without mention of very recent data that failed to find experimental evidence supporting a large stereoelectronic effect in systems analogous to these tetrahedral intermediates.[101]

D. Reactions at Glycosidic Centers

In view of the role that sugars played in the development of stereoelectronic theory, it is appropriate to ask how well stereoelectronic theory accounts for the reactivity of glycosides in the active sites of enzymes. The magnitude of the anomeric effect (1–2 kcal/mol) at the anomeric center of glycosides as well as a substantial amount of confusion in correlating reactivity and structure have made the arguments difficult.

First to be considered are the facts regarding stereoelectronic effects in glycosides in solution. β-Glucosides hydrolyze more rapidly than α-glucosides (Figure 2-27).[102] In the more stable chair conformer, the β anomer has an equatorial leaving group at the anomeric center; in the α anomer, the anomeric substituent is axial. This fact might be viewed as contradicting stereochemical precepts, as these precepts would argue that the axial leaving group, with two lone pairs of electrons antiperiplanar, is better suited to depart.

To reconcile stereoelectronic arguments with the known more rapid hydrolysis of the β anomer, suggestions have been made that the hydrolysis proceeds via a boat conformation. In its simplest form, such a proposal does not clearly explain the difference in rates between the two anomers. The barrier for the departure of an axial group from the α anomer in a chair conformation and from a β anomer in a boat conformation is naively not expected to be different, and the boat conformer of the β anomer is likely to be 6–10 kcal/mol higher in energy[103] than the ground state chair conformation. Further, the chair and boat conformers are most likely in rapid equilibrium. Thus, several additional hypotheses are required if this interpretation is used to explain the enhanced reactivity of the β anomer over the α anomer.

A stereoelectronic analysis can be extended to a discussion of the hydrolysis of glycosides in the active site. In lysozyme, the classical model for hydrolysis involves the distortion of the β-glycoside.[104] This distortion was proposed, in part, in order to meet stereoelectronic requirements.

However, experimental evidence for such a distortion is lacking.[105–107] Together with a computer simulation of the dynamic properties of lysozyme, the presumption of an essential stereoelectronic interpretation and the absence of data supporting the presumably required distortion was strong enough to lead Post and Karplus to suggest an alternative mechanism for lysozyme action where the pyranose ring opens as part of the reaction mechanism (Figure 2-28).[108]

Thus, lysozyme appears either to violate stereoelectronic principles or to

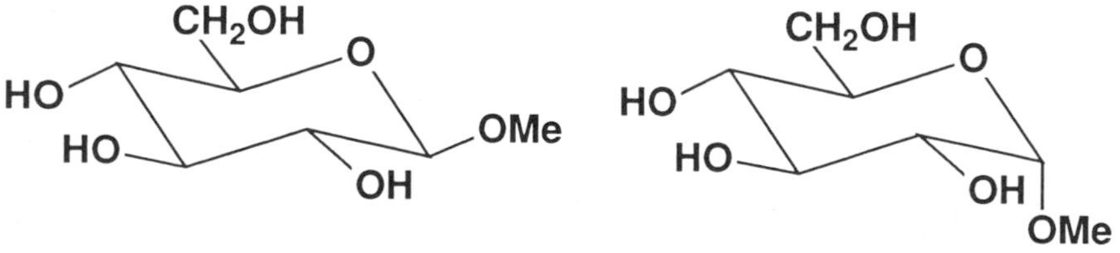

Figure 2-27. The β anomer (left), with an equatorial leaving group, hydrolyzes faster than the α anomer (right). Yet, according to naive stereoelectronic expectations, the α anomer with the axial leaving group is expected to hydrolyze faster.

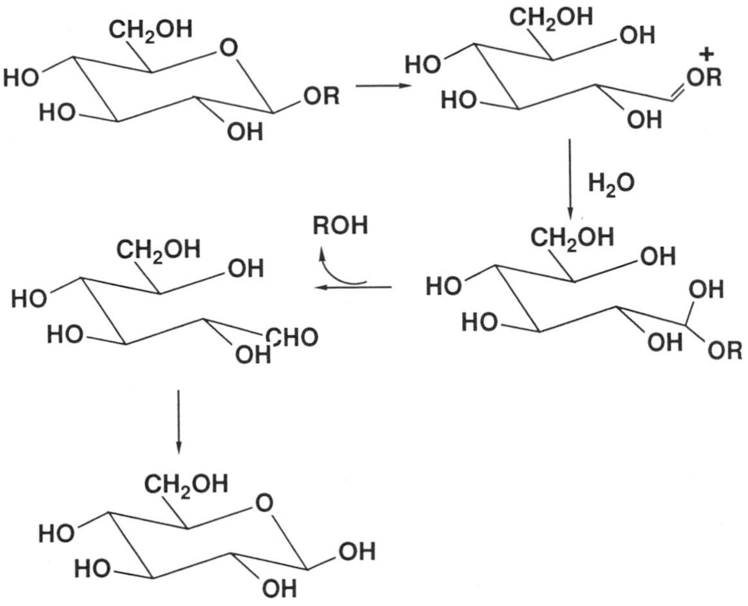

Figure 2-28.   The Post–Karplus mechanism for the action of lysozyme on β-glycosides. The mechanism was formulated following molecular dynamics calculations specifically to avoid violating stereoelectronic rules.

follow an unusual and certainly unexpected mechanism, as suggested by Post and Karplus.

However, the formulation of stereoelectronic principles in the hydrolysis of glycosides has been challenged by a penetrating analysis by Sinnott and co-workers.[109–111] In model studies of the reactivity of cyclic acetals in solution, Sinnott found no evidence that the β anomer reacts through a boat conformer. The conclusion was drawn from the results of a series of measurements of kinetic isotope effects based on the fact that the magnitude of the β deuterium secondary isotope effect is a function of the dihedral angle between the carbon deuterium bond and the departing group at the anomeric center.

These data strongly support the proposition that the *equatorial* substituent in a chair conformation leaves more rapidly than the *axial* substituent. This appears a priori to violate stereoelectronic principles.

Sinnott suggested the following explanation. First, the transition state for the hydrolysis of glycosides appears to be very late. The transition state resembles the oxonium ion intermediate; the anomeric substituent is essentially gone before the transition state is formed. Therefore, this transition state has lost the information that indicates whether it arose from an axial or an equatorial species. In other words, essentially the same transition state is accessible from the chair conformers of both the α and β anomers.

In this view, the difference in reactivity of the α and β anomers is explained simply by the fact that the α anomer is more stable (the anomeric

Figure 2-29. While the bridgehead acetal cannot become a planar oxonium ion, the equatorial isomer of the unconstrained acetal can. Thus, the fact that the constrained acetal hydrolyzes much slower than the unconstrained acetal does not, a priori, mean that the unconstrained compound must convert to a boat conformer before reacting.

effect). The transition state for the departure of the anomeric substituent from the β anomer has already undergone substantial geometric reorganization and, therefore, is essentially as stable as the transition state arising from the α anomer. Thus, the β anomer reacts faster.

Clearly, it is difficult to tell whether lysozyme obeys stereoelectronic rules in the absence of clear model studies that show what these rules are for the molecules in question in solution. Sinnott's challenge has placed the kinetic anomeric effect at the level of the aldehyde oxidation state in some danger. It is clear that the mechanism of Post and Karplus is not *required* by stereoelectronic considerations. Of course, only experimental work will allow us ultimately to conclude whether the mechanism is correct.

As is illustrated by this example, stereoelectronic analysis can be complicated by all of the problems common to physical organic analysis. For example, it has been argued that the fact that a conformationally locked acetal undergoes hydrolysis 13 orders of magnitude slower than an analogous conformationally flexible acetal demonstrates that a conformational change is an essential preliminary to the hydrolysis of β-glycosides (Figure 2-29).[112,113] This argument is not correct. These data only show that the bridgehead cation that results from the hydrolysis of the conformationally locked acetal is quite high in energy and, by analogy, that a twisted ion in the flexible system is presumably also quite high in energy. However, the required conformational change can be part of the reaction coordinate. The flexible acetal can achieve an untwisted conformation (with the correct overlap) as it moves along the reaction coordinate toward the transition state. As long as the transition state is late, the molecule need not achieve this correct overlap by first undergoing a preliminary conformational change.

E. Aldopyranose Dehydrogenases

While hydrolyses of glycosides that proceed via carbonium ion intermediates are not obviously constrained by stereoelectronic principles, stereoelectronic hypotheses remain popular in other reactions at the anomeric center in sugars. Here, we consider one such hypothesis involving aldopyranose dehydrogenases, enzymes that oxidize cyclic hemiacetals to lactones.

For more than a quarter of a century, it has been known that certain aldose dehydrogenases act on only one anomer of their sugar substrate. In 1952, Strecker and Korkes[114] showed that an NAD$^+$-dependent D-glucose dehydrogenase selectively catalyzes the oxidation of the β anomer. (Anomers are named in accordance with conventions[115] that are different in sugar and steroid chemistry). Six other enzymes have been examined since (Table 2-1).[116–120]

TABLE 2-1. Aldopyranose Dehydrogenase Transfer of Axial Hydrogens

Enzyme	Source	Presumed reactive conformation	Anomeric specificity		Absolute anomeric configuration
			Expected	Found	
D-Glucose dehydrogenase (EC 1.1.1.47)	Ox liver	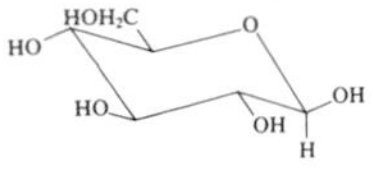	β	β	R
D-Glucose-6-phosphate dehydrogenase (EC 1.1.1.49)	Yeast	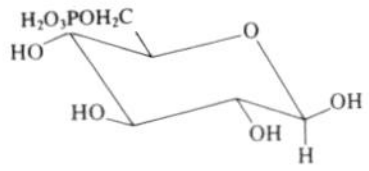	β	β	R
D-Galactose dehydrogenase (EC 1.1.1.48)	*Pseudomonas saccharophila* *Pseudomonas fluorescens*	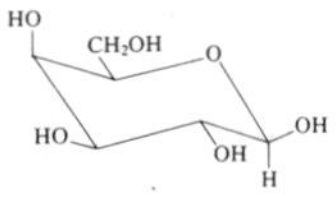	β	β	R
L-Fucose Dehydrogenase (EC 1.1.1.122)	Pig liver	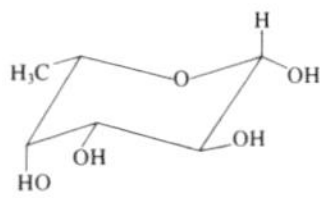	β	β	S
D-Arabinose dehydrogenase (EC 1.1.1.117)	*Pseudomonas*	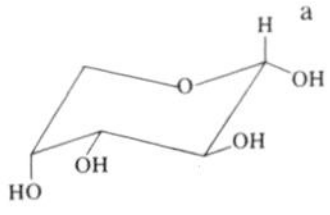	α	α	S
D-Abequose dehydrogenase	*Pseudomonas putida*	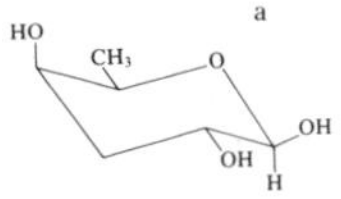	β	β^b	R

In 1976, Benkovic commented that "the observed β-anomeric specificity [of D-glucose-6-phosphate dehydrogenase] is not readily rationalized at present in terms of a mechanism for the redox . . . reaction. It would appear that either anomer should have sufficed."[121]

Recently, Berkowitz and Benner[122] suggested a set of stereoelectronic hypotheses that rationalize the anomeric preferences of aldopyranose dehydrogenases dependent on nicotinamide cofactors (Table 2-1). The hypotheses were based on the presumption that a hydride leaving the anomeric center is facilitated by two lone pairs of electrons antiperiplanar, a presumption that was based on nonenzymatic model oxidations studied extensively by Deslongchamps.[2] In aldopyranoses, a hydrogen in the axial position is antiperiplanar to two lone pairs of electrons when the pyranose is in the chair conformation. One lone pair comes from the ring oxygen; the other comes from the exo hydroxy substituent. In contrast, in the chair conformation a hydrogen in the equatorial position can be antiperiplanar to at most one lone pair (Table 2-1).

Thus, if aldopyranose dehydrogenases have evolved to catalyze reactions along the stereoelectronically favored pathway, they will transfer an axial hydrogen. A second hypothesis, that aldopyranose dehydrogenases act on the lowest energy chair conformation of their substrate, was then suggested. The prediction based on these two hypotheses was that aldopyranose dehydrogenases act on the anomer that has the C-1 hydrogen axial in the lowest energy conformer.

These ideas were formulated as a set of hypotheses that included considerations regarding the adaptation of enzymes:

1. The most efficient catalysis of the expulsion of hydride requires that the transition state have two oxygen lone electron pairs oriented antiperiplanar to the departing hydride. This is the conformation that can transfer hydride via a stereoelectronically favored pathway.
2. The optimal mode of obtaining two antiperiplanar lone pairs is by binding the anomer that provides them in the ground conformational state, as opposed to binding a high-energy conformation of an anomer that does not provide the requisite lone pairs in the ground conformational state.
3. Aldopyranose dehydrogenases have evolved to be optimally efficient catalysts.

These hypotheses predict the anomeric specificity of aldose dehydrogenases that satisfy the following conditions:

1. A strong argument can be made regarding the evolutionarily relevant "natural substrate(s)" of the enzyme.
2. The cofactor is a pyridine nucleotide.
3. The initial product of the reaction is the aldono-1,5-lactone.

The first restriction is necessary because natural selection is the only mechanism for obtaining functional behavior in living systems. Therefore, a functional theory based on the properties of the natural substrate can be predictive only to the extent that the evolutionarily relevant natural substrate is defined. The second restriction constrains us to examine enzymes acting with similar mechanisms, presumably a hydride transfer. The third restriction excludes simple aldehyde oxidation, where the α and β anomers of the substrate are not relevant to the microscopic reaction mechanism. Further, it constrains us to examine enzymes acting on pyranoses. Although the oxidation of furanoses might also conform to stereoelectronic hypotheses, the flexibility of furanoses makes it difficult to analyze the energetics of their different conformations.

These hypotheses account for the observed stereospecificity of all aldopyranose dehydrogenases with known stereoselectivity (Table 2-1) and predicted as well that enzymes oxidizing L-fucose (6-deoxy-L-galactose) to L-fucono-1,5-lactone will act on the β anomer, as it is the β anomer of L-fucose that has a ground state conformer with the requisite lone pairs antiperiplanar to the leaving hydride. Fucose dehydrogenase from pig was then shown to act on the β anomer, consistent with the prediction of the theory.

The analysis of the stereospecificity of aldopyranose dehydrogenase in stereoelectronic terms can be contrasted with alternative functional and historical explanations for selectivity at the anomeric position. For example, it might be that enzymes evolve to act on the anomer which predominates at equilibrium.[121] While this explanation is consistent with data on aldose dehydrogenases in Table 2-1, this explanation is not consistent with data for several aldose kinases and aldose isomerases.[123–127]

Further, in several cases, the equilibrium populations of the two anomers are nearly identical; it seems unlikely that the survival of the host depends on selecting the anomer with 60% abundance at equilibrium. Therefore, this argument was considered to be less reasonable as a working hypothesis.

Alternatively, one might propose that glucose dehydrogenase and glucose-6-phosphate dehydrogenase are related. Their stereochemical preferences are therefore identical because they are conserved. However, with fucose, the absolute configuration at C-1 is the opposite of that for the other hexoses whose dehydrogenases have been examined. Therefore, the anomeric specificity of the enzyme is less likely to be explainable in terms of a simple argument based on pedigree.

The most reasonable historical model divides the enzymes in Table 2-1 into two classes, each representing a separate line of descent. L-Fucose dehydrogenase and D-arabinose dehydrogenase are in one class, the remaining dehydrogenases are in the other. Anomeric specificity is the result of separate historical accidents in the two ancestral dehydrogenases. Once established, anomeric specificity is highly conserved. Thus, the model assumes that the identical preference for presumed axial hydride transfer in the two classes is coincidental (ie, not functional).

This historical model has several weaknesses. First, it is somewhat ad hoc. Any collection of data can be explained historically by postulating a sufficient number of historical accidents and independent pedigrees. Further, the explanation presumes that there are nonfunctional constraints on the drift of anomeric specificity that are more stringent than constraints on the drift of substrate specificity.

So far, this model has been tested only with fucose dehydrogenase. Presumably, the transition state for the oxidation reaction is not extremely late, making the application of stereoelectronic principles to the oxidation of aldopyranoses less problematical than their application to the hydrolysis of glycosides. However, further tests are necessary, both in model systems and in enzymes.

6. CONCLUSIONS

We hope to leave the reader with a sense of potential rather than a sense of conviction. Few examples are known where an enzyme clearly violates a stereoelectronic rule which is associated with a high energetic price tag. The enzyme catalyzing the synthesis of p-hydroxyphenylpyruvate (Figure 2-20) remains the most serious challenge to intuitive pictures of orbital overlap.

Even for stereoelectronic rules with intermediate energetic consequences, reasonably comprehensive working hypotheses can be constructed based on stereoelectronic principles that make testable predictions. The hypotheses advanced to explain the stereochemical course of addition/elimination reactions (Figure 2-14) perhaps constitute a good paradigm for such a working hypothesis.

However, for stereoelectronic effects that have small energetic consequences, stereoelectronic principles do not appear to serve as good guides for predicting reactivity in the active site in every case. Working hypotheses based on stereoelectronic principles now exist both for enzymatic reactions involving nicotinamide cofactors and for enzymes catalyzing the attack of nucleophiles on a carbonyl groups. However, much experimental work needs to be done to determine whether these hypotheses are general, either in dehydrogenases or in β-lactamases. At the other extreme, no pattern of reactivity consistent with stereoelectronic principles has been observed in enzyme-catalyzed vinylogous eliminations.

Nevertheless, this analysis generates an especially exciting prospect for future work. Normally, when studying enzymatic reactions, organic structural theory developed on small molecules is used to design experiments to help understand reactivity in the active site. Rarely do ideas and information flow in the reverse direction, where experiments on enzymes suggest ideas concerning the reactivity of small molecules. However, in the discussion just completed, we have encountered many instances where the analysis of enzymatic reactions suggests that fundamental organic reactivity is incompletely understood and demands studies of small organic molecules.

It has been hoped for some time that enzymatic behavior, presumably optimized by several billion years of biological evolution, might prompt a deeper understanding of organic reactivity in general. The interaction between enzymology and stereoelectronic theory may well be an area where this hope becomes reality.

ACKNOWLEDGMENTS

I am indebted to Professors Duilio Arigoni, Arthur Greenberg, Jeremy Knowles, Joel Liebman, Dieter Seebach, Dr. Simon Moroney, and Rudolf Allemann, Arthur Glasfeld, and Christian Schneider for reading the manuscript and offering many helpful extremely valuable comments and suggestions.

REFERENCES

1. Kirby, A. J. "The Anomeric Effect and Related Stereoelectronic Effects at Oxygen"; Springer-Verlag: Heidelberg, **1983.**
2. Deslongchamps, P. "Stereoelectronic Effects in Organic Chemistry"; Pergamon Press: New York, **1983.**
3. Romers, C.; Altona, C.; Buys, H. R.; Havinga, E. *Top. Stereochem.* **1969,** *4,* 39–97.
4. Lemieux, R. U. "Molecular Rearrangements", Vol 2; de Mayo, P., Ed. Wiley Interscience: New York, **1964,** p. 709.
5. Anderson, C. B.; Sepp, D. T. *Tetrahedron,* **1986,** *24,* 1707.
6. Lemieux, R. U.; Pavia, A. A.; Martin, J. C.; Watanabe, K. A. *Can. J. Chem.* **1969,** *47,* 4427–4439.
7. Eliel, E. L.; Giza, C. A. *J. Org. Chem.* **1968,** *33,* 3754–3758.
8. Stoddart, J. F. "Stereochemistry of Carbohydrates"; Wiley: New York, 1971.
9. Bonner, W. A. *J. Am. Chem. Soc.* **1959,** *81,* 1448–1452.
10. Lemieux, R. U.; Hayami, J. *Can. J. Chem.* **1965,** *43,* 2162–2173.
11. Lemieux, R. U. *Pure Appl. Chem.* **1971,** *25,* 527–548.
12. Paulsen, H.; Gyoergydeak, Z.; Friedmann, M. *Chem. Ber.* **1974,** *107,* 1590–1613.
13. Lemieux, R. U.; Morgan, A. R. *Can. J. Chem.* **1965,** *43,* 2205–2213.
14. Astrup, E. E. *Acta Chem. Scand.* **1971,** *25,* 1494–1495.
15. Bentrude, W. G.; Tan, H.-W.; Yee, K. C. *J. Am. Chem. Soc.* **1975,** *95,* 4659–4665.
16. Petrzilka, M.; Felix, D.; Eschenmoser, A. *Helv. Chim. Acta* **1973,** *56,* 2950–2960.
17. Corey, E. J.; Sneen, R. A. *J. Am. Chem. Soc.* **1956,** *78,* 6269–6278.
18. Perrin, C. L.; Arrhenius, G. M. L. *J. Am. Chem. Soc.* **1982,** *104,* 2839–2842.
19. Cieplak, A. S. *J. Am. Chem. Soc.* **1981,** *103,* 4540–4552.
20. Tenud, L.; Farooq, S.; Seibl, J.; Eschenmoser, A. *Helv. Chim. Acta* **1970,** *53,* 2059–2069.
21. (a) Coward, J. K.; Lok, R.; Takagi, O. *J. Am. Chem. Soc.* **1976,** *98,* 1057. (b) King, J. F.; McGarrity, M. J. *J. Chem. Soc. Chem., Commun.* **1982,** *104,* 175–176.
22. Bennett, A. J.; Sinnott, M. L. *J. Am. Chem. Soc.* **1986,** *108,* 7287–7294.
23. Kimura, M., Ed. "Molecular Evolution, Protein Polymorphism, and the Neutral Theory"; Springer-Verlag: New York, 1982.
24. Benner, S. A.; Nambiar, K. P.; Chambers, G. K. *J. Am. Chem. Soc.* **1985,** *107,* 5513–5517.
25. Lewontin, R. C. *Sci. Am.* **1979,** *239,* 156.
26. Benner, S. A.; Ellington, A. *CRC Crit. Rev. Biochem.* **1988,** in press.

27. (a) Petsko, G. A.; Davenport, R. C., Jr.; Frankel, D.; Raibhandary, U. L. *Biochem. Soc. Trans.* **1984**, *12*, 229–232. (b) Prof. J. R. Knowles has informed the author that this observation may have first been made by Prof. A. Eschenmoser in the mid-1970s following a lecture of his.

28. Iyengar, R.; Rose, I. A. *Biochemistry* **1981**, *20*, 1229–1235.

29. Westheimer, F. H. *Proc. Robert Welch Found. Conf. Chem. Res.* **1971**, *15*, 7–50.

30. Benner, S. A.; Morton, T. H., *J. Am. Chem. Soc.* **1981**, *103*, 991–993.

31. Buchanan, G. L.; Kean, N. B.; Taylor, R. *Tetrahedron* **1975**, *31*, 1583.

32. Sharpless, K. B.; Snyder, T. E.; Spencer, T. A.; Mahesnwari, K. K.; Gahn, G.; Clayton, R. B. *J. Am. Chem. Soc.* **1968**, *90*, 6874–6875.

33. Sharpless, K. B.; Snyder, T. E.; Spencer, T. A.; Mahesnwari, K. K.; Nelson, J. A.; Clayton, R. B. *J. Am. Chem. Soc.* **1969**, *91*, 3394–3396.

34. Rahman, R.; Sharpless, K. B.; Spencer, T. A.; Clayton, R. B. *J. Biol. Chem.* **1970**, *245*, 2667–2671.

35. Ghisalberti, E. L.; de Souza, N. J.; Rees, H. H.; Goad, L. J.; Goodwin, T. W. *J. Chem. Soc., Chem. Commun.* **1969**, 1403–1405.

36. Barton, D. H. R.; Lewis, D. A.; McGhie, J. F. *J. Chem. Soc.,* **1957**, 2907–2915.

37. Knapp, F. F.; Goad, L. J.; Goodwin, T. W. *Phytochemistry* **1977**, 1677–1681.

38. Dunathan, H. *Proc. Natl. Acad. Sci. U.S.A.* **1966**, *55*, 712–716.

39. Kirsch, J. F.; Eichele, G.; Ford, G. C.; Vincent, M. G.; Jansonius, J. N.; Gehring, H.; Christen, P. *J. Mol. Biol.* **1984**, *174*, 497–525.

40. March, J. "Advanced Organic Chemistry", 3rd ed.; Wiley: New York, 1985, p. 657ff.

41. Anet, F. A. L. *J. Am. Chem. Soc.* **1960**, *82*, 994–995.

42. Englard, S. *J. Biol. Chem.* **1958**, *233*, 1003–1009.

43. Jones, V. T.; Lowe, G.; Potter, B. V. L. *Eur. J. Biochem.* **1980**, *108*, 433–437.

44. Takagi, J. S.; Tokushige, M.; Shimura, Y.; Kanehisa, M. *Biochem. Biophys. Res. Commun.* **1986**, *138*, 568–572.

45. Hoberman, H. D.; Havir, E. A.; Rachovansky, O.; Ratner, S. *J. Biol. Chem.* **1964**, *239*, 3818–3820.

46. Englard, S. *J. Biol. Chem.* **1960**, *235*, 1510–1516.

47. Schroepfer, G. L., Jr. *J. Biol. Chem.* **1966**, *241*, 5441–5447.

48. Miller, R. W.; Buchanan, J. M. *J. Biol. Chem.* **1962**, *237*, 491–496.

49. Englard, S.; Britten, J. S.; Listowsky, I. *J. Biol. Chem.* **1967**, *242*, 2255–2259.

50. Cohn, M.; Pearson, J.; O'Connell, E. L.; Rose, I. A. *J. Am. Chem. Soc.* **1970**, *92*, 4095–4098.

51. Givot, I. L.; Smith, T. A.; Abeles, R. H. *J. Biol. Chem.* **1969**, *244*, 6341–6353.

52. Havir, E. A.; Hanson,K. R. *Biochemistry* **1975**, *14*, 1620–1625.

53. Cornforth, J. W.; Cornforth, R. H.; Popjak, G.; Yengoyan, L. *J. Biol. Chem.* **1966**, *241*, 3970–3987.

54. Hanson, K. R.; Rose, I. A. *Proc. Natl. Acad. Sci. U.S.A.* **1963**, *50*, 981–988.

55. Avigad, G.; Englard, S. *Fed. Proc., Fed. Am. Soc. Exp. Biol.* **1969**, *28*, 345.

56. Willadsen, P.; Eggerer, H. *Eur. J. Biochem.* **1975**, *54*, 247–252.

57. Widlanski, T. S.; Bender, S. L.; Knowles, J. R. In "Stereochemistry of Organic and Bioorganic Transformations", Vol. 17; Bartman, W.; and Sharpless, K. B. Eds.; New York: VCH, 275–282.

58. Schwab, J.; Klassen, J. B.; Habib, A. *J. Chem. Soc., Chem. Commun.* **1986**, 357–358.

59. Sedwick, B.; Morris, C.; French, S. J. *J. Chem. Soc. Chem. Commun.* **1978**, 193–194.

60. Messner, B.; Eggerer, H.; Cornforth, J. W.; Mallaby, R. *Eur. J. Biochem.* **1975**, *53*, 255–264.

61. Hanson, K. R.; Rose, I. A. *Proc. Natl. Acad. Sci. U.S.A.* **1963**, *50*, 981–988.

62. Aberhart, J.; Tann, C.-H. *J. Chem. Soc., Perkin Trans. 1* **1979**, 939–942.

63. Scharf, K. H.; Zenk, M. H.; Onderka, D. K.; Carroll, M.; Floss, H. G. *J. Chem. Soc., Chem. Commun.* **1971**, 765–766.

64. Kirby, G. W.; O'Loughlin, G. J.; Robins, D. J. *J. Chem. Soc., Chem. Commun.* **1975**, 402–403.

65. Hanson, K. R.; Rose, I. A. *Acc. Chem. Res.* **1975**, *8*, 1–14.
66. Schwab, J.; Klassen, J. B. *J. Am. Chem. Soc.* **1984**, *106*, 7217–7227.
67. Caspi, E.; Ramm, P. J. *Tetrahedron Lett.* **1969**, 181–185.
68. (a) J. Schwab, personal communication. (b) Mortimer, C. E.; Niehaus, W. G., Jr. *J. Biol. Chem.* **1974**, *249*, 2833–2842.
69. Kozarich, J. W.; Chari, R. V. J.; Ngai, K.-L.; Ornston, N. L. In "Mechanisms of Enzymatic Reactions: Stereochemistry", Frey, P. A., Ed.; Elsevier: Amsterdam, **1986**, pp. 233–246.
70. Yeh, W. K.; Ornston, N. L. *Proc. Natl. Acad. Sci. U.S.A.* **1980**, *77*, 5365–5369.
71. Bright, H. J.; Ingraham, L. L.; Lundin, R. E. *Biochim. Biophys. Acta* **1964**, *81*, 576–584.
72. Turner, M. J.; Smith, B. W.; Haslam, E. *J. Chem. Soc., Perkin Trans. 1* **1975**, 52–55.
73. Anke, H.; Diekmann, H. *FEBS Lett.* **1971**, *17*, 115–117.
74. Hermes, J. D.; Weiss, P. M.; Cleland, W. W. *Biochemistry* **1985**, *24*, 2959–2967.
75. Havir, E. A.; Hanson, K. R. *Biochemistry* **1975**, *14*, 1620–1626.
76. (a) Stork, G.; Kreft, III, A. F. *J. Am. Chem. Soc.* **1977**, *99*, 3850–3851. (b) Stork, G.; White, W. N. *J. Am. Chem. Soc.* **1956**, *78*, 4609–4615.
77. (a) Dobbie, A. A.; Overton, K. H. *J. Chem. Soc., Chem. Commun.* **1977**, 722–723. (b) Oritani, T.; Overton, K. H. *J. Chem. Soc., Chem. Commun.* **1978**, 454–455. (c) Uebel, J. J.; Milaszewski, R. F.; Arlt, R. E. *J. Org. Chem.* **1977**, *42*, 585–591. (d) Vogel, E.; Caravatti, G.; Frank, P.; Aristoff, P.; Moody, C.; Becker, A.-M.; Felix, D.; Eschenmoser, A. *Chem. Lett.* **1987**, 219–222.
78. (a) Drenth, W. *Rec. Trav. Chim. Pays Bas* **1967**, *86*, 319–320. (b) Hill, R. K.; Newkomb, G. R. *J. Am. Chem. Soc.* **1969**, *91*, 2893–2894.
79. Onderka, D. K.; Floss, H. G. *J. Am. Chem. Soc.* **1969**, *91*, 5894–5896.
80. Rose, I. A. *The Enzymes* **1970**, *2*, 281–320.
81. Hermes, J. D.; Tipton, P. A.; Fisher, M. A.; O'Leary, M. H.; Morrison, J. F.; Cleland, W. W. *Biochemistry* **1984**, *23*, 6263–6275.
82. Benner, S. A. *Experientia* **1982**, *38*, 633–636.
83. Nambiar, K. P.; Stauffer, D. M.; Kolodziej, P. A.; Benner, S. A. *J. Am. Chem. Soc.* **1983**, *105*, 5886–5890.
84. Rossmann, M. G.; Liljas, A.; Branden, C. I.; Banaszak, L. J. *The Enzymes* **1975**, *11*, 61–102.
85. Pai, E. F.; Schulz, G. E. *J. Biol. Chem.* **1983**, *258*, 1752–1757.
86. Benner, S. A.; Nambiar, K. P.; Chambers, G. K. *J. Am. Chem. Soc.* **1985**, *107*, 5513–5517.
87. Oppenheimer, N. J. In "Mechanism of Enzymatic Reactions: Stereochemistry"; Frey, P. A., Ed.; Elsevier: Amsterdam, **1986**, pp. 15–28.
88. Oppenheimer, N. J. *J. Biol. Chem.* **1986**, *261*, 12209–12212.
89. Oppenheimer, N. J. *J. Am. Chem. Soc.* **1984**, *106*, 3032–3033.
90. Dixon, D. A.; Lindner, D. L.; Branchaud, B.; Lipscomb, W. N. *Biochemistry* **1979**, *18*, 5770–5775.
91. Ohta, K.; Closs, G. L.; Morokuma, K.; Green, N. J. *J. Am. Chem. Soc.* **1986**, *108*, 1319–1320.
92. Hegazi, M. F.; Borchardt, R. T.; Schowen R. L. *J. Am. Chem. Soc.* **1979**, *101*, 4359–4365.
93. Westheimer, F. H. *Acc. Chem. Res.* **1968**, *1*, 70–78.
94. Gorenstein, D. G.; Findlay, J. B.; Luxon, B. A.; Kar, D. *J. Am. Chem. Soc.* **1977**, *99*, 3473–3479.
95. Gerlt, J. A.; Youngblood, A. V. *J. Am. Chem. Soc.* **1980**, *102*, 7433–7438.
96. Douzou, P.; Petsko, G. A. **1984**, *36*, 246–361.
97. Bürgi, H. B.; Dunitz, J. D.; Shefter, E. *J. Am. Chem. Soc.* **1973**, *95*, 5065–5067.
98. Bizzozero, S. A.; Zweifel, B. O. *FEBS Lett.* **1975**, *59*, 105–108.
99. Bizzozero, S. A.; Dutler, H. *Bioorg. Chem.* **1981**, *10*, 46–52.
100. Knox, J. R.; Kelley, J. A.; Moews, P. C.; Murthy, N. S. *J. Mol. Biol.* **1976**, *104*, 865–875.
101. Perrin, C. L.; Nunez, O. *J. Am. Chem. Soc.* **1986**, *108*, 5997–6003.
102. Edwards, J. T. *Chem. Ind. (London)* **1955**, 1102–1104.

103. Anet, F. A. L.; Brown, J. R. *J. Am. Chem. Soc.* **1967,** *89,* 760–768.
104. Blake, C. C.; Mair, G. A.; North, A. C. T.; Phillips, D. C.; Sarma, V. R. *Proc. R. Soc. London Ser. B* **1967,** *167,* 378–388.
105. Schindler, M.; Aesaf, Y.; Sharon, N.; Chipman, D. M. *Biochemistry* **1977,** *16,* 423–431.
106. Kelly, J. A.; Sielecki, A. B.; Sykes, B. D.; Phillips, D. C. *Nature (London)* **1979,** *282,* 875–878.
107. Platt, S.; Baldo, J. H.; Boekelheide, K.; Weiss, G.; Sykes, B. D. *Can J. Biochem.* **1978,** *56,* 624.
108. Post, C. B.; Karplus, M. *J. Am. Chem. Soc.* **1986,** *108,* 1317–1319.
109. Bennet, A. J.; Sinnott, M. L. *J. Am. Chem. Soc.* **1986,** *108,* 7287–7294.
110. Sinnott, M. L. *Biochem. J.* **1984,** *224,* 817–821.
111. Hosie, L.; Marshall, P. J.; Sinnott, M. L. *J. Chem. Soc., Perkin Trans. 2* **1984,** 1121–1131.
112. Briggs, A. J.; Evans, C. M.; Glenn, R.; Kirby, A. J. *J. Chem. Soc., Perkin Trans. 2* **1983,** 1637–1640.
113. Kirby, A. J. *Acc. Chem. Res.* **1984,** *17,* 305–311.
114. Strecker, H. J.; Korkes, S. *J. Biol. Chem.* **1952,** *196,* 769–784.
115. Kennedy, J. F.; White, C. A. "Bioactive Carbohydrates"; Wiley: New York, 1983.
116. Salas, M.; Vinuela, E.; Sols, A. *J. Biol. Chem.* **1965,** *240,* 561–568.
117. Wallenfels, K.; Kurz, G. *Biochem. Z.* **1962,** *335,* 559–572.
118. Ueberschar, K.; Blachnitzky, E.; Kurz, G. *Eur. J. Biochem.* **1974,** *48,* 389–405.
119. Schiwara, H. W.; Domagk, G. F. *Hoppe-Seyler's Z. Physiol. Chem.* **1968,** *349,* 1321–1329.
120. Cline, A. L.; Hu, A. S. L. *J. Biol. Chem.* **1965,** *240,* 4493–4497.
121. Benkovic, S. J.; Schray, K. J. *Adv. Enzymol.* **1976,** *44,* 139–164.
122. Berkowitz, D. B.; Benner, S. A. *Biochemistry* **1987,** in press.
123. Howard, S. M.; Heinrich, M. R. *Arch. Biochem. Biophys.* **1965,** *110,* 395–400.
124. Heinrich, M. R.; Howard, S. M. *Methods Enzymol.* **1966,** *9,* 407–412.
125. Feather, M. S.; Deshpande, V.; Lybyer, M. J. *Biochem. Biophys. Res. Commun.* **1970,** *38,* 859–863.
126. Cori, C. F.; Colowick, S. P.; Cori, G. T. *J. Biol. Chem.* **1937,** *121,* 465–477.
127. Rose, I. A.; O'Connell, E. L.; Schray, K. J. *J. Biol. Chem.* **1973,** *248,* 2232–2234.
128. I am indebted to Prof. D. Arigoni for bringing this to my attention.
129. The author has been informed by Prof. D. Arigoni that this observation was independently made about ten years ago by the late Prof. R. B. Woodward.

CHAPTER 3

Intramolecularity: Proximity and Strain

Anthony W. Czarnik

The Ohio State University, Columbus, Ohio

CONTENTS

1. Introduction 75
2. Definitions 77
3. Rate Accelerations Due to Covalently Enforced Proximity 79
4. Rate Accelerations Due to Noncovalently Enforced Proximity . 95
5. Rate Accelerations Due to Covalently Enforced Strain 102
6. Rate Accelerations Due to Noncovalently Enforced Strain 104
7. Conclusion 112
Acknowledgments 112
References 113

1. INTRODUCTION

Imagine yourself a chemist responsible for the design of a living organism. One of the most important tasks ahead is making sure the reactions that must occur quickly to sustain life proceed fast enough. Being classically trained in organic chemistry, you consider simply increasing the temperature, but find one of the project guidelines calls for an almost perfectly uniform 37°C reactor temperature. You suggest choosing only reactions with low activation barriers, but find the reaction mixture required to be so complex that undesired side reactions predominate. In frustration, you reluctantly concede that a solvent more polar than chloroform must be used, only to discover that the choice of reaction medium has already been made: water.

Nature's solution, namely, the evolution of thousands of water-soluble catalysts, most apparently specific for a single chemical transformation, has caused chemists to reevaluate their own powers. We have, in fact, yet to devise any catalytic scheme that combines all the highly desirable characteristics of enzymes: high substrate specificity, chemoselectivity, stereoselectivity, stereospecificity, and large (sometimes enormous) rate accelerations. A detailed understanding of enzymatic modes of catalysis will inexorably lead to an ability to design synthetic catalysts for abiotic reactions; it is largely for this reason that the study of enzyme mechanisms has become central to bioorganic chemistry. Work in this area over the past 30 years or so has been described in now classic texts by Bruice and Benkovic,[1] Jencks,[2] and Bender;[3] more recent texts[4-9] have built upon and added to these volumes. To summarize drastically, enzymatic catalysis is today explained largely on the basis of chemical catalysis (catalysis by functional groups) together with an ability of enzymes to reroute *intermolecular* processes through *intramolecular* pathways by binding substrates to preorganized active sites.

It is appropriate here to emphasize that enzymes do not formally catalyze reactions, but rather catalyze state changes that can be described as processes, conversions, transformations, etc. The term *reaction* implies a particular set of reactants transversing a particular reaction coordinate. Enzymes, rather than accelerating a particular reaction, provide an alternate pathway for achieving a change from state A to state B; because the energy of activation is lower in the alternate pathway, the state change proceeds via that pathway. While the terminology "catalyzed reaction" is widely accepted and understood, it is not accurate, and we will attempt to avoid it.

This chapter deals with the topic of *intramolecularity*. Intramolecular reactions usually occur more quickly than do equivalent intermolecular reactions, and this rate enhancement is often attributable to an increase in $\Delta S^{\ddagger}$. The rationale follows. Collision theory predicts that a given bimolecular reaction will occur when two molecules collide with enough kinetic energy and in a productive orientation for bond formation to occur. Thus, the rate of a reaction is determined by a purely energetic term, a purely statistical term, and an orientation term that may be alternately defined as either energetic or statistical. Choosing the latter, the statistical term is therefore determined by how often reactants randomly collide with each other (concentration) and how often those collisions involve the reactive regions of each molecule (orbital alignment). Enzymes are able both to increase the effective concentrations of reactants and/or catalytic groups and to orient reactive regions of substrates in productive spatial relationships; each of these affects primarily the entropy of activation, and each is a result of transient but enforced proximity.

Increasingly, common wisdom is coming to include the likelihood that enzymes catalyze processes by utilizing binding energy to decrease $\Delta H^{\ddagger}$. Of course, it has long been recognized that participation of a catalytic group (eg, general acid catalysis by a carboxylic acid) can decrease $\Delta H^{\ddagger}$, whether on an enzyme or in free solution. However, the binding event itself can lower $\Delta H\ddagger$ even in the absence of catalytic groups if some of the binding energy is transformed to strain energy. The reader must be cautioned, however, that this effect is fruitful only if the strain energy is appropriate to move a bound substrate along the reaction coordinate leading to product. Still, it seems likely that this method of accelerating conversions remains at once both the brightest light and the least examined aspect of synthetic catalyst design. We will present examples that point to its utility.

The functions of this chapter, then, are to differentiate between proximity and strain effects that accelerate conversions, to cite and briefly summarize pertinent theoretical discussions about these ideas, and to present a few well-documented examples of each. The author has chosen to group examples into two categories: *covalent* and *noncovalent* associations. This is a natural distinction to make inasmuch as the simpler enzyme *models* generally enforce intramolecularity by way of a contiguous chain of covalent bonds, while enzyme *mimics* do so using some other type of substrate–catalyst association (eg, hydrophobic binding, hydrogen bonding, ion pairing, etc.). Note that our definition of "intramolecularity" includes anything that is, at a given moment, a single species, and formally includes (if you will) intracomplexity. The distinction is obviously somewhat arbitrary in that either type of association (covalent or noncovalent) may be stronger than the other under appropriate circumstances, although most covalent models are stable enough that binding equilibria need not be considered. In addition, we make the equally arbitrary (and occasionally unpopular) choice of classifying metal–ligand interactions as noncovalent.

Finally, the traditional disclaimers. It is not the function of this chapter to present an unabridged review of work on the topic of intramolecularity, and I apologize ahead of time to my colleagues whose work, although equally relevant, was not included for space reasons. Some of this work and thought has been reviewed previously, and I will make every attempt to alert readers to contemporary reviews of individual topics rather than trying to improve upon them with a retelling in different words. It is sometimes said that an artist's "style" results not so much from their abilities as from their inabilities; this thought likely holds for authors as well.

2. DEFINITIONS

A. Strain

In order to distinguish between proximity effects and strain effects, we must first try to define them. Operationally, it is easier to define strain first and to define proximity reflexively. Insofar as the term *strain* refers to an elevated ground state energy, we will adopt the convention used by molecular mechanicians; that is, by comparing the enthalpy of formation of a "strained" molecule to that of a "strain-free" molecule having the same constitutive units. DeTar has described[10] a transferable method of breaking down the enthalpy of formation into four terms: formal bonding enthalpy (FBE), formal steric enthalpy (FSE), formal polar enthalpy (FPE), and formal medium (or solvent) enthalpy (FME). The term *formal steric enthalpy* is furthermore described as "the energy arising from the deformation of bonds, angles, and torsions together with the energy due to intramolecular nonbonded interactions." The repulsion of methyl groups in *gauche*-butane will therefore increase the energy relative to that of *anti*-butane by both forcing the molecule away from an ideal staggered geometry and by residual methyl–methyl repulsions in that geometry; that is as compared to cyclopropane, which has a higher ground state energy (per CH_2 unit) than does cyclohexane because of nonoptimal orbital overlap (FBE) and eclipsing H—H interactions (FSE) that do not exist in cyclohexane. The anomeric effect in carbohydrates has been interpreted as the result of dipole–dipole repulsion, which can be classified as FPE. And finally, the increased nucleophilicity of anions in nonpolar solvents may be considered to be a ground state elevation owing to the FME term. This subdivision of strain effects, although somewhat arbitrary, is useful in delineating the modes of destabilization that are likely operative in enzymatic catalysis. The reader must use caution in applying these terms with unwarranted authority. It would be comforting to ascribe the relative thermodynamic acidities of substituted benzoic acids in water (the Hammett σ constant) entirely to the formal polar enthalpy term, when it is well known to result from both enthalpic and entropic changes.[11]

Defining the strain present in an enzyme–substrate complex similarly requires comparison to a strain-free reference. As has been noted often before, this is a conceptually imprecise definition because enzyme–substrate complexes *do* form; it therefore follows that a localized strain ($\Delta H^{\ddagger} > 0$) together with the unfavorable entropy of binding ($\Delta S^{\ddagger} < 0$) must be compensated for by other localized regions of favorable binding ($\Delta H^{\ddagger} << 0$). Of course, one can measure only a single $\Delta H^{\ddagger}$ for the binding event, so the separation of localized enthalpic forces is hypothetical. Still, such an approach offers a readily grasped visualization of binding-induced strain and will be used here. Therefore, we will adopt as a reference the ground state conformations of the uncomplexed substrate and enzyme in aqueous solution. Localized strain may thereby be induced in the substrate by binding it in such a way that a group at the active site is forced into juxtaposition with a group on the substrate (FSE); or a dipole at the active site may be forced to lie in close proximity with a dipole on the substrate (FPE); or the solvating properties of the active site may more closely resemble those of DMSO as opposed to water, making negatively charged substituents less stabilized (FME). One can, in an imperfect way, model these effects by considering them separately and carrying out equivalent operations between the substrate and an isolated juxtaposing group, or an isolated dipole, or by changing the solvent. It should be noted that this approach ignores the possible interplay of these effects and can be used only qualitatively.

The notion of ground state destabilization (strain) used here turns out to be almost completely equivalent to the idea of transition state stabilization, as has been espoused and described in a scholarly fashion by the "Fundamentalist" camp.[12] In brief, $\Delta G^{\ddagger}$ for a reaction may be decreased (and therefore the rate increased) either by increasing the ground state energy via strain or by decreasing the transition state energy via its strong binding to the enzyme. Two additional points must be made. First, these notions are functionally equivalent only for *conformationally immobile* systems. If the strained group is conformationally mobile, the catalytically productive conformation will be closer to a local energy maximum in the ground state and to a local energy minimum in the transition state. As has been stated previously, "The presence of one or more free rotations ensures that a reacting group can move out of an unfavorable conformation so that the fraction of starting material in the high-energy, strained, or desolvated form will be negligible."[13] This distinction becomes especially important when describing enzyme models, which may not exhibit strain-induced rate accelerations if a critical functional group is not forced to interact with the functional group undergoing conversion in an initially destabilizing way. The second addendum, which is not mutually exclusive of the first, is the notion of "distortionless strain" as advanced by Fersht[14] and discussed by Jencks.[15] It is quite possible that a favorable, localized interaction in an enzyme–substrate complex would not be as strong as in the corresponding enzyme–transition state complex. Inasmuch as this favorable interaction would be developed as the transition state is approached, $\Delta G^{\ddagger}$ would be decreased. This developing stabilization (as compared to the enzyme–substrate complex), together with ground state destabilization (as compared to our reference state), have been referred to collectively as the reason that the enzyme–substrate complex is "strained toward the transition state."

B. Proximity

After all this, an intramolecular effect due solely to *proximity* results when reactants are held together in a productive orientation such that the complex is strain free, ie, there is an absence of significant strain effects as listed in the preceding discussion. Rate increases are observed because reactants have already lost the required translational and rotational degrees of freedom prior to the rate-determining step, and the "preorganized" reaction has a less negative value of $\Delta S^{\ddagger}$.

3. RATE ACCELERATIONS DUE TO COVALENTLY ENFORCED PROXIMITY

A. Intramolecular Reactions between Reagents

The reaction between A and B, where A is a reagent and B is either a reagent[16] or a catalyst[17–26], can be faster when A and B are linked covalently

(A—B) than when they are not (A + B). One of many frequently cited examples of such an acceleration is the relative rate of anhydride formation in the series of dicarboxylic semiesters shown in Table 3-1.[27-30] Each of the intramolecular cyclizations (examples B through E) occurs more quickly than does the reference intermolecular reaction (example A) 1 M in reactants, although a problem arises in comparing the rates of these reactions. Example A is a bimolecular reaction, and the units of its rate constant are per M.s; examples B–E are each unimolecular reactions, with rate constants of unit/s. The two constants cannot be compared directly, and the usual

TABLE 3-1. Relative Rate of Anhydride Formation in a Series of Dicarboxylic Semiesters

Reactants	k_1/k_2
A) 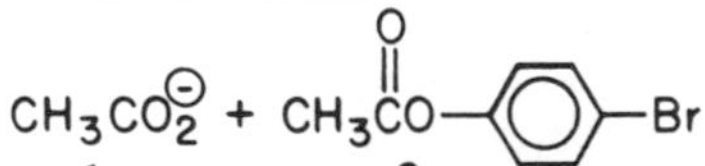 $CH_3CO_2^{\ominus}$ + ...	1.0
B) 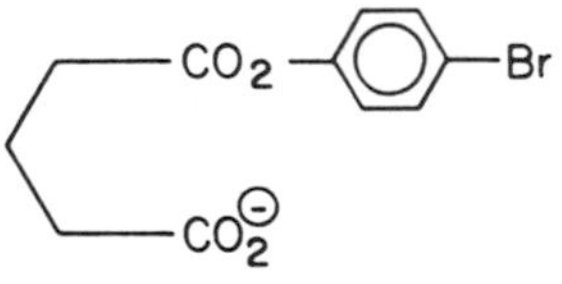 ... 3	1×10^3 M
C) 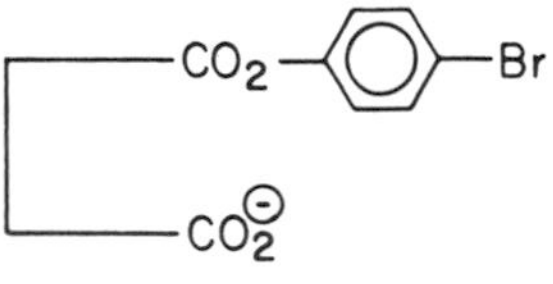 ... 4	~2.2×10^5 M
D) 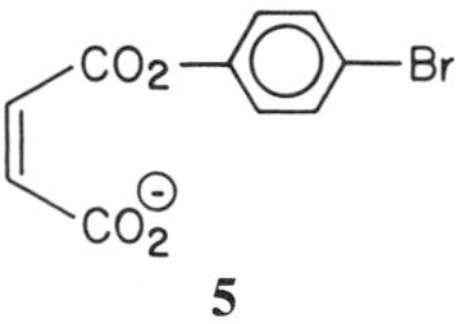 ... 5	1×10^7 M
E) 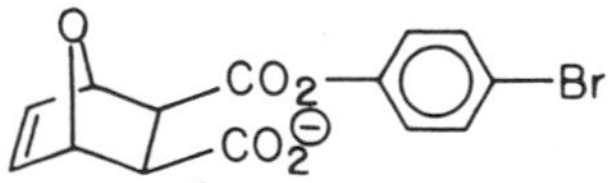 ... 6	~5×10^7 M

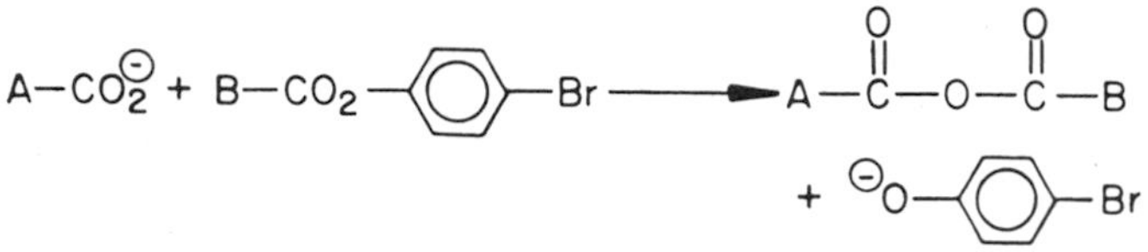

k_1 = first-order rate constant, reaction A
k_2 = second-order rate constants, reactions B-C

solution is to calculate the ratio of k_1/k_2, where the reactions being compared are electronically and sterically as similar as possible near the reacting centers. This ratio must have units $(1/s)/(1/M-S)$, and has been referred to as an *effective molarity* (abbreviated EM). The physical interpretation using reactions A and B, for example, is as follows: given a solution of **1** and a solution of **3** such that $[\mathbf{1}] = [\mathbf{3}]$, $[\mathbf{2}]$ must be 1×10^3 M in order for the initial observed rates to be equal. In other words, the carboxylate ion in **3** "sees" an effective *p*-bromophenyl ester concentration of 1×10^3 M in reaction B. Early attempts to constrain the magnitude of EMs to less than or equal to 55.5 M (the concentration of water) were eventually concluded to be incorrect; such an argument correctly ascribes a rate acceleration due to increasing the probability of collision, but underestimates the acceleration achieved by effectively increasing the number of times such collisions occur with the correct orientation for reaction. The debate over this issue has been summarized in the literature.[31] Examples C–E in Table 3-1 bear out this notion of orientation; the fewer rotational degrees of freedom are available to the molecule, the less organization is required to form the transition state and, consequently, the faster each successive reaction is.

A recent compilation of rate data for the cyclization of ω-bromo carbanions[32] succinctly points out that all intramolecular reactions do not proceed with high EMs and that reactions may even be retarded when carried out intramolecularly (as compared to solutions 1 M in reactants). As summarized in Figure 3-1, only the cyclizations leading to four-, five-, and

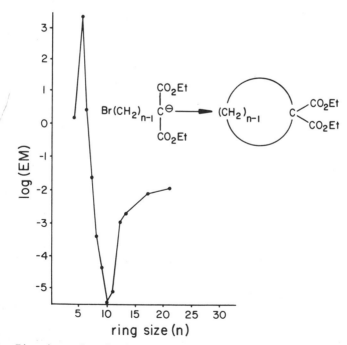

Figure 3-1.   Ring size and cyclization rate data for ω-bromo carbanions. See text for discussion.

six-membered rings had EMs greater than one; for medium-sized rings EM < 1, and at very long chain lengths an asymptote is approached. The enormous sensitivity of $\Delta S^{\ddagger}$ to the number of freely rotating bonds between functional groups is easily demonstrated by the well-documented rapidity of three-membered ring closure.[33] While smaller chains are expected to have a less negative $\Delta S^{\ddagger}$ for cyclization, ring strain expressed in the transition state results in a more positive $\Delta H^{\ddagger}$ and, for many ring sizes, overwhelms the $-T\Delta S^{\ddagger}$ term. Indeed, the authors note a "definite tendency for transition-state strain energies to parallel cycloalkane strain energies."[32] Extrapolating these findings to intermolecular reactions, we expect that the fastest examples will both hold reacting groups in the same region of space ($\Delta S^{\ddagger}$) and decrease energy barriers of enthalpic origin ($\Delta H^{\ddagger}$).

B. Orbital Steering

There have been various attempts to formalize and quantify the importance of orientation factors in achieving rate accelerations. One of the earliest and best known is that of Koshland, termed *orbital steering*, in which it is proposed that for some reaction types the activation energy is very sensitive to deviations from an ideal trajectory angle. If this is the case, enzymes could well accelerate reactions by binding and "steering" reactants into an optimal spatial orientation. The real question, of course, is just how much a moderate deviation of, say, 10°, would increase E_{act}. In support of his thesis, Koshland has presented both theoretical work[34-36] and experimental work[37-41] demonstrating the rate of lactone formation to be a sensitive function of angle between alcohol and carboxylic acid moieties.

The criticism of this theory[31,42-46] and of the experimental evidence supporting it[47-50] began almost immediately, and it is fair to say that the orbital steering theory remains widely unaccepted. Of particular importance is Bruice's observation that the transition state force constants required by orbital steering are much larger than those measured for fully formed covalent bonds.[42] It is often noted, however, that in the face of experimental rate data unambiguously supporting any theory, calculations and prior reasoning could, if history is a guide, be rationalized post facto. A fundamental assumption upon which rests the theoretical basis for the magnitude of orbital steering, that of EM $\leq$ 55.5 M, is clearly incorrect, but as was stated in 1980, "The fact of the matter is that orbital steering has never been proved or disproved."[51] Indeed, literature citings supportive of the precepts of orbital steering can be found.[52,53] The subject of orbital steering has been reviewed,[54] and the interested reader is advised to consult this source for pertinent literature citings prior to 1978.

A recent, thoughtful set of experiments by Menger was designed to fill an obvious gap on this topic. In his paper,[51] Menger lists four criteria that should be met in order to establish unambiguously that rapid lactonization between intramolecular alcohol and carboxylic acid groups is due to a stringent requirement of orbital alignment. They are as follows: "(1) The two compounds must have measureable but differing angular relationships between their hydroxy and carboxy groups. Since molecular flexibility would lead to uncertainty in this regard, a rigid carbon framework must support the functionalities. (2) Despite the angular differences, the initial OH/COOH distances must be the same. (Koshland's compounds are seen to disobey this stipulation.) (3) The hydroxy groups must possess identical inherent reactivities (eg, one hydroxy group should not be primary and the other tertiary). The same holds true for the carboxy groups. (4) The lactone products must have identical strain energies." Two sets of hydroxy acids, **7/8** and **9/10** (Table 3-2), were then synthesized and shown to meet each of the Menger's

TABLE 3-2. Hydroxy Acid Sets Synthesized and Shown to Closely Approximate Menger's Criteria

	angle (deg)	$k_H^{\oplus}$(rel)
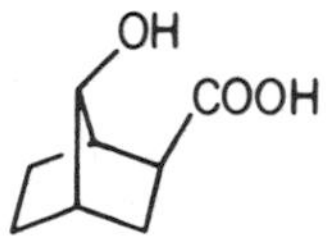 7	70	1
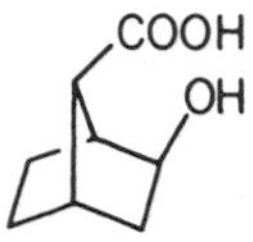 8	80	1.2
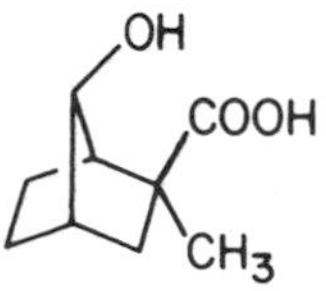 9	76	36
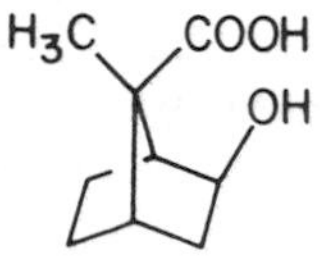 10	85	22

criteria to a close approximation. The rates of acid-catalyzed lactonization (shown in Table 3-2) are seen to vary a negligible amount (relative rate ratios are **7/8**, 0.83, and **9/10**, 1.64) even though a 10–11° difference in alignment exists between the members of each set.[55] This series of experiments does seem to suggest that, at least in the case of lactonization reactions, a misalignment of functional groups by 10° does not result in an enormously slower reaction. The conclusion does, however, rest on the premise that the rate-determining step in each of these reactions is formation of the tetrahedral intermediate and not its decomposition. Houk[56a] has recently described the results of force-field calculations suggesting that tetrahedral intermediate decomposition *is* rate limiting for each of the compounds in Menger's study. In fact, a good correlation was found between rates of reaction and the energy differences between starting hydroxylacids and the *second* transition states (corresponding to decomposition of the tetrahedral intermediate).

C. The Spatiotemporal Hypothesis

Due, in part, to the pervasive formulation of novel postulates and terminology surrounding the topic of intramolecular rate accelerations, Menger has proposed that a greatly simplified picture be considered. This postulate, known as the *spatiotemporal hypothesis*,[57] asserts that reacting moieties rigidly fixed within a certain critical distance with respect to each other will react with a single inherent rate constant. For nonfixed groups, therefore, the bimolecular rate is a function simply of how much time the reacting moieties spend within that critical distance. An experimental example that seems to differentiate this idea from either orbital steering or simple proximity is shown in Figure 3-2. The fact that dehydrohalogenation of **11** occurs intramolecularly insures that proton transfer does not require a linear transition state[58,59] (which has also been demonstrated by other groups).[60–62] However, phenoxide **13** does not react using the intramolecular pathway; rather, dehydrohalogenation occurs intermolecularly. While recognizing that one is comparing a secondary alkoxide with a phenoxide as base and that the O/H/C trajectory angles differ (106° in **11** and 82° in **13**), a major difference is the O---H distance (2.2 Å in **11** and 2.9 Å in **13**). Menger argues that the difference in intramolecular reactivities can be explained if the critical distance for proton transfer is somewhere between 2.2 and 2.9 Å.

Several aspects of this theory warrant discussion. The spatiotemporal hypothesis, as described, requires that activation energy be a step function (or at least a very steep sigmoidal function) of distance. Exactly what the critical distance is for various reactions remains the subject of study, and Menger has rightly stated that this value need not be predictable for the theory to be useful. In its present form, one might conclude from this theory that once the critical distance is passed the activation energy no longer decreases with decreasing interatomic distance. This is a conclusion insepa-

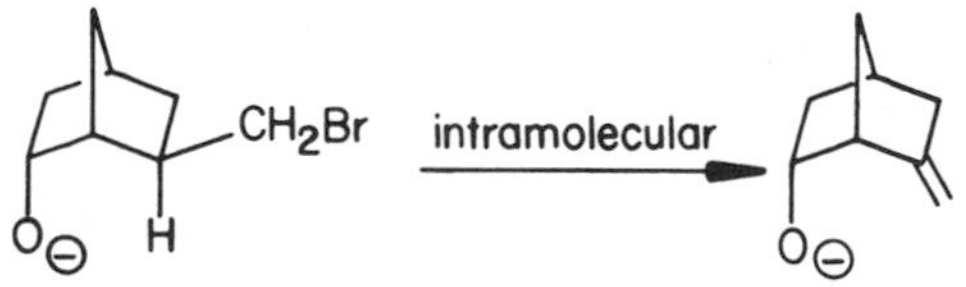

Figure 3-2. Experimental example proposed by Menger to distinguish critical distance from orbital steering and simple proximity. See text for discussion.

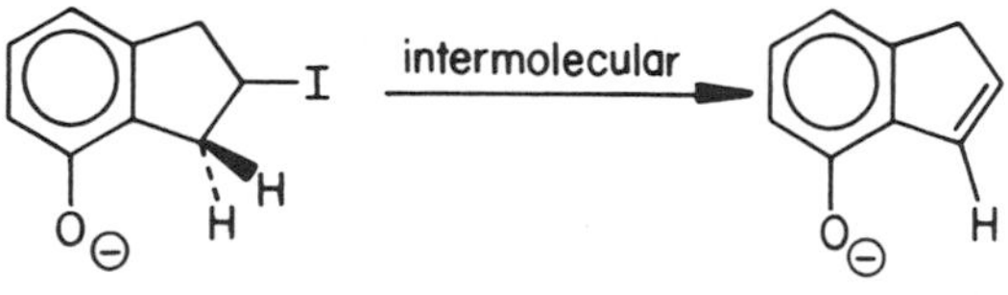

rable from the idea of an intrinsic rate constant and would seem to preclude strain effects. For example, suppose a model compound like **11** could be made in which the enforced distance between O^- and H was intermediate between an O—H covalent bond length (0.96 Å) and an O---H hydrogen bond length (2.07 Å). It seems intuitively reasonable that such a model compound would undergo elimination even faster than **11**. As with all intuitively reasonable conjectures, this one must be tested experimentally before it has merit.

Menger has also stated that "alternative enzyme mechanisms (electrostatic stabilization, rack effects, transition-state stabilization, etc.) are intriguing but unnecessary".[57] This statement comes across as restrictively broad. We might, for example, consider the possibility of strong hydrogen bonding in hydroxyacids **7–10** of the type O—H---O=C. Such an interaction would be expected to be more favorable at an enzyme's active site than in water and would hold the functional groups in a nonproductive orientation. Thus, whether or not there is a narrow angular trajectory (orbital steering) or critical distance (spatiotemporal hypothesis), the hydrogen bond must be broken for reaction to occur if the alcohol itself is acting as nucleophile. By analogy to chymotrypsin, an adjacent imidazole group could serve this funtion, which might be described as either electrostatic or transition-state stabilization. In addition, the closer a critical distance for a reaction comes to the sum of appropriate van der Waals radii, the more converted binding energy will be required to force the reactants to reside at this distance. Inasmuch as such binding energy can be channeled into either "reactant-state destabilization" or "transition-state stabilization," effects other than enforced proximity at an unstrained distance must be considered.[56b]

D. Intramolecular Reactions between a Reagent and a Catalytic Group: The Evolution of Enzyme Models for Chymotrypsin

The vast majority of *enzyme models* fall into the category of reactions accelerated due to covalently enforced proximity. Such compounds possess both functional and catalytic groups attached covalently to a (usually) carbon framework and are designed to determine the potential effectiveness of the catalytic group in the absence of complicating binding equilibria. Many examples of enzyme models have been reported, and it is likely that every known enzymic active site has been modeled, albeit to various degrees of sophistication. We will not attempt to provide an overview of enzyme model chemistry, which has been done previously.[17-24] Rather, we will look at the evolutionary improvement in the modeling of one specific enzyme, chymotrypsin, which will serve both to illustrate the general premise and to point out the extreme technical feat embodied in any single molecule that is a truly accurate model for an enzyme active site.

a. Chymotrypsin: The Enzyme

Requisite to the modeling of any enzyme is a detailed knowledge of its interactions with substrate molecules. The three-dimensional structures of α-chymotrypsin and of its active site have been well documented both by chemical studies with model substrates[63] and by X-ray diffraction analysis.[64-66] The enzyme consists of three chains designated as A, B, and C, and derives a considerable degree of its tertiary structure from five disulfide bridges. The catalytic groups at the active site are located on the B and C chains; these groups are generally considered to be a histidine residue on the B chain (labeled His-57) and a serine residue on the C chain (labeled Ser-195). That these groups are, in fact, active in catalysis has been shown directly by inactivation of the enzyme upon exposure to functional group-specific reagents[67] and indirectly by kinetic studies at various pH values.[68] These histidine and serine residues have been shown to lie in close proximity to each other[69] as well as to an aspartic acid residue (labeled Asp-102), which is now widely believed to be involved in the catalytic activity of the enzyme.[70] The prevailing mechanism[71] is that illustrated in Figure 3-3. In this scheme, the substrate initially binds at the active site through a combination of hydrogen bonding, dipole–dipole interactions, and hydrophobic bonding, which cause the peptide (or ester) to be oriented near the catalytic groups (**15**). Serine-195, hydrogen bonded to the adjacent His-57, affects a nucleophilic attack on the carbonyl group and releases the nitrogen end of the peptide bond as the first product, leaving an acylated enzyme intermediate (**16, 17, 18**). Finally, the acyl enzyme is hydrolyzed, displacing the aminoacyl group from the enzyme with participation of the now deprotonated imidazole, and the second product of the hydrolysis is released from the active site (**19**). Evidence for this two-step mechanism has been compiled

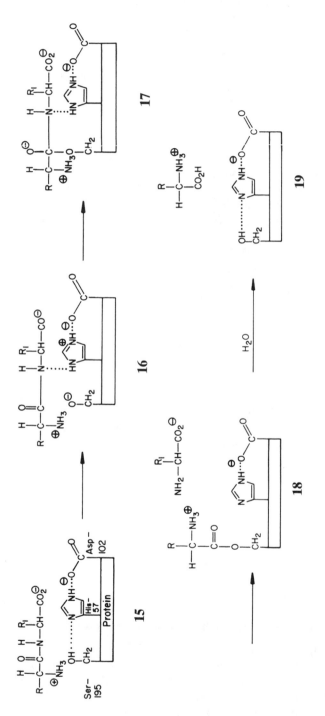

Figure 3-3. Mechanism of chymotrypsin catalysis. See text for discussion.

from a number of sources, including the isolation and even recrystallization of some aminoacyl enzymes.[72]

b. Models for Chymotrypsin: One Catalytic Group Spatially Unconstrained

The action of imidazole as a general base catalyst in the hydrolyses of activated esters has been studied by Jencks[73,74] and by Bruice.[75,76] Similarly, Hardman and co-workers[77] have evaluated imidazole catalysis of the hydrolysis of N,O-diacetylserinamide (**20**) as a model for the deacylation of aminoacyl-α-chymotrypsins. They calculated the activation parameters for this reaction and found a reduction in the $\Delta H^{\ddagger}$ of at least 7.4 kcal/mol relative to 20.8 kcal/mol for the uncatalyzed hydrolysis. The rate was not greatly enhanced because of an unfavorable change in the $\Delta S^{\ddagger}$ of about -35.8 eu; however, the authors concluded that, as the entropy change for the deacylation of aminoacyl-α-chymotrypsins has been observed to be substantially less than this value ($\Delta S^{\ddagger} = -21.3$ eu), imidazole catalysis could accelerate the rate of reaction 10,000-fold.

A study that investigated the feasibility of imidazole-facilitated reaction of an amide with a neighboring hydroxyl group was performed by Shafer and co-workers.[78] The kinetics of lactonization of 2-hydroxymethylbenzamide (**21**) and of its N-benzyl derivative were determined at various concentrations of imidazole buffer solution, and the apparent second-order rate constant for **21** was found to be 4.5×10^{-3}/min/M at 25°C. The authors reported that imidazole was clearly more efficient in catalyzing cyclization of the hydroxyamides than imidazolium ion; in fact, in comparison to specific base catalysis, they observed the second-order rate constant for imidazole-catalyzed lactonization, assuming that a properly oriented un-ionized hydroxyl group is present, to be approximately seven times that of hydroxide-ion-catalyzed hydrolysis of benzamide.

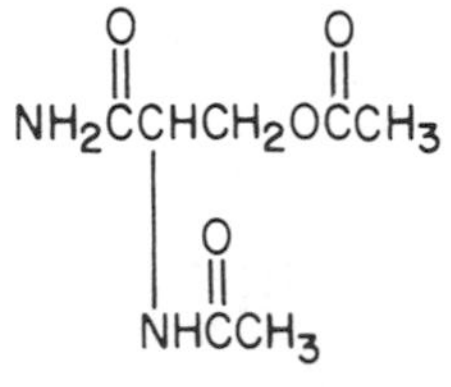

20 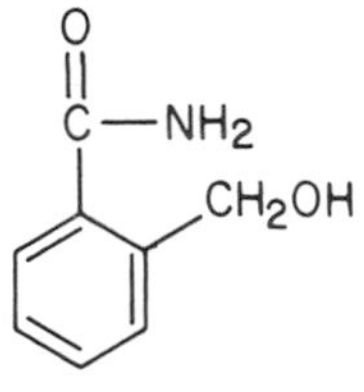   21

c. Models for Chymotrypsin: One Catalytic Group Spatially Constrained

In an attempt to test the validity of nucleophilic attack by a deprotonated serine residue (**16** --→ **18**), Menger and Brock[79] studied the hydrolytic properties of N-n-butyl-8-hydroxy-1-naphthoamide (**22**) under conditions in which the naphthol was largely ionized. They reasoned that since the naphtholate anion was rigidly held in close proximity to the amide, it should act as a general base and assist in the hydrolysis of the amide; however, after 48 h at 25°C and pH 11.48, less than 5% of the starting material had disappeared. Consequently, they argued that, if monofuncitonal general acid–general base catalysis was the sole source of rate acceleration in α-chymotrypsin, then the overall rate of hydrolysis of **22** should have been of the same order of magnitude as that of the enzyme. The rates were found to differ by a minimum of 10^5, and on the basis of this evidence they postulated that at least a second catalytic functionality must be present at the enzyme's active site.

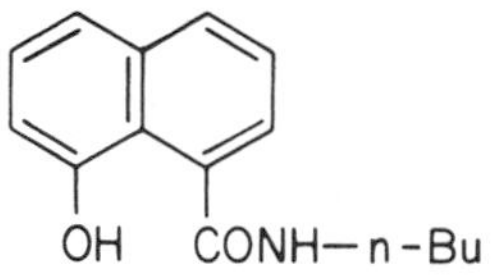

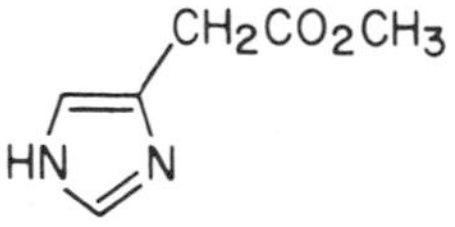

While general base catalysis by imidazole of unactivated alkyl esters has been reported to be quite small,[76,80] Jencks has observed the existance of a slight general base catalysis of methyl acetate at high imidazole concentration.[81] Therefore, approximation of a substrate and imidazole in an active site, which would be expected to increase the EM, might be expected to result in a substantial contribution to the rate acceleration. This idea was tested by several investigators utilizing model compounds with catalytic groups spatially constrained to the vicinity of the functional group. One such study was carried out by Koshland,[82] in which the rates of hydrolysis for a series of imidazole-substituted esters (**23–27**) were measured (Table 3-3). It is important to note that the rates of hydrolysis for **26** and **27** are essentially proportional to the hydroxide concentration, while **23**, **24**, and **25** each show rate enhancements of 3- to 8-fold at pH 7 as compared to **26** and **27**. That this catalytic effect is intramolecular was shown by experiments at 0.005, 0.010, and 0.015 M concentrations of ester, which indicated the rate constants were unchanged. A similar study was done by Utaka,[83] who measured hydrolysis rate constants for related model compounds in which internal rotations of

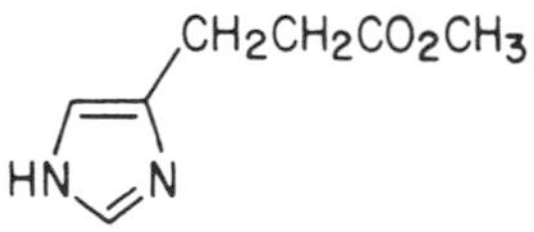

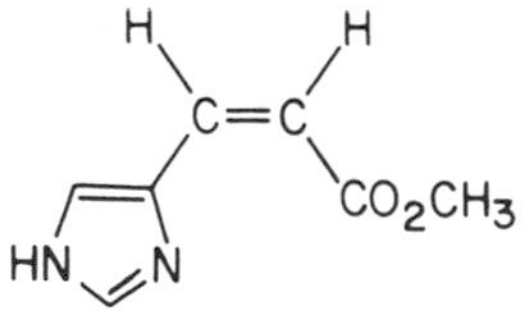

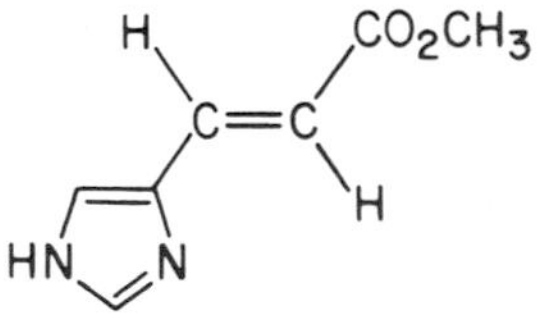

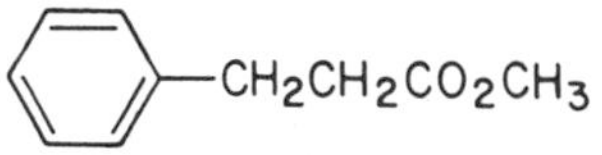

TABLE 3-3. Rates of Hydrolysis of Some Spatially Constrained Substituted Esters

$$10 \cdot k \text{ in } s^{-1} \text{ at } 105°C$$

Compound	pH 6.0	pH 7.0	pH 8.5
23	0.030	0.170	1.36
24	0.011	0.064	0.72
25	0.039	0.14	1.00
26	0.0019	0.022	0.69
27	0.0028	0.028	0.94

the esters were frozen, a circumstance likely to exist at the active site of chymotrypsin. He found that the rate increased by factors as high as 11.5 over the rotationally unrestricted ester as the degree of spatial constraint was increased.

Rogers and Bruice[84] have examined the rate data for another series of spatially constrained enzyme models, o- and p-imidazoylphenyl acetates 28 and 29. Their findings implicate general acid assistance to H_2O attack of the o-disubstituted compound at low pH (2–4) by the neighboring imidazolium cation. At neutral pH, general base catalysis by imidazole was observed, and at alkaline pH (9–11) evidence was obtained to support intramolecular acetyl group transfer to the imidazolyl anion with subsequent hydrolysis of the acetylimidazole intermediate.

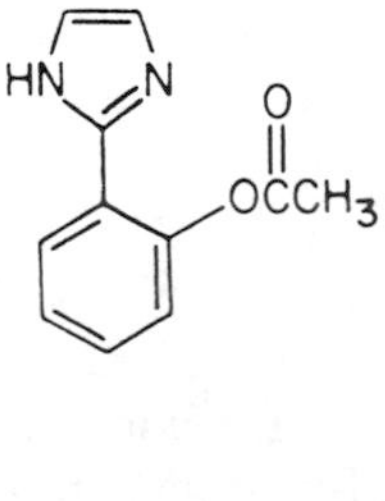

28 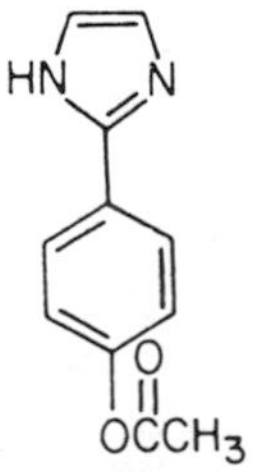   29

Bender and co-workers have provided evidence which seems to support the importance of the charge-relay system. In one study[85] they found an eight-fold acceleration due to cooperation of the imidazolyl and carboxyl groups in the general base-catalyzed hydrolysis of ethyl chloroacetate by 2-benzimidazole acetic acid (30), which has both the imidazolyl and carboxyl groups in the same molecule.

In another investigation[86] Bender reported the synthesis of *endo*- and *exo*-

5-[4(5)-imidazolyl]bicyclo[2.2.1]hept-*endo*-2-yl *trans*-cinnamates **31** and **32** as potential enzyme models. In 1977,[87] Bender reported that the imidazole in **31** functions as an intramolecular general base catalyst and, furthermore,[88] that the rate acceleration attained at a benzoate concentration of 0.5 *M* was 2500 times that attained at zero benzoate concentration. No intramolecular activity was found for **32**.

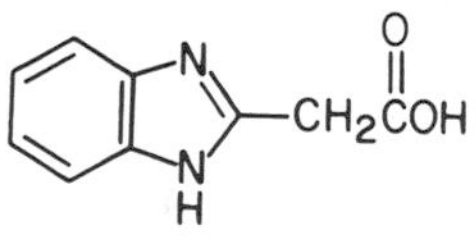

30 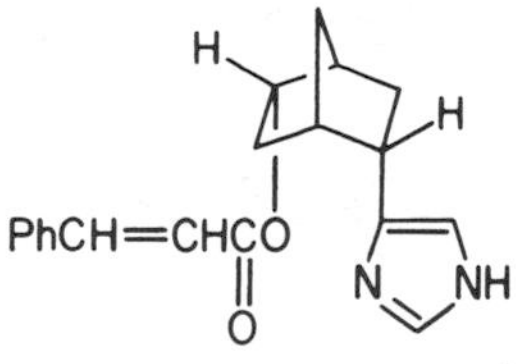   **31** 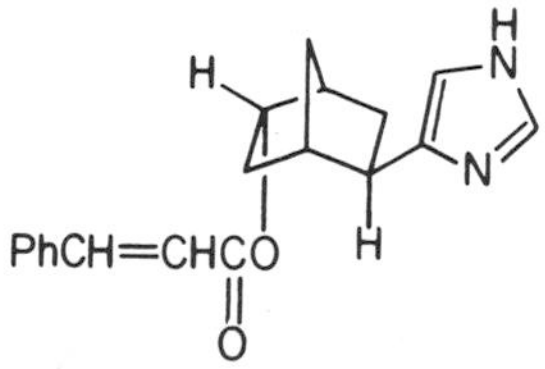   **32**

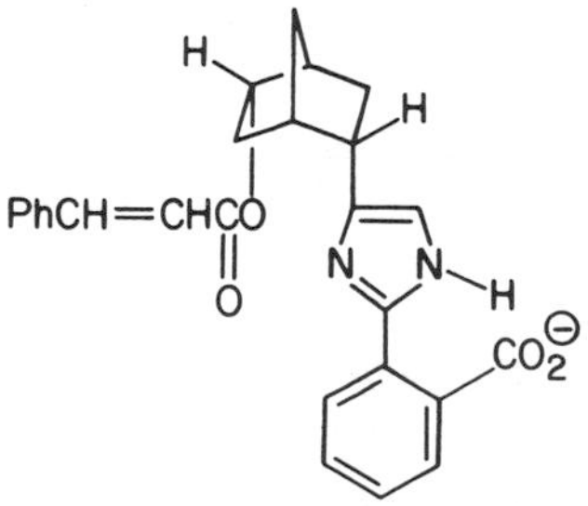

33

d. Models for Chymotrypsin: Two Catalytic Groups Spatially Constrained

A series of experiments designed to test the charge-relay hypothesis for chymotrypsin[89] was carried out by Rogers and Bruice,[90] in which the rates of hydrolysis for several spatially constrained phenyl esters were examined as shown in Table 3-4. It was observed that rate enhancement upon the introduction of a hydrogen-bonded carboxylate was only threefold, a negligible increase from the standpoint of enzymatic catalysis. In order to approximate the partially hydrophobic environment of the enzyme's active site more closely, the hydrolyses were also conducted in acetonitrile containing 3.3 *M* H_2O. The investigators found that transferring from H_2O to a solvent of limited H_2O concentration had virtually no effect on the catalytic role of the carboxylate group in the intramolecular reactions.

However, as is often the case, the effectiveness of a model system may be critically dependent on its structure in as yet unpredictable ways. Ten years

TABLE 3-4. Rate of Hydrolysis for Some Model Compounds of the Charge—Relay System

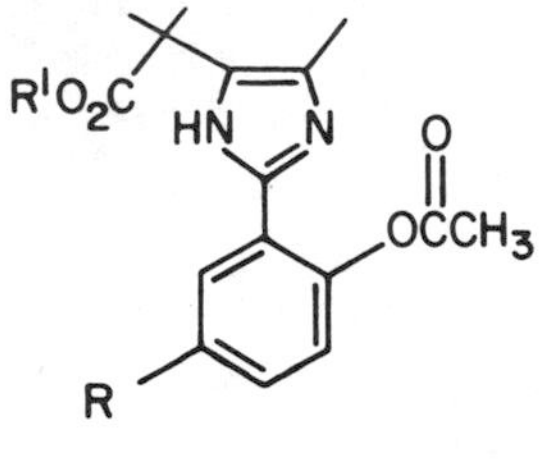

Compound	R^1	R^2	$k(min^{-1})$
28a	—	H	1.00×10^{-2}
28b	H	H	2.75×10^{-2}
28c	CH_3	H	9.50×10^{-3}
28d	H	$SO_3^{\ominus}$	3.60×10^{-2}
28e	—	$SO_3^{\ominus}$	1.20×10^{-2}

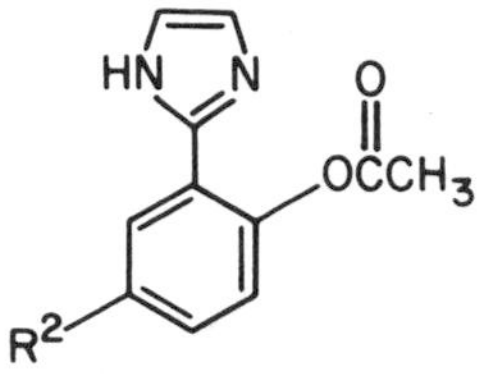

after this work, Bender[91,92] reported the synthesis of enzyme model **33** and studied the kinetics of its ester hydrolysis reaction. This compound possesses an esterified alcohol group, an adjacent imidazole group, and a neighboring carboxylate, and is a totally intramolecular combination of the models discussed earlier in this section. In this respect, it is designed to model the deacylation of chymotrypsin (**18** $\longrightarrow$ **19** in Figure 3-3) as was Bruice's compound, but in this case the effect of the carboxylate group is clearly evident. Figure 3-4 summarizes the rate data for the hydrolyses of models **31** (curve A), **31** plus added benzoate ion (curve B), and **33** (curve C). The sloped line on the right side of the graph reflects the rate of specific base-(hydroxide) catalyzed hydrolysis, which, as expected, is directly proportional to the hydroxide concentration. Curves A, B, and C, which may be considered to be pH-independent regions, exist because in these respective compounds a hydrolysis pathway exists that is faster than the specific base reaction, at least until the curve intersects with the line. At pH 7, all three systems hydrolyze by the intramolecularly catalyzed pathway, and the relative pseudo-first-order rate constants for these reactions may be compared directly. By extrapolation of the sloped line to pH 7, one can estimate that compound **33** deacylates over 100,000 times faster than it would have in the absence of catalytic groups.

The evolution of this model for the esterase activity of chymotrypsin, therefore, nicely demonstrates the power of preorganization—intramolecularity—in amplifying the effectiveness of catalytic groups. Whereas free

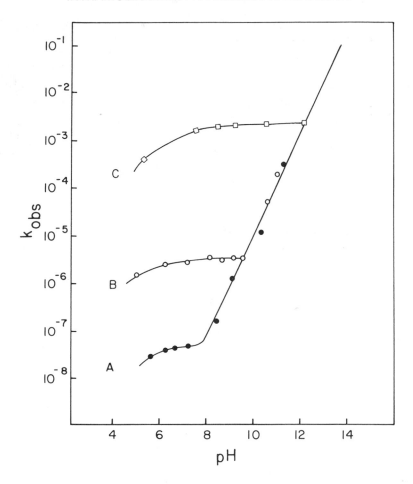

Figure 3-4. pH dependence of the hydrolysis rates for enzyme models **31** (●), **31** + 0.5 *M* benzoate (○), and **33** (□) at 60°C.

imidazole and/or a free carboxylate are not particularly effective catalysts in the hydrolysis of esters, at least one intramolecular model affords a rate acceleration approximating that for the deacylation of chymotrypsin itself.

E. Catalysis via Reversible Covalent Bond Formation: Intramolecular Reactions between a Reagent and a Catalytic Group with the Potential for Turnover

Enzyme models, as described earlier, are covalent versions of enzyme–substrate complexes in which catalytic effects may be studied in the absence of complicating binding equilibria. Such intramolecular reactions have not, to date, been considered as candidates for synthetic catalyst development.

Rather, the biomimetic origins of artificial enzyme research have, quite logically, resulted in focusing attention on binding guests out of aqueous solution; enzymelike association mechanisms (eg, hydrophobic binding, hydrogen bonding, ion pairing, ion-dipole interactions) have been successfully utilized by the productive efforts of many groups. The covalent association of catalytic and functional groups may lead to turnover catalysis, however, if the covalent reaction is itself fast and reversible. There are, of course, many such examples in the organic literature; for example, disulfide exchange, hemiacetal formation, and cycloaddition may occur in both forward and reverse directions. Perhaps even more importantly than the fact that such equilibria can take place in organic solvents, the catalyst–substrate "complexes" generated in these schemes can be rigid with well-defined and predictable geometries.

One such approach to synthetic catalyst design has been recently described.[93] As shown in Figure 3-5, cycloaddition of pyridylanthracene **34** to

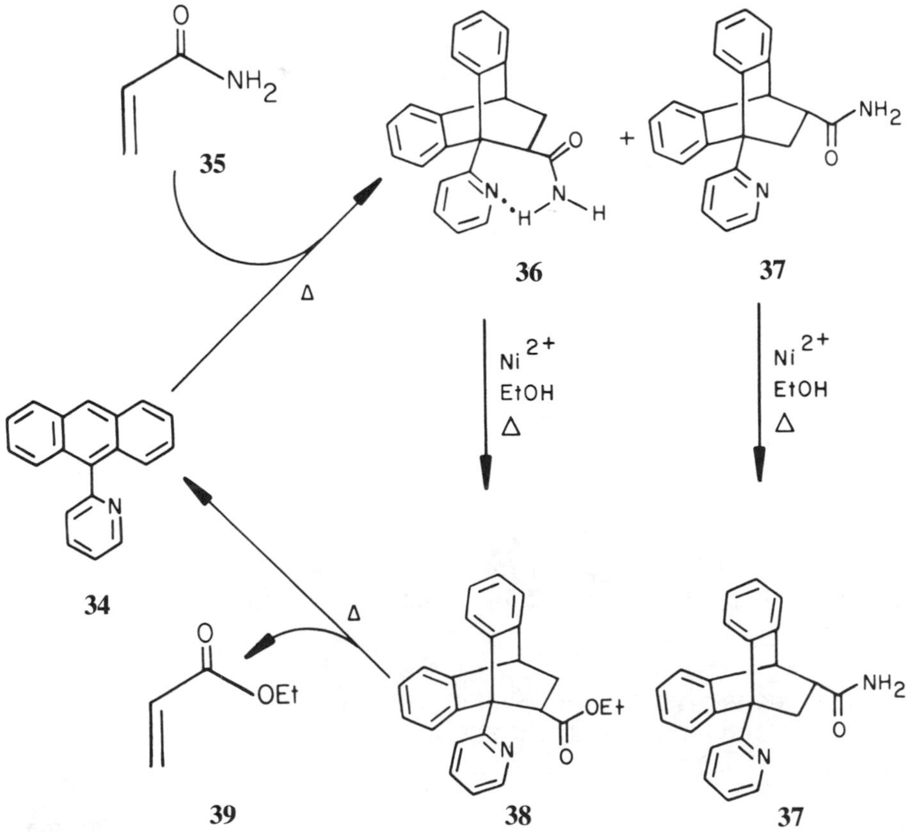

Figure 3-5. Cycloaddition cycloreversion cycle for the conversion of acrylamide (**35**) to ethyl acrylate (**39**).

acrylamide (35) affords a mixture of two isomeric adducts, 36 and 37. In one of these adducts (36), the amide and pyridine groups are close enough that a metal ion chelate with Ni^{2+} may be formed; as a result, the metal-catalyzed conversion of amide 36 to ester 38 is observed under conditions that are extremely mild for the alcoholysis of an amide. The lack of reactivity in amide 37 attests to the catalyzed nature of the amide to ester conversion. Finally, thermal cycloreversion of ester 38 affords ethyl acrylate (39) and the starting anthracene (34), completing the formal cycle. The same general scheme has been used to convert acrylonitrile to acrylamide and ethyl acrylate to acrylic acid.[94] Examples of Diels–Alder reactions that are fast and reversible at room temperature have been reported,[95-97] and work aimed at making such reversibility accessible in the acrylate series is being actively pursued.[98]

4. RATE ACCELERATIONS DUE TO NONCOVALENTLY ENFORCED PROXIMITY

A. Intramolecular Reactions between Reagents

The reaction between A and B, where A is a reagent and B is either a reagent or a catalyst, is sometimes faster in A : : B than in A + B. The symbol A : : B is meant to imply a noncovalent association between A and B, and will hereafter be referred to as *binding*. Among the noncovalent forces that may cause molecules to associate are "hydrophobic bonding," hydrogen bonding, electrostatic interactions, metal–ligand interactions, charge transfer, π-system stacking, and the weaker London dispersion forces; each of these topics has been discussed previously.[2,7,9] The great advantage of noncovalent complex formation from the perspective of enzyme catalysis is that activation energies for noncovalent complex formation and dissociation are very often less than those required for covalent bond formation and cleavage, which allows for fast and reversible associations with substrates. And although correct spatial orientation is essential for obtaining rate accelerations, the interplay of many weak interactions can position a substrate functional group just as efficiently as the most rigid covalent enzyme models— probably better.

A recent example of a conversion that can be accelerated by noncovalent association is Diels–Alder adduct formation. Many common dienes and dienophiles are relatively nonpolar and might logically be expected to self-associate in water as a result of hydrophobic bonding. Such an associative phenomenon would be expected to result in increased local concentrations of the reactants, thereby increasing the observed rates as compared to reactions in nonpolar solvents. This is, in fact, what is found; the reaction of cyclopentadiene with butenone in water at 20°C exhibits a second-order rate constant more than 700 times that of the same reaction in 2,2,4-trimethylpentane.[99] However, as Jencks has pointed out,[100] it is also common to observe rate *decreases* as a result of binding to an added molecule. Addition of β-cyclodextrin to the aqueous Diels–Alder reaction cited above decreases the rate constant, presumably because the dienophile can bind while the diene cannot.[99] Caffeine binds to benzocaine (ethyl *p*-aminobenzoate) in water with an association constant of about $60/M$, but in so doing dramatically decreases benzocaine's hydrolysis rate.[101] Chelation of a metal ion adjacent to an amide may facilitate its hydrolysis,[102] but may also inhibit it.[103] Furthermore, the magnitude of the chelation effect will depend on whether or not the amide carbonyl is physically able to coordinate to the metal ion center; hydrolysis may be faster if the carbonyl and metal *cannot* directly bind![104] Clearly, approximation of reacting groups is not sufficient to favor intramolecular reaction pathways; the approximation must permit, at least, association in a productive orientation.

An interesting macromolecular version of an intracomplex reaction was reported by Webb and Matteucci.[105] The helical complex formed by two (or three) strands of DNA results from both hydrogen-bonding and base-stacking effects, and the affinity of one strand for another is optimal when they are exactly complementary [adenosine (A) with thymidine (T) and guanosine (G) with cytosine (C)]. The authors reasoned that if in one strand of DNA a single base was modified to incorporate an electrophilic reagent but could still base pair, then the event of complex formation would bring that reagent into very close proximity to the corresponding complementary base on the opposite strand. The oligonucleotide $d(T_6\text{-}X\text{-}T_{14})$ was synthesized, in which X is cytosine with an aziridine group in place of the natural exocyclic amine, and was allowed to interact with $d(A_{14}\text{-}N\text{-}A_6T)$, in which N is C, G, T, and A. Gel electrophoresis clearly revealed a cross-linking reaction between $d(T_6\text{-}X\text{-}T_{14})$ and $d(A_{14}\text{-}C\text{-}A_6T)$, and sequencing of the product indicated that the reaction had taken place at the site required by helix formation. Cross-linking was not observed in the other cases, probably because they do not possess strong nucleophiles at the part of the complex exposed to the aziridine group. The potential of such site-specific reagents is clear in that the specificity can be predicted and readily modified using the known base-pairing selectivities in any oligonucleotide of interest. This result warrants a moment for reflection; how far the utility of intramolecularity has come since its humbler origins!

B. Comparing the Rates/Specificities of Enzymes and Enzyme Mimics

In associations that occur with rapid "on" and "off" rates, it becomes no longer appropriate to compare simple unimolecular rate constants for related reactants. If, for example, a reaction proceeding by the scheme

$$\text{A} + \text{B} \underset{}{\overset{K_{eq}}{\rightleftharpoons}} \text{A} :: \text{B} \xrightarrow{k_2} \text{C} \qquad (3\text{-}1)$$

had a large k_2 but an extremely small K_{eq}, the observed rate could be quite slow. The k_{obs} for this scheme is

$$k_{obs} = \frac{K_{eq}k_2}{K_{eq} + 1} \tag{3-2}$$

If K_{eq} is much greater than 1 (A and B exist entirely as A : : B) and k_2 is slow compared to the association rate, then $k_{obs} = k_2$ and

$$\text{rate} = k_2[\text{A : : B}] \tag{3-3}$$

This limiting case is exemplified by association via a strong covalent bond, as in the enzyme model compounds, where [A] = [B] = [A : : B]. On the other hand, if $K_{eq} \ll 1$, then $k_{obs} = K_{eq}k_2$, meaning that the observed rate is determined both by the microscopic rate constant and by how much time the reactants spend in an associated state. In this limiting situation

$$\text{rate} = k_{obs}[\text{A}] [\text{B}] \tag{3-4}$$

in which k_{obs} equals ($K_{eq}k_2$), and has the units of a second-order rate constant.[106]
As mentioned previously, enzymes must *not* bind their substrates or products too tightly or there will be no turnover; therefore, under physiological conditions, enzymes (and many enzyme mimics) are best described by the K_{eq} much less than 1 limiting case. However, the simplified kinetic scheme for enzymatic catalysis looks quite different from that of associated reagents (Equation 3-1), as shown in Equation 3-5:

$$\text{E} + \text{S} \underset{k_{-1}}{\overset{k_1}{\rightleftharpoons}} \text{E : : S} \underset{k_{-2}}{\overset{k_2}{\rightleftharpoons}} \text{E : : P} \underset{k_{-3}}{\overset{k_3}{\rightleftharpoons}} \text{E} + \text{P} \tag{3-5}$$

where E is the enzyme, S is the substrate, E : : S is an enzyme–substrate complex, P is the product, and E : : P is an enzyme–product complex. There is no general analytical solution to the rate equations for this mechanism, but rate expressions can be derived if one of several limiting assumptions is imposed.[107] If (1) the kinetics are determined at low substrate concentration such that the enzyme exists mostly uncomplexed and a steady-state concentration of E : : S exists and (2) dissociation of the E : : P complex is fast, then the rate equation (Equation 3-6) becomes quite similar to that for the simple bimolecular case in Equation 3-4:

$$\text{rate} = \{[k_1/(k_{-1} + k_2)]k_2\}[\text{E}][\text{S}] \tag{3-6}$$

where k_2 is the first-order rate constant for the catalytic step (k_{cat}) and [$k_1/(k_{-1} + k_2)$] may be treated as an apparent association constant (analogous to

K_{eq}). The reciprocal of $[k_1/(k_{-1} + k_2)]$ under these constraints is K_m, the Michaelis constant, which may likewise be treated for some purposes as an apparent dissociation constant.[108] With these conventions, the rate equation becomes simply

$$\text{rate} = k_{obs}[E]\ [S] \tag{3-7}$$

in which $k_{obs} = (k_{cat}/K_m) = (``K_{eq}"\ k_{cat})$, and has the units of a second-order rate constant. The analogy to Equation 3-4 is thereby complete, and k_{cat}/K_m yields an apparent second-order rate constant that incorporates both the catalytic rate constant for conversion of E : : S to E : : P and how much time enzyme and substrate species spend in the productive E : : S form. The quantity k_{cat}/K_m relates the energy of free enzyme and substrate to that of the ES transition state leading to EP. It is k_{cat}/K_m that determines the specificity of an enzyme for competing substrates, and it is the quantity used for comparing the relative rates of one enzyme toward many substrates or of several enzymes toward the same substrate.

While the same quantity should ideally be used for similar comparisons of enzyme mimics, in practice it almost never is. The main reason is that most enzyme mimics do not show turnover kinetics due to slow dissociation rates, so it is more practical to compare initial reaction rates instead.

C. Intramolecular Reactions between a Functional Group and a Catalytic Group: Enzyme Mimics

The vast majority of *enzyme mimics* fall into this category of reactions accelerated due to noncovalently enforced proximity of a functional group to a catalytic group. Such compounds possess one or more catalytic groups attached to a (usually) covalent framework and are designed to determine the potential effectiveness of the catalytic group in the conversion of a functional group brought into proximity by way of a noncovalent binding event. Enzyme mimicry is a more recent pursuit than is enzyme modeling and, consequently, reviews presenting an overview of the area are, to date, few.[109] Recent reviews of the most extensively studied enzyme mimics, such as the cyclodextrins,[110] crown ethers,[111] micelles,[112] dicarboxylic acid binding sites,[113] and metal ions,[114,115] may be found.

a. Systems with the Potential for Turnover

To exemplify the concept of intramolecularity via noncovalent complex formation, we may examine a few very recent examples. β-Cyclodextrin that has been modified on its secondary side with a pyridoxamine moiety (**40**) will catalyze the conversion of α-keto acids (**41**) to α-amino acids (**43**),[116] in complete analogy to the process catalyzed by transaminases (Figure 3-6). The selectivity of enzyme mimic **40** for keto acids depends strikingly on the

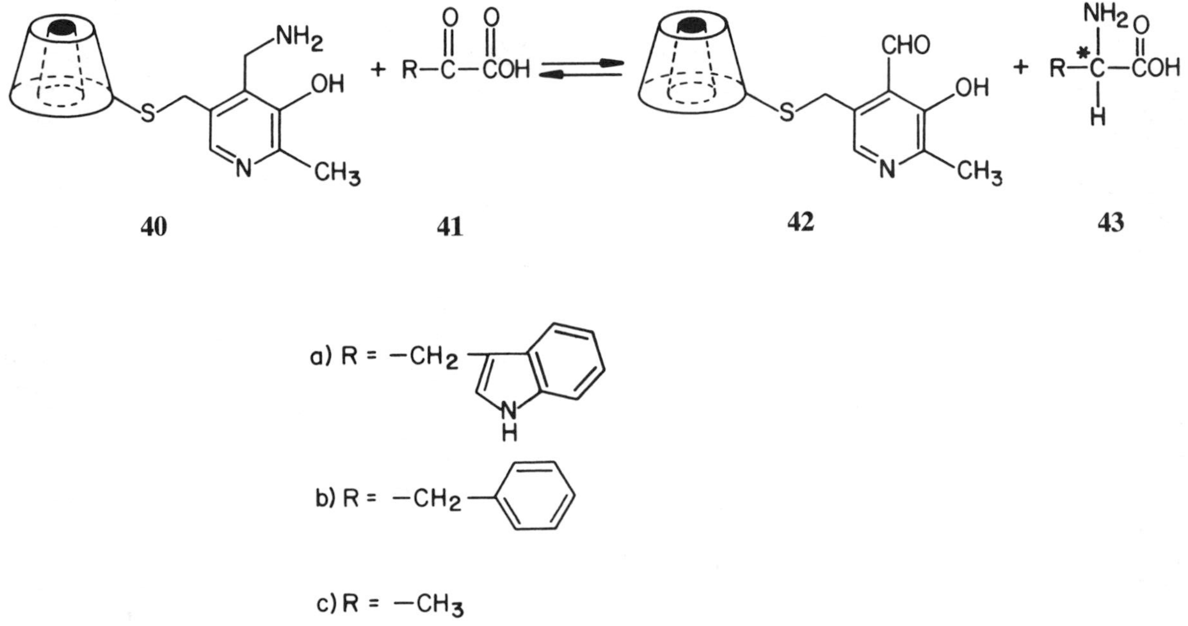

Figure 3-6. Modified β-cyclodextrin catalysis of α-keto acids (41) to α-amino acids (43).

structure of the R group: aromatic-ring-containing substrates such as indolyl (**41a**) and phenyl (**41b**) reacted much more quickly than did the control (**41c**). Such selectivity is due to the known ability of cyclodextrin derivatives to bind aromatics out of aqueous solution; as a result of the chiral binding environment, the amino acid products formed are often chiral [29% enantiomeric excess (ee) in the production of **43a** depicted in Figure 3-6]. In essence, **41c** reacts with pyridoxamine **40** via a simple bimolecular pathway, while **41a** and **41b** bind initially to the cyclodextrin pocket followed by an intramolecular reaction with the pyridoxamine group. As with several other examples in this section, the reaction, while in principle reversible with the potential of turnover, was studied as an irreversible process, and initial rates were used for comparison.

In 1984, Cram[117,118] reported the ability of various crown ether and polyurea hosts to act as "transacylase partial mimics," that is, they have the ability to bind *p*-nitrophenyl esters of alanine (perchlorate salts) and to transfer the amino acid to a hydroxyl group on the host molecule. In these examples, the binding event was very exothermic as would be expected for a chelating ion-dipole association in an organic solvent, effectively converting a bimolecular transacylation into an entirely unimolecular (intracomplex) one. By comparing observed rate constants and calculating what the corresponding second-order rate constant would be, Cram found one of the complexing systems to have a second-order rate constant that was an impressive 10^{11} times faster than a control noncomplexing system. While the method of calculating relative rate enhancements is subject to interpretation,[119] this "incremental strategy" for the design of molecular receptors/reagents/catalysts will surely guide synthetic catalyst design efforts for the foreseeable future.

Dervan has recently examined what may be a particularly useful type of intracomplex catalysis. As shown in Figure 3-7 (with only a bit of literary

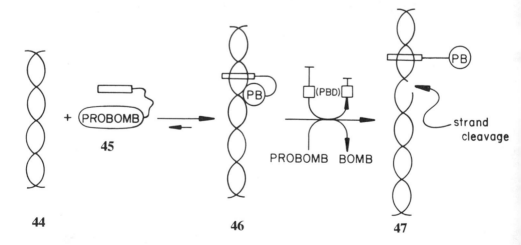

Figure 3-7. Selective delivery of a catalyst to a site on dsDNA using an intercalating agent.

licence applied), a catalyst (45) can be delivered selectively to a site on double-stranded DNA (dsDNA, 44) using an intercalating agent as the delivery vehicle.[120,121] Under the conditions reported to date, the association of 45 with 44 is kinetically irreversible. Once the association is complete, the "probomb" may be activated with an external reagent to generate a "bomb" near the site of intercalation; the probomb is a chelated Fe(II) ion, the "detonator" (PBD) is an oxygen/reducing agent mixture, and the bomb is probably a hydroxyl radical formed in the reaction between the two. Hydroxyl radicals formed in situ diffuse away from the site at which they were formed but, in the process, occasionally cleave the DNA in the vicinity. An alternate cleaving mechanism utilizing a site-specific alkylating reagent has also been employed.[122] This general scheme has been used in several intriguing ways to cleave DNA with selectivity and has already generated an enormous amount of new research in other laboratories.

b. Systems that Exhibit Turnover

Intracomplex reactions that can utilize the "catalyst" more than once are more closely related to enzymes themselves than are previously mentioned

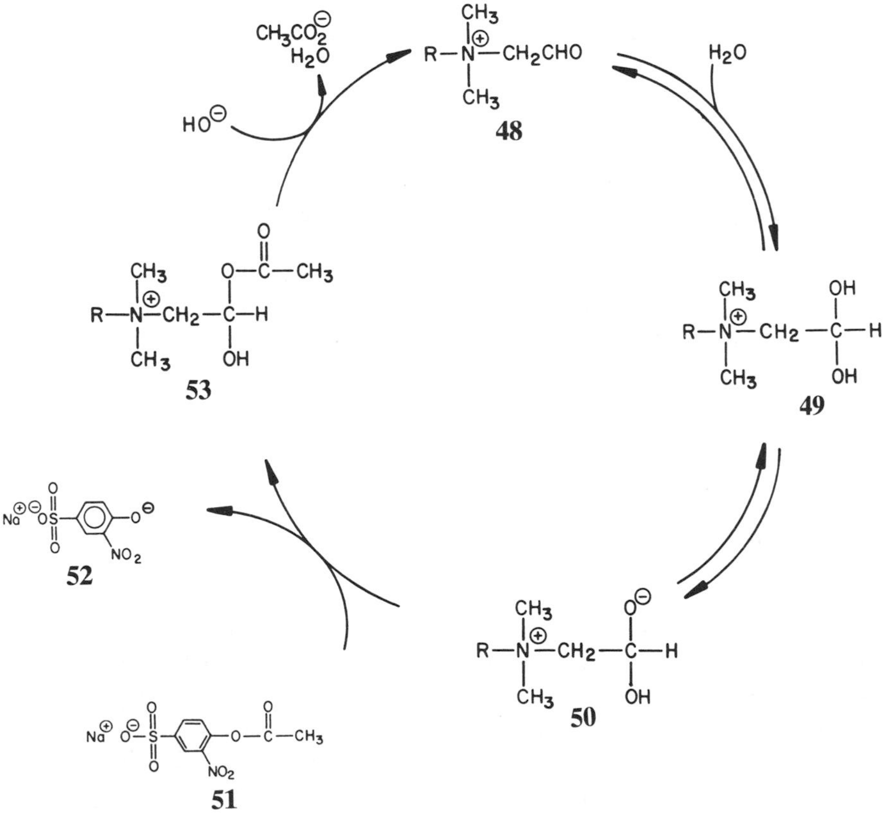

Figure 3-8.   Micellar system catalytic for the hydrolysis of nitrophenylacetate (15).

examples. Perhaps best studied is catalysis by metal ions, which has been well documented in the forms of metalloenzymes, metal-containing enzyme mimics, metal-containing enzyme models, and free metal ions.[114,115] However, while many metal-free enzymes capable of enormous turnover numbers exist, one finds few metal-free enzyme mimics that are actually catalytic. A micellar system catalytic for the hydrolysis of nitrophenylacetate (51) has been reported by Menger and is shown in Figure 3-8.[123] Hydration of micellar 48, in which R is dodecyl, is facilitated by the adjacent ammonium ion to afford hydrate 49. At pH 9.0, some of the hydrate exists in the deprotonated form (50), which can act as a nucleophile in the transacylation reaction with 51. This is the stage at which many "artificial enzymes" stop. However, acetylated micelle 53 has a facile β-elimination pathway available to it that quickly releases acetate ion and regenerates the starting aldehyde (48). Experiments with as high as a 5 : 1 ratio of 51 to 48 indicated complete hydrolysis of all added 51, verifying that turnover occurs; if it had not, only a stoichiometric amount (one-fifth) of 51 would have been hydrolyzed.

5. RATE ACCELERATIONS DUE TO COVALENTLY ENFORCED STRAIN

A. Introduction

The reaction between A and B may be faster in A—B than in A + B if A—B is significantly strained *and* if that strain is at least partially relieved on achieving the transition state. Such an effect is normally manifested as a decrease in the $\Delta H^{\ddagger}$; our working definition for strain (see Subsection A of Section 2) compares enthalpies of formation in strained and strainless states, so observation of the effect in $\Delta H^{\ddagger}$ is to be expected.

The general notion of a reaction affected by strain is one familiar to organic chemists. One may refer to the book *Steric Effects in Organic Chemistry*, published in 1956 by Professor Newman, for the state of the art at that time.[124] Strain effects may be invoked to explain why a reaction proceeds more slowly than in related compounds, as in the unusually low reactivity of cyclohexyl halides toward S_N2 substitution as compared to the corresponding isopropyl halide.[125] In this case, 1,3-diaxial interactions impede approach of the nucleophile or the departure of the leaving group and thus increase the energy of the transition state relative to the ground state. For similar reasons, *t*-butoxide ion is a strong base but not a strong nucleophile. Strain effects may also be invoked to explain particularly fast reactions. Epoxides undergo S_N2 substitution unusually quickly as compared to five-membered and larger cyclic ethers, the result of a strained ground state that loses some of its strain in the transition state. Care must be taken not to assume that loss of strain on achieving the product necessarily means loss of strain in the transition state; some Diels–Alder reactions are slowed in the forward direction with increasingly bulky dienophiles, but the rate of cycloreversion of the corresponding cycloadducts does not increase.[98] These examples, of course, are intermolecular reactions, but they provide the necessary background for intramolecular cases.

B. Intramolecular Reactions

In order to observe a strain-accelerated reaction in a model compound, there must not be any way that the compound can relieve that strain by a simple conformational flexure. Such compounds are, by necessity, rigid structures with energy barriers for bond rotation that are higher than the energy barriers for whatever reaction is being studied *or* for which a bond rotation does not relieve the strain. If ground state strain is present, it is reasonable to expect that bond lengths and/or angles will be different from those found in strain-free models. In part for this reason, such compounds may be difficult to synthesize and must react slowly enough so that they can be isolated.

Rüchardt and Beckhaus[126] have published a scholarly investigation of strain effects in carbon–carbon bond homolysis reactions. By comparing the experimentally determined thermal stabilities of substituted ethanes with the calculated strain enthalpies (using a molecular mechanics method), the authors established for the first time that a linear relationship exists between strain and reactivity. While this notion has been appreciated for many years, their analysis attests to the simple and direct correlation between the two and allows for a quantitative separation of steric and electronic effects.

An important (and still somewhat controversial) set of data (Figure 3-9) has been described by Cohen[127,128] and the various interpretations reviewed thoroughly by Gandour.[129] The rate of acid-catalyzed lactonization of hydroxyacid **57** is 3×10^{11} times faster than in the electronically similar **55**. It was originally argued that the rate acceleration was due to a "stereopopulation control" such that the two functional groups were always in a productive, relative spatial orientation. However, it eventually became clear that this number was extraordinarily high for a simple proximity effect, which is currently thought to be about 10^4 in the best cases. Rather, the rate enhancement is probably due in large part to relief of steric compression that exists as a result of polymethyl substitution. The X-ray structure analysis of hydroxyacid **57** suggests this to be the case; significant deviations exist in the bond angles of the benzene ring from the idealized value of 120° in order to accommodate the methyl groups.[130] A particularly interesting experiment by Loudon[131] provides further support for the strain-acceleration argument. Perdeuteration of the *gem*-dimethyl group of a model compound related to **57** resulted in a remote secondary deuterium isotope effect (k_H/k_D) of 1.09 due to the larger effective radius of hydrogen as compared with deuterium and, therefore, a presumably stronger repulsion in the protio compound.

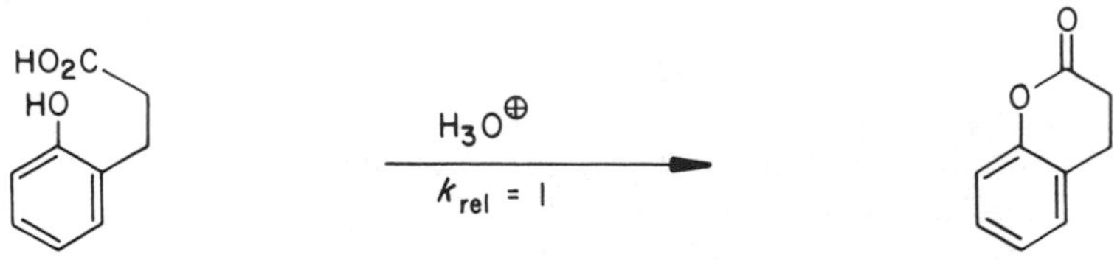

Figure 3-9.   Acid-catalyzed lactonization of hydroxyacids **57** and **55**.

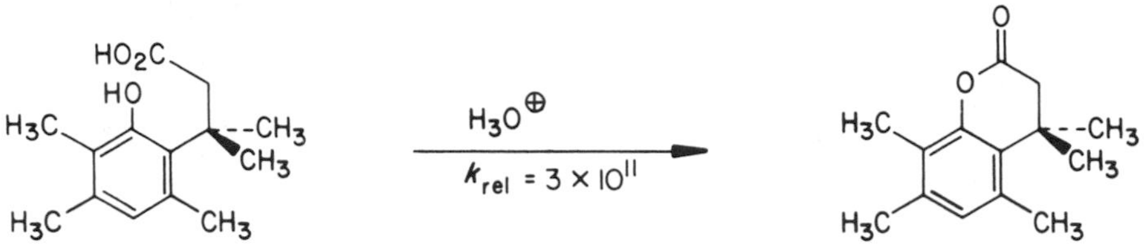

6. RATE ACCELERATION DUE TO NONCOVALENTLY ENFORCED STRAIN

A. Introduction

The reaction between A and B may be faster in A : : B than in A + B if A : : B is significantly strained *and* if that strain is at least partially relieved on achieving the transition state. As discussed previously, such strain may be exhibited as either (1) the binding of a molecule in a geometry that would, in the absence of the binding site, relax spontaneously to a lower energy conformation (strain with distortion)[132] or (2) binding in such a way that some attractive interactions are realized, but all attractive interactions are realized only on achieving the transition state (distortionless strain).[14,133–135] The distinction between these two possibilities is fuzzy at best and it has been suggested that only the second possibility need be discussed since both are mathematically equivalent. It has been argued that proteins are too flexible to generate strain with distortion for small molecules,[136] and the term *stress* has been suggested to describe forces exerted by the distorted enzyme on the bound substrate that do not distort that substrate.[137,138] Such arguments based on computational methods and on "intuition" must be often reevaluated as experimental methods improve; in at least one case, evidence for substrate distortion by the enzyme has been obtained.[139]

B. Lysozyme

The enzyme most closely associated with hypotheses of rate accelerations due to strain is lysozyme. It has been the subject of enzyme modeling perhaps more than any other enzyme, save only chymotrypsin, and most bioorganic textbooks are identical in their representation of its mechanism of action. Unfortunately, it does not appear that any unified review of mechanistic studies on lysozyme has appeared since Chipman's excellent compilation in 1969;[140] individual topics, however, may be found in virtually all recent texts. A widely disseminated mechanism for catalysis of the rate-determining step, cleavage of the glycosidic linkage, is summarized in Figure 3-10. The substrate oligosaccharide binds in such a way that six carbohydrate residues, labeled A–F, are in contact with the enzyme. While five of the six residues bind exothermically, residue D (as shown in Figure 3-10) binds endothermically and is distorted into a half-chair conformation. This half-chair conformation resembles the conformation of the oxonium ion intermediate (and, therefore, the transition state), thus lowering the energy barrier leading to its formation. In addition, a protonated carboxylic acid (Glu-35) at the active site serves as a general acid to protonate the leaving group; an ionized carboxylic acid (Asp-52) serves to stabilize the oxonium ion intermediate. The remaining steps, such as addition of water and diffusion of the products from the enzyme, are relatively fast.

The proposed mechanism, therefore, boils down to three testable components: general acid catalysis, electrostatic catalysis, and strain-induced bond cleavage. The first of these, general acid catalysis by an adjacent protonated carboxylic acid, has been verified by the model studies of several groups (especially by Capon, Fife, Bruice, and Loudon) using acetals bearing covalently attached carboxylic acid moieties. The second, electrostatic stabilization of the positively charged transition state, has been modeled, but without large rate accelerations. The work of Fife[141] does set an upper limit of 100-fold even in a mixed solvent (50% dioxane/50% water), which may, as the author points out, be due entirely to an S_N2 process (anchimeric assistance) rather than to electrostatic stabilization of the S_N1 process.[142]

The third component, due to strain, has been the subject of intensive investigation. A history of the topic up to 1977 has been nicely summarized by Chipman[143] and is reproduced here:

"The role of strain in the lysozyme-catalyzed hydrolysis of oligosaccharides has been under active investigation ever since the proposal by Phillips and co-workers of a model for a hen egg-white lysozyme (HEWL)–substrate complex.[144,145] According to this model, the enzyme can interact with up to six [N-acetyl-D-glucosamine] moieties, so long as the residue in the fourth subsite from the nonreducing end (subsite D) is distorted towards a partially planar conformation, in order to prevent prohibited steric interactions of its C^5-hydroxymethyl group with the protein.[144,146] Since the transi-

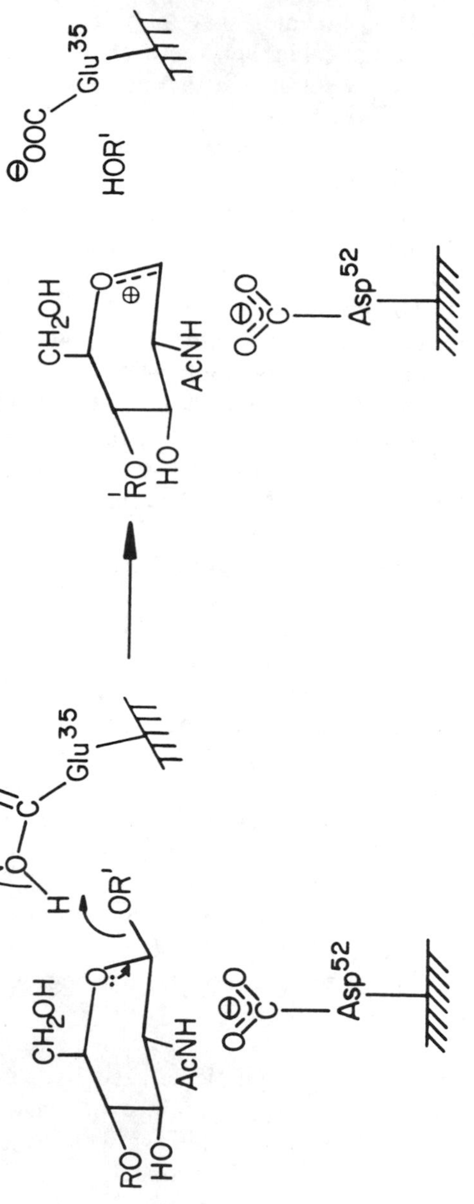

Figure 3-10. Mechanism of lysozyme catalysis.

tion state for cleavage of the glycosidic bond is expected to have oxo-carbonium ion-like character,[147] planar about the C-1—O-5 bond,[148] and since site D is the site of bond cleavage,[144,149] it was proposed that the distortion of the substrate towards the transition state geometry plays an important role in lysozyme catalysis.[144,150,151]

"Experimental evidence for strain in the catalytic mechanism of lysozyme has in the main been concerned with binding interactions in subsite D. Studies with bacterial cell wall oligosaccharides have shown that the introduction of an N-acetylmuramic acid residue into subsite D destabilizes the HEWL–oligosaccharide complex.[152] On the other hand, analogues of N-acetyl-D-glucosamine, expected to fit subsite D *without* prohibited interactions, appeared to stabilize the enzyme–saccharide complex.[153,154,155] Largely on the basis of such data, it has been suggested that strain in subsite D is responsible for a reduction of the activation energy for bond cleavage in the enzymatic reaction by 5–10 kcal/mole.[140,150,153,156,157,158]"

However, in 1977, both Capon[159] and Chipman[160] reported new binding data (and a reevaluation of prior binding data) that strongly suggest binding at subsite D may be much less endothermic than previously thought. This new estimate, about +2.3 kcal/mol,[160] implies that there is little or no ground state strain on binding at subsite D and, therefore, little or no distortion of substrate residue D. Both authors conclude that strain must still be important, but that it is likely to be distortionless strain that increases as the transition state is approached. Specifically, earlier kinetic results were reinterpreted to imply that the C-5 hydroxymethyl group (and probably several other groups) of residue D conspire to afford the greatest interaction with subsite D only at the transition state. This conclusion received support from analysis of an X-ray structure of an HEWL–trisaccharide complex in which subsite D is occupied.[161] In this complex, which is stable to hydrolysis, the residue occupying subsite D shows little distortion of the glucopyranose ring away from a ground state $^4C_1$ chair conformation. Most recently, binding data have been reported that suggest subsites D–F are bound only weakly in the ES complex, but more strongly in the transition state.[162]

It is, in this author's opinion, important to emphasize that each of the above binding studies were carried out with oligosaccharides that, when bound to HEWL, do not span subsite D *on both sides*; in other words, no complexes have residues bound at subsites C, D, and E simultaneously. If residue D were to experience a repulsive interaction at subsite D, then that repulsion must surely be greater in the hexasaccharide (in which residue D is "anchored" on both ends) than in tetra- or trisaccharide models (in which only one end of residue D is anchored). Indeed, hexasaccharide substrates hydrolyze *much* more quickly than do substrates that do not span both sides of subsite D; binding studies utilizing such oligosaccharides are effectively prohibited by the rapidity with which they are hydrolyzed by the enzyme.[140] The fact that the residue occupying subsite D in the most recent HEWL–

trisaccharide X-ray study does not bind as deeply into the enzyme as does the previously reported tetrasaccharide lactone (a transition state analogue)[161] implies that the chair conformation of a D residue cannot bind to subsite D as well as can the so-called "sofa" conformation. Whether or not this inexact complementarity of structures results in "strain with distortion" or "strain without distortion" depends only on the magnitude of the force applied by the rest of the substrate by way of favorable binding. The question of whether or not hexasaccharide substrates bound to HEWL exist with residue D distorted from a chair geometry remains open to inquiry.

While the function of electrostatic stabilization has been suggested for Asp-52 by many investigators, its potential role in substrate destabilization has also been considered. The rather modest effect of an ionized carboxylate group in the hydrolyses of acetal models in aqueous solution[141,142] has prompted some to propose that Asp-52 may not play a catalytic role at all;[163] it is at least secure to say that that the role of Asp-52 has not been verified experimentally. It is simplistic to presume that stabilization of an oxonium ion by Asp-52 is not well modeled in aqueous solution; the active site region near Asp-52 has been described as hydrophobic,[163] slightly hydrophobic,[164] and polar.[144] In any case, the difficulty of forming an ion in an environment of low dielectric suggests that the documented[165] hydrogen bonding by Asp-52 to other active site residues will delocalize its charge proportionate to the nonpolarity of its environment.

The active site of lysozyme, hydrated in free solution, is partially dehydrated upon binding to an oligosaccharide.[161,166] Computational studies by Warshel and Levitt[167] and Scheraga[168] of the complete lysozyme–substrate complex together with the surrounding solvent deemphasize the significance of closed shell (steric) strain in the energetics of lysozyme catalysis. However, the importance of electrostatic stabilization of the transition state by an interaction with Asp-52 was strongly indicated. Both these and other[12,169,170,171] studies have supported the idea proposed by Vernon[151] that ground state desolvation of Asp-52 could result in reactant destabilization, affording a driving force for catalysis that has been referred to as "electrostatic strain."[167] There is, of course, a potential conflict between this theory and the recent downward estimate of binding energy at subsite D. Destabilization via desolvation should be observable after binding but before the transition state for glycolysis. If the binding of a residue at subsite D does require the displacement of a high-energy water molecule, such an event ought to show up in the binding constant. In any case, while the use of calculations to study enzymatic reactions has its critics, the method does allow one to isolate portions of the enzyme for study that cannot yet be isolated experimentally.

A possible resolution to the conundrum described above has been suggested by Czarnik.[172] Asp-52, in addition to stabilizing the positively charged transition state and destabilizing the enzyme–substrate complex via desolvation, can in principle induce an electrostatic repulsion with a lone pair on the ring oxygen that "pushes" the ground state toward the transition state. Such a repulsion between lone pairs of electrons is fully comparable to "steric repulsion" between functional groups; each results from an attempt to bring electron densities into the same region of space and shows up in the exchange integral of the system's wave function. The main difference between this "lone pair exchange repulsion" (LPER) and electrostatic strain via desolvation is that only the LPER effect can be induced as a result of small atomic motions, such as those involved in tight binding of a transition state. In this scenario, the substrate binds to the active site as before; however, the contacts between enzyme and substrate at the subsite D region are not optimal, differing by perhaps a few tenths of an angstrom from the ideal. As the residue D carbohydrate begins to "stretch" toward the transition state, these contacts become more favorable; at the same time, the ribose ring oxygen is forced toward the ionized Asp-52, which is held rigidly by a set of hydrogen bonds. The electrostatic repulsion results in a distortion of the carbohydrate toward the transition state for bond cleavage, thereby coupling many small favorable energy changes with a local strain effect directed along the reaction coordinate for glycolysis. Support for this mechanism lies, to date, entirely on the results of ab initio calculations on acetal model compounds, which indicate (1) forcing a negatively charged species to interact with the oxygen lone pairs increases the system energy dramatically after the 4 Å distance (the distance in lysozyme is 2.5–3 Å at 2 Å resolution[166]), and (2) reoptimizing the carbohydrate structure with a negatively charged group fixed at a distance of 2.5 Å yields a new geometry with distortions in all bond lengths and angles toward those of the oxonium ion intermediate.[183] If this idea stands up to experimental verification, it would provide a reasonable explanation for the failure of flexible organic acetal models to demonstrate large rate accelerations by a neighboring carboxylate.

C. Proteins that Catalyze Conversions

An understanding and utilization of intramolecular rate accelerations due to noncovalently enforced strain effects stands at the intellectual and experimental frontiers of chemistry. The notion of binding-induced strain, which can be traced to Haldane[173] and Pauling,[174] is today summarized by saying that the transition state for a reaction binds to an enzyme more tightly than either the substrate or product. Until recently, experimental evidence was limited to findings that transition-state analogs—molecules more closely resembling presumed transition-state geometries than substrate geometries—often bind more tightly to enzymes than the natural substrates themselves. For the past several years, Fersht[175,176] has been studying transition-state binding by modifying the enzyme instead. In the case of tRNA[Tyr] synthetase, all of the microscopic rate and equilibrium constants may be determined experimentally, making it an ideal subject for systematic study (Fersht refers to it as "reverse evolution") using site-directed mutagenesis techniques. The energetics of substrate binding to a mutated enzyme may be determined from ES dissociation constants, while the energetics of transition state binding to a mutated enzyme may be calculated from rate data. In this way, Fersht is able to establish how an individual enzymatic residue affects binding energies; if a change in one residue was found to decrease slightly the binding of substrate but decrease significantly the binding of the transition state, it would at the same time provide strong support for the likely importance of transition-state binding and would implicate the involvement of that particular residue.

The results obtained with some 15 mutant enzymes were, in part, supportive of the transition-state binding theory and, in part, unexpected. In the reaction of E + ATP + Tyr to give E : : Tyr − AMP + PP$_i$, mutations of Thr-40 and His-45 (which are removed from the site of bond cleavage) to non-hydrogen-bonding residues (Ala and Gly, respectively) did not appreciably change the binding constant of ATP and Tyr to the enzyme, but did dramatically decrease the rate of reaction. Several other residues also had more of an effect on the rate than on the binding constant. These results are in accord with the supposition that tight binding of the transition state was occurring. However, the decrease in binding of the high-energy intermediate (E : : Tyr-AMP) was even more pronounced than that of the transition state. While this finding can be rationalized with some success,[175] the fact that the intermediates bind with increasing strength as the reaction proceeds is an important confirmation of the original idea by Haldane and Pauling. The method of enzyme point mutation via site-directed mutagenesis promises to be a very powerful tool in the study of binding effects on catalysis.

Pauling's conjecture that "enzymes are antibodies to transition states" has recently received an unusually direct verification. In 1986, two groups (Tramontano, Janda, and Lerner;[177] Pollack, Jacobs, and Schultz[178]) simultaneously reported that antibodies to tetrahedral phosphonates and phosphates catalyze hydrolysis reactions of esters and carbonates, respectively. Inasmuch as these tetrahedral antigens are structurally and electronically similar to the tetrahedral intermediates in ester and carbonate hydrolyses, they may be viewed as transition-state analogs for those reactions. (Because the intermediates are "high energy" relative to the starting materials, Hammond's postulate predicts that the intermediates will resemble the corresponding transition states.) It is not yet possible to ascertain whether catalysis occurs because of strain with or without distortion toward the transition state; in fact, it is possible that the antibodies act by holding the bound reactant near a catalytic group on the antibody's surface and that the effect is largely entropic. Regardless, the now accessible technology to produce monoclonal antibodies suggests wide-ranging possibilities for this line of research. Both groups have raised the additional possibility of modifying the antibody binding sites with chemical catalytic groups in much the same way that Kaiser's group has pioneered the study of semisynthetic enzymes.[179]

Because the synthetic problems associated with the construction of substrate-encompassing binding sites are so formidable, the potential of such combined biological/chemical approaches is substantial.

D. Nonprotein Systems that Demonstrate Strain-Induced Rate Accelerations

Sargeson and co-workers[180] have reported an unusually fast electron transfer reaction in a Co(II)/Co(III) couple that is likely due to strong binding of the transition state. The Co(II)–Co(III) self-exchange rate in sepulchral complex **59** is about 10^5 times faster than that in ethylenediamine complex **60**. Except for the enforced binding possible in **59**, complexes **59** and **60** would seem to be very similar; the author proposes that the "hole" inside the sepulchrate ligand may be a little too large for Co(III) and too small for Co(II), but close to the right size for the transition state for electron transfer. The Co—N bond lengths do not differ significantly between **59** and **60**, so that any strain must be manifested in nonoptimal bond angles and torsion angles. It should be noted that Jencks described the possibility of just such an effect in 1975.[181]

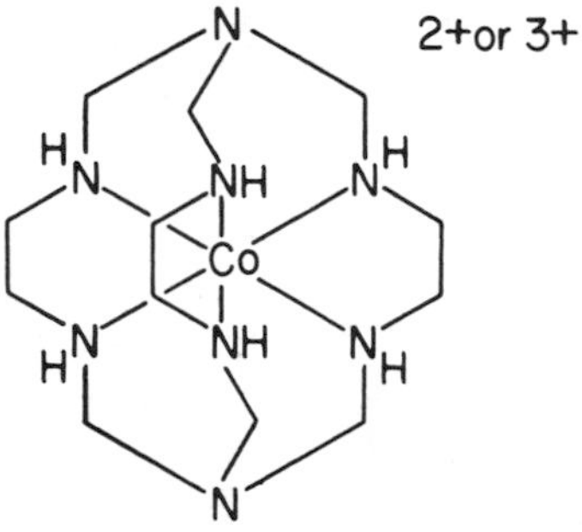

59 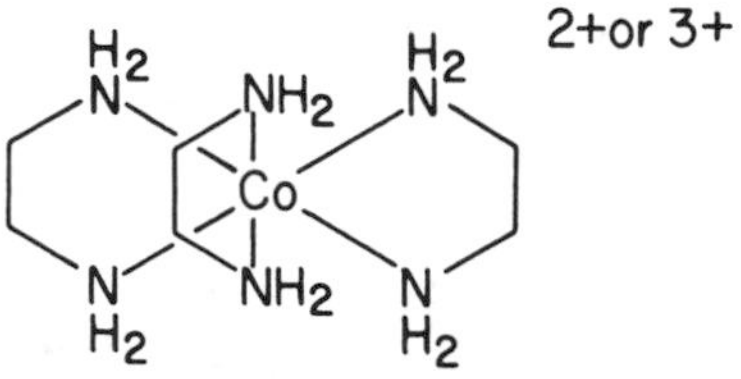   **60**

The use of bipyridyls as vehicles to test the idea of transition state binding has been studied and reviewed by Rebek.[182] As shown in Figure 3-11, chiral biphenyl (**61**) can racemize through a planar (excluding the *gem*-dimethyl group) transition state (**62**); the activation energy (calculated from the temperature required for methyl group coalescence) for the uncatalyzed process at room temperature is 14.5 kcal/mol. It was predicted that chelation of a Lewis acid, which is expected to be strongest in the planar form (ie, **62**), would decrease the activation energy in that the binding event and racemization each lie on the same reaction coordinate. This was, in fact, the case; addition of one equivalent of $HgCl_2$, $ZnCl_2$, or H^+ resulted in a lowering of the activation energy by about 4 kcal/mol. Interestingly, addition of two equivalents of H^+ increased the activation energy by 2.3 kcal/mol, as would be expected if each biphenyl nitrogen were protonated. Rebek has also

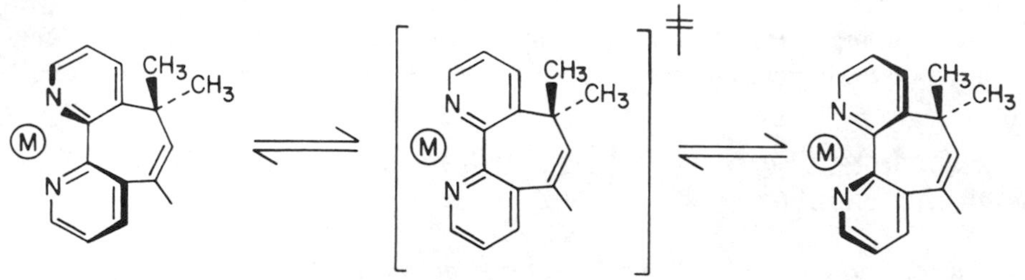

Figure 3-11. Chiral biphenyl (**61**) racemization through a planar transition state.

shown that reactions other than racemization, such as dehydrohalogenation, can be accelerated by metal binding in biphenyl systems. His group's studies on the experimental verification of rate accelerations due to strong binding of the transition state are among the earliest and most definitive; they are, probably, the clearest examples that can be used to explain this notion to an as yet uninitiated chemical community.

7. CONCLUSION

Reactions that occur intramolecularly often are faster than reactions that must occur intermolecularly, sometimes enormously faster. The rate accelerations observed result from proximity and strain effects, and can be seen in both covalently and noncovalently associated reacting systems. Enzymes, which at least initially form noncovalent associations with substrates, certainly harness both effects. The relative contributions from approximation, orientation, and strain (both with and without distortion) remain estimatable but unquantified. Work aimed at understanding enzyme mechanisms has also laid the groundwork for the design of synthetic catalysts based on the tenets of enzymatic catalysis. The production of such catalysts that actually work and solve real-life problems (or, more likely, generate solutions to problems as yet unrecognized) will be viewed as a focus for vanguard chemical research in the coming years.

ACKNOWLEDGMENTS

The author would like to thank many of the primary authors cited in this chapter for their constructive comments. In addition, the efforts of Ronald Breslow, Robert Hanson, Sam Gellman, and Joel Liebman in their critical reading of the entire manuscript are acknowledged with gratitude.

REFERENCES

1. Bruice, T. C.; Benkovic, S. J. "Bioorganic Mechanisms," Vols. I and II; Benjamin: Reading, MA, 1966.
2. Jencks, W. P. "Catalysis in Chemistry and Enzymology"; McGraw-Hill: New York, 1969.
3. Bender, M. L. "Mechanism of Homogeneous Catalysis from Protons to Proteins"; Wiley Interscience: New York, 1971.
4. Walsh, C. "Enzymatic Reaction Mechanisms"; Freeman: San Francisco, 1979.
5. Bender, M. L.; Bergeron, R. J.; Komiyama, M. "The Bioorganic Chemistry of Enzymatic Catalysis"; Wiley (Interscience): New York, 1984.
6. Dugas, H.; Penney, C. "Bioorganic Chemistry"; Springer-Verlag: New York, 1981.
7. Fersht, A. "Enzyme Structure and Mechanism"; Freeman: San Francisco, 1977
8. Gandour, R. D.; and Schowen, R. L., Eds. "Transition States of Biochemical Processes"; Plenum Press: New York, 1978.
9. Page, M. I., Ed. "The Chemistry of Enzyme Action"; Elsevier: Amsterdam, 1984.
10. DeTar, D. F.; Binzet, S.; Darba, P. *J. Org. Chem.* **1985,** *50,* 2826–2836.
11. Bolton, P. D.; Fleming, K. A.; Hall, F. M. *J. Am. Chem. Soc.* **1972,** *94,* 1033.
12. The "Fundamentalist" view of enzyme–substrate binding interactions has been put forward by R. L. Schowen in Ref. 8, pp. 90–101.
13. Page, M. I.; Jencks, W. P. *Proc. Natl. Acad. Sci. U.S.A.* **1971,** *68,* 1678.
14. Fersht, A. R. *Proc. R. Soc. London Ser. B.* **1974,** *187,* 397.
15. Jencks, W. P. In "Molecular Biology, Biochemistry and Biophysics", Vol. 32; Chapeville, F.; and Haenni, A.-L., Eds.; Springer-Verlag: New York, 1980, pp. 3–25 (p. 12). Jencks goes on to say that "This raises a difficult challenge for their [destabilization mechanisms] experimental demonstration, but should not be allowed to lead to neglecting their potential significance for catalysis."
16. Descriptions of this effect abound, but one particularly valuable summary of comparative rate data exists: Kirby, A. J. In "Advances in Physical Organic Chemistry", Vol. 17; Academic Press: New York, 1980, pp. 183–278.
17. Fife, T. H. In "Advances in Physical Organic Chemistry", Vol. 11; Academic Press: New York, 1975, pp. 1–122.
18. Kirby, A. J.; Fersht, A. R. *Prog. Bioorg. Chem.* **1971,** *1,* 1.
19. Capon, B. *Essays Chem.* **1972,** *3,* 127.
20. Gandour, R. D.; Schowen, R. L. *Annu. Rep. Med. Chem.* **1972,** *7,* 279.
21. Balakrishnan, M.; Rao, G. V.; Venkatassubramina, N. *J. Sci. Ind. Res.* **1974,** *33,* 641.
22. Capon, B. In "Proton-Transfer Reactions"; Caldin E. F.; and Gold, V., eds.; Chapman and Hall: London, 1975, pp. 339–384.
23. Ref. 3, Chapter 9.
24. Gandour, R. D. In Ref. 8, pp 540–548.
25. Page, M. I. *Chem. Soc. Rev.* **1973,** *2,* 295.
26. Kirby, A. J. In "Studies in Organic Chemistry, Vol. 10—Chemical Approaches to Understanding Enzyme Catalysis: Biomimetic Chemistry and Transition-State Analogs"; Green, B. S.; Ashani, Y.; and Chipman, D., Eds.; Elsevier: Amsterdam, 1982, p. 219.
27. Bruice, T. C.; Pandit, V. K. *Proc. Natl. Acad. Sci. U.S.A.* **1960,** *46,* 402.
28. Bruice, T. C.; Pandit, V. K. *J. Am. Chem. Soc.* **1960,** *85,* 5858.
29. Bruice, T. C.; Bradbury, W. C. *J. Am. Chem. Soc.* **1965,** *87,* 4846.
30. Bruice, T. C.; Bradbury, W. C. *J. Am. Chem. Soc.* **1968,** *90,* 3808.
31. Bruice, T. C. In "Annual Review of Biochemistry", Vol. 45; Annual Reviews, Inc.: Palo Alto, 1976, pp. 331–373 (esp. pp. 352–357).
32. Casadei, M. A.; Galli, C.; Mandolini, L. *J. Am. Chem. Soc.* **1984,** *106,* 1051.
33. For example, see Knipe, A.; Sterling, C. J. *J. Chem. Soc. B* **1968,** 67.
34. Dafforn, A.; Koshland, D. E., Jr. *Proc. Natl. Acad. Sci. U.S.A.* **1971,** *68,* 2463.

35. Dafforn, G. A.; Koshland, D. E., Jr. *Bioorg. Chem.* **1971**, *1*, 129.
36. Dafforn, A.; and Koshland, D. E., Jr. *Biochem. Biophys. Res. Commun.* **1973**, *52*, 779.
37. Storm, D. R.; Koshland, D. E., Jr. *Proc. Natl. Acad. Sci. U.S.A.* **1970**, *66*, 445.
38. Koshland, D. E., Jr.; Carraway, K. W.; Dafforn, G. A.; Gass, J. D.; Storm, D. R. *Cold Spring Harbor Symp Quant. Biol.* **1971**, *36*, 13.
39. Storm, D. R.; Tijan, R.; Koshland, D. E., Jr., *J. Chem. Soc. D.* **1971**, *15*, 854.
40. Storm, D. R.; Koshland, D. E., Jr. *J. Am. Chem. Soc.* **1972**, *94*, 5805.
41. Storm, D. R.; Koshland, D. E., Jr. *J. Am. Chem. Soc.* **1972**, *94*, 5815.
42. Bruice, T. C.; Brown, A.; Harris, D. O. *Proc. Natl. Acad. Sci. U.S.A.* **1971**, *68*, 658.
43. Bruice, T. C. *Cold Spring Harbor Symp. Quant. Biol.* **1971**, *36*, 21.
44. Bruice, T. C. *Nature (London)* **1972**, *237*, 335.
45. Jencks, W. P.; Page, M. I. *Biochem. Biophys. Res. Commun.* **1974**, *57*, 887.
46. Scheiner, S.; Lipscomb, W. N.; Kleiner, D. A. *J. Am. Chem. Soc.* **1976**, *98*, 4770.
47. Moriarty, R. M.: Adams, T. *J. Am. Chem. Soc.* **1973**, *95*, 4070.
48. Moriarty, R. M.; Adams, T. *J. Am. Chem. Soc.* **1973**, *95*, 4071.
49. Herschfeld, R.; Schmir, G. L. *J. Am. Chem. Soc.* **1972**, *94*, 6788.
50. Capon, B. *J. Chem. Soc. B* **1971**, 1207.
51. Menger, F. M.; Glass, L. E. *J. Am. Chem. Soc.* **1980**, *102*, 5404. The trajectory angles and O/C distances reported in this contribution have been modified by the author. Compound **1**: angle = 69°, not 70°; distance = 3.02 Å, not 2.83 Å. Compound **2**: angle = 79°, not 80°; distance = 2.95 Å, not 2.81 Å. Compounds **3** and **4** have not been recalculated (private communication, F. Menger).
52. Hoare, D. G. *Nature (London)* **1972**, *236*, 437.
53. Bürgi, H. B.; Dunitz, J. D.; Shefter, E. *J. Am. Chem. Soc.* **1973**, *95*, 5065.
54. Gandour, R. D., In Ref. 8, pp. 529–552 (especially pp. 535-539).
55. This author would like to take a stab at differentiating between the usages of *enzymic* and *enzymatic*, which have to date been used interchangably. We propose that the adjective *enzymic* be limited to the modification of "passive" nouns, ie, nouns that do not imply motion. For example, one could refer to the "enzymic Asp-52 residue" or to the "enzymic substrate." Alternately, it is proposed that the adjective *enzymatic* be reserved for the modification of "active" nouns, ie, nouns that imply action. For example, one could refer to "enzymatic catalysis" or to an "enzymatic process." Such a usage is, unfortunately, a true convention inasmuch as the Greek origins of these terms do not suggest a differentiation.
56. (a) Dorigo, A. E.; Houk, K. N. *J. Am. Chem. Soc.* **1987**, *109*, 3698. (b) For a computational treatment of the spatiotemporal hypothesis, see Dorigo, A. E.; Houk, K. N. "On The Relationship Between Proximity and Reactivity. An ab Initio Study of the Flexibility of the OH⁻ + CH₄ Hydrogen Abstraction Transition State, and a Force-Field Model for the Transition States of Intramolecular Hydrogen Abstractions"; submitted to *J. Am. Chem. Soc.* **1987**. In this article, Houk argues that, ". . . there is no simple relationship between calculated reaction rates and distances between the reacting atoms in [the] starting materials."
57. Menger, F. M. *Acc. Chem. Res.* **1985**, *18*, 128.
58. Menger, F. M.; Chow, J. F.; Kaiserman, H.; Vasquez, P. C. *J. Am. Chem. Soc.* **1983**, *105*, 4996.
59. Menger, F. M. *Tetrahedron* **1983**, *39*, 1013.
60. Hine, J.; Cholod, M. S.; King, R. A. *J. Am. Chem. Soc.* **1974**, *96*, 835.
61. Menger, F. M.; Grossman, J.; Liotta, D. C. *J. Org. Chem.* **1983**, *48*, 905.
62. Bednar, R. A.; Jencks, W. P. *J. Am. Chem. Soc.* **1985**, *107*, 7135.
63. Pattabiraman, T. N.; Lawson, W. B. *J. Biol. Chem.* **1971**, *247*, 3029.
64. Steitz, T. A.; Henderson, R.; Blow, D. M. *J. Mol. Biol.* **1969**, *46*, 337.
65. Sigler, P. B.; Blow, D. M.; Matthews, B. W.; Henderson, R. *J. Mol. Biol.* **1968**, *35*, 143.
66. Blow, D. M. *Biochem. J.* **1969**, *112*, 261.
67. Nakagawa, Y.; Bender, M. L. *J. Am. Chem. Soc.* **1969**, *91*, 1566.

68. Ref. 2, pp. 45–59 and pp. 218–226.
69. Matthews, B. W.; Sigler, P. B.; Henderson, R.; Blow, D. M. *Nature (London)* **1967**, *214*, 652.
70. Blow, D. M.; Birktoff, J. J.; Hartley, B. S. *Nature (London)* **1969**, *221*, 337.
71. Bender, M. L.; Kezdy, F. J. *Annu. Rev. Biochem.* **1965**, *34*, 49.
72. McDonald, C. E.; Balls, A. K. *J. Biol. Chem.* **1957**, *227*, 727.
73. Jencks, W. P.; Carriuolo, J. *J. Biol. Chem.* **1959**, *234*, 1272.
74. Jencks, W. P.; Carriuolo, J. *J. Am. Chem. Soc.* **1961**, *83*, 1743.
75. Pandet, U. K.; Bruice, T. C. *J. Am. Chem. Soc.* **1960**, *82*, 3386.
76. Bruice, T. C.; Benkovic, S. J. *J. Am. Chem. Soc.* **1963**, *85*, 1.
77. Boland, M. J.; Hardman, M. J.; Watson, I. D. *Bioorg. Chem.* **1974**, *3*, 213.
78. Belke, C. J.; Su, S. C. K.; Shafer, J. A. *J. Am. Chem. Soc.* **1971**, *93*, 4552.
79. Menger, F. M.; Brock, H. T. *Tetrahedron* **1968**, *24*, 3453.
80. Bruice, T. C. *Brookhaven Symp. Biol.* **1962**, *15*, 52.
81. Kirsch, J. F.; Jencks, W. P. *J. Am. Chem. Soc.* **1964**, *86*, 837.
82. Lukton, A.; Blackman, D.; Koshland, D. E., Jr., *Biochim. Biophys. Acta* **1964**, *85*, 512.
83. Utaka, M.; Koyama, J.; Takeda, A. *J. Am. Chem. Soc.* **1976**, *98*, 984.
84. Rogers, G. A.; Bruice, T. C. *J. Am. Chem. Soc.* **1974**, *96*, 2463.
85. Komiyama, M.; Bender, M. L. *Bioorg. Chem.* **1977**, *6*, 13.
86. Utaka, M.; Takeda, A.; Bender, M. L. *J. Org. Chem.* **1974**, *39*, 3772.
87. Komiyama, M.; Roesel, T. R.; Bender, M. L. *Proc. Natl. Acad. Sci. U.S.A.* **1977**, *74*, 23.
88. Komiyama, M.; Bender, M. L.; Utaka, M.; Takeda, A. *Proc. Natl. Acad. Sci. U.S.A.* **1977**, *74*, 2634.
89. Blow, D. M.; Birktoff, J. J.; Hartley, B. S. *Nature (London)* **1969**, *221*, 337.
90. Rogers, G. A.; Bruice, T. C. *J. Am. Chem. Soc.* **1974**, *96*, 2473.
91. Mallick, I. M.; D'Souza, V. T.; Yamaguchi, M.; Lee, J.; Chalabi, P.; Gadwood, R. C.; Bender, M. L. *J. Am. Chem. Soc.* **1984**, *106*, 7252.
92. D'Souza, V. T.; Bender, M. L. *Acc. Chem. Res.* **1987**, *20*, 146.
93. Czarnik, A. W. *Tetrahedron Lett.* **1984**, *25*, 4875.
94. Nanjappan, P.; Czarnik, A. W. *J. Am. Chem. Soc.* **1987**, *109*, 1826.
95. Sauer, J. *Angew. Chem.* **1966**, *5*, 211 (p. 230).
96. Kiselev, V. D.; Konovalov, A. I.; Veisman, E. A.; Ustyugov, A. N. *Zh. Org. Khim.* (English translation) **1978**, *14*, 128.
97. Sauer, J. *Angew. Chem.* **1964**, *3*, 150.
98. Nanjappan, P.; Czarnik, A. W. *J. Org. Chem.* **1986**, *51*, 2851.
99. Rideout, D. C.; Breslow, R. *J. Am. Chem. Soc.* **1980**, *102*, 7816.
100. Ref. 2, p. 400.
101. Higuchi, T.; and Lachman, L. *J. Am. Pharmacol. Assoc.* **1955**, *44*, 521.
102. For example, see Collman, J. P.; Kimura, E. *J. Am. Chem. Soc.* **1967**, *89*, 6096.
103. For example, see Fife, T. H.; Squillacote, V. L. *J. Am. Chem. Soc.* **1977**, *99*, 3762.
104. Groves, J. T.; Chambers, R. R., Jr. *J. Am. Chem. Soc.* **1984**, *106*, 630.
105. Webb, T. R.; Matteucci, M. D. *J. Am. Chem. Soc.* **1986**, *108*, 2764.
106. Essentially this same treatment, but for unimolecular reactions, has been described in Ref. 8, pp. 17–18.
107. A convenient summary of these solutions may be found in Cantor, C.R.; and Schimmel, P. R. "Biophysical Chemistry, Part III"; Freeman: San Fransisco, 1980, pp. 887–904.
108. For a clear presentation of simple enzyme kinetics, see Ref. 7, pp. 84–102.
109. (a) Ref. 5, pp. 277–305. (b) Ref. 6, pp. 253–328.
110. (a) Bender, M. L.; Komiyama, M. "Cyclodextrin Chemistry"; Springer-Verlag: New York, 1978. (b) Breslow, R. *Science* **1982**, *218*, 532. (c) Ref. 9, pp. 505–527.
111. Ref. 9, pp. 529–562.
112. (a) Fendler, J. H.; Fendler, E. J. "Catalysis in Micellar and Macromolecular Systems"; Academic Press: New York, 1975. (b) Ref. 9, pp. 461–505.
113. Rebek, J., Jr., *Science* **1987**, *235*, 1437.

114. Ref. 6, pp. 329–386.
115. Ref. 5, pp. 191–215.
116. Breslow, R.; Czarnik, A. W. *J. Am. Chem. Soc.* **1983,** *105,* 1390.
117. Cram, D. J.; Katz, H. E.; Dicker, I. B. *J. Am. Chem. Soc.* **1984,** *106,* 4987.
118. Cram, D. J. In "International Symposium on Bioorganic Chemistry"; 1986, Breslow, R.,
     Ed.; New York Academy of Sciences: New York, 1986, pp. 22–40.
119. See (a) The opening shot: Breslow, R. *Chem. Eng. News,* April 11, 1983, p. 4. (b) The
     return volley: Cram, D. J. *Chem. Eng. News,* April 11, 1983, p. 4. (c) The arbitrated cease-
     fire: Schowen, R. L. *Chem. Eng. News,* May 23, 1983, p. 39. For a more detailed discus-
     sion of this topic, see Katz, H. E. "Molecular Structure and Energetics", Volume 4;
     Liebman, J. F.; and Greenberg, A., Eds.; VCH: Deerfield Beach, FL, 1987, Chapter 5.
120. For a recent overview, see Dervan, P. B. *Science* **1986,** *232,* 464.
121. Dervan, P. B., in Ref. 118, pp. 51–59.
122. Baker, B. F.; Dervan, P. B. *J. Am. Chem. Soc.* **1985,** *107,* 8266.
123. Menger, F. M.; Whitesell, L. G. *J. Am. Chem. Soc.* **1985,** *107,* 707.
124. Newman, M. S. "Steric Effects in Organic Chemistry"; Wiley: New York, 1956.
125. Streitweiser, A., Jr.; Heathcock, C. H. "Introduction to Organic Chemistry", 3rd ed.
     MacMillan: New York, 1985, p. 182.
126. Rüchardt, C.; Beckhaus, H.-D. *Angew. Chem.* **1980,** *19,* 429.
127. Milstein, S.; Cohen, L. A. *J. Am. Chem. Soc.* **1972,** *94,* 9158.
128. Borchardt, R. T.; Cohen, L. A. *J. Am. Chem. Soc.* **1972,** *94,* 9166.
129. Ref. 8, pp. 531–535.
130. Karle, J. M.; Karle, I. L. *J. Am. Chem. Soc.* **1972,** *94,* 9182.
131. Danforth, C.; Nicholson, A. W.; James, J. C.; Loudon, G. M. *J. Am. Chem. Soc.* **1976,**
     98, 4275.
132. Jencks, W. P. *Curr. Aspects Bio. Chem. Energ.* **1966,** 273–298.
133. Jencks, W. P. *Adv. Enzymol. Relat. Areas Mol. Biol.* **1975,** *43,* 219.
134. Jencks, W. P. *Mol. Biol. Biochem. Biophys.* **1980,** *32,* 3.
135. Ref. 9, pp. 27–34.
136. Levitt, M. In "Peptides, Polypeptides, and Proteins"; Blout, E. R.; Bovey, F. A.; Good-
     man, M.; Lotan, N., Eds.; Wiley: New York, 1974, p. 99.
137. Fersht, A. R.; Kirby, A. J. *Chem. Br.* **1980,** *16,* 136.
138. Ref. 7, pp. 269–270.
139. Stein, R. L.; Romero, R.; Bull, H. G.; Cordes, E. H. *J. Am. Chem. Soc.* **1978,** *100,* 6249.
140. Chipman, D. M.; Sharon, N. *Science* **1969,** *165,* 454. See also the very recent review of
     lysozyme model chemistry by Kirby: Kirby, A. J. *CRC Critical Reviews in Biochemistry*
     1987, *22,* 283.
141. Fife, T. H.; Przystas, T. J. *J. Am. Chem. Soc.* **1977,** *99,* 6693.
142. For an authoritative review of both topics, see Sinnott, M. L., in Ref. 9, pp. 389–431.
143. Schindler, M.; Assaf, Y.; Sharon, N.; Chipman, D. M. *Biochemistry* **1977,** *16,* 423.
144. Blake, C. C. F.; Johnson, L. N.; Mair, G. A.; North, A. C. T.; Phillips, D. C.; Sarma,
     V. R. *Proc. R. Soc. London Ser. B* **1967,** *167,* 378.
145. Phillips, D. C. *Sci. Am.* **1966,** *215* (5), 78.
146. Ford, L. O.; Johnson, L. N.; Machin, P. A.; Philips, D. C.; Tjian, R. *J. Mol. Biol.* **1974,**
     88, 349.
147. Dahlquist, F. W.; Rand-Meir, T.; Raftery, M. A. *Biochemistry* **1969,** *8,* 4214.
148. Capon, B. *Chem. Rev.* **1969,** *69,* 407.
149. Rupley, J. A.; Gates, V. *Proc. Natl. Acad. Sci. U.S.A.* **1967,** *57,* 496.
150. Imoto, T.; Johnson, L. N.; North, A. C. T.; Phillips, D. C.; Rupley, J. A. In "The
     Enzymes", 3rd ed., Vol. 7; Boyer, P. D., Ed.; Academic Press: New York, 665.
151. Vernon, C. A. *Proc. R. Soc. London Ser. B* **1967,** *167,* 389.
152. Chipman, D. M.; Grisaro, V.; Sharon, N. *J. Biol. Chem.* **1967,** *242,* 4388.
153. Secemski, I. I.; Lehrer, S. S.; Lienhard, G. E.; *J. Biol. Chem.* **1972,** *247,* 4740.
154. van Eikeren, P.; Chipman, D. M. *J. Am. Chem. Soc.* **1972,** *94,* 4788.

155. Capon, B.; Dearie, W. M. *J. Chem. Soc., Chem. Commun.* **1974**, 370.
156. Chipman, D. M. *Biochemistry* **1971**, *10*, 1714.
157. Thoma, J. A. *J. Theor. Biol.* **1974**, *44*, 305.
158. Reprinted with permission from the author and *Biochemistry,* copyright 1977, American Chemical Society.
159. Ballardie, F. W.; Capon, B.; Cuthbert, M. W.; Dearie, W. M. *Bioorg. Chem.* **1977**, *6*, 483.
160. Schindler, M.; Assaf, Y.; Sharon, N.; Chipman, D. M. *Biochemistry* **1977**, *16*, 423.
161. Kelly, J. A.; Sielcki, A. R.; Sykes, B. D.; James, M. N. G.; Phillips, D. C. *Nature (London)* **1979**, *282*, 875.
162. Chipman, D. M.; Schindler, M., in Ref. 26, p. 227.
163. See the discussion on p. 479 in Loudon, G. M.; Smith, C. K.; Zimmerman, S. E. *J. Am. Chem. Soc.* **1974**, *96*, 465.
164. Parsons, S. M.; Raftery, M. A. *Biochemistry* **1972**, *11*, 1623.
165. Blake, C. C. F.; Mair, G. A.; North, A. C. T.; Phillips, D. C.; Sarma, V. R. *Proc. R. Soc. London Ser. B* **1967**, *167*, 365.
166. Ford, L. O.; Johnson, L. N.; Machin, P. A.; Philips, D. C.; Tijan, R. *J. Mol. Biol.* **1974**, *88*, 349.
167. Warshel, A.; Levitt, M. *J. Mol. Biol.* **1976**, *103*, 227.
168. For a review of work on this topic, see Pincus, M. R.; Scheraga, H. A. *Acc. Chem. Res.* **1981**, *14*, 299.
169. Levitt, M., quoted in Ref. 166.
170. Warshel, A. In "Peptides: Proceedings of the 5th American Peptide Symposium"; Goodman, M.; and Meienhofer, J., Eds.; Halsted Press: New York, 1977, p. 574.
171. Warshel, A. *Pontif. Acad. Sci. Scr. Varia* **1984**, *55*, 59.
172. Bakthavachalam, V.; Czarnik, A. W. *Tetrahedron Lett.* **1987**, *28*, 2925.
173. Haldane, J. B. S. In "Enzymes"; Longmans, Green: London, 1930, p. 182.
174. Pauling, L. *Chem. Eng. News* **1946**, *24*, 1375.
175. Fersht, A. R.; Leatherbarrow, R. J.; Wells, T. N. C. *Trends Biochem. Sci.* **1986**, *11*, 321, and references to earlier work cited therein.
176. Wells, T. N. C.; Fersht, A. R. *Biochemistry* **1986**, *25*, 1881.
177. Tramontano, A.; Janda, K. D.; Lerner, R. A. *Science* **1986**, *234*, 1566.
178a. Pollack, S. J.; Jacobs, J. W.; Schultz, P. G. *Science* **1986**, *234*, 1570.
178b. Shultz, P. G.; Jacobs, J. W. In "Environmental Influences and Recognition in Enzyme Chemistry" (*Molecular Structure and Energetics*), Liebman, J. F. and Greenberg, A., Eds., VCH Pub., New York, 1988, Chapt. 11.
179. For a recent review, see Kaiser, E. T.; Lawrence, D. S. *Science* **1984**, *232*, 505.
180. Creaser, I. I.; Geue, R. J.; Harrowfield, J. MacB.; Herlt, A. J.; Sargeson, A. M.; Snow, M. R.; Springborg, J. *J. Am. Chem. Soc.* **1982**, *104*, 6016.
181. Ref. 133, p. 248.
182. Rebek, J., Jr., *Acc. Chem. Res.* **1984**, *17*, 258.
183. Bakthavachalam, V.; Czarnik, A. W. *Tetrahedron Lett.* **1987**, *28*, 2925.

CHAPTER 4

Structural and Energetic Aspects of Protolytic Catalysis by Enzymes: Charge-Relay Catalysis in the Function of Serine Proteases

Richard L. Schowen

University of Kansas, Lawrence, Kansas

CONTENTS

1. Introduction.   119
2. Crystallographic and NMR-Spectroscopic Studies   136
3. Chemical Models for Charge-Relay Catalysis   141
4. Theoretical Models for Charge-Relay Catalysis.   146
5. Proton-Inventory Studies.   156
6. Summary and Prospects   162
7. Appendix: Some Points of Detail   163
   References.   165

I. INTRODUCTION

A. Proton Transfer and Protonic Bridging in Enzyme Catalysis

The transfer of protons, and more generally the formation of proton bridges in transition states, occurs in the course of many enzymatic reactions and is variously involved in the generation of enzyme catalytic power. Only a limited class of enzymatic proton-transfer and protonic-bridging phenomena will be addressed in this chapter. The cases we shall consider all involve transition-state protonic bridging among electronegative atoms such as oxygen, nitrogen, or sulfur in the catalysis of such reactions as ester or amide cleavage.

This is the equivalent of the protolytic mode of general acid–base catalysis of these reactions in nonenzymatic systems. (Although the phrase *general acid–base catalysis* is sometimes used to mean catalysis by means of transition-state protonic bridging, it is properly used in a broader sense, including the nucleophilic mode of general catalysis.) Protolytic general catalysis *may* involve the actual transfer of a proton (ie, motion of a proton from a donor atom to an acceptor atom as a part of the reaction-coordinate motion of the transition state). However, there are also models of protolytic general acid–base catalysis which do not require net transfer of the proton, and these will be included in our considerations.

In addition to this restriction to protolytic general catalysis, our attention in this chapter will be further limited to a smaller aspect of enzymatic general acid–base catalysis, namely the concept that has come to be called "charge-relay catalysis."

B. Protolytic Catalysis by Enzymes and the Charge-Relay System

Protolytic catalysis (exergonic proton bridging in the transition state, the bridge linking a catalyst functional group that can act as a proton donor or acceptor to a center in the reacting substrate that has high or low electron density, respectively) is an effective means of stabilizing transition states and accelerating reactions near neutral pH.[1-7] Since there has been no catalytic technique yet described in any situation that does not seem to be employed, in some sense, by enzymes, it is to be expected that protolytic catalysis will be a feature of enzyme action. Indeed, the pH dependence of enzymatic rates led early investigators to the equivalent of this idea.[8] Then the solvent-isotope-effect studies of Bender and co-workers[9] in the 1960s showed that rate reductions in deuterium oxide (DOD) as solvent were around two- to three-fold for the amide-cleaving and ester-cleaving reactions of the enzyme α-chymotrypsin. This is in exact agreement with solvent isotope effects for protolytic catalysis in similar nonenzymatic reactions.[10] The concept of protolytic catalysis by active-site functional groups, such as the imidazole/imidazolium side chain of His or the carboxylate/carboxyl side chains of Asp and Glu, thus became an established feature of mechanistic enzymological thought.

The crystallographic elucidation of the structure of chymotrypsin was expected to confirm the presence of machinery in the active site for protolytic catalysis as well as other features of the then-hypothesized[6] enzyme mechanism. It was known, for example, that the $HO—CH_2—$ side chain of a

Ser residue of the active site served as a nucleophile in the first ("acyla-tion") stage of the enzymatic reaction, expelling the leaving group X of an acyl substrate RCOX and becoming acylated in the process. In the second ("deacylation") stage of the reaction, the acyl enzyme RCO—O—[Ser]—[enzyme] was hydrolyzed by reaction with water:

Acylation: RCOX + E—OH ----→ RCO—O—E + HX

Deacylation: RCO—O—E + HOH ----→ RCOOH + E—OH

Protolytic catalysis by an active-site His was believed to occur in both stages. Each stage, acylation and deacylation, should consist of at least two steps: the formation and decomposition of the tetrahedral intermediate T common in acyl-transfer reactions. In an intriguing kind of dialectic, each step in the sequence prepares the catalytic group for its role in the subse-quent step *if* one assumes that net transfer of the proton occurs in the catalytic process:

Acylation, formation of T:

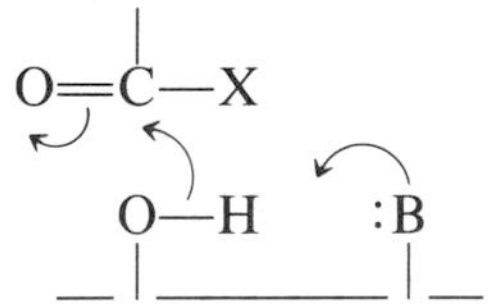

Acylation, decomposition of T:

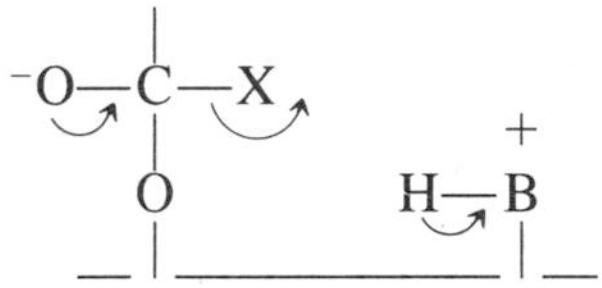

Deacylation, formation of T:

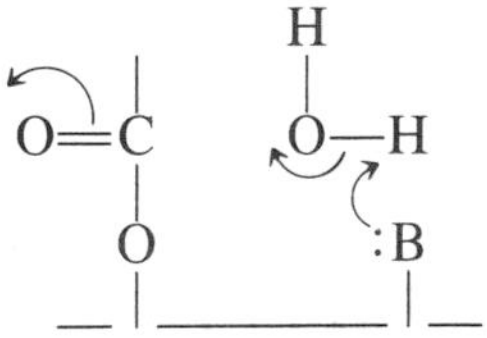

Deacylation, decomposition of T:

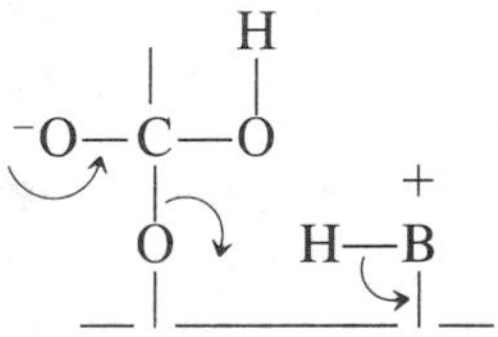

Which stage, acylation or deacylation, actually limits the rate in a given case depends mainly on the reactivity of X as the leaving group (high reactivity leaving the deacylation rate limiting, low reactivity favoring the acylation).

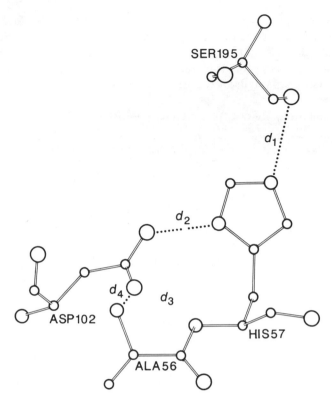

Figure 4-1. The charge-relay system in its current incarnation.[15] This structure comes from the data for native α-chymotrypsin, unperturbed by tosylate, which had the effect in the original report[11,12] of producing a mistake in the position of the Ser-195 O$^\gamma$.

As expected, the actual crystallographic enzyme structure[11] in fact showed the active-site Ser-195 hydrogen bonded to His-57 (Figure 4-1; note that this conclusion has had its historical vicissitudes, but that the pendulum currently rests with the common existence of this hydrogen bond). The surprise[12] lies at the other side of the imidazole ring, where a further hydrogen bond linked the N—H bond of the His to the carboxylate of Asp-102. This chain of two successive H bonds is the entity called the *charge-relay system*[12,13] or the *catalytic triad*.[14]

Such a structure has now been widely documented[15] in serine hydrolases. It is clear that an evolutionary pressure exists which has brought this structure into being in the course of both convergent and divergent lines of evolution. Organismic populations with enzymes possessing this structure apparently enjoy a sufficient competitive advantage that the catalytic triad has come to dominate in the serine proteases over other chemical solutions to the catalytic problem of ester and amide cleavage.

The subsequent history of the charge-relay concept has been complex, to a degree disputatious, to a greater degree confused and contradictory, and the detailed role of the entity in catalysis remains unsettled. Nevertheless, the evolutionary conservation of the catalytic triad clearly indicates that the structure is a valuable discovery of molecular evolution. It is apparently a quintessential element of the overall acid–base catalytic capability of the serine proteases. How much wider its importance is remains to be seen.

This chapter presents an account of the study of the charge-relay concept by several experimental and theoretical tools, and a critical assessment of the findings and their significance is attempted. The tools for consideration are listed in Table 4-1. They are considered in the order given, following an introductory review of basic issues in protolytic catalysis.

TABLE 4-1. A Selection of the Methods Used for the Study of the Charge-Relay System as an Element of the Acid–Base Catalytic Entity in Serine Proteases

Technique	Methodological design
Crystallography	X-Ray crystallography is used to characterize crystalline enzymes and their complexes or derivatives structurally. The active-site functional groups are located with as much precision as possible in the native enzyme. The charge-relay concept originally arose from the observation of a chain of hydrogen bonds in the chymotrypsin active site, connecting Ser-195 to His-57 to Asp-102 and perhaps farther into the structure. Protein inhibitors or product molecules are used as simulators of the substrate, or transition-state or intermediate analogs may be employed. Neutron diffraction can locate the proton positions with greater precision. From these studies of crystalline forms of stable states of the enzyme and its derivatives, conclusions about the mechanism and events in the transition states are sometimes drawn by inference.
NMR spectroscopy	Chemical shifts and coupling constants are determined, sometimes for specifically isotopically labeled enzymes or their derivatives, under various conditions of pH, binding of inhibitors or other substrate simulators, etc. From the observations, structural features, such as the active-site functional group which bears a proton at a particular pH, of the enzyme or its derivatives in solution are inferred. From these inferred structures of stable states, conclusions about the mechanism and events in the transition states are drawn by further inference.
Chemical modeling	Small molecules are constructed by synthetic chemical techniques. These molecules include functional groups identical to or related to those in the enzyme active site, with relative positions similar to (or instructively different from) those derived from crystallographic studies. These model molecules are evaluated kinetically as catalysts of hydrolytic reactions. From these observations, conclusions about the probable roles of the functional groups and their interactions in the transition states for enzyme action are drawn by inference.
Theoretical modeling	Theoretical techniques (molecular mechanics or other potential-function methods, semiempirical quantum-mechanical methods, ab initio

(continued)

TABLE 4-1. Cont.

Technique	Methodological design
	quantum-mechanical methods, or some combination of these applied either in the calculation of static structures or in dynamic simulations) are applied to models of the enzyme. The models vary in the fraction of the enzyme structure which is simulated, the precision with which it is simulated structurally, the number of degrees of freedom which is allowed to relax, the degree to which the solvent inside and outside of the active site is simulated, the precision with which the substrate structure is simulated, and the care with which the reaction paths are explored and the transition states characterized. In principle, theory has accurate access to the enzymatic transition-state events in a way no other technique can match. Practically, the limits of current technology dictate that conclusions are being drawn by inference from simplified models studied by approximate methods.
Proton inventories	Kinetic studies of enzyme action are conducted in HOH, DOD, and binary mixtures of the two solvents. Solvent isotope effects are evaluated for each kinetic term. Each kinetic term links a reactant state to a transition state. From the Gross–Butler treatment for each kinetic term, the number of sites that generate an isotope effect and the magnitude of each isotope effect are inferred. While the method gives direct experimental access to transition-state information, the structural location of the sites that generate the isotope effects must be inferred from other information.

C. Basic Issues in Protolytic Catalysis

a. PT and HAR Components[10,16]

The fundamental idea of protolytic catalysis is that the transition state for some reorganization of nonhydrogen atoms [heavy-atom reorganization (HAR)] can be stabilized by bridging to a hydrogen donor/acceptor. In many cases, although not necessarily in all or even most cases, the bridging proton will be transferred during catalysis. The interaction is, therefore, often described as the *process* of proton transfer (PT). It is also true that, in some cases, the mere presence of the acid–base catalyst is sufficient for catalysis to occur without direct transition-state interaction (preassociation or spectator catalysis: see later in this section).

For the case of R—OH addition to $\diagup^{\diagdown}C{=}O$, as in the nucleophilic attack of an enzyme active-site Ser on the carbonyl group of an ester or amide substrate, the overall process can be represented by Equation 1.

$$B + HO— + \overset{\diagdown}{\underset{\diagup}{C}}{=}O \dashrightarrow BH^+ + —O—\overset{|}{\underset{|}{C}}—O^- \qquad (4\text{-}1)$$

Here, the HAR process consists of O—C bond formation and the concomitant fission of the π bond of the C=O group. The PT process consists of removal of the H from the nucleophilic O to the catalytic base B.

Various degrees of coupling between PT and HAR can be imagined. The two processes may occur separately in time, or exactly simultaneously, or roughly simultaneously with one of the processes lagging to some extent. In nonenzymatic systems, information on the degree of coupling between HAR and PT in a given example of the reaction may be obtained from the kinetic order in B, the dependence of rate constant on the basicity of B (Brønsted β coefficient), the dependence of the rate constant on structure in —OH and $\overset{\diagdown}{\underset{\diagup}{C}}{=}O$ (Hammett ρ and other structure–reactivity coefficients), the way in which the various structure–reactivity coefficients themselves change with substrate structure (Cordes–Jencks coefficients[17]), or from kinetic isotope effects.[7,10,18] In enzymatic systems, only kinetic isotope effects[19] may be used in general, although limited applications of other techniques are possible in special cases.[20]

In the limit of *zero* coupling between PT and HAR, one possibility is that the proton transfer occurs rapidly before a slow and rate-limiting HAR (in a bimolecular reaction of B and HO—R) or rapidly after a slow and rate-limiting HAR (in a bimolecular reaction of B and the protonated product) (this zwitterionic tetrahedral adduct is often called $T^{\pm}$). In nonenzymatic systems, these reactions are both zero order in B and, thus, kinetically distinct from a general-catalyzed reaction. A general-catalyzed reaction must be first order in B since, if proton bridging to B is to occur in the transition state, B must be present in the transition state. Rapid, prior proton transfer in the sense described, therefore, produces the kinetically identifiable specific, rather than general, catalysis; rapid, subsequent proton transfer produces an "uncatalyzed" reaction.

Nevertheless, zero coupling can also occur in general-catalyzed reactions. This can happen if a reaction involves rapid prior or rapid subsequent proton transfer *but in a kinetically significant manner only when all three reactants have assembled into the same solvent cage and remain assembled throughout the reaction sequence.* The rate-limiting HAR transition state, then, has B present, but either completely protonated (rapid-prior PT) or completely unprotonated (rapid-subsequent PT). One way this can occur is if either the conjugate base of R—OH (R—O$^-$) or $T^{\pm}$ is too unstable to be formed in sufficient concentration to serve as a reaction intermediate in a bimolecular reaction. In the ternary collision complex, however, $\overset{\diagdown}{\underset{\diagup}{C}}{=}O$ can trap

R—O⁻ immediately after its formation by proton transfer to B. Likewise, $T^\pm$ can be trapped by immediate loss of a proton to B if the base is present when $T^\pm$ is generated. This is the phenomenon called *preassociation catalysis* because all reacting fragments must associate in the same solvent cage before any fruitful reaction ensues.[21] It is also called *spectator catalysis* because in one or more of the component steps of the reaction sequence at least one of the reactants is not interacting with the others but is present merely as a spectator.[22]

For enzymatic reactions, the importance of spectator catalysis is that active site functional groups can accelerate substrate reactions by the same type of trapping mechanism. For an acceleration to be produced, it is therefore not necessary that the catalytic functional group be directly interacting in the rate-limiting transition state with the fragments undergoing HAR.

b. Use of MAR Diagrams[23,24]

For further discussion of coupling between PT and HAR in protolytic catalysis, a useful tool is the diagram called by Bruice[23] a "map of alternate routes," (MAR; Figure 4-2). The axes of such plots measure the *fractional reaction progress*, $\hat{I}$, for the PT process along the ordinate and for the HAR process along the abscissa. Values of $\hat{I}$ can be translated into structural terms: for example, a given value of $\hat{I}_{PT}$ corresponds to particular lengths of the bonds connecting the proton to B and to O. Similarly, a value of $\hat{I}_{HAR}$ implies certain values for the bond lengths of the heavy-atom bonds. Any point on the MAR has as its coordinates $(\hat{I}_{HAR}, \hat{I}_{PT})$ so that each point specifies a complete structure of a transition state *if* it is assumed that the generally less drastic structural reorganizations occurring elsewhere in the

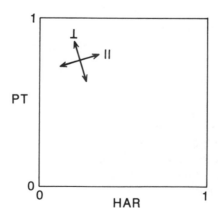

Figure 4-2. A "map of alternate routes"[23] or MAR. The point drawn indicates a transition-state structure with proton transfer (PT) complete to a fractional degree of 0.75 and heavy-atom reorganization (HAR) complete to a fractional degree of 0.25. The parallel and perpendicular vectors indicate a reaction-coordinate motion that is mostly HAR, with a small component of PT.

molecule can be related to either $\hat{I}_{HAR}$, $\hat{I}_{PT}$ or both. Any actual reaction will have a particular transition-state structure which will then correspond to a particular point on the MAR.

Through each point on the MAR, two vectors may be constructed.[25] One of these describes the relative contribution of changes in PT and HAR to the reaction-coordinate (transition-coordinate[26]) motion of the transition state, the motion which leads to decomposition of the activated complex to reactants or products. If this vector is parallel to the PT axis, then there is no contribution of HAR to the reaction-coordinate motion; this corresponds to a transition state for proton transfer with the degree of HAR remaining constant. In some sense, one would say "proton transfer is rate determining," but this can be confusing since some degree of HAR may have occurred prior to the reaction-coordinate motion and the remainder could occur afterward, which does not require a separate transition state for HAR.[27] If the vector is parallel to the HAR axis, then HAR alone constitutes the reaction-coordinate motion ("HAR determines the rate," a statement that must, just as above, be interpreted with care) and the degree of PT remains constant during passage through the transition state.

If the MAR was a potential energy surface, which it assuredly is not,[28] the vector just described would be tangent to, and thus parallel to, the reaction-coordinate vector at the equilibrium structural coordinates of the activated complex (transition structure).[25] The vector on the MAR is, therefore, known as a *parallel vector*.

A second vector on the MAR can be constructed perpendicular to the parallel vector; it is, therefore, called the *perpendicular vector*. For an activated complex, the energy profile along the "perpendicular coordinate" on the actual potential energy surface must form a potential minimum at the equilibrium structure.[25,29]

c. Coupling Modes of PT and HAR

Several modes in which protolytic catalysis can occur are shown in Figure 4-3 (there are other possible modes but these are the most important for our purposes). As illustrated for the first mode, the reaction-coordinate motion can be depicted in two ways: first, as often written with "curved arrows," and second, as the vibrational normal mode. An MAR is shown for each mode. Finally, the shapes of the energy profiles are plotted along coordinates of the potential energy surface which correspond to the parallel and perpendicular vectors of the MAR. Table 4-2 shows the experimental expectations for each of the coupling modes depicted.

The fully coupled mode is shown in Figure 4-3a. If PT and HAR are "balanced"[30] then the transition-state structure will lie on the MAR diagonal. If the coupling between HAR and PT is exact, then the parallel vector will point along the diagonal. This corresponds to one kind of transition state, one with exact coupling of PT and HAR. Other transition states with other coupling schemes are also possible, corresponding to parallel vectors

PT, HAR FULLY COUPLED

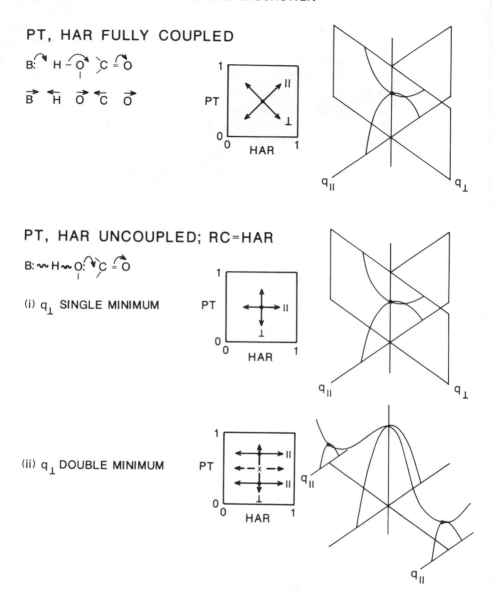

PT, HAR UNCOUPLED; RC=HAR

(i) $q_\perp$ SINGLE MINIMUM

(ii) $q_\perp$ DOUBLE MINIMUM

Figure 4-3. Modes of coupling between PT and HAR in transition states for protolytic cataly-sis. See text for discussion.

"tilted" so that PT or HAR plays a larger role in the reaction-coordinate motion. This kind of coupled PT and HAR is the most commonly accepted model of protolytic catalysis. It corresponds to Jencks's rule[16] that the pK of a protolytic catalyst B must lie between the pK of the reactant nucleophile —OH and the protonated product $T^\pm$. If this is so, then PT is unfavorable in the reactant state, favorable in the product state, and also favorable at some

TABLE 4-2. Experimental Expectations for Reactions in Which Various Degrees of
Coupling between HAR and PT Exist in the Transition State (Figure 4-3)

Degree of coupling	Experimental expectations
PT, HAR fully coupled	The situation is analogous to E2 eliminations, in which substantial coupling of PT and HAR is considered to prevail.[58] Primary isotope effects are expected for the transferring proton, the nucleophilic oxygen, the carbonyl carbon, and the carbonyl oxygen.
PT, HAR uncoupled, RC = HAR	
Single-minimum perpendicular coordinate	Primary isotope effects are expected for the nucleophilic oxygen, the carbonyl carbon, and the carbonyl oxygen. The isotope effect at the protonic center depends on the nature of the single-minimum hydrogen bond.[32] Eliason and Kreevoy[59] have suggested that the kind of AHA⁻ hydrogen bond studied by Kreevoy and co-workers, which is formed with equilibrium isotope effects of 2–5, could be present in these transition states.
Double-minimum perpendicular coordinate	In this case, the point labeled x on the MAR in Figure 4-3 is not a saddle point on the potential-energy surface and cannot be an activated complex.[29] One or both of the minima along the perpendicular coordinate must serve as activated complexes.[a]

[a] There are two types of mechanisms that may be important. (1) Semiclassical mechanism: Pathways from reactants to products pass through both minima of the perpendicular coordinate and the reaction thus occurs by these two competing pathways. If both minima are of equal free energy, then the competing reactions will occur with equal rates; otherwise reaction across the lower free-energy minimum will dominate. Isotope effects are expected to be weighted averages of those for the competing paths. Primary effects are expected for heavy atoms. For the proton, isotope effects should be small if the H bond is of the ordinary type, large for the AHA⁻ type (see above). (2) Resonant-tunneling mechanism:[34–39] The "bottom" minimum on the MAR connects only to reactants, the "top" minimum connects only to the products. Reaction occurs by resonant tunneling of molecules from the "bottom" minimum into the "top" minimum. Resonant tunneling demands that the two minima be of very similar energy, and so the mechanism is restricted to this circumstance.

point in between. If the transition-state structure corresponds to this or a later point, then the transition-state energy will be reduced by a transfer of the proton coupled to HAR. The greatest problem[31] in reconciling this model with experimental data is that, while the analogous E2 elimination reactions commonly show quite large isotope effects (five to seven, for example) for the transferring proton, protolytically catalyzed reactions of the types involved in ester and amide hydrolysis rarely show effects larger than three.

The other two models shown in Figure 4-3 are for HAR as the sole reaction-coordinate motion, with the perpendicular coordinate therefore being that for proton transfer. This is the catalytic model[10,31] called "solvation catalysis." Since the PT coordinate here is not coupled into the reaction coordinate, the system B---H---O is a stable hydrogen bond in the simple sense. The two versions of the model correspond to a single-minimum hydrogen-bond potential function and a double-minimum hydrogen-bond potential function. The issues involved in which this sort of description is correct for hydrogen bonds under various circumstances have been reviewed recently by Hibbert.[32]

The single-minimum situation is that originally[31] envisioned for solvation catalysis. It is to be expected particularly if the two bases binding the proton are of nearly equal strength and if the distance across the B---H---O system is not large. The model is readily reconciled with the modest hydrogen isotope effects for protolytic catalysis, but is inconsistent with the interpretation of Cordes–Jencks coefficients in some cases.[33]

The double-minimum model, depending on the nature of the actual potential surface, can correspond to either of two situations as explained in Table 4-2. Whether resonant tunneling[34-36] enters the picture should depend on the length of the B---H---O system (shorter distances will favor tunneling) and on the "match" between energy levels in the two minima. One important characteristic of the resonant tunneling model is that it is capable of accounting for temperature-independent isotope effects of substantial magnitude.[37-39] This phenomenon will be observed if there is little isotope effect for reaching the state from which tunneling occurs; the isotope effect then arises wholly from the relative isotopic tunneling probabilities, which will be temperature independent. Temperature-independent isotope effects of the magnitude of two or more are otherwise very hard to explain.[40] The competing-reaction (semiclassical) version of the double-minimum model must account for isotope effects in the ordinary way and will generate the usual sort of temperature dependence.

There are possibilities for further models (those in which PT and HAR are uncoupled but the reaction coordinate consists wholly of PT are the most clearly evident). These models, although of probable general importance, are not required for our considerations here and will not be described.

D. Transition-State Stabilization in Charge-Relay Catalysis

The evolutionary conservation of Asp-102 (and its congener in other enzymes) indicates its mechanistic importance and, thus, the mechanistic importance of the charge-relay system in terms of rate acceleration and the corresponding savings to the organism in the costs of protein synthesis. There are several ways in which the charge-relay system could produce rate accelerations, and it is the choice among these that has stimulated debate over the past few years. Among the important acceleration schemes, now to be described in detail, are (1) two-proton protolytic catalysis with coupled PT and HAR, (2) two-proton protolytic catalysis with uncoupled PT and HAR, and (3) one-proton protolytic catalysis with electrostatic or conformational stabilization by Asp.

a. Two-Proton Protolytic Catalysis with Coupled PT and HAR

In this model, the two protons of the charge-relay system could undergo PT in concert with each other and with the HAR of the Ser–substrate system, thus producing the rate acceleration associated with an extraordinary form of protolytic catalysis.[13] The degree of coupling among the component events of this process could vary, but the concept is that some degree of coupling be present among HAR, PT(1) for the His–Ser proton bridge and PT(2) for the His–Asp proton bridge. This type of protolytic catalysis would be more effective than one-proton protolytic catalysis in both base-catalytic and acid-catalytic modes. In base catalysis, the removal of the effective positive charge associated with the proton to a greater distance would permit stronger bond formation by the nucleophilic electrons being liberated by proton abstraction. In acid catalysis, the negative charge neutralized by proton donation would be relayed a greater distance away, leading to smaller repulsions between this charge and other negative charges in the reacting system. In both cases, the transition-state energy would be reduced by two-proton, charge-relay protolytic catalysis compared to simple, one-proton protolytic catalysis. In this model for charge-relay catalysis, actual net proton transfer along both of the H bonds occurs. In the state just preceding the transition state, both protons are on one "side" (eg, Ser and His), and in the state just succeeding the transition state, both are on the other side (eg, Asp and His).

This model of charge-relay catalysis is reasonable for most of the steps in the enzymatic mechanism for amide hydrolysis in terms of one of the popular views of protolytic catalysis,[16] which holds that PT and HAR are substantially coupled for cases that obey Jencks' rule (that the pK of the catalyst lies between the pK's of reactant and product—see the subsection entitled "Coupling Modes of PT and HAR"). This rule will be obeyed for attack of Ser on substrate carbonyl, attack of water on acyl–enzyme carbonyl, and expulsion of Ser from adduct of water to acyl–enzyme carbonyl. It will not be obeyed for the expulsion of the amine leaving group from the tetrahedral adduct if ordinary pK's remain correct in the active site at this point; the amine nitrogen should initially be more basic than the His imidazole *or* the Asp carboxylate. PT is thus expected to occur rapidly, exergonically, and prior to HAR.[41]

The coupled, two-proton protolytic mechanism is predicted to give primary heavy-atom isotope effects (or secondary isotope effects diagnostic of HAR) and to give hydrogen isotope effects of a magnitude between 1.5 and 12 (typical primary hydrogen isotope effects) at each of the coupled protonic positions.[10]

b. Two-Proton Protolytic Catalysis with PT and HAR Uncoupled

In this model, HAR is considered to constitute the reaction-coordinate motion. The protolytic bridges are taken to be the type of strong, polarizable hydrogen bonds known to give substantial isotopic fractionation[32] (see the appendix for detail). There are several features of this model which distinguish it from the preceding model:

1. These protolytic bridges should form most easily if the two ligands of the bridging hydrogen have equal basicities;[32] the enzyme may have been evolutionarily developed toward adjusting these basicities to match in the transition state, perhaps through alterations of the geometric or electrostatic environment as the transition state is entered.

2. This type of protolytic bridging does *not* require that net transfer of the bridging proton occur.[31] Whether it transfers or not is simply a matter of the relative acidities of the hydrogen-bonding partners in the reactants and products as opposed to the transition state (see the appendix for detail). *This model is therefore consistent with a finding that the His is the protonated functional group (not Asp) in the resting state of the enzyme, the acyl enzyme, tetrahedral intermediates, etc.*

3. This type of protolytic bridging will be affected by compression or extension of the distances across the component hydrogen bonds of the catalytic triad (see Figure 4-3 and the accompanying discussion).[32] If this distance is short, single-minimum potentials will be favored. If it is longer, double-minimum potentials will be preferred. In the latter case, the reaction will proceed by one or both of two competing routes through the two minima or by tunneling from one minimum to the other. [A further modification would allow the parallel pathways to lead not to saddle points (ie, transition states) but into "blind alleys," one connected only to reactants, the other only to products, with reaction occurring by tunneling between these stable states.] These particular tunneling mechanisms are not transition state theoretical mechanisms and constitute a departure from current thinking about enzyme mechanisms.[34–39,42] While of great interest, these mechanisms will not be further considered in this article, since the questions involved go far beyond the realm of charge-relay catalysis.

4. This mechanism is predicted to give primary heavy-atom or substrate-secondary isotope effects diagnostic of HAR and hydrogen isotope effects of 1.5–5 (corresponding to isotope fractionations in short, strong hydrogen bonds[32,59]) at each of the protolytic bridges.[10,32]

c. One-Proton Protolytic Catalysis with Electrostatic or Conformational Stabilization by Asp

According to this model, the Asp of the catalytic triad does not engage in protolytic hydrogen bridging of either of the two types just considered. Only the bridge from His to other partners is considered to be a protolytically catalytic bridge. There are actually two versions of this model, depending on whether the protolytic bridge is considered to be a proton-transfer bridge (coupled to HAR) or a "solvation" bridge (not coupled to HAR). However, this distinction is less important than the other differences between this overall one-proton model and the preceding two-proton models. The evolutionary conservation of the Asp is interpreted in terms of another role. The two simplest of these are (1) electrostatic stabilization and (2) conformational stabilization.

1. Electrostatic stabilization by a negative Asp of the adjacent His partial-imidazolium ion, as the latter accepts a proton bridge, would reduce the energy of the transition state and thus contribute to catalysis. The electrostatic interaction would require no close approach of the Asp and the His and would generate no isotope effect at the protonic center between His and Asp. This stabilization differs from the electrostatic component of hydrogen-bonding stabilization in that it lacks the directional character and geometrical requirements of hydrogen-bond formation.

2. Asp could conformationally stabilize the His imidazole ring by weak hydrogen bonding, holding it in the best position to serve as a protolytic catalyst. This stabilization, if already present in the reactant state, would reduce the entropy requirement for formation of the transition state. Thus (by preventing the entropic relative stabilization of the stable states over the transition state that would occur if the His were freely rotating in the stable states but bound in one conformation in the transition state), the Asp would contribute to catalysis. If the stabilizing hydrogen bond were of the ordinary type, which generates very small isotope effects, this mechanism predicts no isotope effect at the Asp–His site. If the stabilizing hydrogen bond were of the strong, polarizable type that generates isotope effects, then an isotope-effect contribution from this site is expected. In that case, however, the hydrogen bond would itself contribute to catalysis, and the mechanism becomes indistinguishable from a two-proton protolytic model.

E. Catalytic Machinery in Reactant and Transition States

The investigation of the charge-relay system is an investigation of enzyme catalysis, and thus an investigation most centrally of events in enzymatic transition states. Information directly deriving from transition states can be obtained at present only from kinetic studies and from theoretical studies. Nevertheless, enzymes frequently have been the objects of crystallographic and spectroscopic investigations. Scientists conducting these investigations would like to draw conclusions about catalysis, which is the most significant feature of enzymes. They are placed in a difficult position; their data bear only on stable states of the enzymes, since only these are accessible to direct crystallographic and spectroscopic observation, while the most interesting conclusions concern exactly the states inaccessible to study by their techniques: the transition states.

In spite of this fact, valuable mechanistic information has been derived from studies of stable states of enzymes. This is because (1) the situation in stable states establishes the baseline from which catalysis begins, and (2) *reliable inference* about transition states from stable states may occasionally be possible. In the latter connection, for example, functional groups can be located in enzyme active sites by crystallography of stable forms of the enzyme. If these states sufficiently resemble states immediately preceding or succeeding transition states, then there is hope for a reliable inference that no extremely large alteration in the relative disposition of these groups occurs as the transition state is entered. Such inferences must be drawn with great care. As the methods of molecular modeling continue to develop, the extent to which such inferences are valid will be clarified.

A useful question can, however, be asked at this point: To what extent do we expect the molecular machinery of a catalyst to change in structure and function between reactant state and transition state? If the answer to this question was "very little," then crystallographic and spectroscopic studies of stable states of enzymes would generate fine and reliable detail about transition states. If the answer was "very much," then crystallographic and spectroscopic studies will give, at best, general leads but not fine detail.

The nineteenth century answer, "very little," was epitomized by the "lock and key" concept of Fischer. The passage of time has generally seen an increased concern with the mobility of enzymes, an important idea being Koshland's induced-fit hypothesis, leading up to the views summarized in a recent book called *The Fluctuating Enzyme*.[43] There seems to be very little doubt that the *potential* for considerable motion is present in enzyme structures, but is there any catalytic advantage in changes in catalyst structure between reactant and transition states?[44-46]

One advantage is *differential complementarity*. An enzyme needs to be sufficiently complementary to reactant-state structure so that the reactant molecule is not excluded from entering the active site, for example. In the transition state, the enzyme needs to be strongly complementary to the structure of the substrate-derived fragments. A simple way to attain this differential complementarity is to have the enzyme undergo a structural change between reactant and transition state. Indeed, structural changes in the enzyme can be coupled to structural changes in the substrate, so that each drives the other.

A second advantage is the generation of *differential properties*. The transition state may impose requirements for stabilization through interaction of enzyme groups with substrate-derived structures that are best met through enzyme reorganization accompanying the formation of the transition state. For example, consider nucleophilic attack by Ser —OH at substrate carbonyl such that, in the absence of protolytic catalysis, this —OH would have attained a pK of 10 in the transition state. Optimal interaction with the Asp–His grouping in a strong single- or double-minimum hydrogen bond (as discussed previously) is favored by the Asp–His also having an effective pK of 10 at the His nitrogen. This can readily be achieved by appropriate location of the Asp for stabilization of the HisH$^+$ either electrostatically or through hydrogen bridging, so that its pK rises from 7 to 10. However, if this change were to be brought about permanently in the reactant state, the Asp–His pair would already be protonated at pH 7–8 and the electron pair of the His nitrogen would not be available for bridging to the Ser —OH in the transition state. The best strategy is thus for the enzyme to maintain a sufficient distance in the reactant state between Asp and His so that the pK of the latter is unperturbed and it can be at least partially unprotonated between pH 7 and 8. Then, as the transition state is formed, the enzyme structure can change so as to bring the Asp and His together and to modulate the pK in exactly the optimal manner.

As a general matter, enzymes and other catalysts must achieve transition-state stabilization without reactant-state stabilization. A near-certain concomitant of this task is an alteration in structure as the enzyme and substrate cooperatively enter the transition state.

F. Some Interesting but Mechanistically Irrelevant Questions

Some problems are posed by the strong likelihood that enzyme structure and properties in reactant and transition states are not the same but differ, perhaps rather sharply, for reasons that are not only germane to catalysis but central to the fact of high catalytic activity. Table 4-3 lists a number of

TABLE 4-3.   Selected Questions to Which the Answers Are Intrinsically Interesting but Irrelevant to the Question of Charge-Relay Catalysis

Question: Where is the proton located, on the His-57 or the Asp-102, in the protonated, low pH forms of the native serine proteases?

Comment: The location of the proton in the free enzyme not only *may* be different from its position in the transition state (which is the relevant location in terms of the mechanism of catalysis) but detailed theories of catalysis indicate that it *must* be different.[44-46]

Question: Where is the proton located in the protonated form of the complexes of these enzymes with their natural polypeptide inhibitors?

Comment: These complexes are often thought to simulate features of various states along the reaction path. The complexes are extremely stable, however, and this shows prima facie that they are very poor simulators of the transition state. The enzyme conformation and substrate structure both are, in the transition state, at points of high energy. The probability is very low that these structures are simulated accurately in stable, isolable complexes. Indeed, there is reason to think that inhibitors may *avoid* simulation of transition-state interactions in order to avoid catalytic turnover.

Question: Where is the proton located in the phosphorylated or otherwise similarly derivatized forms of the enzymes?

Comment: The phosphorylated derivatives possess a polar, tetrahedral moiety bound to the active-site Ser and are, thus, analogs of the tetrahedral intermediates of the catalyzed reaction, bound to the enzyme. Good theories of catalysis again hold that stabilizing features present in the transition state will be avoided in stable states. The phosphorylated enzyme is then expected not to give accurate information about the transition state. There is a certain chance, however, that it could give useful information about nearby states (see Subsection D of Section 2).

Question: Where is the proton located in an enzyme complex with a transition-state analog?

Comment: If the analog were rather exacting, the information might be highly relevant indeed. It is here classified as irrelevant because of the extraordinarily low probability of constructing an analog of the transition state that is at once stable enough to study *and* a sufficiently accurate simulator of the subtle transition-state interactions required to draw the enzyme into its transition-state conformation.

questions that are both susceptible to experimental study and fascinating, sometimes challenging, to answer. They are all "direct" approaches which involve the study of entities isolable (eg, in crystalline form) or obtainable in high concentration in solution (eg, for spectroscopic study). The entities are thus materials of *high stability*. As an inevitable result of their high stability, these entities cannot be good simulators of the quintessentially *unstable* transition state. In the most fundamental sense, such studies are therefore *irrelevant to an understanding of transition-state interactions or events in catalysis*.

This is not to say that the studies are devoid of interest. They are often of great interest. The reactant-state characteristics of enzymes, by indicating what the enzyme is *not* doing in catalysis, illuminate the ways in which the enzyme is avoiding the loss of catalytic power through what would be a premature expression of catalytic interactions in the reactant state.

What is erroneous is a conclusion about transition-state interactions in catalysis, unless a powerful chain of inference vigorously supports the way reactant-state characteristics are to be translated into conclusions about transition states.

2. CRYSTALLOGRAPHIC AND NMR-SPECTROSCOPIC STUDIES

A. Results from Earlier Studies: the Situation in 1982

The charge-relay system is a crystallographic discovery, and crystallographic investigations have continued to play a role in thinking about charge-relay catalysis. A problem quickly recognized by crystallographers and others is that the situation in a crystal may not prevail in solution. Spectroscopic approaches, preeminently nuclear magnetic resonance (NMR), were thus applied to examine the status of serine proteases and their complexes in solution. In 1982, the crystallographer Steitz and the NMR spectroscopist Shulman combined forces to review the results from these two techniques up to that point. They wrote, "Our conclusions, subject to further test, are that the triad does not exist in the resting state of the enzyme because the serine–histidine bond is not formed. However, we suggest that this bond does form upon complexation with substrates allowing the triad to play an important role in catalysis."

Steitz and Shulman are led to this conclusion by the following arguments:

1. In the absence of substrate, the most careful crystallographic studies suggest that water molecules in the active sites of the serine proteases compete for hydrogen-bonding sites and interrupt the catalytic triad hydrogen-bond chain. This is a revision of the original crystallographic finding, based on more highly refined structures and reevaluation of the older data. The view that the absence of the charge-relay system in the resting, native enzyme excludes charge-relay catalysis is "not generally warranted since

. . . a Ser–His hydrogen bond is observed in the structure of trypsin–pancreatic trypsin inhibitor (PTI).'' The NMR evidence on the existence of the Ser–His hydrogen bond in the resting enzymes is ''ambiguous but does not disagree with the structural [ie, crystallographic] conclusion.'' The NMR evidence clearly indicates the existence of the Asp–His hydrogen bond: a proton in this position has a residence time greater than 3 ms, which is too long unless a hydrogen bond exists. Further, N-15 spectra show that the His–imidazole ring is in what would normally be the (slightly) less stable tautomeric form, ie, the N—H is on the Asp side of the ring. Presumably, the formation of the Asp–His hydrogen bond more than compensates for the unfavorable tautomerization. Titration of the native enzymes is shown by a variety of NMR probes to result in a transition of the His imidazole to imidazolium (charge zero to charge unity) at the pK of 6.9, so that no charge is being relayed to Asp.

2. The trypsin–trypsin inhibitor complexes are ''[g]ood analogs of a pretransition state Michaelis complex'' and exhibit a Ser–His hydrogen bond. Binding of the inhibitors appears to have excluded the active site water molecules, and it is presumed that this results in formation of the catalytic triad hydrogen-bond chain. The NMR results strongly suggest (on the basis of C_2—H proton resonance of the His imidazole) that the His is fully protonated in the PTI complex. This means the Ser —OH in this complex is ionized to alkoxide unless an extra proton is present (a possibility that is rendered somewhat unlikely by the small space seen crystallographically around the Ser oxygen).

3. MIP-trypsin [which bears on the Ser oxygen a negatively charged, tetrahedral monoisopropylphosphoryl (MIP) group and is, therefore, used as an analog of the tetrahedral intermediate of the enzymatic acylation or deacylation step] is found by neutron diffraction studies of the protonated material to have ''the proton between histidine 57 and aspartic acid 102 . . . firmly bound to the histidine nitrogen.'' The conclusion drawn is ''that whatever the role of Asp 102, it does not result in the transfer of a proton from histidine to the carboxylate group.'' The NMR studies of the low-pH, protonated form of these phosphorylated enzymes suggest that His–imidazolium hydrogen bonds to one of the phosphoryl oxygens, as would occur in the course of acid catalysis of leaving-group departure from a tetrahedral intermediate. Steitz and Shulman remark, ''On this basis, we see a role for the His–Asp section of the charge relay in providing the His-tetrahedral N—H proton with a larger electronic buffering system than that provided by the imidazole, alone.''

B. Some Recent Crystallographic Results: The Work of Tsukada and Blow[15]

Crystallographic studies of the charge-relay system since 1982 have generally had the effect of chipping away at the evidence that the charge-relay system is not intact in the free enzyme. The current picture indicates that the absence of the charge-relay system in a free, crystalline enzyme is the exception. Most crystalline serine proteases appear to possess both Ser–His and His–Asp hydrogen bonds in the active site. One example of the progress in this direction is the refinement at 1.68 Å resolution of the structure of α-chymotrypsin, accomplished by Tsukada and Blow.[15]

The previous structure, on which the charge-relay hypothesis had originally been based, was constructed from data on a tosylate-containing crystal, the tosylate having been added to assist in the isomorphous-replacement procedure required for the method of real-space refinement necessarily employed in 1972. The new method of "phase-free" refinement allowed Tsukada and Blow to omit the perturbing tosylate. This turned out to be critical for the charge-relay system. Tsukada and Blow write, "This refinement corrects the angle χ_1 (195) which fixes the position of O^γ of the active serine, and substantially alters the distance of this atom from $N^{\varepsilon 2}$ (57), the atom to which it is linked in the charge-relay system. The structure at the active site agrees closely with other members of the trypsin family whose structures have been refined at high resolution. . . ."

Tsukada and Blow comment on the prevailing impression that the charge-relay system is not present in free-enzyme crystals. Kraut's statement that the Ser–His hydrogen bond is distorted to the point where it is nonexistent or very weak, "which remains true for subtilisin but has been disproved for trypsin, S. griseus proteases A and B, and α-chymotrypsin by high-resolution refinements, is still widely quoted." The conclude their article, "The present results confirm that a strong hydrogen bond is formed in the free enzyme between His57 and Ser195, providing a pathway for the proposed proton transfer, in α-chymotrypsin, as well as in trypsin, S. griseus proteases A and B, and kallikrein. The structures of the various trypsin inhibitor complexes . . . emphasize that when a substrate or inhibitor is bound, the whole charge relay system becomes buried, and this will certainly enhance the polarizing influence of the system."

C. Some Recent NMR Results: The Work of Bachovchin[49]

Bachovchin has reviewed the crystallographic and spectroscopic evidence on the charge-relay system and provided important new evidence on the N-15 NMR spectroscopy of the native enzyme and the sulfonated (PMSF) and phosphorylated [diisopropyl fluorophosphate (DFP)] derivatives of α-lytic protease. This enzyme is of importance for NMR studies because it comes from a bacterial source (and can thus be cheaply produced with specific isotopic enrichments) and because it contains only a single His, that of the catalytic triad. Feeding of the bacterium with labeled His thus results in a single label of great interest. The significant new results of Bachovchin proceed from the establishment in model systems of the correct chemical shifts in His for (1) the non-proton-bearing (β) nitrogen [128 ppm upfield (positive shift) from external 1 M nitric acid in DOD], (2) for the neutral proton-bearing nitrogen (α, 210 ppm), and (3) for the imidazolium positive proton-bearing nitrogen (α^+, 201 ppm). Further useful results are the establishment of an 8–10 ppm upfield shift for a nitrogen accepting a hydrogen bond and an 8–10 ppm downfield shift for an N—H (neutral or positive) that donates a hydrogen bond. These latter shifts correspond to normal hydrogen bonds; unusual hydrogen bonds with greater amounts of charge transfer should give larger shifts.

Within the framework of these benchmarks, the results for the enzyme and its derivatives can be interpreted.

1. For the free enzyme at high pH (imidazole form), resonances are seen at 199 and 138 ppm. The resonance of 199 ppm is assigned to N—H donating a hydrogen bond to Asp (199 = 210 − 11 ppm). The resonance of 138 ppm is assigned to the β =N— of His accepting a hydrogen bond from Ser (138 = 128 + 10 ppm). The conclusion is that the charge-relay system is intact in the native enzyme in the unprotonated form.

2. For the free enzyme at low pH (imidazolium form), resonances are seen at 192 and 204 ppm. The 192-ppm resonance is assigned to the α^+ N—H donating a hydrogen bond to Asp (192 = 201 − 9 ppm). The 204-ppm resonance is taken as sufficiently near 201 ppm as simply to confirm that the imidazolium bears a full positive charge. The conclusion is that Asp does not become protonated but does hydrogen bond to the protonated His.

3. For the sulfonylated enzyme at high pH (imidazole form), resonances arise at 200 ppm and 128 ppm. The resonance at 200 pm (200 = 210 − 10 ppm) shows the Asp–His hydrogen bond to remain intact. The resonance at 128 ppm is the normal resonance for =N—, now that the Ser is sulfonylated and incapable of donating a hydrogen bond. The corresponding resonances in the DFP-derived enzyme are 206 and 133 ppm, respectively. Bachovchin remarks, "Essentially the same result [as for the sulfonylated enzyme] is obtained when diisopropyl fluorophosphate (DIFP) is used to phosphorylate Ser-195." He also notes that the signals are broader in the phosphorylated than in the sulfonylated enzyme. The conclusion is that in this tetrahedral-intermediate analog, the Asp–His bond remains intact and, as expected, there is no His–Ser hydrogen bond.

4. For both sulfonylated and phosphorylated enzymes, the observed resonances at low pH come at 198–199 ppm and 202 ppm. These are taken as equal essentially to the 201 ppm expected of both nitrogens of an imidazolium moiety not hydrogen bonded at all. The conclusion is that in the protonated form of these tetrahedral-intermediate analogs, the His–Ser hydrogen bond is, of course, absent, but in contrast to the neutral form of the enzyme, *the Asp–His hydrogen bond is also broken.*

Bachovchin's study thus shows that the entire charge-relay chain in the native enzyme is intact, but that the Asp is never protonated at low pH, neither in the free enzyme nor in the analogs of the tetrahedral intermediate. The Asp–His hydrogen bond is intact in all species, except the low-pH forms of the tetrahedral-intermediate analogs. These last are regarded by Bachovchin as simulators of the state immediately following serine addition to substrate carbonyl and immediately preceding leaving-group expulsion. He suggests that the formation of this state triggers disruption of the Asp–His hydrogen bond to allow the His-H$^+$ to function as a better proton donor from both geometrical and electronical viewpoints, in protonating the departing leaving group.

D. Mechanistic Evaluation of the Crystallographic and NMR Results

Mechanistically, the potentially most meaningful result in the crystallographic domain is the finding by Kossiakoff and Spencer[50] that the imidazole is the protonated group in the active site of enzymes bearing an analog of the tetrahedral intermediate. The two most potentially meaningful results in the spectroscopic domain are Bachovchin's findings (1) that the imidazole is also the protonated group in the solution form of these enzyme derivatives, and (2) that the protonated imidazole is not hydrogen bonded in the tetrahedral-intermediate analogs.

The finding that the His imidazole is protonated in these analogs can be interpreted in two ways:

1. It could be assumed that the enzyme species under observation corresponds exactly to the first-formed configuration following the transition state for generation of the tetrahedral intermediate. If this were so, then the presence of the proton on His means that it did not transfer to the Asp in the just-preceding transition state. This means that coupled PT and HAR cannot be the mechanism of charge-relay catalysis. It says nothing about whether any of the models for uncoupled PT and HAR are or are not the mechanism of charge-relay catalysis.

2. It could be assumed that the enzyme species under observation corresponds not to the first-formed configuration after the transition state, but rather to one or a mixture of the myriad forms to which this first-formed configuration doubtless relaxes after emergence from the transition state. This is by far the more likely interpretation. The question then becomes this: Is the position of the proton the same in this configuration as in the first-formed configuration? If the answer is yes, then the coupled PT/HAR version (but no other version) of charge-relay catalysis can be excluded; If the answer is no, the result contains no mechanistic information. At the present time, we have no way of answering this question. Unfortunately, "no" is favored probabilistically.

A similar set of formulations can be set up for the mechanistic significance of the demonstration of a non-hydrogen-bonded imidazolium ring. It may very well be that this observation is an indication for the "moving-imidazole" mechanism, one of two favorites of Satterthwaite and Jencks[41] among the six mechanistic possibilities they considered for the acylation stage of the enzymatic reaction. Whether it can be relied upon is a matter of how accurate the sulfonylated and phosphorylated enzymes are as analogs of the appropriate enzyme species on the reaction pathway. Again, we have no currently available method to decide the exact mechanistic significance of the experiment.

In summary, the crystallographic and NMR studies, while both intrinsically interesting and presenting a fascinating historical pattern of contrasts and reversals, do not offer any basis for evaluating whether charge-relay catalysis of any character (coupled or uncoupled PT/HAR, with or without net proton transfer) is an aspect of the function of the serine proteases.

3. CHEMICAL MODELS FOR CHARGE-RELAY CATALYSIS

It is expected that if a small-molecule array could be built which completely simulates the structure and dynamics of the charge-relay system of the serine proteases, then it would display the same reactivity in the hydrolysis of ester and amide substrates. Such a construct would be very valuable in that it would permit the detailed study of many features of the system which may be obscured by the complexity of the entire enzyme. On the other hand, there is substantial question as to whether a completely accurate model of this kind is in principle possible. If the entire protein structure, or some large part of it, is essential to the effective functionality of the catalytic center of the serine proteases, then an incremental approach to modeling the catalytic function (an approach followed by Bender[51] and Cram[52] and their co-workers) will end in the total synthesis of one or more of the serine proteases. Nevertheless, models which fall far short of mimicking every aspect of enzyme function can be very valuable in elucidating the inherent capacity of small-molecule fragments, *in the absence of the polypeptide framework*, to produce parts of the catalytic effect. Comparative studies, using different arrays of functional groups or different settings for them, can also indicate the degree to which the enzyme structural variability is limited by fundamental structural and mechanistic considerations.

There have been various investigations in the direction of modeling protease action, but two studies have been particularly informative about charge-relay catalysis. These are the study of Rogers and Bruice[53] and the series of investigations of Bender and co-workers,[51] which have led to the compounds shown in Figures 4-4 and 4-5. The fact that contrasting results have emerged from the different models developed by the two groups actually renders both studies more valuable.

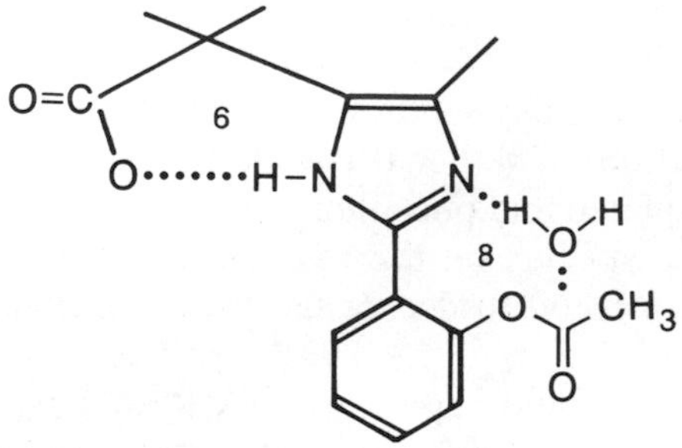

Figure 4-4. The charge-relay model of Rogers and Bruice. The ring sizes of 6 and 8 are shown.

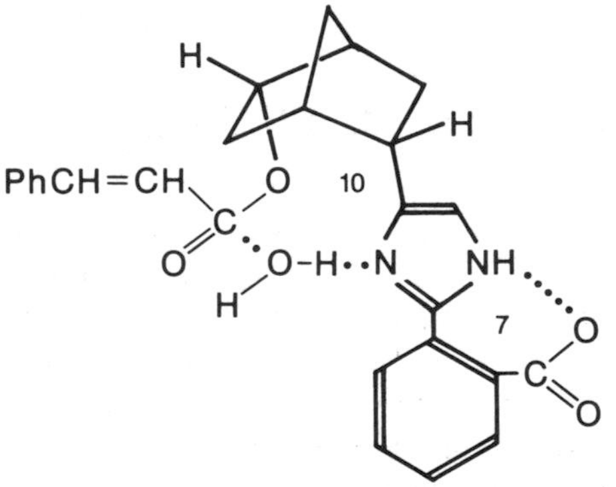

Figure 4-5. The charge-relay model of Bender and co-workers. The ring sizes of 7 and 10 are shown.

A. The 6 + 8 Model of Rogers and Bruice

The model constructed by Rogers and Bruice is notable for showing no sign of charge-relay function, although it appears to deacylate in a manner similar to the hydrolysis of an acylated serine protease. This model we denote a 6 + 8 *model* to describe the ring sizes involved if charge-relay catalysis of water attack at ester carbonyl were to occur.

In all small-molecule models of either stage of the hydrolytic reaction in serine protease active sites, two rings must be present in the transition state (see Figures 4-4 and 4-5). One part of the model must simulate the Asp–His interaction. A carboxylate group must interact with the imidazole N—H bond; in the model, the carboxylate and the imidazole will have to be attached to a connecting fragment of structure so that a ring is formed. The ring must contain the C—O⁻ grouping of the carboxylate (2 atoms) and the

=C—N—H grouping of the imidazole (3 atoms); the connecting fragment will then add more atoms to the ring. In the Rogers and Bruice model, the connecting fragment is a single sp^3 carbon atom, so the "Asp–His" ring is a 6-ring. The second part of the model must simulate His-catalyzed attack of water on the carbonyl of the acyl enzyme (or a geometrically equivalent breakdown of the tetrahedral intermediate with acid catalysis by imidazolium or Asp—H—His). The atoms necessarily present are the C—N= or C=N— bond of His (depending on the direction along which the ring is constructed; two atoms in either case), the H—O of water (two atoms) and the O—C of the ester (2 atoms). The "His–Ser" ring must therefore contain these 6 atoms together with those of the fragment which binds the Ser-like grouping to the His-like grouping. In the Rogers–Bruice model, the connecting fragment consists of the 2 sp^2 carbons of the phenyl ring, producing an 8-ring on the "His–Ser" side. The model is thus a 6 + 8 model.

The 6 + 8 model undergoes deacetylation with a pH-independent term in mildly acidic solution (imidazole protonated) with a rate constant of 1.7 × 10^{-6}/s. The rate increases according to a titration curve (pK 7.05, obviously the imidazole) as the solution is made more basic, yielding a pH-independent rate constant (imidazole as free base) of 4.6 × 10^{-4}/s. The imidazole is thus functioning catalytically. For the related compound not possessing the carboxylate side chain, the solvent isotope effect of 3.8–3.9 was taken to indicate the mechanism of catalysis to be protolytic base catalysis of water attack at the ester carbonyl. The carboxylate (anionic throughout the pH range under consideration) is held by the "trimethyl lock" in a position apparently correct for removal of the imidazole proton in charge-relay catalysis.

However, when the carboxylate is methylated, the magnitude of the pH-independent term at high pH (free imidazole), which should exhibit the effects of charge-relay catalysis, is very slightly reduced from 4.6 × 10^{-4}/s to 1.6 × 10^{-4}/s. This factor of 3 is correctly characterized by Rogers and Bruice: "From the standpoint of enzyme catalysis this is, of course, negligible."

From this elegant work, we may conclude that an array of Asp–His–Ser-like entities in this particular 6 + 8 configuration is incapable of yielding charge-relay function.

B. The 7 + 10 Model of Bender and Co-workers[54]

In this model, the arrangement of the functional groups is different. On the Asp–His side of the model, the transition state ring size will now be 7 atoms because the 5 necessary atoms are now joined by the 2 sp^2 atoms of the linking phenyl ring. On the His–Ser side, if water attack is rate limiting, the 6 necessary atoms are linked through a chain of 4 sp^3 atoms of the norbornyl ring (note that there is no ambiguity: the chain is 4 atoms in length around

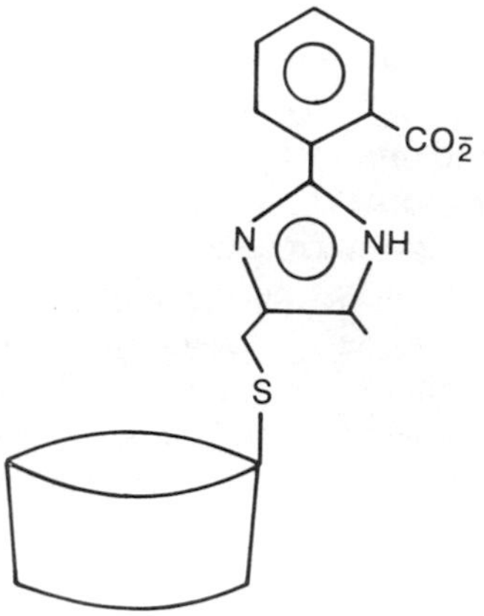

Figure 4-6. The β-cyclodextrin "artificial enzyme" of Bender and co-workers.

either side of the norbornyl ring). If leaving-group expulsion is rate limiting, then the His–Ser ring is an 8-membered ring. Thus, the model of Bender and co-workers is a 7 + 10 model, or if water attack is not rate limiting, then a 7 + 8 model. For simplicity, we use the first designation.

The mechanism of deacylation is again protolytic hydrolysis of the acyl function, as shown by a solvent isotope effect k_{HOH}/k_{DOD} of 3, whether or not the carboxylate is present. In the compound lacking the benzoate fragment, the rate constant for deacylation at 60° and pH 7.9 is 4.5×10^{-5}/s, while for the full model the corresponding rate constant is 690×10^{-5}/s, a value that is over 150-fold larger. This is an entirely credible estimate for the possible contribution of Asp to the protolytic catalysis of serine proteases. Further, the rate constant for the model compound is only 20-fold smaller than the deacylation rate constant for cinnamoyl–chymotrypsin, extrapolated to the same conditions.

In an even more impressive accomplishment, the Bender group has produced a cyclodextrin-based model catalyst which they justifiably call an "artificial enzyme" (Figure 4-6). Assuming that the hydroxyl group adjacent to the attachment of the sulfur is the one which is acylated then this model is a 7 + 9 model in deacylation and a 7 + 7 model for acylation. The model catalyzes the hydrolysis of m-t-butylphenyl acetate, exhibiting full turnover with burst kinetics (ie, a first-order release of m-t-butylphenol corresponding to acylation of the catalyst, followed by zero-order deacetylation), solvent isotope effect of 3, and k_{cat} and k_{cat}/K_m values scarcely different from those for chymotrypsin-catalyzed hydrolysis of p-nitrophenyl acetate.

C. Evaluation of the Chemical Models

These two valuable examples of chemical model approaches to charge-relay function exemplify the importance of posing the correct question to be answered by model studies of any kind, whether the models are experimental chemical models, theoretical models, crystallographic or spectroscopic models, etc. The question to ask of the model is not "Does charge-relay catalysis operate in the serine proteases?" The answer to this question must be either "No, or the model differs from the enzyme," or "Yes, or the model differs from the enzyme." Since the model obviously differs from the enzyme, nothing has been gained. Instead, the questions to ask of the model are "Does charge-relay catalysis operate in this model system? If so, why? If not, why not?" These questions can, in principle, be answered (the first readily, the second and third with vastly greater difficulty), and with great profit.

Comparing the 6 + 8 model with the 7 + 10 model shows that the former system is incompatible with charge-relay catalysis, the latter strongly compatible with charge-relay catalysis. It does not, of course, mean that the 7 + 10 model more closely approximates the enzymatic situation, nor does it mean that the ring sizes themselves are the cause of the difference.

The ring sizes were considered by Rogers and Bruice not to prevent charge-relay catalysis in the 6 + 8 model simply by excluding the nucleophilic water molecule or preventing hydrogen bonding between the Asp-like carboxylate and the His-like imidazole. The solvent isotope effect of 3.8–3.9 seemed to show that the nucleophilic water is present in the deacylation transition state. Rogers and Bruice provide spectroscopic evidence for the Asp–His-like hydrogen bond.

Among the reasons, consistent with these assumptions, that charge-relay may not be manifested in the 6 + 8 model are (1) the Asp–His-like hydrogen bond may be rendered very strong in the reactant state through compression engendered by the trimethyl lock; no increase in strength, then, would be observed on entering the transition state and, thus, no catalysis would occur; and (2) the 6 + 8 geometry may produce an unfavorable alignment of the two component hydrogen bonds of the relay chain, and this may prevent their coupling in the transition state.

A further complication is suggested by the magnitude of the solvent isotope effect, 3.8–3.9. This clearly controverts simple nucleophilic catalysis; Rogers and Bruice give arguments against acetyl migration to imidazole, followed by rate-limiting hydrolysis of the acylimidazole. However, the studies of one-proton protolytic catalysis by imidazole of water attack at ester carbonyl, carried out by Hogg and co-workers,[55] give values of 2.8 and 3.3. The larger value in the Rogers and Bruice system may simply indicate a different transition state structure for one-proton catalysis. If, for example, the imidazole is stabilizing an adjacent positive moiety as water, rather than imidazole itself, catalyzes water attack (an "immature hydronium ion" mechanism[56]), then a large, multisite isotope effect would be expected.[56] This mechanism might have been forced by the small ring size and would probably not be susceptible to remote acceleration by carboxylate.

A notable feature of both these models is that they violate Gandour's Rule[57] for protolytic base catalysis (preferred formation of the s-trans carboxylic acid) while the enzymic charge-relay system in fact obeys it (Figure 4-1). Gandour has pointed out that carboxyl groups prefer to exist in the s-trans configuration. Proton transfer to Asp in the serine protease active site produces this stable configuration. Proton transfer to the Asp-like carboxyl in either the 6 + 8 or 7 + 10 models produces the unstable s-cis configuration and is thus expected to be slower than corresponding reactions which would produce the stable conformer. If the 7 + 10 model were reconfigured to permit Gandour's Rule to govern, it might be even more active as a catalyst than chymotrypsin.

4. THEORETICAL MODELS FOR CHARGE-RELAY CATALYSIS

In principle, theoretical approaches to the exact function of the charge-relay system are the most powerful possible. In no other way can one get a clear picture of both the structural and dynamical features of the transition states along the enzymatic reaction pathway. Unfortunately, at the present point in history the gap between principle and practice remains very great. Nevertheless, informative beginnings have been made in theoretical treatments of the charge-relay system. Selected examples are outlined in Table 4-4 and will be discussed here.

A. The EVB/SCSSD/LDPD Approach

The most ambitious attempt on the problem has been reported recently by Warshel and Russell[60] and is based on substantial earlier work from the same group. An unusual feature is that, because this group wanted to develop an account of catalysis, not only is the enzyme-catalyzed reaction treated but an uncatalyzed, or standard, reaction is also studied.

As in almost all approaches, the problem is simplified by dividing the overall system into two parts: the core (in which bond-making and bond-breaking processes are at work) and the surroundings. The core is treated quantum mechanically, and the effect of the surroundings is introduced in a manner not requiring as much computation as a quantum-mechanical method.

The quantum-mechanical method used is the "empirical valence bond" or EVB technique developed by this group.[61] For each configuration of the reacting system, an array of resonance-contributing structures is set up, such that ionic and covalent contributions are considered for each bond that will be made or broken. The energy, including bonding, nonbonding, and strain terms, corresponding to each of these contributors is estimated (partly with the use of empirical data). The bonding terms, for example, are set

TABLE 4-4. Selected Theoretical Studies of Charge-Relay Catalysis[a]

Authors	System studied	Techniques employed	Results and conclusions
Warshel and Russell, 1986	Catalyzed reaction: active Ser in trypsin active site (X-ray coordinates) on $>C=O$ as substrate; 22 water molecules included. Standard reaction: HCO_2H, CH_3OH, imidazole, and dimethylformamide reacting in water cage	EVB calculation of PE for gas phase reaction, incorporation of ΔG for environmental interaction by SCSSD method for aqueous media and PDLLD method for active site	"The enzyme stabilizes the ionic resonance form ($A^- $ Im-H$^+$ t$^-$) (where t$^-$ designates the negatively charged tetrahedral intermediate) by about 5 kcal/mol more than water does; this results in a reduction of the activation energy in the enzyme $\Delta G_{cat}^{\ddagger}$ by ~5 kcal/mol relative to the corresponding activation energy in a solvent cage, $\Delta G_{cage}^{\ddagger}$." "The charge-relay mechanism is not an important factor"
Stamato and co-workers, 1986; Longo and co-workers, 1985; Stamato and co-workers, 1984	HCO_2^-, imidazole, CH_3OH at α-chymotrypsin X-ray coordinates for Asp, His, Ser, and a bridging water molecule, methyl acetate as substrate with X-ray coordinates from AcTrpOMe: α-chymotrypsin complex, protein surroundings represented by array of Poland-Scheraga atomic charges; twelve-step mechanism for AcOMe $\rightarrow$ MeOH + AcOH	At each step, deformations introduced with standard geometries, then optimized with REFINE; three-step ISCRF calculation: (1) in vacuum, with protein surroundings omitted; (2) permanent protein charge distribution; (3) reaction-field polarization	"The reactive hydroxyl protein proton of Ser-195 is transferred to the Nε_2 atom of His-57, thus originating the multicharged structure Asp$^-$-His$^+$-Ser$^-$." ". . . our model seems to have incorporated the most relevant aspects of the problem"
Dewar and Storch, 1985	$HCO_2^- + HOCH_3 + CH_3NH-CH=O \rightarrow CH_3NH_2 + HCO_2^- + CH_3OCHO$ $(+H_2O) \rightarrow HCO_2^- + CH_3OH + HOCHO$ "Since the intervening imidazole ring remains virtually unchanged during the charge relay, it can be ignored"	AM1 "because it alone reproduces hydrogen bonds in a reasonably satisfactory manner"	"The relevant anions are generated by proton transfer from HX to the carboxylate group of Asp-102 via the imidazole ring of His-57 ('charge-relay mechanism')"

(continued)

TABLE 4-4. Cont.

Authors	System studied	Techniques employed	Results and conclusions
Nakagawa and Umeyama, 1984	$-CO_2^-$ is introduced near HIm in reactant and "transition state": $H-Im$ $HOCH_3$ $NH_2CH=O \rightarrow$ $H-ImH^+$ $CH_3O-CH{\Large\langle}^{O^-}_{NH_2}$	4-31G basis set, ab initio calculations of "quantum mechanical region." In related calculations, a point-charge model of the surrounding "classical region" of protein structure was used	The 22-kcal/mol reduction in activation energy by $-CO_2^-$ was chiefly electrostatic (24 kcal/mol). ". . . the 'charge-relay' mechanism is denied"
Kollman and Hayes, 1981	all at X-ray coordinates of trypsin/PTI complex (reactant), MIP-trypsin ("transition state") HCO_2^- HIm $HOCH_3$ A B C (linear H bonds)	Optimized structure with STO-3G, point calculations with 4-31G with corrections for minimum basis-set error, for AB, BC, and AC; then, nonadditivity correction for ABC applied	HCO_2H ImH $^-OCH_3$ (31.9 kcal/mol) less stable than HCO_2^- $HImH^+$ $^-OCH_3$ (25.3 kcal/mol). "We conclude that Asp 102 . . . is likely to stay unprotonated during catalysis"
Allen, 1981	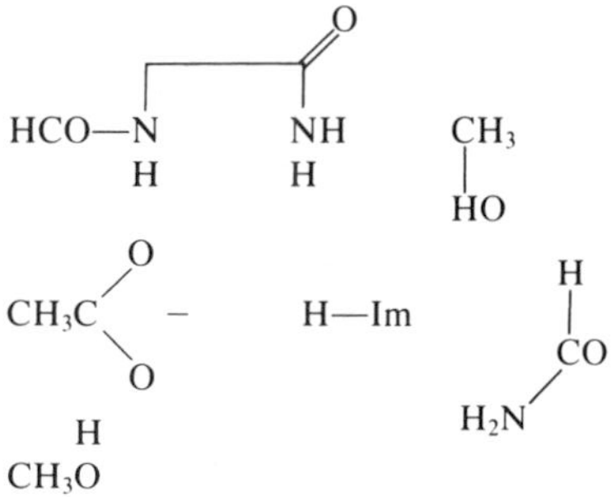 (all at chymotrypsin X-ray coordinates)	4-31G with Boys–Bernardi counterpoise correction for active site; point-charge model for protein surroundings; calculation of proton affinities of Asp, His in absence of substrate	". . . a one-proton transfer process; . . . neither a 'charge relay' nor a 'proton relay' exists in the serine proteases"

Kitayama and Fukutome, 1976	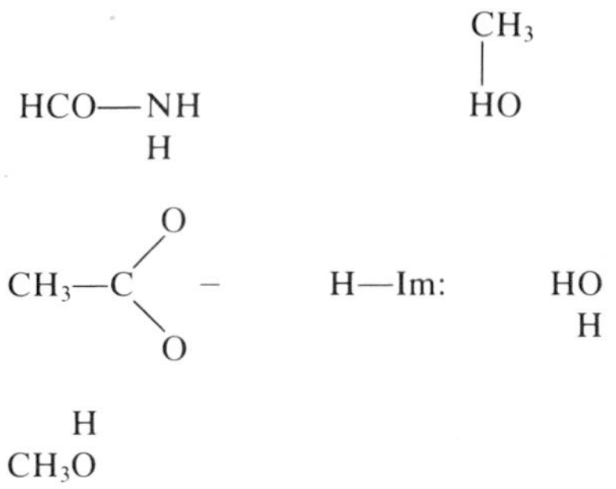 HCO—NH H CH₃ — HO H—Im: CH₃—C with O, –O, and H/CH₃O; HO—H (small molecule X-ray coordinates with relative locations at α-chymotrypsin X-ray coordinates)	CNDO/2 semiempirical calculations establish a linear relation of E for proton transfer to HO^- with experimental pK_a of acids; apply this to active site functional groups	"The neutral state is lower in energy than the zwitter ionic state . . . which leads to coupled proton transfer"
Scheiner, Kleier, and Lipscomb, 1976, 1975	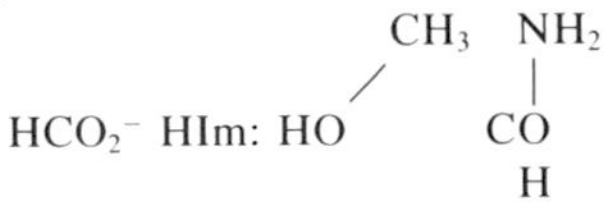 HCO_2^- HIm: HO — CH₃, NH₂, CO, H	PRDDO with minimum basis-set corrections; potential energy surface for formation and breakdown of tetrahedral intermediate	Proton is relayed to HCO_2^- in all cases but relay is synchronous in formation of tetrahedral intermediate and stepwise in breakdown
Gandour, Maggiora, and Schowen, 1974	X H—Y H—X |←R→|←R→| X = H₂O, NH₃ Y = F, OH	INDO	Proton transfers uncoupled for $R = 3.0$ Å, coupled for $R = 2.75$ Å, single-minimum potential for 2.5 Å

equal to Morse functions based on measured bond energies and force constants. Next, the Schrödinger equation is solved to obtain the energy of the actual resonance hybrid in terms of the energy formulas for the contributors. Up to this point, the calculation includes only the reacting structures (in drastically truncated form, since the energy formulas take into account only reacting bonds and immediately surrounding atoms) and, at best, accounts for the situation in the gas phase. However, with this approach the potential-energy surface for the gas-phase reaction can be constructed.

To account for environmental effects, the environmental-interaction ("solvation") energies of the resonance-contributing forms are estimated, using one approach for the reaction in aqueous solution and a different approach for the reaction in the protein environment. Both approaches are microscopic–dielectric modeling methods.

For aqueous solution, the surface constrained soft sphere dipoles (SCSSD) method is used, "which represents the water molecules as point dipoles attached to the centers of soft spheres and minimizes the solute–solvent and solvent–solvent energies with respect to the orientations and positions of these dipoles."[62] The authors calibrated this method to reproduce the solvation enthalpies of ions at 300 K. Here the method is employed to estimate the solvation energy of a given resonance-contributing form.

For the protein environment, the protein dipoles–Langevin dipoles (PDLD) approach is taken: the active-site environment is represented by several residues of the protein structure at their crystallographic coordinates and an array of about 22 water molecules. The nearest water molecules are treated as "real," ie, they are ascribed atomic charges and van der Waals parameters. More distant water molecules are represented as time-averaged point dipoles ("Langevin dipoles"), and both permanent dipoles and dipoles induced in the protein structure are calculated. Then the interaction energy between the environment and a given resonance form is estimated. This method was tested by using it to calculate successfully the pK's of active site functional groups.

At this point, environmental effects had been calculated only for the contributing structures, not for the actual resonance hybrid. Now the Schrödinger equation is again solved, but all terms are not corrected for the effect of environment: only the energies of the individual contributors (diagonal elements of the Hamiltonian) are corrected, while the terms that describe the mixing of the resonance forms (off-diagonal elements of the Hamiltonian) are assumed to retain their gas-phase values. This approximate solution to the problem is used to calculate parts of the potential-energy surface for the reaction occurring either in aqueous solution or in the enzyme active site. In proceeding to compare the results with experiment, the difference between potential energy and Gibbs free energy is neglected except insofar as it was empirically introduced in the "calibration" of the techniques.

The first important initial result reported by Warshel and Russell is the free energy of activation (about 28–29 kcal/mol) for the reference (standard or "uncatalyzed") reaction, the reaction of "formic acid, imidazole, methanol, and dimethylamide [a misprint for dimethylformamide] in a solvent cage." This calculation does not disagree strongly with an estimate of 25 kcal/mol made from experimental observations. This estimate was made by the authors as follows:

1. A value of 3×10^{-8}/s for "the pseudo-first-order rate constant . . . for 1 M OH^-" for the hydrolysis of urea was taken from an article by K. R. Lynn (*J. Phys. Chem.* **1965,** *69,* 687).
2. Multiplication by 55 "which guarantees the presence of the OH^- in the solvent cage of the amide" yields 1.5×10^{-6}. From this value, ΔG^* was calculated to be 26 kcal/mol.
3. Heats of formation and relative pK's were used to estimate that the free energy of formation of the tetrahedral adduct from reactants should be more favorable for CH_3O^- than for HO^- by 14 kcal/mol. This same quantity was assumed to apply to the free energies of activation, so the estimated ΔG^* is reduced: $26 - 14 = 12$ kcal/mol.
4. Since the initial reactants in the cage are assumed to be imidazole, methanol, and the amide, the free energy of proton transfer from methanol to imidazole is estimated as 13 kcal/mol and is added to the estimate of 12 to obtain a final estimate of 25 kcal/mol.

Some elements in this procedure are unrealistic, although it should certainly be recognized that good data for a proper estimate are difficult to locate and that the estimation is hard to make. At least the following criticisms would, however, seem to apply:

1. The assumption is introduced from the beginning, and was apparently confirmed by the EVB/SCSSD calculation, that the mechanism of imidazole-catalyzed addition of methanol to amide carbonyl is rapid prior proton transfer, followed by reaction of the alkoxide ion. The expectation on experimental grounds is that protolytic bridging from imidazole to oxygen will occur in the transition state.[41] Thus, the mechanism calculated for the standard reaction within the solvent cage is unlikely to be correct. There is nothing improper about *arbitrarily choosing* a hypothetical, unrealistic mechanism for the standard reaction, but if, as seems to be the case, the calculational method has independently suggested that this mechanism is physically correct, then it is in disagreement with experiment.
2. The value of 14 kcal/mol for both the kinetic and equilibrium advantage of methoxide over hydroxide in carbonyl addition is too large. Alcohols do form adducts more readily at carbonyl than does water but with an advantage[63] of only about 10-fold (1.4 kcal/mol), while alkoxide nucleophilicities toward ester carbonyl are typically around 50-fold that of hydroxide ion,[64] not 10^{10}-fold. Attack at amide carbonyl might show a somewhat greater

selectivity, but an error of well over 10^6-fold (8–9 kcal/mol) is likely to be involved here.

3. Urea is not a good model for dimethylformamide. Correcting the Lynn second-order rate constant for urea by a more realistic factor of 10^3 for changing hydroxide to methoxide gives a value of $3 \times 10^{-5}/M/s$ while Fersht[65] gives $3.4 \times 10^{-1}/M/s$ for methoxide attack on N-formylmorpholine, an error of around 10^4-fold (5–6 kcal/mol).

These considerations suggest that the estimate of the apparent experimental activation energy for comparison with the calculated activation energy in the standard reaction may be in error by a substantial amount and that the calculation has generated an incorrect reaction mechanism.

The second important finding reported by Warshel and Russell is the free energy of activation for trypsin-catalyzed attack of serine hydroxyl on amide carbonyl, a value around 21–22 kcal/mol or 7 kcal/mol smaller than ΔG^* for the reference reaction. This was felt to agree well with the fact that the "rate constant for amide hydrolysis in the trypsin active site, k_{cat}, is about 7 orders of magnitude larger than the rate constant for the reference reaction." A rate acceleration of 7 orders of magnitude would produce a 9–10 kcal/mol difference between what Warshel and Russell call ΔG^*_{cage} and ΔG^*_{cat}.

The following criticisms may be made:

1. As in the standard reaction (Figures 5 and 6 of Reference 60), the reaction coordinate for the enzymatic reaction in the rate-limiting transition state (Figure 6 of Reference 60) is calculated to be purely HAR ("displacement of the serine oxygen toward the peptide carbonyl"), with the PT component complete well before the rate-limiting transition state. This mechanism corresponds to complete ionization of the Ser–His pair before nucleophilic attack. We consider that this is inconsistent with isotope-effect measurements (as shown later) that indicate protonic bridging in the transition state for enzymatic reaction. Professor Warshel has kindly informed me that he has repeatedly found the same energetic results for concerted and nonconcerted mechanisms of this type.

2. The calculated ΔG^* of 21–22 kcal/mol corresponds to a value of k_{cat} of about 0.4–$2 \times 10^{-3}/s$ for dimethylformamide. While this may be considered an encouraging value for further work, it is not a basis for detailed mechanistic conclusions because experimental evidence (summarized in Reference 66) indicates that the specific side chain interactions of substrate and enzyme, some at sites remote from the scissile residue, which produce variations in k_{cat} also produce variations in catalytic mechanism. In particular, these interactions are thought to lead to activation of the charge-relay system. Since these interactions are neglected in the calculation, the calculation cannot be expected to give the correct, detailed catalytic mechanism. Warshel and Russell have striven to avoid this error by locating the small substrate in the correct position for good substrates, but they still fail to simulate critical remote-subsite contributions.

3. The catalytic-acceleration equivalent of 7 kcal/mol vs 9–10 kcal/mol can be considered a most encouraging and impressive achievement in terms of pursuing this approach. However, to draw detailed mechanistic conclusions on the basis of this agreement is, in view of the disputable features of both the nonenzymatic and enzymatic models, inadvisable.

The authors have made an effort to establish the reliability of their approach and reach the judgment that "the present approach has an error limit of about 5 kcal/mol for the *absolute* [their emphasis] value of the catalytic free energy." It is first of all difficult to understand how the calculation of any free-energy *difference* can be this reliable: Warshel and Russell's tables cite many calculated quantities, all carrying error estimates of 2–10 kcal/mol, with the most common value being around 5 kcal/mol. To an experimentalist, it seems that differences between quantities having uncertainties of 5 kcal/mol will themselves have uncertainties larger than 5 kcal/mol. It is furthermore not clear whether "catalytic free energy" in their statement quoted above means ΔG_{cat}^{*} or $(\Delta G_{cat}^{*} - \Delta G_{cage}^{*})$. If the latter is meant and if we accept the error estimate of 5 kcal/mol, then Warshel and Russell estimate the free-energy equivalent of catalytic acceleration as 7 ± 5 kcal/mol or a catalytic-acceleration factor of $10^{5\pm4}$. If the former is meant, the error should propagate to $5 \times (2)^{1/2} = 7$ kcal/mol, so that the catalytic-acceleration factor has been estimated at $10^{5\pm5}$. Finally, if we return to the tabulated results and take the reasonable view that the energy of any state can be estimated to about ±5 kcal/mol, then both the reference and catalytic ΔG^{*} values should be estimated to about ±7 kcal/mol and their difference to about $7 \times (2)^{1/2} = \pm10$ kcal/mol. The free-energy equivalent of the catalytic acceleration would then be calculated as 7 ± 10 kcal/mol and the catalytic-acceleration factor as $10^{5\pm7}$.

Thus, the overall evaluation of the EVB/SCSSD/PDLD approach indicates that it can calculate that the trypsin catalytic-acceleration factor is not distant from the experimental value. This is encouraging for further work, but certainly no detailed mechanistic conclusions can be drawn. The citations in the table, including the conclusion that the charge-relay mechanism is not an important factor, can hardly be supported by the calculations. This is especially true in view of the several different versions of charge-relay catalysis which could be at work. Not all were considered in the calculations.

B. The ISCRF Approach[67–69]

These authors also have an approach (inhomogeneous self-consistent reaction field) that may eventually yield mechanistic information. Its spirit and methodology are not extremely different from those of Warshel and his collaborators (the quantum-mechanical method used for the reacting system is more traditional but still semiempirical, a complete-neglect-of-overlap-

(CNDO)-based technique). The calculation includes the catalytic triad and finds His as proton acceptor rather than Asp, thus excluding overall charge relay as a mechanism. This is not claimed as a finding by the authors, however, who did not explore alternatives: "Since the H-bond between serine and imidazole is extremely deformed [ie, in the X-ray coordinates they used], no attempt is made to establish a proton pathway. In a dynamical view, the position fluctuation of these groups may allow for favorable orientation for proton translocation." Until the exploration of various roles for the catalytic triad is included in the calculations, this approach cannot be considered to have provided detailed information on charge-relay or other aspects of acid–base catalysis.

C. Alternative View of Enzyme Reactions[70]

In an article of this title published in 1985, Dewar and Storch[70] consider a number of issues in enzyme catalysis, including charge-relay catalysis. They conclude, in contrast to the authors already cited, that the system functions to relay the proton. A somewhat startling feature of the calculation is that, for the reason included in the table, the imidazole ring is omitted from the calculation! The idea that, since it begins and ends in the same state, the imidazole can be ignored is one of those propositions that might be classified as possessing "dangerous verisimilitude" (cf. D. R. Hofstadter and D. C. Dennett, *The Mind's I*, Basic Books, New York, 1981, p. 403: "There is a famous puzzle in mathematics and physics courses. It asks, 'Why does a mirror reverse left and right, but not up and down?' It gives many people pause for thought . . ."). The difficulty is that much of the argument has been about *whether* the imidazole does begin and end in the same state or, if it does, what it is doing in the middle. The AM1 calculation thus might be taken to suggest that formate-ion general-base catalysis of *N*-methylformamide methanolysis is feasible but not that the charge-relay mechanism is functional in the action of serine proteases.

D. Ab-Initio-Plus-Classical Approaches

Three groups of authors in the 1980s have employed variations of the same general idea, one which is similar to the approach used by Warshel and Russell and treated in detail previously (although no other group has treated environmental effects so thoroughly). Umeyama and co-workers,[71] Kollman and Hayes,[72] and Allen[73] divide the enzyme–substrate complex into a reacting array of structure, which they treat by ab initio quantum mechanics, and a surrounding region, which they treat by classical models. The approach is quite promising in terms of eventually producing useful information, but the conclusion drawn by all three sets of authors (that the charge-relay system does not function in catalysis) is not, at this point, warranted by the calculation models. In all cases, the conclusion rests on the greater stability of a system having the proton on His than on Asp. If this were a firm conclusion, it would exclude the version of charge-relay catalysis which leads to overall shift of the proton to Asp, but not other versions. In none of the calculations was the transition-state structure sufficiently well characterized to give detail on protolytic bridging per se.

Nakagawa and Umeyama, as did Allen, employed a static enzymatic framework to treat the effect of the enzyme structure. This further neglects the fact that reorganization of the enzyme structure in the course of catalysis may change the result of the calculation. Of course, to introduce such reorganization at the present development of these theoretical methods is not a realistic expectation. However, the conclusion which is permitted is that *if the enzyme is a static entity, frozen at its X-ray coordinates*, then charge-relay catalysis is not indicated. Nothing can be said about the highly likely situation that the enzyme is mobile. Kollman and Hayes, although they used the dual approach in other problems addressed in their paper, in fact omitted all consideration of the enzyme structure in their treatment of the charge-relay question. In this case, then, the permissible conclusion is that *if the enzyme structure is unimportant*, charge-relay catalysis will not occur. A fortiori, no conclusion about the enzyme mechanism is justified.

E. Simple Semiempirical Approaches

The three last entries in the table correspond to approaches using semiempirical methods of treating active-site reactions or related questions without representations of the enzyme structure being included at all. The only conclusions permissible thus correspond to the circumstance that the enzyme structure has no effect on the events in question. The study of Kitayama and Fukutome[74] models proton transfer in a charge-relay system and leads to the conclusion that the proton is relayed to the carboxylate. The conclusion may be questioned on the grounds of the difficulty many semiempirical methods have in describing hydrogen-bonding systems.[75] The study of Scheiner and Lipscomb[76] attempts to model protolytic catalysis of amide hydrolysis across a charge-relay chain. The treatment of the corrections for the small basis set were later followed by Kollman and Hayes, who reached the opposite conclusion about charge relay; Kollman and Hayes merely dealt with motion of the proton, however, while Scheiner and Lipscomb include the reacting substrate system. In neither case was the enzyme surrounding region and its influence included.

The final study of Gandour et al.[77] did not attempt to model enzyme reactions but only the general behavior of hydrogen-bond chains as a function of geometry. The salient finding is that shortening of the distance across a chain of hydrogen bonds leads to a coupling of their motions and, eventually, to a single-minimum structure in which all protons are in the centers of their respective H bonds. The quantum-mechanical study can be criticized because of problems of this technique in the treatment of hydrogen bonds.[75] The conclusion, however, seems reasonable on simpler grounds and may be correct. It can readily be reached by the consideration of simple overlapping Morse potentials for the individual hydrogen bonds. The utility of the study is that it suggests that an enzyme might modulate the mechanism of proton motion, and the degree of coupling of proton motions, along a chain of hydrogen bonds (such as the charge-relay system) by altering the distances across the component hydrogen bonds. This could readily be accomplished through substrate-induced changes in detailed enzyme conformation.

F. Some Other Studies

A number of theoretical investigations have been omitted from detailed treatment. Three in particular may be mentioned briefly here. Ressler[78] applied the qualitative concept of Fajans charge configurations to the question of the charge-relay system. He reached the conclusion that enzyme motions might indeed modulate coupling of the proton motions in the system. Naray-Szabo[79] and co-workers have conducted a series of investigations on serine-protease problems which, while not definitive for charge-relay function, offer promise for the future. Warshel's molecular dynamics simulation (Warshel, A. *Proc. Natl. Acad. Sci. USA*, **1984**, *81*, 444) demonstrates how to remove the problems associated with static enzyme-structural models.

G. Evaluation of Theoretical Models of Charge-Relay Catalysis

Theoretical approaches offer the best entry to the solution of the charge-relay problem: the actual depiction of the protolytic interactions in the transition state. The state of advancement of current methods is, however, incapable of treating enzymatic reactions with the degree of detail and accuracy needed to solve the problem. Current approaches, for example, have frequently treated the enzyme structure statically and quite locally and have not achieved high levels of demonstrated accuracy. Dynamic studies are now appearing, and more powerful and sophisticated calculations can be anticipated.

5. PROTON-INVENTORY STUDIES

A. The Proton-Inventory Technique[56,80–83]

Whether charge-relay catalysis functions in serine hydrolase action and, if so, how it functions are questions of the nature of protonic interactions in the enzymic transition state. This is an ideal problem for investigation by deuterium isotope effects, and since the hydrogens in question exchange rapidly on the scale of minutes,[47] the sites of the charge-relay system are labeled by immersion of the enzyme in deuterium oxide. Partial labeling is achieved in mixtures of HOH and DOD (the deuterium is actually randomly distributed in these mixtures so that an equimolar mixture of HOH and DOD contains 25% each of HOH and DOD and 50% DOH). The rate as a function of the atom fraction of deuterium, n, in binary mixtures of HOH and DOD is called a *proton inventory*.

Experimentally, use of the proton-inventory method requires the measurement of a rate constant, k, in mixtures of HOH and DOD having n ranging from 0 to 1. For reactions in which one and only one step determines the rate, the shape of the function $k(n)$ can be interpreted and in terms of (1) the number of sites which contribute to the isotope effect and (2) the isotope effect generated at each site. In enzymatic reactions, particularly, more than one step frequently contributes to limiting the rate. Then the interpretation must take this into account.

For serine hydrolases, two rate constants can be determined: k_{cat} and k_{cat}/K_m. For k_{cat}/K_m, the initial state is the free enzyme and free substrate, and the final state is the transition state for acylation of the enzyme. For k_{cat}, the initial state and final state depend on the nature of the substrate. For substrates which have the acylation of the enzyme as the rate-determining step, the initial state is the noncovalent enzyme–substrate complex, ES, and the final state is the transition state for acylation. For substrates which have the deacylation stage rate limiting, the initial state is the acyl enzyme and the final state is the transition state for deacylation.

If a single step limits the rate for a parameter k, then the function $k(n)$ is given by $k(n) = TSC(n)/RSC(n)$ where $TSC(n)$ is the transition-state contribution, the contribution of the final state for that parameter, and $RSC(n)$ is the reactant-state contribution, the contribution of the initial state for that parameter. Both $RSC(n)$ and $TSC(n)$ are functions of the equilibrium isotope effects K_D/K_H, called fractionation factors, ϕ, for each site in the reactant-state or transition-state structure, respectively. The fractionation factor for each site in the reactant or transition state measures the equilibrium isotope effect for conversion of a bulk-water average site into that particular site. If, as seems commonly to be the case for serine hydrolases, the various initial states (free enzyme and substrate, ES, acyl enzyme) have, as mechanistically significant sites, only those sufficiently similar to an average bulk-water site that $\phi = 1$, then $RSC(n) = 1$. (For most free OH and NH sites, as well as those engaged in ordinary hydrogen bonding, $\phi = 1$). When $RSC(n) = 1$, the proton inventory probes the transition state directly: $k(n) = TSC(n)$.

$TSC(n)$ will have a factor $(1 - n + n\phi)$ for each site in the transition state. Clearly sites with $\phi = 1$ will give no isotope effect. The three common situations of interest here are

1. $k(n) = 1 - n + n\phi$ (linear proton inventory, one site generates the isotope effect, "one-proton catalysis"),
2. $k(n) = (1 - n + n\phi_1)(1 - n + n\phi_2)$ (quadratic proton inventory, two sites generate isotope effects, "two-proton catalysis"), and
3. $k(n) = Z^n$ (Z a constant; exponential proton inventory, many sites generate isotope effects, "generalized solvation change").

These equations are for simplified models and often the experimental distinction between two-proton catalysis or three-proton catalysis cannot be drawn with confidence. However, if the overall solvent isotope effect is larger than about 1.5–2, and if data of reasonable precision can be obtained, then it is commonly possible to distinguish one-proton catalysis (linear function), catalysis by two or a few protons (roughly quadratic function), and generalized solvation effects (roughly exponential dependence).

In the investigation of the charge-relay question, $k_{cat}(n)$ can be used to probe events in either acylation or deacylation transition states (depending on the choice of substrate), and $k_{cat}/K_m(n)$ can be used to probe events in the acylation transition states *if* the transition states for the actual chemical events, as opposed to processes such as binding, conformation changes or product release, totally determine the rates.

B. Substrate Recruitment of Multiproton Catalysis

In a review of the proton-inventory technique, published in 1984, Venkatasubban and Schowen[83] reviewed the evidence, from this kind of experiment, that enzyme–substrate interactions in the serine hydrolases, if sufficiently extensive, activate or *recruit* a catalytic entity of the enzyme which exhibits multiproton catalysis (ie, $k(n)$ is approximately quadratic). Their summary (Table 7 of Reference 83) for the serine proteases contains 18 entries for five enzymes: chymotrypsin, trypsin, porcine pancreatic elastase (PPE), human leukocyte elastase (HLE) and α-lytic protease. Most entries are for k_{cat}, and most of these are for substrates with rate-limiting deacylation. One case has an isotope effect too small (1.4) for confident interpretation; another shows signs that more than one step determines the rate. We shall omit these two cases here.

Of the remaining 16 cases, 11 give linear proton inventories indicative of one-proton catalysis. Commonly, these are for "truncated" substrates which have minimal interaction of enzyme with substrate in the catalytic transition state. Examples are Cbz-Phe-ONp (ONp, p-nitrophenyl ester) with chymotrypsin, Bz-Arg-OEt with trypsin, Ac-ONp with PPE and α-lytic protease. An exception is the relatively large substrate Suc-Ala-Ala-Ala-NHNp (p-nitroanilide) with HLE; an important point (to which we return) is that although the peptide is long, its first residue does not have the specific structure required by HLE.

These linear proton inventories are what would be expected if the charge-relay system did not function to generate an isotope effect at the Asp-His site, but only at the His–Ser site. In the deacylation reactions, for example, the His could serve as a simple one-proton protolytic catalyst, bridging to the oxygen of the attacking water molecule if formation of the tetrahedral intermediate is rate limiting, bridging to the oxygen of the departing serine if the decomposition of the tetrahedral intermediate is rate limiting. In principle, the Asp residue could serve to orient the His or to stabilize the HisH$^+$, or, indeed, it could serve no function at all; these substrates do not simulate the substrates which guided the molecular evolution of the enzymes and, thus, the evolved mechanism may not be simulated.

Five of the cases cited by Venkatasubban and Schowen show quadratic proton inventories. These are Cbz-Gly-ONp with chymotrypsin, Bz-Phe-Val-Arg-NHNp with trypsin, Ac-Ala-Pro-Ala-NHNp with PPE, MeOSuc-Ala-Ala-Pro-Val-NHNp with HLE, and Ac-Ala-Pro-Ala-NHNp with α-lytic protease. The first of these is an exception and will not be discussed; the original report[84] may be consulted for further information. The other four cases are (1) oligopeptides of at least three residues and (2) oligopeptides of which the first residue meets the specificity requirements of the enzyme (Arg for trypsin, Ala for PPE and α-lytic protease, Val for HLE).

The quadratic proton inventories indicate two-proton catalysis and are what would be expected if the charge-relay chain were functioning with protolytic bridges at both the Asp-His and the His-Ser sites. The finding gives no information on whether PT and HAR are coupled (although the small magnitude of the isotope effects suggests they are not) or whether the bridging involves net transfer of the proton. If net transfer occurred, then the Asp would be at least momentarily protonated; if no net transfer occurred in catalysis, the proton could remain—in effect—throughout the sequence on the His.

Venkatasubban and Schowen summed up the situation this way:

"Considered together, the entire data seem to us to suggest that the charge-relay apparatus of the enzyme is susceptible to activation through an extensive set of enzyme–substrate interactions, involving not only those at the scissile residue, but also those at remote subsites. When these interactions are absent, the enzyme functions as a simple one-site general catalyst. The activating interactions may have an internally balancing, or conflicting character so that not only the correct interaction at one site, but the correct *combination* of interactions at various sites is required for the reliable activation of the catalytic machinery.

"We suspect that this is one manifestation of a general phenomenon. In this view, enzymes and their substrates experience mechanical alterations of structure, which are mutually brought about in both enzyme and substrate as the two enter the catalytic transition state together. Here, the activation of the charge-relay system may be brought about by structural compression of the entire hydrogen-bond chain, as a result of attractive interactions between the residues along the polypeptide tail of the substrate and the appropriate subsites of the enzyme. The substrate polypeptide tail being anchored at the scissile residue, these attractions would produce a compressive mechanical force which might suffice to engender a spatial contraction. The shortening of the distances across the individual hydrogen bonds of the chain might then have a multiple effect. The sites could be coupled together, allowing for charge displacement. The effective pK of the chain might be altered, enhancing its catalytic effect. Finally, in a shortened hydrogen bond, the protons might be unusually mobile, producing isotope effects of 1.5 to 2 at each site.''

C. Recent Studies of Charge-Relay Recruitment: The Work of Stein and Strimpler[85]

A much clearer picture of substrate recruitment of multiproton catalytic function in the serine proteases has recently emerged in a publication of Stein and Strimpler,[85] the latest contribution in a series of elegant applications of the proton-inventory technique by Stein and co-workers. The crucial findings of Stein and Strimpler are summarized in Table 4-5.

Three enzymes were studied: PPE, HLE, and α-chymotrypsin (CT). Previous workers[86] had shown the three enzymes to possess different specificities for the scissile residue: a small side chain (eg, Ala) for PPE, the same or a larger alkyl side chain (Ala, Val) for HLE, and an aromatic side chain (Phe) for CT. Tri- and tetrapeptide analogs were used by Stein and Strimpler to measure proton inventories for the deacylation process for each enzyme, substrates being included which did and did not meet the scissile-residue specificity requirements for each of the enzymes.

TABLE 4-5. Results of Stein and Strimpler[85] for the Role of the Scissile Residue in Oligopeptide Recruitment of Multiproton Catalysis by Serine Proteases (Peptidyl–Thiobenzyl Esters, pH 7.65 and Equivalent, 25°C)

Peptidyl	$10^{-5}k_{cat}/K_m$ (per M/s)	k_{cat}/s	Solvent isotope effect	Active protons
Porcine pancreatic elastase				
MeOSuc-Ala-Pro-Ala-	12.0	47.0	3.08	2
MeOSuc-Ala-Ala-Pro-Ala-	26.0	38.0	2.64	1
Suc-Ala-Ala-Pro-Phe-	5.8	34.0	2.41	1
Human leukocyte elastase				
MeOSuc-Ala-Ala-Pro-Ala-	41.0	53.0	3.56	2
MeOSuc-Ala-Ala-Pro-Val-	56.0	13.0	2.84	2
Suc-Ala-Ala-Pro-Phe-	0.6	3.0	2.52	1
α-Chymotrypsin				
MeOSuc-Ala-Ala-Pro-Ala-	12.0	6.0	2.79	1
MeOSuc-Ala-Ala-Pro-Val-	3.0	0.2	3.17	1
Suc-Ala-Ala-Pro-Phe-	67.0	56.0	2.44	2

Table 4-5 shows that the expected specificities are found for these substrates in the values of k_{cat}/K_m (for the thiobenzyl esters), although the selectivity is small in some cases. Similar specificities are, in some cases, only weakly reflected in k_{cat}, which here refers to deacylation.

With one exception, the proton inventories indicate multiproton catalysis when the scissile residue meets the specificity requirements of the enzyme, and they indicate one-proton catalysis when this residue does not meet the specificity requirements. Since it has already been shown that "short" substrates generally do not activate multiproton catalysis, even if they meet the scissile-residue specificity requirement, Stein and Strimpler have demonstrated that *at least* a combination of (1) correct scissile residue and (2) two to three residues for interaction at more remote sites are required for recruitment of multiproton catalysis. It is not clear whether the linear proton inventory observed for MeOSuc-Ala-Ala-Pro-Ala-PPE is an exception to the rule for activation; Stein and Strimpler suspect the proton inventory may be rendered artifactually linear by the incursion of a step other than deacylation as partially rate limiting.

The general, provisional conclusion from these results is that the anchoring at scissile residue (presumably in the "specificity pocket") is a prerequisite for activation of charge-relay catalysis. If, in addition, there are sufficient remote-site interactions to compress the charge-relay system into its catalytically functional state, then multiproton catalysis is observed. Otherwise, it is not.

D. Is Multiproton Catalysis Charge-Relay Catalysis?

The preceding discussion suggests that enzyme–substrate structural interactions in the catalytic transition state, if the requirements of scissile-residue specificity and several remote-site interactions are fulfilled, are capable of activating a multiproton catalytic entity. This is indicated by the proton-inventory experiments, but there is no direct way to deduce from these experiments the location in the transition state of the centers that are generating the observed isotope effects. It seems reasonable to imagine that the entity generating these effects is the charge-relay system for these reasons:

1. The His of the charge-relay system is almost certainly the one-proton catalyst which is involved in those cases that display one-proton catalysis.
2. In many stable forms of the enzyme, the Asp–His hydrogen bond is intact so that it seems reasonable to imagine it could be coupled to the already active His–Ser part of the charge-relay system under appropriate circumstances.
3. Activation of the charge-relay system by interactions of the polypeptide tail of the substrate is easy to envision because the remote binding sites are spatially close to the catalytic triad.

An alternative considered previously[84] that has been favored by Fink[87] is the protonic sites of the "oxyanion hole." This seems less likely than the charge-relay system for the reasons given above and because the generation of large isotope effects of the order of 1.5–2.0 requires the formation not of the usual kind of hydrogen bonds but of the type involved in protolytic bridging. This seems not as likely between the oxyanion of the transition state and the N—H bonds of the oxyanion hole as between the members of the charge-relay system. However, this is not a firm conclusion because the solvation of fully formed anions (rather than the partially formed transition-state oxyanion) involves isotope effects of around 1.4 from ROH donors, and the value of the isotope effect is strongly dependent on environment.[88] It can therefore not be excluded that the sites being activated are in the oxyanion hole.

E. Evaluation of Proton-Inventory Studies

The proton inventories offer probes directly into the transition-state events and are, in this sense, superior to all other approaches except for future theoretical studies (at a point when the level of detail involved is accessible to theory). They indicate that the serine proteases have the intrinsic potential for multiproton catalysis, but that its activation requires not just the product of enzyme evolution but the product of the coevolution of enzyme and substrate, namely, the correct mutual interactions of enzyme and natural substrate. The assumption not yet tested is that the multiproton entity detected by the proton inventories is indeed the charge-relay system and not another structure.

If the charge-relay system is under observation, the dependence of its function on fine details of enzyme–substrate interaction furnishes an explanation of the difficulty of reaching reliable conclusions about its function from inference based on model systems, whether chemical models, theoretical models, or spectroscopic or crystallographic models. Any model which fails to simulate a sufficient number of aspects of the activating interactions will fallaciously indicate the absence of charge-relay catalysis.

6. SUMMARY AND PROSPECTS

A number of approaches of potentially great value were not covered here, among them the modification of the enzymes, eg, chemically by methylation of histidine[89] or through molecular–biological techniques in which aspartate has been changed to asparagine.[90] Both kinds of experiments have given results in general agreement with the conclusions presented here, but the future of both methods is promising. For example, proton-inventory studies with various substrates and with modified and mutant enzymes will be important.

In addition, as the resolution and speed of the crystallographic experiments increase and as the power of spectroscopic methods grows, particularly as they are linked more closely to theory in the simulation of actual transition-state events (especially in dynamic simulations), greater understanding of the significance and detailed function of charge-relay catalysis should emerge. This will challenge chemists to make use of this information in constructing synthetic systems which not only achieve enzymatic rates (which has now been done) but which match or surpass biological enzymes in features such as substrate-activated catalysis.

7. APPENDIX: SOME POINTS OF DETAIL

The MAR and the PES

Each point on a potential-energy surface (PES) corresponds to a specific set of molecular coordinates for a particular molecular system. One or more of these points, but usually only a few, are saddle points (they possess one and only one coordinate along which the curvature of the potential-energy function is negative) and are, thus, activated complexes (transition structures, transition-state structures). The contours on a PES contour map are contours of constant energy (ie, each contour traces a set of molecular configurations that share the same potential energy).

In contrast, each point on an MAR (map of alternate routes) specifies a possible transition-state structure. If the MAR refers to a single reaction, then, if it is sufficiently general, it displays all possible transition-state structures for that reaction. Since most reactions have only one or, at most, a few transition states, there will be one or a few points on the MAR which correspond to physical reality (ie, which have the coordinates of the real transition states in nature). All of the remaining points on the MAR then describe hypothetical transition states, which can be imagined but do not actually exist in nature. Experimental results can sometimes be used to discover which points on an MAR describe the real transition states (or within what range of points on the MAR the real transition states lie).

Often, an MAR refers not to a single reaction but to a large collection of reactions that may have different transition-state structures. One can imagine a very large reaction set in which the structural characteristics can be so finely tuned that each point on the MAR represents the real transition-state structure of one particular reaction from the set. Then, each point on the MAR corresponds to a real transition state, but each point refers to a different molecular system from every other point. Now, several points may be located by the use of experimental data on related reactions. This is very frequently done.

In neither of these formats is an MAR like a PES. Every point on a PES represents a physically real configuration of a single, particular molecular system. Every point on an MAR represents a transition state, but either the MAR describes a single system and only a few points represent real transition states *or* the MAR describes a collection of systems and all points represent real transition states but each for a different reacting system.

The contours on a PES trace sets of real molecular configurations of equal energy. Contours on an MAR cannot have any simple, concrete meaning. If it should be insisted that an MAR possess "energy contours," then they would have to be constructed differently for the two types of MAR. If the MAR represents a single reaction, the points corresponding to the real transition states would, presumably, have lower energies than all other points, which correspond to imaginary or hypothetical transition states. These other points will not be quantum-mechanical stationary states, but their energy could be calculated. Contours could be drawn, but their utility would not be obvious. If, on the other hand, the MAR refers to a collection of reactions with each point representing the transition-state structure for a real reaction, then, of course, each of the points could be associated with the energy of the corresponding real transition state. The energy contours would then connect transition-state structures having equal energy. These contours would not connect in general transition states of reactions with equal activation energy because the reactant energies for the individual reactions would differ and the reactant energies would not be represented on the MAR.

REACTION COORDINATE IN PROTOLYTIC CATALYSIS

Protolytic catalysis produces isotope effects for deuteration at the hydrogen bridge, so there must be some change in binding at this hydrogenic site when the reactant state is converted to the transition state. Catalysis itself can arise from stabilization of the transition state through hydrogen bonding, but most hydrogen bonds of the ordinary, asymmetric, double-minimum type do not exhibit isotope effects of more than 10–20 percent[82]. There are hydrogen bonds, however, which form particularly in non-aqueous media between bases of nearly equal basicity and which seem to be unusually strong, perhaps with a single potential minimum or a low barrier in a symmetrical double minimum; they generate large isotope effects (factors of 2–5)[32,59]. They are sometimes characterized as "highly polarizable," because of the ready mobility of the proton in the broad single-minimum potential well (or symmetrical potential well with very low barrier). These bonds may be the stabilizing interactions between catalyst and substrate in transition states for protolytic catalysis. Their formation in such transition states may be favored by (a) the opportunity for transient generation—just at the transition state— of a center in the substrate of nearly equal basicity to the catalyst, producing the required symmetrical potential; (b) the simulation of a nonaqueous environment by the transition-state solvation shell, which may well lag behind the charge reorganization and bonding changes in the transition state, so that the water molecules (in their non-equilibrium, "unsuited" orientations) do not induce asymmetry in the hydrogen-bridge potential well, as may usually occur with hydrogen bonds in a relaxed aqueous solvent environment. Thus HAR could constitute the entire reaction coordinate.

APPENDIX: NET PROTON TRANSFER IN PROTOLYTIC CATALYSIS

Net proton transfer often occurs in protolytic catalysis because protolytic catalysis commonly involves a "crossover" of acid-base properties: an effective base catalyst, for example, has a basicity intermediate between that of the deprotonated reactant and the deprotonated product[16]. This favors net proton transfer, but the fact of net proton transfer may in effect be coincidental. The critical point is probably to achieve an approximate match of basicities between catalyst and substrate at the transition state, so as to promote strong interaction through a hydrogen bridge. The probability that such a match of equal basicities will occur in the transition state is maximized if the basicity of the substrate site begins as greater than the catalyst basicity in the reactant state and ends in the product state as greater than the catalyst basicity. In simple molecules, achieving the requisite match of basicities at the transition state may be possible ONLY by this kind of "crossover." With an enzyme, the situation is totally different. The enzyme environment may actively modulate the basicity of any functional group in the reactant state or transition state. Thus the match of basicities required for catalysis may be produced transiently in the transition state through modulation of the pK of the His, or of the Asp or of the substrate site—or of all of these simultaneously. Furthermore, as the enzyme relaxes out of the transition state into the subsequent state(s), these pK's may themselves relax to values not necessarily near those achieved in the transition state.

REFERENCES

1. Bell, R. P. "Acid-Base Catalysis"; Oxford University Press: Oxford, 1941.
2. Bell, R. P. "The Proton in Chemistry"; 1st ed., Cornell University Press: Ithaca, New York, 1959; 2nd ed., Chapman and Hall: London, 1973.
3. Caldin, E.; and Gold, V. Eds. "Proton-Transfer Reactions"; Chapman and Hall: London, 1975.
4. Jencks, W. P. "Catalysis in Chemistry and Enzymology"; McGraw-Hill: New York, 1969.
5. Bender, M. L. "Mechanisms of Homogeneous Catalysis from Protons to Proteins"; Wiley (Interscience): New York, 1971.

6. Bruice, T. C.; Benkovic, S. J. "Bioorganic Mechanisms", Vols. 1 and 2; Benjamin: New York, 1966.

7. Gandour, R. D.; Schowen, R. L., Eds. "Transition States of Biochemical Processes"; Plenum Press: New York, 1978.

8. See the discussion in Bruice, T. C.; Schmir, G. L. *J. Am. Chem. Soc.* **1959,** *81,* 4552.

9. Bender, M. L.; Clement, G. E.; Kezdy, F. J.; Heck, H.d'A. *J. Am. Chem. Soc.* **1964,** *86,* 3680. Bender, M. L.; Hamilton, G. A. *J. Am. Chem. Soc.* **1962,** *84,* 2570. Schonbaum, G. R.; Zerner, B.; Bender, M. L. *J. Biol. Chem.* **1961,** *236,* 2930.

10. Schowen, R. L. *Prog. Phys. Org. Chem.* **1972,** *9,* 275.

11. Matthews, B. W.; Sigler, P. B.; Henderson, R.; Blow, D. M. *Nature (London)* **1967,** *214,* 652.

12. Blow, D. M.; Birktoft, J. J.; Hartley, B. S. *Nature (London)* **1969,** *221,* 337.

13. Blow, D. M. *Acc. Chem. Res.* **1976,** *9,* 145.

14. Stroud, R. M. *Sci. Am.* **1974,** *231(1),* 24.

15. Tsukada, H.; Blow, D. M. *J. Mol. Biol.* **1985,** *184,* 703.

16. Jencks, W. P. *Chem. Rev.* **1972,** *72,* 705. *J. Am. Chem. Soc.* **1972,** *94,* 4731.

17. Structure-reactivity coefficients like the Hammett ρ are first derivatives of a reactivity measure (eg, log k) with respect to a structure measure (eg, σ). Thus $\rho = \partial(\log k)/\partial\sigma$ and $\beta = [\partial(\log k)]/[\partial(pK_a)]$. Cordes–Jencks coefficients are the next higher derivatives: $[\partial^2(\log k)]/(\partial\sigma^2)$ or $[\partial^2(\log K)]/[\partial\sigma\partial(pK_a)]$, for example. A complete treatment is given by Jencks, D. A.; Jencks, W. P. *J. Am. Chem. Soc.* **1977,** *96,* 6967.

18. Melander, L.; Saunders, W. H., Jr. "Reaction Rates of Isotopic Molecules"; Wiley (Interscience): New York, 1980.

19. Cleland, W. W.; O'Leary, M. H.; Northrop, D. B. Eds. "Isotope Effects on Enzyme-Catalyzed Reactions"; University Park Press: Baltimore, 1977.

20. Kirsch, J. F. In "Advances in Linear Free Energy Relationships"; Chapman, N. B.; and Shorter, J. Eds.; Plenum Press: New York, 1972, Chapter 8.

21. Jencks, W. P.; Salvesen, K.; *J. Am. Chem. Soc.* **1971,** *93,* 1419. Jencks, W. P. *Acc. Chem. Res.* **1976,** *9,* 425.

22. Kershner, L. D.; Schowen, R. L. *J. Am. Chem. Soc.* **1971,** *93,* 2014.

23. Bruice, T. C. *Annu. Rev. Biochem.* **1976,** *45,* 331.

24. Jencks, W. P. *Chem. Rev.* **1985,** *85,* 511.

25. Thornton, E. R. *J. Am. Chem. Soc.* **1967,** *89,* 2915.

26. Dewar, M. J. S. *J. Am. Chem. Soc.* **1984,** *106,* 209.

27. Maggiora, G. M.; Schowen, R. L. In "Bioorganic Chemistry", Vol. 1; van Tamelen, E. D., Ed.; Academic Press: New York, 1977, Chapter 9.

28. There is a strong temptation to consider MAR's identical to potential-energy surfaces (PES's) when the latter are displayed as contour maps. They are very different. Useful discussions are given by E. K. Thornton and E. R. Thornton[7] and by Maggiora and Schowen.[27] Further comments are presented in the appendix.

29. Murrell, J. H.; Laidler, K. J. *Trans. Faraday Soc.* **1968,** *64,* 371. McIver, J. W., Jr. *Acc. Chem. Res.* **1974,** *7,* 72. McIver, J. W., Jr.; Stanton, R. E. *J. Am. Chem. Soc.* **1972,** *94,* 8618. Stanton, R. E.; McIver, J. W. *J. Am. Chem. Soc.* **1975,** *97,* 3632. Pechukas, P. *J. Chem. Phys.* **1976,** *64,* 1516.

30. Bernasconi, C. *Tetrahedron* **1985,** *41,* 3219.

31. Swain, C. G.; Kuhn, D. A.; Schowen, R. L. *J. Am. Chem. Soc.* **1965,** *87,* 1553.

32. Hibbert, F. *Adv. Phys. Org. Chem.* **1986,** *22,* 113.

33. Palmer, J. L.; Jencks, W. P. *J. Am. Chem. Soc.* **1980,** *102,* 6466.

34. Dogonadze, R. R.; Krishtalik, L. I. *Russian Chem. Rev.* (transl. from *Us. Khim.*) **1975,** *44,* 11. Ulstrup, J. "Charge Transfer Processes in Condensed Media"; Springer-Verlag: Berlin, 1979.

35. Brickmann, J.; Zimmermann, H. *Ber. Bunsenges. Phys. Chem.* **1966,** *70,* 157, 521.

36. Verhoeven, J. W.; Koomen, G. J.; van der Kerk, S. M. *Rec. Trav. Chim.* **1986,** *105,* 343.

37. Khoshtariya, D. E.; Topolev, V. V.; Krishtalik, L. I. *Bioorg. Khim.* **1978,** *4,* 1341.

38. Khoshtariya, D. E. *Bioorg. Khim.* **1978,** *4,* 1673.
39. Khoshtariya, D. E.; Topolev, V. V.; Krishtalik, L. I.; Reizer, I. L.; Torchilin, V. P. *Bioorg. Khim.* **1979,** *5,* 1243.
40. The concept that large temperature-independent isotope effects indicate bent transition-state structures (Kwart, H. *Acc. Chem. Res.* **1982,** *15,* 401) was fundamentally incorrect: Anhede, B.; Bergman, N. A. *J. Am. Chem. Soc.* **1984,** *106,* 7634.
41. Satterthwait, A. C.; Jencks, W. P. *J. Am. Chem. Soc.* **1974,** *96,* 7018.
42. Mordy, C. W.; Schowen, R. L. In "Proteases and Peptidases: Recent Advances"; Barth, A.; and Schowen, R. L., Eds. Pergamon Press: Oxford, 1987, pp. 273–280.
43. Welch, G. R., Ed. "The Fluctuating Enzyme"; Wiley (Interscience): New York, 1986.
44. Fersht, A. R. *Proc. R. Soc. London Ser. B* **1974,** *B187,* 397; "Enzyme Structure and Mechanism"; 1st ed., Freeman: San Francisco, 1977; pp. 271–272; 2nd ed., 1985.
45. Jencks, W. P. *Adv. Enzymol.* **1975,** *43,* 219.
46. Schowen, R. L. In "Transition States of Biochemical Reactions"; Plenum Press: New York, 1978, Chapter 2.
47. Steitz, T. A.; Shulman, R. G. *Annu. Rev. Biophys. Bioeng.* **1982,** *11,* 419.
48. Kraut, J. *Annu. Rev. Biochem.* **1977,** *46,* 331. Matthews, D. A.; Alden, R. A.; Birktoft, J. J.; Freer, S. T.; Kraut, J. *J. Biol. Chem.* **1977,** *252,* 8875.
49. Bachovchin, W. W. *Biochemistry* **1986,** *25,* 7751.
50. Kossiakoff, A. A.; Spencer, S. A. *Biochemistry* **1981,** *20,* 6462.
51. D'Souza, V. T.; Bender, M. L. *Acc. Chem. Res.* **1987,** *20,* 146.
52. Cram, D. J.; Lam, P. Y.-S.; Ho, S. P. *J. Am. Chem. Soc.* **1986,** *108,* 839.
53. Rogers, G. A.; Bruice, T. C. *J. Am. Chem. Soc.* **1974,** *96,* 2473.
54. Mallick, I. M.; D'Souza, V. T.; Yamaguchi, M.; Lee, J.; Chalabi, P.; Gadwood, R. C.; Bender, M. L. *J. Am. Chem. Soc.* **1984,** *106,* 7252.
55. Patterson, J. F.; Huskey, W. P.; Venkatasubban, K. S.; Hogg, J. L. *J. Org. Chem.* **1978,** *43,* 4935.
56. Alvarez, F. J.; Schowen, R. L. *Isot. Org. Chem.* **1987,** *7,* 1.
57. Gandour, R. D. *Bioorg. Chem.* **1981,** *10,* 169.
58. Saunders, W. H., Jr. *J. Am. Chem. Soc.* **1985,** *107,* 164 and references cited therein.
59. Eliason, R.; Kreevoy, M. M. *J. Am. Chem. Soc.* **1978,** *100,* 7037.
60. Warshel, A.; Russell, S. *J. Am. Chem. Soc.* **1986,** *108,* 6569.
61. Warshel, A.; Weiss, R. M. *J. Am. Chem. Soc.* **1980,** *102,* 6218.
62. Warshel, A. *Biochemistry* **1981,** *20,* 3167.
63. Sander, E.; Jencks, W. P. *J. Am. Chem. Soc.* **1968,** *90,* 6154.
64. Hupe, D. J.; Jencks, W. P. *J. Am. Chem. Soc.* **1977,** *99,* 451.
65. Fersht, A. *J. Am. Chem. Soc.* **1971,** *93,* 3504.
66. See discussions in Fersht's books (Reference 44), data and discussions in the articles cited in Reference 86, and the discussion in Section V of this chapter.
67. Stamato, F. M. L. G.; Longo, E.; Ferreira, R.; Tapia, O. *J. Theor. Biol.* **1986,** *118,* 45.
68. Longo, E.; Stamato, F. M. L. G.; Ferreira, R.; Tapia, O. *J. Theor. Biol.* **1985,** *112,* 783.
69. Stamato, F. M. L. G.; Longo, E.; Yoshioka, L. M.; Ferreira, R. C. *J. Theor. Biol.* **1984,** *107,* 329.
70. Dewar, M. J. S.; Storch, D. M. *Proc. Natl. Acad. Sci. U.S.A.* **1985,** *82,* 2225.
71. Nakagawa, S.; Umeyama, H. *J. Mol. Biol.* **1984,** *179,* 103, and preceding contributions from Umeyama's group.
72. Kollman, P. A.; Hayes, D. M. *J. Am. Chem. Soc.* **1981,** *103,* 2955.
73. Allen, L. C. *Ann. N.Y. Acad. Sci.* **1981,** *367,* 383.
74. Kitayama, H. P.; Fukutome, H. *J. Theor. Biol.* **1976,** *60,* 1.
75. Dewar, M. J. S.; Zoebisch, E. G.; Healy, E. F.; Stewart, J. J. P. *J. Am. Chem. Soc.* **1985,** *107,* 3902.
76. Scheiner, S.; Lipscomb, W. N. *Proc. Natl. Acad. Sci. U.S.A.* **1976,** *73,* 432. Scheiner, S.; Kleier, D. A.; Lipscomb, W. N. *Proc. Natl. Acad. Sci. U.S.A.* **1975,** *72,* 2606.
77. Gandour, R. D.; Maggiora, G. M.; Schowen, R. L. *J. Am. Chem. Soc.* **1974,** *96,* 6967.

78. Ressler, N. *Physiol. Chem. Phys. Med. NMR* **1985,** *17,* 183. ibid. *J. Theor. Biol.* **1982,** *97,* 195.
79. Naray-Szabo, G.; Nagy, P. *Enzyme* **1986,** *36,* 44, and preceding contributions from this group.
80. Schowen, R. L. In "Isotope Effects on Enzyme-Catalyzed Reactions"; University Park Press: Baltimore, 1977, pp. 64–99.
81. Schowen, K. B. In "Transition States of Biochemical Processes"; Plenum Press: New York, 1978, Chapter 6.
82. Schowen, K. B.; Schowen, R. L. *Methods Enzymol.* **1982,** *87C,* 551.
83. Venkatasubban, K. S.; Schowen, R. L. *CRC Crit. Rev. Biochem.* **1984,** *17,* 1.
84. Stein, R. L.; Elrod, J. P.; Schowen, R. L. *J. Am. Chem. Soc.* **1983,** *105,* 2446.
85. Stein, R. L.; Strimpler, A. M. *J. Am. Chem. Soc.* **1987,** *109,* 4387.
86. Bauer, C. A.; Thompson, R. C.; Blout, E. R. *Biochemistry* **1976,** *15,* 1296. Kasafirek, E.; Eric, P.; Slaby, J.; Malis, F. *Eur. J. Biochem.* **1976,** *69,* 1. Zimmerman, M.; Ashe, B. M. *Biochim. Biophys. Acta* **1977,** *480,* 241. Harper, J. W.; Cook, R. R.; Roberts, J.; McLaughlin, B. J.; Powers, J. C. *Biochemistry* **1984,** *23,* 2995.
87. Fink, A. In "Enzyme Mechanisms"; Page, M. I.; and Williams, A., Eds.; Royal Society of Chemistry: London, 1987.
88. Kresge, A. J.; More O'Ferrall, R. A.; Powell, M. F. *Isot. Org. Chem.* **1987,** *7,* 177.
89. Henderson, R. *Biochem. J.* **1971,** *124,* 13. Byers, L. D.; Koshland, D. E., Jr. *Bioorg. Chem.* **1978,** *7,* 15. Fastrez, J. *J. Chem. Soc., Perkin Trans. 2* **1980,** 1067.
90. Craik, C. S.; Roczniak, S.; Largman, C.; Rutter, W. J. *Science* **1987,** *237,* 909. Sprang, S.; Standing, T.; Fletterick, R. J.; Stround, R. M.; Finer-Moore, J.; Xuong, N.-H.; Hamlin, R.; Rutter, W. J.; Craik, C. S. *Science* **1987,** *237,* 905.

CHAPTER 5

Electron Transfer in Cytochromes *c* and *b*₅

Dabney White Dixon

Georgia State University, Atlanta, Georgia

CONTENTS

1. Introduction... 169
2. Structure, Theory, and Experimental Methods.............. 171
3. Factors Controlling Electron Transfer..................... 184
4. Intramolecular Electron Transfer 195
5. Intermolecular Electron Transfer 204
6. Miscellaneous Topics................................... 215
   Acknowledgments....................................... 217
   References 217

1. INTRODUCTION

Electron transfer is one of the most basic of biological reactions, important in photosynthesis,[1-3] oxidative phosphorylation,[4,5] oxidation of endogenous and exogenous substrates, and maintenance of enzymes in active states.[6-9] Our understanding of electron transfer in biological systems has increased greatly in the last ten years. Perhaps the most dramatic strides have been the demonstrations of the Marcus inverted region[10] and of long-distance electron transfer through proteins.[11-13]

Current work in this field is focused on a number of aspects of electron transfer. Substantial progress is being made in determining the distance dependence of electron transfer.[11-14] The role of the material between the donor and acceptor is important. Although it is difficult to assess this role both theoretically and experimentally, progress is being made in both areas.[15-18] Site-directed mutagenesis,[19,20] a very valuable tool in this regard, is beginning to be used for both cytochrome c[11,21,22] and cytochrome b_5.[23] Site-directed mutagenesis complements investigation of different species,[24-31] other mutants[32-34] and proteins made via semisynthesis,[24-31] or chemical modification of native cytochromes.[42-45]

The relative orientation of the donor and acceptor is likely to be an important controlling factor in electron transfer.[46,47] Many groups are synthesizing models to separate the three variables of distance, orientation, and intervening material in controlling electron transfer.[48-58] The role of conformational change is the subject of considerable speculation.[59-62] Although there is as yet little evidence on the importance of protein conformational change in electron transfer, the widespread availability of high-field nuclear magnetic resonance (NMR) spectrometers and rapid advances in two-dimensional NMR allow structural and motional information on proteins unobtainable only a few years ago.[25,26,64-69] Finally, for bimolecular electron transfer, work continues on the importance of electrostatics in determining the geometry of the electron transfer pair and, hence, the rate constant for electron transfer.[13,25,26,31,70-73] There has also been a sharpened awareness of the possibility of more than one electron transfer complex in bimolecular electron transfer.[74-77]

Much recent work has focused on cytochromes c[25,26,27,65,78] and b_5.[78-80] These are small, well-characterized, soluble proteins which participate in a variety of electron transfer reactions. Cytochrome c is involved in mitochondrial electron transport, carrying electrons from complex III to the terminal oxidase, complex IV. Cytochromes c are also involved in photosynthetic electron transfer, as well as a number of other metabolic pathways. Cytochrome b_5 participates in electron transfer in stearyl-CoA desaturation and in the reduction of in cytochrome P-450 and methemoglobin.[78] This review is concerned with these proteins and their electron transfer partners cytochrome c peroxidase (ccp), cytochrome P-450, and the blue copper proteins. Cytochrome c peroxidase is a heme-containing protein which catalyzes the oxidation of reduced cytochrome c using hydrogen peroxides and alkyl hydroperoxides.[6,79] Cytochrome P-450, another heme-containing protein, oxidizes both endogenous and exogenous substrates in a wide variety of metabolic processes.[7,8] Blue copper proteins, for example, azurin and plastocyanin, are electron transfer proteins found in bacteria and plants.[82]

The great interest in biological electron transfer and the extensive progress in both theory and experiment have led to a large number of recent reviews. To cite only those since 1983, recent overviews include those by Mayo et al.,[11] Scott et al.,[83] and Pielak et al.[59] Guarr and McLendon[12] have provided a review of many aspects of theory along with descriptions of experiments on both biological and nonbiological systems. Bertrand[84] and Dreyer[85] have also discussed electron transfer in biological systems. De-Vault[86] has discussed quantum-mechanical tunneling in biological systems in a recent book. Jortner and Pullman's book[87] on tunneling gives papers presented at a symposium in 1986.

Marcus and Sutin[13] have given a comprehensive review of theory and experiment with particular emphasis on Marcus theory as applied to biological systems. This extends earlier reviews by Newton and Sutin[88] and Sutin.[89] Creutz, Linck, and Sutin[90] have provided a detailed treatment of electron transfer in inorganic systems. Isied[91] has discussed both theory and experiment in electron transfer with emphasis on both biological and inorganic/organometallic systems. Bowden, Hawkridge, and Blount[92] have reviewed heterogeneous electron transfer of heme proteins. A very interesting discussion of electron tunneling in solid state reactions has been given by Mikkelsen and Ratner.[14] Hush[93] has also reviewed electron transfer in a variety of systems. Haim[94] has reviewed electron transfer within ion pairs.

Williams and co-workers[25] and Moore and co-workers[26] have written two very useful reviews of the structure, spectroscopy, and electron transfer properties of cytochrome c. Comprehensive descriptions of the properties and spectra of cytochromes c have been given by Meyer and Kamen.[27] The structure and reactions of heme proteins have been reviewed by Poulos and Finzel[6] and by Mathews;[78] Senn and Wüthrich[65] have reviewed the structure and coordination of cytochromes c. Greenwood[79] has discussed cytochromes with particular emphasis on cytochrome c peroxidase and nitrate reductase. Tollin et al.[95] have reviewed the work of the groups of Cusanovich, Meyer, and Tollin on electron transfer measured via flash photolysis. Margoliash and Bosshard[70] have reviewed their groups' work on electron transfer in cytochrome c and its derivatives. Both Satterlee[66] and Bertini and Luchinat[67] have provided extensive discussions of nuclear magnetic resonance (NMR) spectroscopy of paramagnetic heme proteins. Peterson-Kennedy et al. have reviewed the studies of Hoffman and co-workers on intramolecular electron transfer.[96]

In addition to the recent reviews cited above, important earlier books in the field include those by Cannon on electron transfer,[97] Chance et al. on tunneling,[98] Lever and Gray on iron porphyrins,[99] and the invaluable seven-volume series on the porphyrins edited by Dolphin.[100] The literature search for this review was completed through the end of 1986; selected references to works in 1987 have also been added.

2. STRUCTURE, THEORY, AND EXPERIMENTAL METHODS

A. Protein Structure

In cytochromes c and b_5, electron transfer takes place to and from the heme itself. The heme is a tetrapyrrole macrocycle, with a highly conserved set of

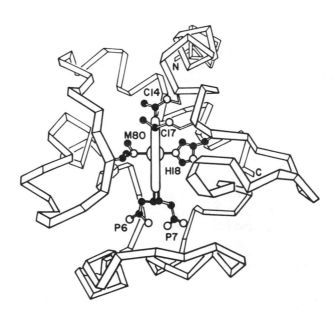

Figure 5-1. Sructure of the heme. The peripheral positions are numbered 1–8. The four protons directly attached to the ring are the meso protons and are designated with Greek letters.

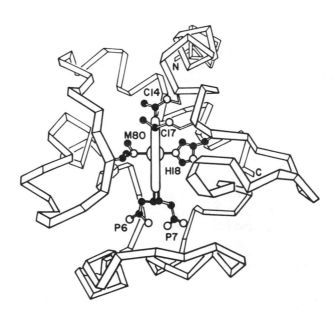

Figure 5-2. A ribbon drawing of cytochrome c. The heme is on edge, the propionates extend down. The axial ligands are His-18 and Met-80; the heme is covalently bound to the protein through thioether linkages at positions 2 and 4 on the heme. Reproduced with permission.[78]

substituents around the macrocycle periphery including two propionic acids which extend out from the same side of the heme (Figure 5-1). For cytochromes *c* and *b*$_5$, there are two oxidation states of importance: Fe(III), or ferric (which has 5 d electrons), and Fe(II), or ferrous (which has 6 d electrons). In vivo, cytochromes *c* and *b*$_5$ are hexacoordinate, and the iron centers are low spin; the ferric low-spin iron is paramagnetic, and the ferrous low-spin iron is diamagnetic.

A ribbon drawing of tuna cytochrome *c*, taken from the crystal structure coordinates,[101] is shown in Figure 5-2. The protein has 103 amino acids; the heme is attached covalently to the polypeptide backbone by thioether bridges from cysteines at positions 14 and 17 to the α-carbons of the heme periphery at positions 2 and 4. The axial ligands are Met-80 and His-18. The two propionic acids extend toward the bottom of the protein in the figure (P6 and P7). The heme is slightly exposed to solvent, with about 1% of the surface area of the protein being heme.[102] The isoelectric point of eukaryotic cytochromes is about 10. The surface charges are asymmetrically placed; in particular, there are a number of lysines around the exposed heme edge. In general, these lysines are highly conserved in eukaryotic cytochromes *c*, but cytochromes from other organisms have a wider variety of charge distribu-

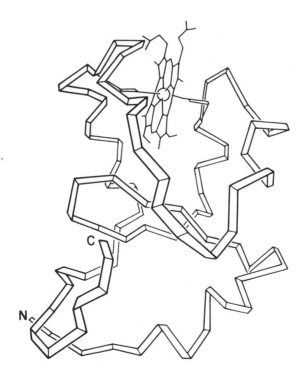

Figure 5-3. A ribbon drawing of the heme binding domain of cytochrome *b*$_5$. One of the heme propionates extends into solution, the other is hydrogen bonded to Ser-64. Reproduced with permission.[78]

tions. Insight into the structure and spectroscopy of cytochromes c is provided in a number of recent reviews.[25–27,65,66,78]

As mentioned previously, cytochrome b_5 participates in electron transfer in stearyl-CoA desaturation and in the reduction of in cytochrome P-450 and methemoglobin.[78] In the first of these roles, the protein is found in a membrane-bound form. The heme-binding portion can be solubilized by treatment of microsomes with a number of different proteases. The structure of the heme-binding fragments from bovine cytochrome b_5 (93 residues)[103] is shown in Figure 5-3. The heme is not covalently bound to the polypeptide in this protein, but held in only by the bonds from the two axial histidines to the iron atom. The protein is less spherical than cytochrome c. It has an overall negative charge of approximately -7.5 (calculated from the sequence); the charges are again asymmetrically distributed, with a number of negatively charged amino acids near the exposed heme edge. As can be seen, one of the heme propionates extends into solution, whereas the other is hydrogen bonded internally (to the peptide nitrogen and side chain OH of Ser-64).

B. Electron Transfer Theory

Electron transfer theory has been reviewed recently by Marcus and Sutin,[13] Newton and Sutin,[88] Hush,[93] Guarr and McLendon,[12] Creutz, Linck, and Sutin,[90] and Mikkelsen and Ratner.[14] The following discussion is, therefore, an abbreviated treatment.

To transfer an electron, the oxidant and reductant must undergo fluctuations in their atomic coordinates such that the oxidant attains the geometry of its reduced form and the reductant attains the geometry of its oxidized form. Because so many nuclear coordinates are important both on the reactants and in the surrounding medium, atom motion is a complicated multidimensional problem. This is generally simplified by treating the multidimensional potential energy surface as a simple one-dimensional surface (for related discussions see References 86 and 88). This is illustrated in Figure 5-4, in which the left-hand parabola represents the potential surfaces of reactants plus the surrounding medium, and the right-hand parabola represents those of products plus surrounding medium. The bottoms of the wells are the equilibrium conformations of reactants (left) and products (right).

Electron transfer occurs at or near the crossing point of the two parabolas. This follows the Franck–Condon principle. That is, the electron and nuclear motion can be separated and the nuclei, being far heavier than the electrons, do not have time to alter their positions during the electron transfer event. It follows that electron transfer can occur when thermal fluctuations bring the system from the reactant well to the crossing point.

Electron transfer need not occur each time the system reaches the crossing point. This is expressed as the transmission coefficient, κ, the probability the electron transfer will occur as the nuclei pass through the crossing point.

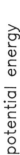
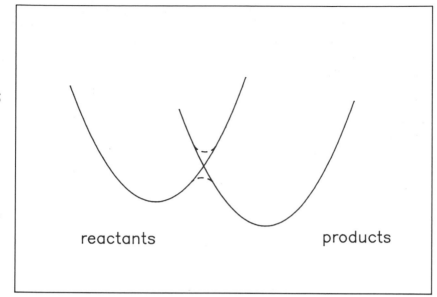

nuclear configuration

Figure 5-4.   Potential surface for an electron transfer reaction. The left-hand parabola represents the reactants, the right-hand parabola the products. The dotted lines indicate electronic interaction between the reactants.

When the system moves smoothly from reactants to products with unit probability, the reaction is said to be adiabatic. The probability for any given system depends upon the coupling between the orbitals of the two reactants. This coupling, in turn, depends on the orientation of the molecular orbitals and distance between the two species. The value of κ is usually taken to fall off exponentially with distance.

The intramolecular (first-order) rate constant is given by

$$k_{et} = \kappa\nu \, \exp(-\Delta G^*/RT) \qquad (5\text{-}1)$$

where ν is an effective frequency for movement along the reaction coordinate, generally taken as approximately equal to a bond vibration, or 10^{13} s^{-1}. ΔG^* is the free energy of activation of the reaction which can be expressed

$$\Delta G^* = \frac{(\Delta G^\circ + \lambda)^2}{4\lambda} \qquad (5\text{-}2)$$

where λ is the reorganization energy. When the free energy change from reactants to products (ΔG°) is zero, $\Delta G^* = \lambda/4$. ΔG^* is closely related to the conventional free energy of activation, $\Delta G^{\ddagger}$, differing by a term that involves the free energy necessary to bring the two electron transfer partners together in solution. The form of the latter term depends upon the way in which the association constant is calculated.[90,104] Equation 5-2 predicts that the observed rate will be fastest when $\Delta G^* = 0$, which occurs when the reorganization energy equals the free energy of the reaction, that is, when $\Delta G° = -\lambda$. It also predicts that the rate will decrease when the free energy of the reaction is greater than the reorganization energy. This is usually called the "Marcus inverted region".[105]

The reorganization energy is generally separated into two terms, that for reorganization of the bonds within the molecule (inner-sphere reorganization) and that for reorganization of the solvent (outer-sphere reorganization). The former is given by

$$\lambda_{in} = \sum_i \frac{f_1 f_2}{f_1 + f_2} (\Delta a°)_i^2 \tag{5-3}$$

where f_1 and f_2 are the force constants of the ith bond in the two reactants, $\Delta a°$ is the difference in equilibrium bond distance between the reactant and product, and the summation is over all of the normal modes of the intramolecular vibrations.

The outer-sphere reorganization term has been derived using dielectric continuum theory as

$$\lambda_{out} = \frac{(\Delta \varepsilon)^2}{4\pi\varepsilon_0} \left(\frac{1}{2a_1} + \frac{1}{2a_2} - \frac{1}{r} \right) \left(\frac{1}{D_{op}} - \frac{1}{D_s} \right) \tag{5-4}$$

where $\Delta \varepsilon$ is the charge transferred in the reaction, a_1 and a_2 are the radii of the species and $r = a_1 = a_2$, D_{op} is the optical dielectric constant (square of the refractive index), and D_s is the static dielectric constant.

Bimolecular reactions can be treated as a collision followed by electron transfer within the collision complex. Thus

$$k_{et} = \kappa A \exp(-\Delta G^*/RT) \tag{5-5}$$

where A has the dimensions of a collision frequency. The value of ΔG^* (which is often broken into terms for the inner- and outer-sphere reorganization, ΔG_{in}^* and ΔG_{out}^*) must be corrected for the work necessary to bring the two reactants together in solution (w^r):

$$\Delta G^* = w^r + \Delta G_{out}^* + \Delta G_{in}^* \tag{5-6}$$

When the rate constant of the electron transfer reaction approaches diffusion control, then both electron transfers and diffusion must be considered:

$$\frac{1}{k_{obsd}} = \frac{1}{k_{diff}} \frac{1}{k_{et}} \qquad (5\text{-}7)$$

Quantum-mechanical theories for electron transfer have been developed in some detail.[12–14,86,87,90,93] Very briefly, the electron transfer rate constant is given as

$$k = (2\pi/h)H_{AB}^2\{FC\} \qquad (5\text{-}8)$$

where H_{AB} is the electronic matrix element for coupling of the reactants and products electronic states and $\{FC\}$ is the Franck–Condon factor. The value of H_{AB} is determined by the distance between the donor and the acceptor and the energy of the tunneling electron. When the donor–acceptor distance is large, this term decays approximately exponentially with distance. However, the decay length may change with the driving force for the reaction and with the energy of the orbitals involved. The terms "hole" and "electron" transport are coming into use. If the weighted average of the donor and acceptor orbitals is closer to the highest occupied molecular orbit (HOMO) than to the lowest unoccupied molecular orbit (LUMO) of the bridge between the two centers, then electron transfer is termed "hole transport." However, if this average is closer to the LUMO of the linker, then the electron transfer is termed "electron transport."[106,107]

Solvent relaxation dynamics are coming under increasing scrutiny as a determining factor in electron transfer reactions. These include studies of photochemical processes,[108] electrochemical reactions,[109] organometallic and inorganic electron transfer,[109–112] and theoretical treatments.[113] The importance of these in biological reactions is not yet established.

C. Measurement of Electron Transfer

a. Techniques

Electron transfer in biological systems is generally measured by flash photolysis, pulse radiolysis, stopped-flow, and NMR techniques. The following summary is very brief; the reader is referred to the original articles for details of the measurements.

Using flash photolysis, measurements can be made over a very wide time scale—picoseconds to seconds.[114] In general, flash photolysis does not demand much sample. However, in those instances where it is necessary to flow the sample being photolyzed, large amounts of protein are necessary. Heme proteins per se are not good candidates for studies of photoinduced electron transfer because photoexcited Fe(II) and Fe(III) porphyrins deactivate very rapidly (<1 ps) to the ground state.[115] Also, excited Fe(II) porphyrins can undergo ligand dissociation,[116] which may compete with electron transfer and complicate analysis of the results. For this reason, flash photolysis studies utilize other chromophores, often a free base porphyrin (no metal) or Zn porphyrin, inserted into the apoprotein after the native heme has been removed. These chromophores have relatively short excited singlet state lifetimes and relatively long excited state triplet lifetimes (usually <15 ns and >1 ms, respectively). Long-lived ionic product states (D^+, A^-) are generally observed only after electron transfer from the triplet state. This is because reverse electron transfer within the initially produced triplet radical pair (to form the ground state) is spin forbidden and, therefore, slow. The spin-allowed decay of the singlet radical pair to the singlet ground state can be very fast, often on the picosecond time scale.[117]

Recently, much effort has centered on measuring both "forward" and "reverse" electron transfer within a complex.[16,60–62,121–125] Figure 5-5 shows a schematic of electron transfer in both the forward and reverse direction. Initial photoexcitation of the donor produces the excited singlet, some fraction of which undergoes intersystem crossing to the triplet. As noted previously, either the singlet or the triplet can undergo electron transfer, depending on the relative rate constants of electron transfer and other deactivation routes of the excited state. Reverse electron transfer from the acceptor to the donor then regenerates the starting complex. Analysis of the observed kinetics can be complicated, however. Unless the forward and reverse rate constants are quite different (with forward at least ten times that of the reverse), one must deal with both of the rate constants together. Of course, if the forward rate constant is substantially slower than the reverse, the elec-

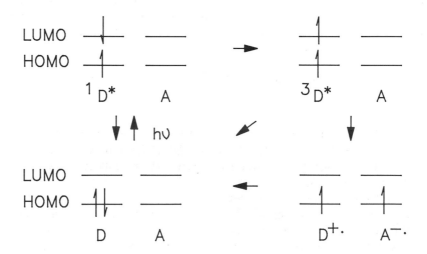

Figure 5-5. A schematic of electron transfer in both the forward and reverse direction. Initial photoexcitation of the donor produces the excited singlet, some fraction of which undergoes intersystem crossing to the triplet. Either the singlet or the triplet can undergo electron transfer, but most studies observe the triplet (see text). Reverse electron transfer from the acceptor to the donor regenerates the starting complex.

tron transfer reaction will appear as a deactivation directly from the excited state to ground. In this case one must distinguish the electron transfer reaction from other processes which take the excited state to the ground state.[125]

Even if the intermediate, B, can be seen, its maximum concentration, $[B]_{max}$, can be small and, hence, difficult to measure. For the reaction $A \rightarrow B \rightarrow C$, the maximum concentration of B is given by[126]

$$[B]_{max} = [A_o](k_2/k_1) \exp[(k_2/(k_1 - k_2)]$$ (5-9)

Thus, if $k_1 = 1$ and $k_2 = 100$, $[B]_{max}$ is only 1% of $[A_o]$. The appearance and disappearance of B is most easily measured at an isosbestic point of A and C. For the system

$$A^*_{red}/B_{ox} \rightleftharpoons A_{ox}/B_{red} \rightleftharpoons A_{red}/B_{ox}$$

this corresponds to isosbestic points between B_{ox} and B_{red}. One can measure the spectrum of B_{ox} in the complex A_{red}/B_{ox} because the complex is stable. However, measurement of the spectrum of B_{red} in the complex A_{ox}/B_{red} is difficult because this complex is reacting via electron transfer to form A_{red}/B_{ox}. The approximation is usually made that B_{red} has the same spectrum in a complex as free in solution. This is a good approximation, but it may not be adequate for small changes, which would be found either because the A complexes do not absorb strongly at the $B_{ox}-B_{red}$ isosbestic point or because $k_2 \gg k_1$. Small absorbance changes can be seen upon complexation of the proteins in some instances, as discussed in Subsection A of Section 4. Of course, understanding of a system depends, in large part, upon the ability to see intermediates; two or more spectroscopically identical intermediates will confuse interpretation of the results.

Measurements of forward and reverse electron transfer are important, because the forward and reverse electron transfer reactions should be similar for similar driving forces in the absence of changes in spin multiplicity and conformation. This is so because electron transfer is over the same distance, with the same intervening medium, and the same orientation of electron transfer partners. The two reactions differ in the driving force for the electron transfer and in the orbitals involved. Measurement of both rates should, in principle, allow separation of the effects of driving force and orbital occupation from those of distance and orientation.

Pulse radiolysis[114,127] is similar in concept in many ways to flash photolysis. In pulse radiolysis, the reducing equivalents are received from a high-energy electron pulse. Intramolecular electron transfer is measured in compounds or complexes with a donor and an acceptor held at a defined distance. The system is prepared with both centers in the oxidized state; the pulse is not selective and reduces both centers (Figure 5-6). The intramolecular electron transfer rate constants are calculated from the approach to equilibrium of the state produced initially (a mixture of both donor and

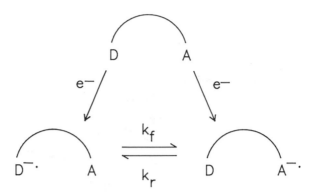

Figure 5-6. A kinetic scheme for measurement of electron transfer in pulse radiolysis studies. Initially, both centers are in the oxidized state. The pulse is not selective and reduces both centers. The forward and reverse rate constants can be calculated from the approach to equilibrium.

acceptor reduced by the electrons). To measure intramolecular electron transfer, it is necessary that the initial electron capture process not be rate determining. It should be noted that the electrons which add to the acceptors are not the initial high-energy electrons, but electrons derived from solvent or additives. Also, loss of an electron from a solvent molecule results in a positive "hole," and recombination of holes with the negative ions produced may complicate the kinetic analysis. Closs, Miller, and co-workers note that ion pairs generated by high-energy radiation are generally not contact ion pairs.[48] This makes data interpretation easier because rates of electron transfer can be controlled by counterion motion when the ions are close.

A great deal of work has come from stopped-flow measurements; these are appropriate for reactions with half-lives in the millisecond-to-second range.[127,128] Concentrations can often be arranged such that electron transfer is a pseudo-first-order reaction between the reduced donor and the oxidized acceptor (one in substantial excess), although complex formation may be observed. One of the difficulties with these experiments is maintaining the donor in the reduced state without an excess of reducing agent (which would itself reduce the acceptor).

NMR techniques are used increasingly to measure electron transfer.[25,26,64,69,129] This is due largely to the increasing accessibility of high-field NMR instruments with provision for data analysis. NMR has the considerable advantage that one can obtain a great deal of information from the experiment. In particular, both structural and motional information are available. Structural information includes the orientation of groups near the heme, often determined via nuclear Overhauser techniques.[25,66,69] NMR is also used to determine which groups contribute to observed pKas of the protein.[130] Motional information includes the rate constants of motions of various amino acid residues, most notably the flipping of aromatic rings.[25,26]

Protein motion on longer time scales can be determined by measuring the rate of exchange of the amide NH protons in the molecule.[131,132,133]

Although NMR can be used to determine the rate constants of electron transfer between two different species, it is most commonly used to measure electron self-exchange, electron transfer between the oxidized and reduced forms of the same molecule:[134]

$$cyt_a^{ox} + cyt_b^{red} \rightleftharpoons cyt_a^{red} + cyt_b^{ox}$$

$\Delta G°$ is zero in this reaction because the reactants and products are the same. Measurement of the electron self-exchange rate constant therefore allows estimation of the importance of factors other than the free energy of the reaction. Offsetting the advantages of NMR, however, are (1) the necessity for large amounts of protein, (2) the possibility of complex formation at the high concentrations necessary for NMR (0.5–5 mM), (3) the difficulty of accurate assessment of the ionic strength, and (4) the possibility of nonunity activity coefficients at high protein concentrations. Large amounts of protein will be increasingly available through site-directed mutagenesis techniques,[11,19–23,135–137] though protein availability will still remain a problem. The ionic strength of solutions of protein at NMR concentrations is difficult to assess because the concentration of the protein is so high that a considerable fraction of the ions come from the protein itself and not from the added salt.[138] Proper theoretical treatment of this problem is difficult. High solute concentrations may also result in activity coefficients which are not unity. NMR concentrations are indeed high; a 6 mM solution of cytochrome b_5 (MW 16,000) is 6% by volume protein (approximately 100 g/L). Calculations from osmotic pressure data on sheep hemoglobin in saline solution give an activity coefficient of about 2 at 100 g/L protein,[139] indicating that nonunity activity coefficients are indeed a possibility at NMR concentrations of protein.

Electron transfer rate constants can be measured by NMR by saturation transfer, line-broadening, and inversion recovery methods in one dimension,[64] and by volume integration or transfer of magnetization in two dimensions.[64,129,140] The technique used depends on the relationship between the rates of electron transfer, spin-lattice relaxation of the resonances of interest, and the difference in chemical shift (in Hz) between the peaks in the oxidized and reduced species. When electron transfer is fast with respect to the frequency difference, the system is in fast exchange and the exchange contribution to the line width is measured. When electron transfer is slow on the NMR time scale, saturation transfer, inversion recovery or line broadening can all be used. The technique of choice depends on the system. Complications arise when the electron–nuclear dipolar relaxation is not fast compared to other processes under consideration,[141] but this is not a problem for ferric heme proteins. Although one can fit a spectrum in the intermediate exchange regime to appropriate equations to calculate the rate constant, judicious choice of NMR field strength, protein concentration, ionic strength, and temperature usually allows one to force the system into either slow or fast exchange. In addition, it should be noted that saturation transfer in either one or two dimensions has proved very useful in correlating the resonances of the oxidized protein with those of the reduced protein.[66–69,140]

b. The Role of Protein Conformational Change

Protein conformational change[25,26,132,133,142] is important in a number of contexts. This section considers conformational change that is slow on the time scale of electron transfer (conformational change that occurs on the time scale of electron transfer is discussed in Subsection C of Section 3 under reorganization energy). Such slow conformational change is important because it can alter the redox potential of the electron transfer partners, their distance, or their orientation, all of which are important in determining the rate constant of electron transfer. Conformational change can also be important because it is not always easy to differentiate conformational change from other processes. Observed spectral changes can be due not to rate-limiting electron transfer but to a rate-limiting conformational change.

The simplest scheme that accommodates electron transfer and slow conformational change for an intramolecular reaction is shown in Figure 5-7a. In this figure, electron transfer takes place horizontally and conformational change vertically. Two conformations are represented, one by a rectangle and a second by a circle. Thermodynamic balance requires that $K_{12}K_{23}K_{34}K_{41} = 1$. Consider a situation in which the reactants are substantially in one conformation (1) and the products substantially in the second conformation (2). In this case, both K_{23} and K_{41} are much greater than one. Because $K_{12}K_{34} = 1/(K_{23}K_{41})$, for a given K_{12}, the larger the product of K_{23} and K_{41}, the smaller the value of K_{34}. In other words, the proposal that the reactants are in one conformation and the products in a second requires a redox potential change of one of the partners upon complexation. If a conformational change is proposed, it is important to identify the features of the protein which cause the change in the redox potential. Factors controlling heme protein redox potentials are discussed in more detail in Subsection B of Section 3.

c. Association and Dissociation of Complexes

Many of the current studies of long-distance electron transfer utilize protein/protein complexes. This technique allows measurement of electron transfer in a system of ostensibly fixed distance and geometry. However, association and dissociation of protein/protein complexes can complicate some analyses in two ways. First, if the rate constant for complex dissociation is similar to that of electron transfer, it would be possible to mistake the optical density changes associated with complex dissociation for those of electron transfer because absorbance changes measured are generally small. Second, if com-

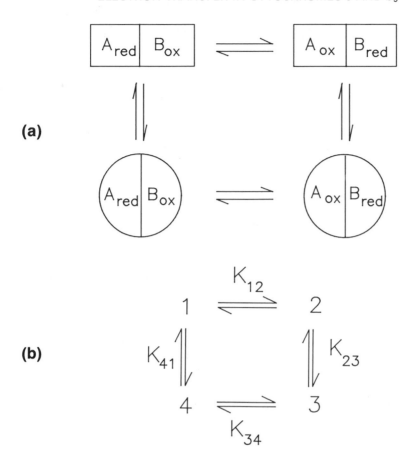

Figure 5-7. (a) Kinetic scheme for intramolecular electron transfer and conformational change. The rectangles represent one conformation of the protein, the circles a second conformation. Electron transfer takes places horizontally, conformational change vertically. (b) The same scheme, annotated for discussion in the text.

plex dissociation is very fast with respect to electron transfer, then there are not two but four complexes present in the reaction mixture. Consider two heme proteins and the desired reaction:

$$Fe^{II}_A/Fe^{III}_B \rightarrow Fe^{III}_A/Fe^{II}_B$$

If complex dissociation is fast with respect to electron transfer and if the association constants are independent of the oxidation state of the protein, then when electron transfer is half complete there will be an equimolar mixture of four species: the expected $Fe(II)_A/Fe(III)_B$ and $Fe(III)_A/Fe(II)_B$, and also $Fe(II)_A/Fe(II)_B$ and $Fe(III)_A/Fe(III)_B$. The latter are unreactive themselves but sequester reactive species.

3. FACTORS CONTROLLING ELECTRON TRANSFER

A. Distance

Both the classical and quantum-mechanical theories predict that the rate constant for electron transfer will fall off exponentially with distance between the two centers (r), $k_{et} \propto \exp(-\alpha r)$. This has been observed in many instances, with most studies having been done in the solid state. The constant, α, has been measured in conduction, photoconduction, fluorescence quenching, and pulse radiolysis experiments. Values of α range from 0.3 to 3.3 with most systems having values in the 0.7–1.2 range. Studies in inorganic and organometallic systems have been reviewed recently by Mikkelsen and Ratner,[14] Hush,[93] and Marcus and Sutin,[13] and the reader is referred to these articles for lists of experimental data. Other recent studies include that of Isied et al.,[143] who have measured electron transfer in a series of binuclear complexes of the form Os(III)—L—Co(III), where L is a spacer of variable length. For *trans*-polyproline spacers [iso(Pro)$_n$, n = 0–2], α = 2.0 Å^{-1}. Guarr et al. have investigated electron tunneling in rigid polymers and found a weak dependence of α on binding energy.[144]

Determination of α in biological systems is more difficult because it is difficult to create systems with all factors equal except the distance of the electron transfer. However, great progress is being made via the study of intramolecular electron transfer reactions. Mayo et al. have plotted data from seven intramolecular electron transfer reactions as a function of distance and found a intramolecular electron transfer reaction as a function of distance and found a value of α of 0.7–0.9 Å^{-1}.[11] There is some indication that the value of α may depend on the driving force for the reaction (see also Subsection E of Section 3). This is a very active field, and more data are to be expected soon.

B. Driving Force

a. Control of Redox Potential

As discussed in the previous section, the rate of electron transfer is expected to depend both on the driving force (ΔG^*) and on the reorganization energy (λ) of the reaction. This section discusses how proteins control the redox potential of the heme and, hence, the driving force for the reaction. Experiments to determine the reorganization energy of the reaction are discussed below in Subsection C of Section 3.

The factors which control the redox potentials of an electron transfer protein can be divided into changes in conformation (either bonding at the redox center or changes in the conformation of protein residues) and changes in the electrostatic interactions between the charge at the redox center and other charges in the protein or surrounding medium.[145] These are the factors which might change upon complexation, incorporation into a membrane, etc. The nature of the axial ligands is very important in determining the redox potential of the protein,[146] but there are apparently no examples of a cytochrome c or b_5 with one set of axial ligands in the ferric state and a different set in the ferrous state (see, however, the subsection entitled "The Axial Ligands" in Subsection B of Section 4). Changes in the axial ligand are important for myoglobin and hemoglobin; these proteins bind a water molecule to the iron in the ferric state but not in the ferrous state.

One factor which controls the redox potential is a change in the geometry at the redox center. However, these changes are usually small in electron transfer heme proteins. X-Ray studies of tuna cytochrome c[101] and *Pseudomonas aeruginosa* cytochrome c_{551} (Reference 147) in the oxidized and reduced forms show only small changes in the protein as a function of the redox state; similar results are found for cytochrome b_5.[103] Extended X-ray fine structure (EXAFS) studies on cytochrome c also showed only a small change.[148] Calculations of the energy required to reduce and oxidize cytochrome c have been performed by Warshel and co-workers; these are discussed below in the subsection entitled "The Rate as a Function of Temperature" in Subsection C of Section 3.

Electrostatic control of the redox potential is substantial. Kassner pointed out many years ago that placement of a charged heme in a nonaqueous environment significantly altered the redox potential.[149] This idea has been extended recently by Churg and Warshel,[150] who looked at the difference in redox potential between cytochrome c and the octapeptide–methionine complex formed by hydrolysis of cytochrome c. The latter has a redox potential that is about 300 mV more negative than the former. Calculations indicated that this was due largely to the destabilization of the oxidized heme by its local environment in the protein. Charge–charge interactions are also important. Moore[151] has postulated that the redox properties of cytochromes c might be governed in part by the electrostatic interaction between the positively charged ferric iron atom and a negatively charged propionate group.[130,152] Rogers et al.[153] have calculated the interaction between the ferric heme and buried heme propionate in cytochrome c_{551} as resulting in a lowering of the redox potential by about 65 mV at 25°C. Rees[154] has suggested that the redox potentials of electron transfer proteins correlate with the overall charge on the protein, with more negatively charged proteins having lower redox potentials. This is the expected direction of the effect, because the negative charges should stabilize the oxidized form of the protein. While many proteins follow this general pattern, there are exceptions.[145] Schejter et al.[155] have estimated the contribution of the interaction between the charge on the ferric heme and the surface charges of the protein. For *Euglena gracilis* cytochrome c_{552} at pH 7 and an ionic strength of zero ($I = 0$), they estimate an interaction of 2.4 kcal mol^{-1} ($E^{\circ\prime}$ of -105 mV); the redox potential of *E. gracilis* cytochrome c_{552} is 275 mV under the same conditions. For horse heart cytochrome c the contribution is smaller. The interaction is estimated to be only -1.1 kcal/mol ($E^{\circ\prime}$ of 48 mV) at pH 7 and zero ionic strength ($I = 0$). Under these conditions the $E^{\circ\prime}$ of the protein is 275 mV. Interesting examples of kinetic control of electron transfer via pH changes have been given by Schaap and Gagnon[156] and by Meyer and co-workers.[157]

The current interest in electron transfer between redox partners within a protein/protein complex focuses our attention again on the question reviewed by Nicholls in 1974 (Reference 158): How much do the properties of cytochromes change on binding to proteins? Changes in the redox potentials or in the distance between the two reaction centers or in the orientation of the two centers should all be important. From the early work of Vanderkooi and Erecinska,[159] one expects that the effect of complexation on the redox potential of cytochromes will be small. These authors studied the binding of cytochrome c to yeast ccp, cytochrome b_5, cytochrome c oxidase, and succinate cytochrome c reductase. No change of the E_m of the cytochrome c was seen for the first two proteins. In the latter two cases the E_m for cytochrome c was between the value in solution and that in membranes. The effect of complexation in the plastocyanin–cytochrome c complex has been studied by King et al.[160] No large effects were noted, but small effects would have been difficult to detect in this system. The next section considers in more detail the factors which control redox potential for cytochrome c, cytochrome b_5, and cytochrome c peroxidase.

b. Cytochrome c

The redox potential of horse heart cytochrome c is approximately 265 mV [$E^{\circ\prime}$ vs normal hydrogen electrode (NHE)].[161] The dependence of $E^{\circ\prime}$ on temperature has been studied under a number of conditions by Koller and Hawkridge.[162] In phosphate buffer (pH 7.0, $I = 0.20$) the formal potential decreased linearly from 267 mV at 5°C to 240 mV at 65°C. In Tris/cacodylic acid buffer (pH 7.0, $I = 0.20$) the potential decreased from 275 mV to 247 mV over the same temperature range.[162] Although the redox potential varied linearly, the heterogenous electron transfer rate constant showed a sharp break, first increasing and then decreasing as the temperature was raised. The effect was observed in both phosphate and Tris/cacodylate buffer. The rate constants were smaller in the former (by approximately a factor of two) and the break came at 41°C (55°C in phosphate). Koller and Hawkridge have postulated that the kinetic plot results from an increased exposure of the heme to solvent as the temperature is raised, which could stabilize the Fe(III) oxidation state. Cytochrome c has also been studied in membrane-bound form.[161] Binding of the cytochrome c to various membranes results in a drop in the $E^{\circ\prime}$ of 15 to 50 mV depending on the phospholipid and other experimental conditions. The states of the membrane also affect the redox potential.

There are several different anion binding sites on horse heart cytochrome c.[25,26,162] For this reason it has been suggested that Tris/cacodylate (a non-binding buffer) be used for kinetic studies.[26] However, although there is no evidence for cacodylate binding, buffer effects are often subtle and use of a nonbinding buffer does not necessarily result in simplified reactions. Koller and Hawkridge[162] observed breaks in their plots of the heterogeneous electron transfer rate constant vs temperature for both phosphate and cacodylate buffers. Kang et al. studied the reaction of yeast iso-1 and horse cytochromes c with yeast cytochrome c peroxidase in various buffers.[163] The ratios of V_{max} were quite dependent on the nature and concentration of the buffer. For example, V_{max}(iso-1)/V_{max}(horse) with cacodylate was 0.18 at 50 mM and 5.3 at 200 mM. With phosphate, the ratio was 2.2 at 50 mM and 17 at 100 mM, and with chloride the ratios were 0.75 at 50 mM and 11 at 200 mM. In all three media, there was an ionic strength at which V_{max}(iso-1) equaled V_{max}(horse).

c. Cytochrome b_5

Reid et al.[164] have measured the redox potential of cytochrome b_5 as a function of pH, temperature, and ionic strength. The dependence of $E°$ on pH was consistent with an ionizable group on the protein with a pK_a of 5.7 in the oxidized protein and 5.9 in the reduced protein. This group was not one of the heme propionates, as shown by reconstitution of apocytochrome b_5 with hemin dimethyl ester.[164] The reconstituted protein (DME-b_5) differed from the native protein in that its redox potential was higher by more than 50 mV [68.8 vs 5.1 mV at pH 7.0 (phosphate), $I = 0.1$ M, 25°C]. The pH and ionic strength dependencies of the redox potential were very similar in the native and reconstituted protein, however, indicating that ionization of the heme propionates is not the reason for the observed pH dependence of the redox potential. It should be noted that cytochrome b_5 undergoes a conformational change to a high-spin form above 45°C; the change is reversible below 55°C.[165] Salerno et al.[166] have pointed out that mitochondrial cytochromes b_5 undergo gradual and only partially reversible changes upon lipid removal.

d. Cytochrome-c Peroxidase (c-cp)

The redox potential of ccp is a function of pH, decreasing linearly with a slope of -57 mV/pH unit.[167] At pH 7 and 4.5, the redox potentials are -194 mV and -52 mV, respectively (25°C). The pH dependence indicates that the ferrous protein has a residue with a pK_a of 7.6 ± 0.1, while the ferric protein does not show a pK_a in this region. Early work by Nicholls and Mochan showed that ccp showed approximately equal binding with ferric and ferrous cytochrome c.[168] More recent work by Erman and co-workers confirms this and shows that the two redox forms differ in their binding only at high ionic strength.[169] Electron transfer kinetics in the cytochrome c–cytochrome c

peroxidase complex[16,118,120,121] are measured at low ionic strength, which promotes complex formation. Under these conditions the binding constants of the two redox forms of the complex are the same.

The purpose of this discussion of redox potentials was to assess the possible magnitude of the redox potential change that might accompany either complexation or a conformational change within the complex. This issue has been under consideration for many years.[158] While a good deal is now known about the factors which control redox potentials in heme proteins, there seems to be no evidence for any dramatic changes of redox potential upon complexation of cytochromes c and b_5 with other proteins. The magnitude of redox changes induced by conformational changes (eg, approach to the transition state) is at present unknown. Experimental evidence would be very welcome but is difficult to obtain. Progress will most likely come with further work on calculations.[72,73,170]

C. Reorganization Energy

As discussed in Subsection B of Section 2, the rate of electron transfer is thought to depend on the reorganization energy of the donor and acceptor according to Equation 5-2. There have been a number of efforts to determine the reorganization energy involved in heme protein electron transfer. These fall into four categories: (1) investigations of model systems, (2) calculations based on crystal structure data, (3) measurements of rate constants as a function of temperature, and (4) measurements of rate constants as a function of the driving force for the reaction. It is somewhat difficult to compare numbers obtained from different techniques. Some calculations give the free energy and some give the enthalpy; if the entropy is not known, no direct comparison can be made. In addition, the ΔG^* of electron transfer theory is similar, but not identical, to the $\Delta G^{\ddagger}$ of Eyring theory. These difficulties aside, one can make two generalizations. Electron transfer between two low-spin hemes has only a relatively small reorganization energy. Electron transfer accompanied by changes of the axial ligands at the heme center has a larger reorganization energy.

a. Model Systems

The first question to be considered is whether the heme itself has any significant reorganization barrier to electron transfer. Model studies show the intrinsic barrier to electron transfer between hemes is very low. Dixon and co-workers measured the electron self-exchange rate constants, using NMR techniques, for a series of iron bis(N-alkylimidazole)-tetraphenylporphyrins.[24] The rate constants ranged from 5×10^7 to $2 \times 10^8 \, M^{-1} \, s^{-1}$ in CD_2Cl_2 at $-21°C$. Although this reaction is fast, it is not diffusion controlled. This is primarily because electron transfer is accompanied by rearrangement of the solvent around the complexes as the electron is transferred (outer-sphere reorganization, Equation 5-4). There are two lines of evidence for this. First, the sterically hindered Fe(II/III)(2,4,6-Me$_3$TPP)(1-MeIm)$_2^{0/1}$ system had an electron self-exchange rate constant approximately twice that of the Fe(II/III)(TPP)(1-MeIm)$_2^{0/1}$ system.[24] This is most easily explained if the methyl groups in the former complex keep the solvent farther away from the iron center, resulting in less outer-sphere reorganization. Second, the solvent dependence of the rate constant for Fe(II/III)(TPP)(CN)$_2^{2-/1-}$ in Me$_2$SO-d_6 and CD$_3$OD was in line with Marcus theory predictions for outer-sphere reorganization.[171] Thus, the intrinsic barrier to electron transfer in low-spin hemes is very close to zero.

b. Calculations from Crystal Structures

Warshel and co-workers[172,173] have computed ΔG^* from the crystal structures of oxidized and reduced tuna cytochrome c. The X-ray structure of tuna cytochrome c shows that the bond distances and angles at the heme iron undergo little change as the heme is oxidized and reduced.[101] Warshel and co-workers have calculated an upper limit to the reorganization energy, λ, of 6.4 kcal/mol^{-1} ($\lambda/4$ of 1.6 kcal mol^{-1}).[172,173] This includes values for changes in the electrostatic potential of the heme atoms (2.4 kcal mol^{-1}), relaxation of the hemes (approximately 1 kcal mol^{-1}) and rearrangement of the surrounding water molecules (3 kcal mol^{-1}). Marcus and Sutin[13] have proposed an alternate calculation assuming the protein and water around the heme can be treated as a continuous dielectric with $D_e = 10$. Given a heme radius of 5 Å and an Fe—Fe distance of 18 Å, they estimate a total $\lambda/4$ of 6 kcal mol^{-1}. Values from these two estimates are different by a factor of four, but both estimates indicate that the reorganization energy is low.

b. The Rate as a Function of Temperature

Determination of the reorganization energy from the electron transfer rate constant as a function of temperature has been done for both intermolecular and intramolecular reactions. For intermolecular electron transfer, the self-exchange rate constants are most useful because $\Delta G^° = 0$ and there is no net thermodynamic driving force for the reaction. Gupta[174] measured the activation energy for electron self-exchange of horse heart cytochrome c as 13 ± 1 kcal mol^{-1} at 0.1 M ionic strength and 7 ± 1 at 1 M ionic strength. Dixon and Barbush found similar values for pigeon cytochrome c, $\Delta H^{\ddagger} = 9.7 \pm 0.9$ kcal mol^{-1} and $\Delta S^{\ddagger} = -6.3 \pm 3.0$ eu.[175] These experimental values are somewhat larger than the reorganization energies calculated from the crystal structures, as expected because factors other than the reorganization energy, primarily the electrostatic work necessary to bring the complexes together in solution, also enter. The dependence of the rate constant on ionic strength indicates the importance of these electrostatic interactions.

    In measurements of intramolecular electron transfer, both Gray and co-workers[176] and Isied and co-workers[177] have measured the temperature dependence of intramolecular electron transfer from Ru(II) to Fe(III) in $Ru(NH_3)_5$(histidine-33)$^{2+}$-ferricytochrome c. The flash photolysis studies of Nocera et al.[176] gave a ΔH^* of approximately 1.5 kcal mol^{-1} while the pulse radiolysis studies of Isied et al. gave $\Delta H^* = 3.5$ kcal mol^{-1}. Using Marcus theory, the standard enthalpy change for the electron transfer reaction ($\Delta H^° = -11.9$ kcal mol$^-$), and the reorganization energy of electron self-exchange in $Ru(NH_3)_5(py)^{3+/2+}$, Nocera et al.[176] calculated that ΔH^* of the cytochrome c self-exchange reaction was approximately 8 kcal mol^{-1}. This is similar to the $\Delta H^‡$ of the self-exchange reaction. For intramolecular electron transfer, $\Delta H^‡$ and ΔH^* are the same, but for bimolecular reactions, the difference between $\Delta H^‡$ and ΔH^* depends on the model which one assumes for the association of the two partners.[104] For bimolecular reactions extrapolated to infinite ionic strength, the two enthalpies should differ by only a few kilocalories.

Gray and co-workers have measured long distance electron transfer in ruthenated myoglobins.[61,124] The redox potentials of the two metal sites in pentaammineruthenium (His-48)-myoglobin differ by only 20 mV, allowing measurement of electron transfer in both directions.[124] The forward (Ru^{2+} to Fe^{3+}) and reverse (Fe^{2+} to Ru^{3+}) had $\Delta H^‡$ values of 7.4 ± 0.5 and 19.5 ± 0.5 kcal mol^{-1}, respectively. A reorganization enthalpy (ΔH^*) of 20 kcal mol^{-1} was calculated assuming a ΔH^* for the ruthenium center of 6.9 kcal mol^{-1}. The reorganization energy of myoglobin was higher than that of cytochromes c; this was as expected because the former undergoes loss of a bound axial water molecule upon reduction.

Peterson-Kennedy et al. have measured the rate constant as a function of temperature for intramolecular electron transfer in [Zn,Fe(III)] hemoglobin hybrids.[96,178] The [α(Zn), β(Fe(III)H$_2$O)] complex had an optical spectrum that did not change upon cooling from 313 to 77 K, consistent with ligation of the Fe(III) center that was invariant over this temperature range. A fit to a quantum-mechanical formulation of electron transfer gave a reorganization energy of 2.06 eV (48 kcal mol^{-1}). Below 160 K, the rate constant was independent of temperature with a value of 9 s^{-1}. Again, the reorganization energy is expected to be large because ferrihemoglobin loses its axial water molecule upon reduction. The authors note that this value for the reorganization energy is the average of those for cytochrome c (λ = approximately 28 kcal mol^{-1}) and that for myoglobin (of approximately 80 kcal mol^{-1} (Reference 124), as might be expected for electron transfer between two centers, one which loses a ligand and the other which does not.

c. The Rate as a Function of Driving Force

Theory predicts that the rate of electron transfer will be at a maximum when the driving force for the reaction equals the reorganization energy (ie, which $\Delta G^* = -\lambda$, Equation 5-2). Thus, measurements of the rate constant as a function of driving force should show a maximum at the reorganization energy of the system. McLendon and Miller have measured rates of electron transfer as a function of the driving force for the reaction in four cytochrome c/cytochrome b_5 complexes.[122] Fitting of the data to Marcus theory gave a maximum rate for electron transfer at a driving force of approximately 17 kcal mol^{-1} (Figure 5-8). This value of λ is between the theoretical value calculated by Warshel and co-workers and the experimental values calculated from self-exchange measurements and measurements of electron transfer in ruthenated cytochrome c. Considering the very different techniques used, the agreement is reasonable. The agreement also indicates that geometry changes at the interface between the two proteins are not a major factor in the reorganization energy.

Intramolecular electron transfer in various substituted cytochrome c–cytochrome c peroxidase complexes has also been measured.[16,118,120,121] The rate constant as a function of driving force is again plotted in Figure 5-8; a good fit is found to a reorganization energy of 1.5 eV. Gray and co-workers have extended their earlier work[124] to measure long-distance electron transfer in a number of ruthenated myoglobins.[11,61] The rate constants as a function of driving force were consistent with a λ of 1–2 eV.

Cho et al.[179] have investigated electron transfer between cytochrome c and

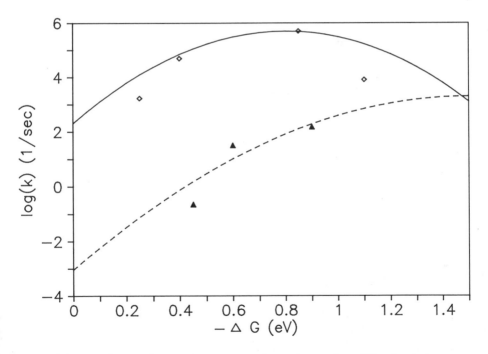

Figure 5-8. Log k vs ΔG for cytochrome c–cytochrome b_5 complexes (◇) and cytochrome c–cytochrome-c peroxidase complexes (▲). The lines are theoretical fits to Equation 5-1 with the preexponential factor $= 5 \times 10^5$, $\Delta G = 0.8$ eV for the c/b_5 complex,[122] and the preexponential factor $= 2 \times 10^3$/s, $\Delta G = 1.5$ eV for the c/ccp complex.[120]

various photoexcited porphyrins. The second-order rate constant increased as the driving force for the reaction increased. They calculated a λ of approximately 1 eV. However, it should be noted that it was not possible to observe a decrease in reaction rate below the maximum with increasing energy, as is common in bimolecular reactions where the observed rate constant levels out at the diffusion-controlled rate constant.[48]

It is encouraging that these different approaches and different biochemical systems all lead to the same general conclusions: that systems which have axial ligand changes have much higher reorganization energies than those which do not, and that electron transfer between two low-spin hemes has a relatively small reorganization energy. A number of laboratories continue extensive work in this area, and progress should be rapid.

D. Orientation

The orientation of the donor and acceptor may be an important controlling factor in biological electron transfer. Good experimental models are very difficult to design because a change in orientation is necessarily accompanied by many other changes in the molecule. Progress to date has, therefore, been mainly in theory, although experimental progress is beginning. Cave et al.[47] have modeled the effects of orbital and potential shape on the rate of electron transfer at defined geometries. The matrix element for electron transfer (H_{AB}) is very sensitive to the orientation and the particular orbitals involved in the transfer. For example, Figure 5-9a shows the relative orientations of two porphyrins which can adopt a face-to-face orientation or swing apart to assume a butterflylike geometry. Cave et al. have considered electron transfer within a complex in which the donor is photoexcited. The forward process involves electron transfer from the lowest excited singlet of the donor to the acceptor to give a donor radical cation and an acceptor radical anion; the reverse process is electron transfer from the LUMO of the acceptor radial anion (now occupied by the single electron) to the HOMO of the donor radical cation. Figure 9b shows the value of H_{AB} for both of these processes as a function of the angle between the porphyrin planes. H_{AB} shows a substantial dependence on the angle between the two porphyrins. For the orbital assumptions of this model, the forward and reverse electron transfers are significantly different for values of less than 20°. This difference becomes negligible for angles greater than 20°. A related theoretical treatment has been presented by Domingue and Fayer.[46] Again, orientation is calculated to influence the rate of electron transfer substantially.

In experimental work, Closs et al.[48] have investigated intramolecular electron transfer in a series of molecules of the form donor–spacer–acceptor, an extension of their seminal work on electron transfer as a function of driving force.[10] The rate constant reached a maximum at approximately 1 eV driving force and decreased as the reaction was made more exothermic. Rate con-

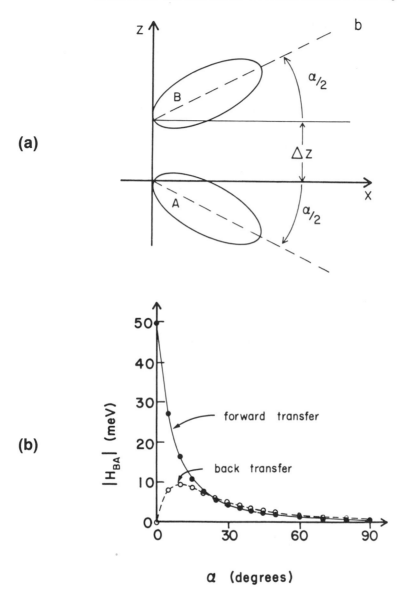

Figure 5-9.   (a) Relative orientation of two porphyrins which can undergo electron transfer. (b) The calculated matrix element H_{AB} as a function of the angle between the two porphyrin planes. Reproduced with permission.[47]

stants were strongly influenced by the geometry of attachment of the donor and acceptor to the spacer. These substantial stereoelectronic effects indicate that through-bond coupling is important. They have proposed that the absence of through-bond connectivities in proteins may substantially slow electron transfer.

E. Intervening Material

The nature of the intervening material between the donor and acceptor may also be very important. Again, this is difficult to study in the proteins themselves because it is difficult to alter the material between two electron transfer centers while maintaining all other aspects of the electron transfer reaction (distance, orientation, driving force) unchanged. However, progress is being made in both experiment and theory.

Beratan[107] has treated electron transfer in the porphyrin-{[2.2.2]-bicyclooctane}$_n$-quinone system. This was studied experimentally by Dervan and Hopfield and their co-workers[55,180] and by Bolton et al.[56] Beratan concludes that, in this system, porphyrins with higher energy excited states should show a steeper dependence of the electron transfer rate constant on distance. In particular, the values for α in H_{AB} [$\propto \exp(-\alpha R)$] in the forward and reverse direction are 0.9 and 0.5 Å, respectively. This translates into a slowing of the forward and reverse rate constants by factors of 1500 and 60, respectively, per bicyclooctane unit. The difference arises from the energies of the orbitals and from their symmetry with respect to the linker orbitals. In very recent work, Onuchic and Beratan[181] discuss the expectation, based on topological considerations, that different bridges are expected to give different electron transfer rate constants even when systems employ the same donors, acceptors, and electron transfer distance. In addition, Beratan et al.[17] have proposed that fluctuation of interactions within the bridging medium influences the dependence of the rate on temperature. Although the protein systems are far more complicated than the models, the idea that the rate constant depends on the orbital energies and topologies needs further exploration.

The idea that aromatic residues between the donor and acceptor can promote electron transfer has been in the literature for many years. Experimental evidence has been difficult to obtain. However, Mauk, Hoffman, and co-workers have recently published a study of electron transfer between yeast cytochrome c and Zn cytochrome-c peroxidase.[16] Both forward [^{3}ZnCcp-Fe(III)Cyt$c$] and reverse [ZnCcp$^+$-Fe(II)Cyt$c$] intramolecular rate constants were measured for yeast cytochrome c with phenylalanine, tyrosine, serine, or glycine at position 87. The forward rate constants for the first three derivatives were all about 200 s^{-1}, while that for Gly-87 was only 13 s^{-1}. The difference in the reverse rate constants was even more dramatic: 1.7×10^4 s^{-1} for the aromatic residues and 1–2 s^{-1} for the aliphatic residues. The molecular basis of this large difference is not yet understood. The authors have speculated that the aromatic residue may enhance "hole" superexchange interactions, although conformational changes or differences in the structure of the complexes cannot yet be ruled out. It should be noted, however, that the rate constants greater than 10^4 s^{-1} are almost surely faster than the rate of dissociation of the complex, even at 1 mM phosphate and

0°C. This is discussed in more detail in the subsection entitled "Cytochrome c/Cytochrome-c Peroxidase" in Subsection B of Section 4. It is interesting that the replacement of Phe-87 does influence these intramolecular electron transfer reactions, but does not have a noticeable effect on bimolecular electron transfer.[21] The lack of a dramatic effect on bimolecular electron transfer has also been observed by Kortenaar et al., who used semisynthesis techniques to alter Phe-82 of cytochrome c to Leu and found that the change neither altered the redox potential of the cytochrome c nor prevented its electron tranfer to cytochrome c oxidase.[39]

In other work on the role of aromatic residues in electron transfer, Kuki and Wolynes[15] have used a Monte Carlo method to calculate electron tunneling pathways from a surface ruthenium atom to the heme in a derivative of myoglobin in which Ru has been attached to the surface of the protein. They found that the pathways followed a reasonably direct line between the two metal centers. Tryptophan-14, which occurs near the "line of flight" of the electron between the two centers, did not receive a disproportionate number of trajectories of the electron. This may indicate that aromatic residues in the pathway do not facilitate electron transfer; further work on this question is necessary.

F. Spin State

Electron transfer is generally faster between two low-spin species than between a low-spin and a high-spin species, although the observed differences probably mainly reflects larger changes in geometry in the latter case than in the former.[11,182] This, in turn, should be seen as a larger reorganization energy in high-spin than in low-spin hemes. This has been observed in the studies to date, as discussed in Subsection C of Section 3. Recent work on cytochrome P-450 should also be mentioned in this context. Fisher and Sligar[183] have measured the rate constants for electron transfer to ferric cytochrome P-450 as a function of the spin state of the iron (controlled by the addition of various substrates).[184] The rate constants increased and the activation energy decreased as the fraction of low-spin protein increased; there was a very good linear free energy relationship between the rate constants and activation energies and the free energy of the spin-state equilibrium ($-RT \ln K_{spin}$). While this may be simply a spin-state effect, the spin state is correlated with the conformation of one tyrosine residue.[185] A correlation between the spin state and reaction rate is not general for cytochrome P-450, however; Lambeth and Kriengsiri observed no correlation in the reaction of cytochrome P-450 and adrenodoxin.[186] The recent X-ray structure of cytochrome P-450$_{cam}$ provides a basis for further work in this area.[187]

4. INTRAMOLECULAR ELECTRON TRANSFER

A. Protein–Protein Complexes

Much of the current work on the dependence of the rate constant for electron transfer on the driving force of the reaction is being done on protein–
protein complexes. These include the tetrameric hemoglobin[96,178,188,189] and
the complexes cytochrome c/cytochrome c peroxidase,[16,118,120,121] cytochrome c/cytochrome b_5 (References 122 and 125), and hemoglobin/cyto-
chrome b_5.[190] In all cases, electron transfer measurements have been made
under conditions which strongly favor the complex of the two proteins.
However, even when the proteins are found entirely in the complexed form,
the rate constants of formation and dissociation of the complex are important if these rate constants compete with that of electron transfer. Finally,
these experiments are also difficult because the complexation cannot always
be forced to greater than 99% at the concentrations of protein and buffer
accessible. This problem should not affect observation of the processes
which contribute substantially to the observed optical density change, but
makes it difficult to observe cleanly those kinetic processes which contribute
little to the observed optical density change.

a. Cytochrome c/Cytochrome b_5

Mauk et al.[74] found equilibrium constants for the cytochrome b_5/cytochrome
c complex of $8 \pm 3 \times 10^4 \, M^{-1}$ in $I = 0.01 \, M$ phosphate (pH = 7.0, 25°C) and
$4 \pm 3 \times 10^6 \, M^{-1}$ in $I = 0.001 \, M$ phosphate (pH = 7.0, 25°C). From the
dependence of the equilibrium constant on ionic strength, they estimated an
association constant of $3 \pm 2 \times 10^7 \, M^{-1}$ at zero ionic strength. The equilibrium constants were calculated from the absorbance changes upon mixing;
at pH 7.0 {$I = 0.001 \, M$ and [cytochrome c]/[cytochrome b_5] = 1}, the
absorbance of the protein at 416 nm is about 2% greater than that of the
proteins before mixing.

   The rate constants for the reaction of cytochrome c with cytochrome b_5
has been measured by stopped-flow techniques as $3.9–4.7 \times 10^7 \, M^{-1} s^{-1}$ (pH
7 phosphate buffer, $I = 0.1 \, M$).[122,191] It has also been calculated from the
kinetics in the NADH–cytochrome b_5 reductase–cytochrome b_5–cyto-
chrome c system as $3 \times 10 \, M^{-1} s^{-1}$ in 0.1 M phosphate, pH 7.0 and 3×10^8
$M^{-1} s^{-1}$ in 0.02 M phosphate, pH 7.0, 0.200 M sucrose.[192,193] Stonehuerner et
al.[193] measured the rate constant as a function of ionic strength in a number
of buffers. Plots of the rate constant as a function of the square root ionic
strength were linear and extrapolated to a rate constant at zero ionic strength
of approximately $5 \times 10^8 \, M^{-1} s^{-1}$. Differences in reaction conditions, calculation of association rate constants, and extrapolation of rate and equilibrium
constants to different ionic strengths all present problems. However, a rough
guess of the off rates can be made by dividing the on rate by the association

equilibrium constant. To the nearest order of magnitude, these are 10 at $I = 0$, 100 at $I = 0.001$ M, and 1000 at $I = 0.01$ M. McLendon and Miller have investigated the dependence of the electron transfer rate constant within the cytochrome c/cytochrome b_5 complex on the exothermicity of the reaction.[122] Their buffer was 1–2 mM phosphate and all rate constants were greater than 10^3 s^{-1}. Thus, in this system intramolecular electron transfer is substantially faster than dissociation of the complex.

b. Cytochrome c/Cytochrome-c Peroxidase

The kinetics and equilibria of the reaction between cytochrome c and ccp are complex. A thorough discussion of the issues is beyond the scope of this review; we focus here only on the association and dissociation rate constants and association equilibrium constant.

The complex between cytochrome c and cytochrome c peroxidase has been studied by Erman and co-workers,[169,194] extending earlier work by Kang et al.[163] and by Leonard and Yonetani.[195] Erman and Vitello investigated the complex formation as a function of ionic strength in cacodylate/KNO$_3$ buffer and found the association constant to decrease from 6.0×10^6 M^{-1} to 2.2×10^3 M^{-1} as the ionic strength increases from 0.01 to 0.20 M.[194] This work was based on the small perturbation of the absorption that occurs when the two proteins form a complex. The maximum change was observed at 408 nm and was an increase of a little over 1% in the absorbance under conditions where 73% of the cytochrome-c peroxidase was bound. More recently, Erman and co-workers have measured the equilibrium constants for cytochrome c/ccp binding as a function of ionic strength in phosphate/KNO$_3$ buffer at pH 7.5.[169a] They found an equilibrium association constant of 2×10^6 M^{-1} at $I = 0.01$. Extrapolation to $I = 0$ and $I = 0.10$ M gave equilibrium constants of approximately 10^8 M^{-1} and approximately 2×10^4 M^{-1}, respectively. Kang et al.,[163] using gel filtration techniques, found evidence for binding of more than one cytochrome c to cytochrome c peroxidase under some experimental conditions; this was not observed by Erman and co-workers.

Kang et al. have investigated the reactions of horse and yeast iso-1-ferrocytochromes c with yeast ccp.[196] For horse heart cytochrome c in phosphate buffer, the apparent second-order rate constant at $I = 0.1$ is approximately 7×10^7 M^{-1} s^{-1} and extrapolation of the rate constant as a function of the square root of ionic strength gives rate constants of 5×10^8 M^{-1} s^{-1} and approximately 10^9 M^{-1} s^{-1} at $I = 0.01$ and $I = 0$, respectively. As above, calculation of the off rates will give only approximate values. To the nearest order of magnitude these are 10 at $I = 0$, 100 at $I = 0.01$, and 1000 at $I = 0.10$. Gupta and Yonetani investigated the cytochrome c–cytochrome c peroxidase complex by NMR and calculated a lower limit to the off rate of 200 s^{-1}.[197] Satterlee et al. have reinvestigated this system very recently. The cytochrome c–cytochrome c peroxidase complex is in rapid exchange on the

NMR time scale at 500 MHz. They estimate a lower limit of 1100 s^{-1} for the off rate (23°C, pD = 6.6; 0.01 M KNO$_3$).[169c]

There have been a number of recent measurements of the rate constant for intramolecular electron transfer in the cytochrome c/ccp complex. Experiments have been done at both 0.010 and 0.001 M phosphate buffer. For example, Cheung et al. have reported a rate constant of 0.23 ± 0.02 s^{-1} for electron transfer within the Fe(II)ccp/Fe(III)cytochrome c complex (generated by rapid mixing, pH 7.0, 10 mm phosphate).[121] If the off rate of 100 s^{-1} estimated above is correct, the complex has dissociated approximately 300 times by the time electron transfer is half complete.

It should be noted that cytochrome c/ccp reactions are quite dependent both on the nature of the buffer and on the species of cytochrome c employed. For example, the apparent second-order rate constant for yeast iso-1-cytochrome c is approximately 10 times that of horse heart cytochrome c in a Tris buffer (pH 6.0) with phosphate counterions at I = 0.2.[196] With chloride counterions, however, the difference vanishes. It is interesting that the factor of 10 difference in on rates of horse and yeast cytochrome c is the same as that observed in the intramolecular electron transfer experiments both in the native proteins[121] and in the quenching of the Zn–ccp triplet by ferric cytochrome c.[118] Kang et al. have found that the apparent second order on rates show substantially more variation with species than do the equilibrium constants.[163] If the difference appears also in the off rate, then the possibility that the observed species dependence of the electron transfer rate constant is connected to the off rate must be explored. In this context a recent study of Bhattacharyya et al. on the kinetics of electron transfer in ferredoxin-ferredoxin-NADP$^+$ reductase complexes is of interest.[193] The off rates of these complexes are within a factor of two of the electron transfer rates. It is not clear whether this is a coincidence or whether the two processes are related in some way.

B. Protein–Organometallic Complexes

Bechtold et al. have investigated intramolecular electron transfer in cytochromes c with various ruthenium derivatives attached to His-33.[60] When the Ru center was an (NH$_3$)$_5$Ru(II) group, electron transfer occurred to the Fe(III) heme with a rate constant of approximately 53 s^{-1} as measured by pulse radiolysis. However, Bechtold et al. observed no intramolecular electron transfer from the Fe(II) to the Ru(III) center in a similar complex in which one of the ammonium moieties was replaced with a nicotinamide to give a similar driving force in the "reverse" direction. Although this result is unexpected because the driving forces for the two reactions are similar and, in a single system, the forward and reverse pathways must be microscopically the same, it is the case that the forward and reverse reactions were measured in somewhat different systems. In very recent work, Lieber et al.

have measured the electron transfer from Fe(II) to Ru(III) in a ruthenium-modified myoglobin.[61] This rate constant was 0.058 ± 0.004 s^{-1}, within experimental error of the reverse electron transfer rate of 0.060 ± 0.004 s^{-1}.[124] This result is in line with expectations that the forward and reverse reaction rates should be similar for systems with similar driving forces in both directions.

C. Conformational Change

As discussed in the subsection entitled "The Role of Protein Conformational Change" in Subsection C of Section 2, protein conformational change can be important because it can alter the redox potential of two electron transfer partners, their distance, or their orientation. In addition, it is possible to mistake conformation change for electron transfer. This section discusses conformational changes in heme proteins and electron transfer reactions which have been discussed in terms of conformational change.

a. Large Conformational Changes

Large conformational changes certainly occur in many heme proteins. A striking example of this is found in studies of reconstituted heme proteins. The heme in cytochrome *c* is attached via thioether linkages at the 2- and 4-positions and cannot be removed without breaking covalent bonds. However, the heme in cytochrome *b*$_5$ is held in only by the two axial histidine linkages. Native cytochrome *b*$_5$ is a mixture of two isomers, as first observed in the hyperfine shifted peaks in the NMR by Keller et al.[199] These isomers are two protein forms with the heme flipped by 180° around the α, γ-meso axis.[200,201] At equilibrium, both forms are present in about a 10 : 1 ratio. However, immediately after reconstitution, the two forms are present in approximately equal amounts.[201] The rate of isomer interconversion is a strong function of pH; rate constants are 6.2×10^3 (pH 5.8), 0.54×10^3 (pH 7.0), 4.5×10^{-3} (pH 8.3), and 5.1×10^3 (pH 8.9) per minute.[202]

The kinetics of interconversion between these two isomers of heme proteins have been studied in a number of other instances as well. In sperm whale myoglobin, the two forms equilibrate on the time scale of hours.[203] The kinetics of yellowfin tuna myoglobin[204] and various forms of hemoglobin A[205] have also been studied; the approach to equilibrium is a sensitive function of the pH and axial ligand in these systems. It is interesting to note that for sperm whale myoglobin heme reorientation is slightly faster than heme replacement with another, similar heme from solution.[206] La Mar et al. have interpreted that as a reorientation occurring in a "protein cage." For hemoglobin, the reorientation and replacement have approximately equal time constants, indicating that the heme dissociates and then reinserts.[205]

The functional significance, if any, of the presence of two isomers of heme proteins is not yet clear. Livingston et al. have found that the two forms of sperm whale myoglobin have oxygen affinities that differ by about a factor of 10, with the minor form having the higher affinity.[207] However, Olson and co-workers[208] have not found any difference in either the kinetic or equilibrium parameters for binding of CO and O_2 to the two forms of myoglobin. This discrepancy remains to be resolved. In another study, Yamamoto and La Mar have indicated that the rate of autoxidation of oxy-Hb has a slight dependence on the orientation of the heme. The azide affinity of the Met-aquo form of the protein is significantly lower for the reversed than for the native orientation.[205] Although the interconversion of the two forms in heme proteins is probably too slow to be of consequence for electron transfer in any of the systems studied to date, it is clear that large motions of the heme protein are occurring.

b. Small Conformational Changes

Many smaller conformational changes have also been measured. Burns and La Mar have showed by NMR that there is conformational heterogeneity around the 3-Me group in many eukaryotic cytochromes.[209] For horse heart cytochrome c, the amounts of the two conformers are somewhat pH dependent, varying from approximately $1:2$ at pD 8.4 to $1:1$ at pD 6.0. At pD 5.8 the amounts of the two conformers are approximately equal and the lifetime in each is approximately 10^{-3} s (interconversion rate constant approximately 1000 s^{-1}). There is no direct evidence that this conformational heterogeneity is reflected in rate constants for electron transfer. However, Dixon and Barbush have observed non-Arrhenius behavior below room temperature in electron self-exchange measurements for some eukaryotic cytochromes c which may be due to slow interchange between two conformations.[175]

Williams, Moore, and co-workers have studied the motions of tuna cytochrome c in some detail.[26,27] The protein has eight tyrosine and phenylalanine residues. Four of these residues are packed together in two pairs and undergo ring flipping only slowly on the NMR time scale. The rate constant for Tyr-67 is also slow at room temperature, although flipping can be seen at high temperature. Phe-36, Try-74, and Phe-82 undergo rapid ring flipping. The flip rate for Phe-82 is greater than 10^4 s^{-1}. This rate constant is substantially faster than most rates that have been measured for intramolecular electron transfer within a cytochrome c–partner complex. Internal valine or leucine residues flip slowly on the NMR time scale (<100 s^{-1}).

Protein motions on a much faster time scale have been calculated. Northrup et al. simulated the molecular dynamics of ferrocytochrome c and compared these with thermal parameters from the X-ray structure.[210] The two types of approaches gave similar results. As expected, residues on the surface of the molecule have greater mobility than those in the interior, and many of the motions are quite anisotropic. Henry et al. have recently reported on protein motions on the fast time scale using molecular dynamics simulations for cytochrome c.[211]

c. The Axial Ligands

Motion of the axial ligand might be expected to be important because the geometry change around the heme, and hence the reorganization energy, will be greater in systems with labile axial ligands. It has been known for many years that the Fe(II)—S bond of cytochrome c is labile. Sutin and co-workers[212–214] first measured the off rate of the methionine in horse heart cytochrome c by adding another ligand which could bind to the five-coordinate Fe(III) atom; rate constants of 30–50 s^{-1} were found. Bechtold et al.[123] have recently reinvestigated this reaction with imidazole as the external ligand and calculated the following values, assuming rate-limiting dissociation of the methionine: $k_{substitution} = 28 \pm 2$ s^{-1}, $\Delta H^{\ddagger} = 13.9 \pm 0.5$ kcal mol^{-1}; $\Delta S^{\ddagger} = 5 \pm 1$ eu. These compare with $k = 55$ s^{-1}, $\Delta H^{\ddagger} = 3.5$ kcal mol^{-1} for electron transfer.[123] Thus, the rate constants for methionine dissociation and electron transfer differ by only a factor of two at room temperature, but the temperature dependence of the loss of the methionine is far greater than that for electron transfer. As Bechtold et al. have pointed out, the difference in activation enthalpies is evidence against the hypothesis that dissociation of Met-80 is required for intramolecular electron transfer.

Bechtold et al. have also measured the rate constant for intramolecular electron transfer in a solution of cytochrome c containing 0.1 M imidazole.[123] The decay after pulse radiolysis ($CO_2^{\cdot -}$ as reductant) was fit to two exponentials which gave rate constants of 55 s^{-1} (native cytochrome c) and 1.2 s^{-1} (presumably the imidazole-ligated species.) Surprisingly, this rate constant for electron transfer to the imidazole-ligated species is the same as that for loss of imidazole $= (1.5 \pm 0.1$ s$^{-1})$. The simplest explanation is that electron transfer and ligand loss occur simultaneously, but this is unexpected in view of the general observation that electron transfer is slowed by reorganization of the bonds at the electron transfer center. One complicating factor in these experiments is that imidazole might bind to the heme in more than one orientation.

If the rate of electron transfer is slower than the off rate of an axial ligand, then the geometry of the redox center at the time of electron transfer is not necessarily well defined. However, in many instances the rate of electron transfer is substantially larger than that of axial ligand loss. For example, McLendon and Miller found rates for intramolecular electron transfer within the cytochrome c–cytochrome b_5 complex of 10^3 to 5×10^5 s^{-1},[122] far faster than the slow conformational reorganization of the methionine of cytochrome c. Meyer et al. have investigated the reduction of tuna cytochrome c by FMN semiquinone.[215] The second-order rate constant was approximately 1×10^8 M^{-1} s^{-1} (ionic strength of approximately 0.03 M). The flavin concentration was 60 μM, giving a first-order rate of 6×10^3 s^{-1}, again far faster than the off rate of the methionine of cytochrome c. Similar observations have been made for the intramolecular oxidation of ferrocytochrome c by bound ferricyanide, although this is a complicated system to analyze.[216]

A related example of ligand loss or reorganization is found in the bacterial cytochromes c_{553} from *Desulfovibrio vulgaris* and *Desulfovibrio desulfuricans*. Senn et al. have interpreted the NMR and circular dichroism (CD) spectra of these proteins as indicating an S chirality of the iron-bound methionine in the ferrous form and an R chirality in the ferric form.[217] This was unexpected because X-ray structures on horse heart cytochrome c and *Ps. aeruginosa* cytochrome c_{551} show very similar geometries in the oxidized and reduced states.[101,147] Senn and Wüthrich have shown that these proteins also have low redox potentials, about 250 mV lower than other c-type cytochromes.[65] For comparison of electron transfer properties between the *Desulfovibrio* and other proteins, one wants the electron self-exchange rate constants to circumvent the problem of different driving forces for the reactions of different proteins. The electron self-exchange rate constants of these *Desulfovibrio* cytochromes are not known. Mixtures of the Fe(II) and Fe(III) oxidation states are in slow exchange on the NMR time scale.[217] However, to achieve slow exchange it was necessary to cool the solutions. Horse heart cytochrome c is in slow exchange at ambient temperature. The NMR properties of horse heart and the *Desulfovibrio* cytochromes are similar, implying that the latter have faster self-exchange rate constants than the former. It appears that movement of the axial ligand does not slow electron transfer dramatically in these cytochromes.

The axial methionine of cytochromes c is important not only because the Fe—S bond is labile on the time scale of many of the electron transfer experiments, but also because the methionine can be derivatized, allowing creation of a protein with an altered active site. The derivative of cytochrome c in which the methionine has been oxidized, methionine-80-sulfoxide cytochrome c, has been known for many years. Recently, Feinberg et al. have been able to prepare material that is chromatographically homogenous by high-performance liquid chromatography (HPLC).[218] This sulfoxide derivative has an $E^{\circ\prime}$ (NHE) of 240 mV, vs 262 mV for the native cytochrome c. Feinberg et al. have studied the electron transfer reactions of this derivative and found that the derivatized protein is very similar to the native protein in its reaction with succinate cytochrome c reductase and cytochrome oxidase. For the former redox partner, the K_m was approximately 9% of that for native cytochrome c, while the maximum turnover number was only about a factor of two of smaller. For cytochrome oxidase in Tris buffer, the K_m were very similar and the turnover number was a factor of 2–3 smaller for the sulfoxide derivative. It is surprising that oxidation of the sulfur to the sulfoxide produces such a small change in the electron transfer rate constant. The difference between the native and sulfoxide protein is not yet easy to assess. Dyer et al. concluded that the heme crevice region is only slightly opened in the sulfoxide derivative.[219] However, this work should be repeated because it was based on earlier preparations, which apparently did not give pure material. Early NMR experiments were interpreted as showing some Fe—S coordination in the reduced state, and probably some in the oxidized state as well.[220] A more quantitative assessment may be possible using modern NMR spectrometers.

One might expect that the sulfoxide would not bind well to the Fe(III) center. This is indicated by model studies—in particular, measurements of the solvent dependence of the overall equilibrium constant for binding of two 1-methylimidazoles (K^{2Im}) to Fe(III)TPP. The K^{2Im} in CHCl$_3$ and Me$_2$SO are 1.1–1.5 × 10^3 (References 221–223) and 1.3 × 10^4 (Reference 224), respectively. Had the Me$_2$SO bound to the hemin significantly, K^{2Im} would have been much smaller in this solvent. The equilibrium constant is larger in Me$_2$SO, however. Recent NMR studies have shown that the di-n-BuNH$_2$ complex of protohemin can be formed with only 4 eq of n-BuNH$_2$ in Me$_2$SO-d_6.[225] Sulfoxides do bind to ferrous porphyrins. Thus, Fe(II)TPP(Me$_2$SO)$_x$ has at least one axial ligand because the spectrum of Fe(II)TPP in benzene [four-coordinate, spin 1 (Reference 226)] is quite different from that in Me$_2$SO-d_6.[227] Although model studies indicate that Fe(III)–sulfoxide binding is weak, the bond may be formed in the protein due to proximity enforced by the structure of the protein.

The axial methionine can be replaced by other residues. This has been studied most extensively in the alkaline isomerization of cytochrome c.[228,229] As the pH of a solution of cytochrome c is raised, a new species appears which reaches a maximum at approximately pH 9. It is low spin, as indicated by its NMR spectrum.[230] The sixth ligand has been postulated to be either hydroxide or lysine. It appears that more than one group can serve as a ligand, although the simple pattern of the hyperfine-shifted ring methyl groups would indicate one predominant species in solution. Gupta et al. have measured electron self-exchange of horse heart cytochrome c as a function of pH.[134] They interpreted their results in terms of an exchange-active species below pH 9 and an exchange-inactive species above that pH. The rate constant of the exchange-active species increased as the pH was taken up to the isoelectric point (~10).

The outlook for further work on the role of the axial ligand is excellent. Hampsey et al.[33] have found a strain of the yeast *Saccharomyces cerevisiae* which makes an iso-1-cytochrome c in which the axial methionine has been replaced with an isoleucine. The protein is "slightly functional at a reduced temperature" in a whole-cell-growth assay on lactate. A more detailed characterization of this protein will be welcome. Sligar and co-workers[23] have recently isolated the gene for rat cytochrome b_5 and made the protein in which one of the axial histidines (His-63) has been replaced with a methionine. This derivative is five-coordinate; the methionine does not bind to the heme. This approach should allow assessment of the role of the axial ligand in electron transfer, particularly if exogenous imidazole will bind to the heme.

d. Other Examples

In other work on the dependence of electron transfer on conformational change, English has observed that electron transfer from Fe(II) cytochrome c peroxidase to horse Fe(III) cytochrome c shows an experimental time constant but is independent of the concentration of the proteins.[231] At 0.2 M sodium phosphate, most of the cytochrome c peroxidase is not in a complex with the cytochrome c. The data has, therefore, been explained as a slow conformational change of the cytochrome c peroxidase, followed by electron transfer. Hazzard et al. have measured electron transfer to and within a horse heart cytochrome c–flavodoxin complex.[232] Reduction of the complexed cytochrome showed a nonlinear concentration dependence, interpreted as a rate limiting conformational change between two forms of the protein complex.

5. INTERMOLECULAR ELECTRON TRANSFER

Studies on intermolecular electron transfer complement those of intramolecular electron transfer because these experiments allow assessment of controlling factors that cannot be measured in intramolecular electron transfer experiments. The two most important of these are electrostatic control of electron transfer and the site of electron transfer on the protein in solution.

A. Electrostatic Control of Electron Transfer

In solution, electrostatic interactions are one of the dominant controlling factors for electron transfer. Indeed, without a reliable way to factor out the electrostatic contribution, it is not possible to determine the importance of other controlling factors for electron transfer. Electrostatic interactions are determined by measuring kinetics and equilibria as a function of ionic strength.

It has been known for some time that the ionic strength dependence of the electron transfer rate constant cannot be explained in terms of the reaction between two uniformly charged spheres. For example, Beoku-Betts and Sykes have explained the temperature and ionic strength dependence of electron transfer between cytochrome f and $[(CN)_5FeCNCo(CN)_5]^{5-}$ in terms of interaction at a positive site, although the overall charge on the protein is negative.[233]

Meyer et al. have investigated the ionic strength dependence of electron transfer reactions between flavin mononucleotide and a large number of cytochromes c.[215] They have used this data to calculate an effective charge at the reaction site for each cytochrome. These data are plotted as a function of the overall charge on the cytochrome in Figure 5-10. For some cyto-

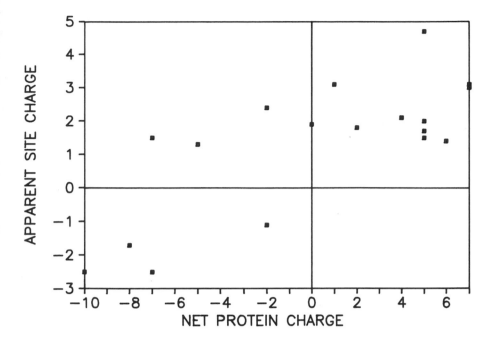

Figure 5-10. The apparent charge at the reactive site vs the net charge on the protein for a series of cytochromes c reacting with flavin mononucleotide semiquinone.[215]

chromes, the sign of the net protein charge is opposite that for the apparent interaction site. For others, the net charge is smaller than that of the apparent interaction site. Meyer et al. have explained their data in terms of electrostatic, and perhaps steric, interactions.

To separate the factors which control electron transfer in cytochromes, it is useful to extrapolate the electron self-exchange rate constant to infinite ionic strength. A number of formalisms have been developed to predict ionic strength dependencies of reaction rates in these types of systems.[31,193,215,234,235] A discussion of the relative merits of these is beyond the scope of this review. However, two points should be made. First, some of these formalisms use dipole moments. This is a very convenient approach because the dipole moments can be calculated from the crystal structure data.[236] It is not yet clear, however, whether characterization of the proteins in terms of only a monopole and dipole is adequate. Second, it is well known that the approximations of the Debye–Hückel law break down at high ionic strength, and, thus, measurements at high ionic strength are more difficult to treat theoretically. Historically, electron transfer rate constant measurements were made by following optical changes, and at these very low concentrations of proteins, low ionic strengths were possible. Measurements are increasingly made with NMR techniques, however, and for these experiments and the protein concentrations are 1–10 mM. In order for the solution's ions to be derived largely from buffer and not from the protein itself, quite high ionic strengths are needed. For example, if we assume a cytochrome c concentration of 10 mM, a charge on the protein of approximately +7, and maximum pairing of the surface charges on the protein,[138] then the protein itself accounts for an ionic strength of 0.07 M.

B. Electron Transfer at the Heme Edge?

Heme proteins generally have one edge of the heme exposed to solvent. It might be expected that electron transfer would take place largely at this exposed heme edge. This is because the two hemes can approach most closely when the proteins are oriented heme-edge-to-heme-edge and the rate of electron transfer is thought to fall off exponentially with distance. Evidence for electron transfer at the heme edge comes from three types of experiments: (1) measurements of electron transfer to derivatized cytochromes, (2) derivatization and cross-linking of cytochrome c–partner com-

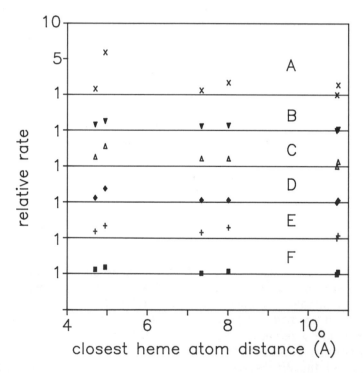

Figure 5-11. Ratio of the rate constant for reaction of various proteins with native cytochrome c divided by that for reaction with a cytochrome c derivative as a function of the distance from the α carbon of the derivatized lysine to the closest heme atom. Lysines are [position (distance in Å)]: 8(10.7), 13(4.9), 27(7.3), 72(8.0), 79(4.7), 100(10.7). Electron transfer partners: A, Adrenodoxin (Reference 45b); B, cytochrome aa_3 (Reference 45c); C, cytochrome c_1 (Reference 45d); D, sulfite oxidase (Reference 45e); E, cytochrome b_5 (Reference 193); F, flavocytochrome c_{552} (Reference 44).

plexes, and (3) computer models of heme protein electron transfer complexes.

a. Experimental Evidence

A number of groups have made derivatives of specific lysines on the cytochrome *c* surface and studied the rates of reaction of these cytochromes with various electron transfer partners.[42,43,45,237,238] In general, the closer the site of derivatization to the exposed heme edge, the more the rate of reaction with the partner is perturbed. However, detailed analysis of these studies shows that other factors are also important. The data from a number of studies on horse heart cytochrome *c* derivatives are shown in Figure 5-11. This figure plots the rate constant ratio [$k_{native\ protein}/k_{derivatized\ protein}$] as a function of the distance from the α carbon of the derivatized amino acid to the closest heme atom. This is a through-space distance, not the distance across the surface of the protein to the exposed heme edge. However, a number of trends can be noted. First, the trend is for derivatives closer to the heme to show greater differences with the native protein. Differences are often small, as seen in Figure 5-11, although much larger effects have been noted in some systems. Examples are the oxidation of horse cytochrome *c* with plastocyanin and stellacyanin,[42] and in the self-exchange reaction of two cytochrome *c* derivatives.[239] Second, there are local electrostatic effects. This is seen in Figure 5-12, which plots the rate of the derivatized protein divided by

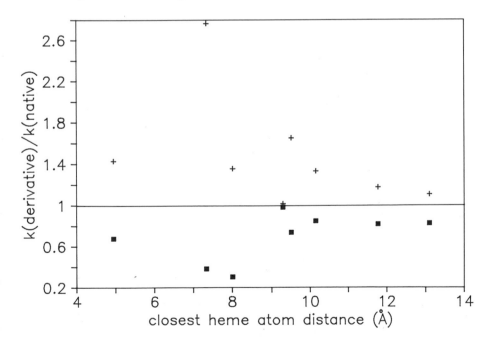

Figure 5-12. Ratio of the rate constant for reaction of [Fe(EDTA)]$^{2-}$ (■) and [Co(sep)]$^{2+}$ (+) with native cytochrome *c* divided by that for reaction with a cytochrome *c* derivative as a function of the distance from the α carbon of the derivatized lysine to the closest heme atom.[42]

that of the native protein for reactions with the negatively charged $[Fe(EDTA)]^{2-}$ and positively charged $[Co(sep)]^{2+}$.[43] The two sets of rate constants show approximate mirror symmetry when plotted in this manner, indicating the importance of local electrostatic effects. Local effects are, in general, important, as seen in Figure 5-11 where all of the reactions show a similar pattern despite the wide variation in redox partners.

Cytochrome c forms tight complexes with its electron transfer partners in a number of instances. For some of these, it has been possible either to cross-link the partners or to derivatize the surface of cytochrome c within the complex.[45,163,240] It is generally found that the area of contact between the two partners is at the exposed heme edge, although a recent article by Rieder et al. presents evidence for a complex between *Rhodospirillum rubrum* cytochrome c_2 and the photosynthetic reaction center from the same organism which involves the side of the protein opposite to the heme edge in the cytochrome.[241]

Computer models have been made of a number of heme protein electron transfer pairs, including cytochrome c–cytochrome b_5,[242] cytochrome c–cytochrome-c peroxidase,[243] cytochrome c–flavodoxin,[244,245] cytochrome b_5–methemoglobin,[246] and cytochrome c–cytochrome c.[13] These models assume a heme-edge-to-heme-edge geometry and minimize the electrostatic energy to obtain a best fit of the two partners. In each case it has been possible to obtain a good geometric and electrostatic fit of the two partner proteins.

b. Geometrical Constraints

If electron transfer in cytochromes takes place only at the exposed heme edge, then there are substantial geometrical constraints on the reaction. The problem can be formulated in terms of electron transfer between two spheres, each with a small reactive spot. There are a number of models which predict the rate constant for the reaction between a sphere with a reactive spot and either a fully reactive sphere or another sphere with a reactive spot. Berg and von Hippel have reviewed the literature;[247,248] other groups have also discussed the issue recently.[249–252]

In a recent formulation of the problem for electron self-exchange, Temkin and Yakobson have derived[253]

$$\frac{1}{k_{calc}} = \frac{1}{k_{el}f^2} + \frac{1}{k_D F} \qquad F = [2R^2/D\tau]^{1/2}f^2 \qquad (5\text{-}10)$$

where k_D is the diffusion-controlled rate constant, R is the sum of the radii of the spheres, D is the relative diffusion coefficient, τ is the orientational relaxation of the two partners, f is the fraction of the sphere that is reactive, and k_{el} is the rate constant for the chemical reaction within the complex (electron transfer in the present case). For horse heart cytochrome c, $f =$

$0.01.$[13,102] The rate constant k_D can be given by the diffusion-controlled encounter of two neutral spheres 16.6 Å in diameter[31] with a relative diffusion coefficient $D = 2.6 \times 10^{-6}$ cm²/s;[254] the calculated value is $6.5 \times 10^9 M^{-1} s^{-1}$. The value of τ has been measured from NMR experiments as $4-6 \times 10^{-9}$ s.[255] The only remaining variables are f and the rate of electron transfer, k_{el}. Figure 13 shows the k_{calc} as a function of k_{el} for $f = 0.01$. Figure 5-13 also shows this calculation for *Ps. aeruginosa* cytochrome c_{551}. This protein has 82 rather than 104 amino acids, a slightly larger percentage of the surface of the protein that is heme ($\sim 3\%,$[24b]) and, hence, a slightly smaller radius (14.4 Å).[31]

In the simplest approximation for a reactive spot on a sphere, the rate constant for electron transfer between two free hemes in solution would be multiplied by the fraction of the sphere that is reactive (f) for each of the two proteins. For a self-exchange reaction, the multiplicative factor would be $f^2.$[256] Indeed, when the intrinsic rate of electron transfer is slow, the calculated rate constant (k_{calc}) is $k_{el}f^2$, as expected. As the rate constant for electron transfer increases, the calculated rate constant becomes less sensitive to this rate constant. This is because diffusion is becoming competitive with electron transfer. When the rate constant for electron transfer is very high ($>10^{11}$ s⁻¹ in the examples shown in Figure 5-13), the calculated rate

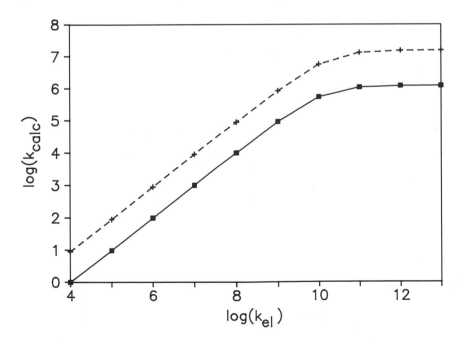

Figure 5-13. The expected observed rate second-order rate constant for electron self-exchange as a function of the first-order rate constant for electron transfer within the heme-edge-to-heme-edge cytochrome complex. +, *Pseudomonas aeruginosa* cytochrome c_{551}; ■, horse heart cytochrome *c*. The calculations are based on the formalism of Temkin and Yacobson.[253]

constant is no longer sensitive to k_{el}; the reaction has become diffusion controlled.

One important aspect of Figure 5-13 is that the observed electron self-exchange rate constant for horse heart cytochrome c (10^4–10^5 M^{-1} s^{-1} at infinite ionic strength) can be explained in terms of a rate constant for electron transfer within the complex of about 10^9 s^{-1}. This is reasonable and in line with predictions by Marcus and Sutin based on a computer model by Weber.[13] The *Ps. aeruginosa* cytochrome c_{551} predictions are somewhat surprising, however. This protein has an electron self-exchange rate constant of approximately 2×10^7 M^{-1} s^{-1} [257] which is independent of ionic strength.[258] To account for this, one must postulate a rate constant within the complex of greater than 10^{10} s^{-1}. This is very fast and would indicate that the hemes are very close to one another. In addition, there could be no other process demanding energy (eg, substantial protein reorganization) for the rate constant to be this fast in this model. Activation parameters for this reaction have not yet been measured.

It is also possible that the model of electron transfer occurring only at the exposed heme edge is inadequate to explain the observed results. Electron transfer can occur over a distance and, as pointed out previously, is thought to fall off exponentially with distance (Equation 5-7). The smaller the protein, the closer an electron transfer partner approaching the surface of the protein can get to the heme. If electron transfer falls off exponentially with distance and electron transfer through the protein is significant, then smaller proteins will have faster electron self-exchange rate constants. Electron self-exchange rate constants have been measured for about 20 cytochromes. Figure 5-14 shows a plot of the log of the observed rate constant vs the size of the protein expressed simply as the number of amino acids in the chain. It can be seen that the electron self-exchange rate constants do indeed correlate with size, with smaller proteins having faster self-exchange rate constants.

Two other studies should be noted in this context. Hasinoff has noted a decrease in the rate constant for reduction of the heme by the hydrated electron as the size of the protein increases.[259] However, the change is much smaller, only about a factor of 4 as the number of amino acids increases from 104 (cytochrome c) to 153 (myoglobin). Tollin et al. have proposed that the extent of solvent exposure correlates with the rate constant for electron transfer in some cases but not in others.[260] They note that the sulfur of the thioether bridge is exposed to solvent in cytochromes c and that there is also electron density on this center. Whether these sulfur atoms, or the axial ligands, serve as "pathways" for electron transfer remains to be determined. In addition, different cytochromes c have different orientations of the axial methionine and different extents of delocalization of the unpaired electron on this ligand. There is as yet no apparent correlation of electron transfer rates with these properties of the methionine, but this may represent a lack of sufficient data.

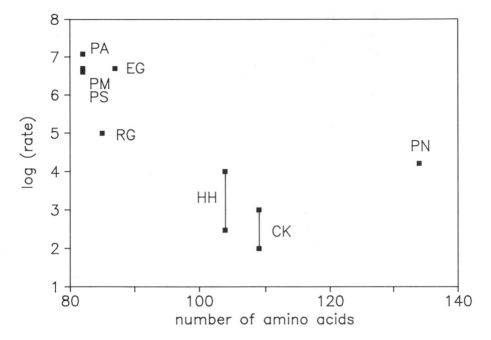

Figure 5-14. The log of the electron self-exchange rate constant as a function of the size of the protein expressed in terms of the number of the residues in the protein. The vertical lines represent the ionic strength dependencies of the rate constants which have been measured for horse heart (HH), *Candida krusei* (CK) cytochromes c,[174] and *Ps. aeruginosa* c_{551} (PA) (independent of ionic strength).[258] Other abbreviations: EG, *Euglena gracilis*; PM, *Pseudomonas mendocina*; PS, *Pseudomonas stutzeri*; RG, *Rhodopseudomonas gelatinosa*; PN, *Paracoccus denitrificans*. Data are given in Reference 24.

C. Complex Formation

Much of the work on electron transfer between two redox partners assumes that there is only one complex and that this complex is the most favorable one for electron transfer. The ideas that electron transfer takes place from more than one complex or that dead-end complexes are important have been largely, although not entirely, ignored. However, recent work in a number of systems has been interpreted in terms of more than one complex.

The existence of a complex between two proteins can be demonstrated via absorption spectroscopy, saturation kinetics, gel filtration, or NMR. When a spectral change can be seen upon complexation, ultraviolet (UV)/visible spectroscopy is convenient; this technique has been used to measure equilibrium constants for complexes between cytochrome c and cytochrome b_5 (Reference 74) and cytochrome c and cytochrome c peroxidase.[194] Various groups have investigated the complexation between the heme protein and second protein by NMR. Complexation is seen in changes in both the line width and chemical shift of the peaks as a function of concentration. Because the paramagnetically shifted resonances are so easily seen, these have been the focus of investigation in all complexes studied to date. The shifts of these resonances are sensitive to minor structural changes, and one generally cannot tell the site of complexation or, indeed, whether more than one complex is formed. Data on most systems has been interpreted in terms of only one complex. NMR studies include those on cytochrome c–cytochrome c peroxidase,[169,197] cytochrome c–cytochrome c oxidase,[261] cytochrome c–cytochrome b_5,[255] plastocyanin–horse cytochrome c,[160] myoglobin–cytochrome b_5,[262] cytochrome c–flavodoxin,[263,264] and *Ps. aeruginosa* cytochrome c_{551}–nitrite reductase cd_1.[265,266] Horse heart cytochrome c shows no evidence of dimerization at NMR concentrations; cytochrome b_5, however, shows slight changes in the chemical shifts of the heme resonances as a function of concentration which are consistent with some dimerization ($K_{eq} = 1$–$10\,M$ at $I = 0.1\,M$),[267] but may simply reflect a viscosity increase as the concentration increases, which results in an increase in the rotational correlation time.[66]

a. Multiple Complexes

Increasingly, equilibrium data is being interpreted in terms of more than one complex. This has been the case for some years for the interaction of cytochrome c with inorganic and organometallic reagents; this work has been summarized. The interaction of cytochrome c with organometallic reagents was one of the first areas to be studied, largely because the complexes can, in some instances, be observed directly by NMR. Early work centered on ferricyanide binding to cytochrome c.[77,138,216] More recently, Williams and Moore have shown that Fe(III)–polyaminocarboxylate complexes bind to four sites on cytochrome c.[268]

Mauk et al. have investigated the interaction of cytochrome c with cytochrome b_5 reconstituted with hemin dimethyl ester.[74] One might have expected that neutralization of the heme propionate would reduce the association constant between the two proteins. However, this was not observed, and computer modeling indicated that there were two different complexes with very similar stabilities in this system.

Very recently, Northrup et al. have investigated the reaction between cytochrome c and cytochrome c peroxidase using Brownian dynamics.[269] They propose that there are a number of encounter complexes which have the correct geometric requirements for electron transfer. They calculate that formation of only one association complex is insufficient to account for the observed rate of association. It should be noted that Northrup et al. have also calculated that dissociation, and not association or electron transfer, is rate limiting at low ionic strengths in this system.[269,270]

Tollin et al. have interpreted the changes in the NMR spectrum upon cytochrome c complexation with *Clostridium pasteurianum* flavodoxin as due to a loosening of the heme crevice in the cytochrome.[264] They note that this is consistent with the observation that complexation of cytochrome c with cytochrome-c peroxidase enhances the rate of reaction of the peroxidase.[271,272]

In this context, it should also be noted that electron transfer to plastocyanin occurs through two distinct sites.[82] Both Mauk et al.[74] and Sykes and co-workers[75,82] have stressed that electron transfer within a number of complexes should be considered. Although it is experimentally difficult to assess this point, electron transfer within a series of complexes of approximately equal energy may, in general, be important, and this possibility must be kept in mind.

b. Nonproductive Complexes

Rieder et al. have investigated the reactivity of surface lysines of *R. rubrum* cytochrome c_2 both free in solution and bound to the photosynthetic reaction center.[241] Three residues are less reactive in the complex than in the free enzyme, and these are all clustered together on the side of the protein away from the exposed heme edge. It is not known whether this is a nonproductive complex, an artifact which is different from the complex in the intact photosynthetic membrane, or, indeed, electron transfer occurs from a complex in which the two hemes are not edge to edge. In this context, Tiede has recently reported that the heme orientations of electron transfer complexes between cytochromes c and c_2 bound to the reaction center of *Rhodobacter sphaeroides* are different.[273] Tiede suggests that the binding sites of the two cytochromes are not the same. The 3 Å crystal structure of *Rhodopseudomonas viridis* reaction center published by Deisenhofer et al.[274] has provided a clear molecular picture of one reaction center and has generated great enthusiasm for further work in this area.

c. Electron Transfer to Complexes

The measurement of electron transfer to complexes allows estimation of the extent to which the complexation blocks or enhances electron transfer. In favorable cases, intramolecular electron transfer may be measured.

Recently, Erman et al. have investigated the 1 : 1 covalent complex between horse heart cytochrome c and yeast cytochrome-c peroxidase.[169b,c] Free Fe(II) cytochrome c reduces the ccp I/Fe(III) cytochrome c complex 10^{-4} to 10^{-5} times more slowly than it does for ccp I. This slow electron transfer may represent simply a greater heme–heme distance between the Fe(II) cytochrome c and ccp I hemes when the latter is in the complex. At the other end of the spectrum of mechanistic possibilities, electron transfer could occur only after a large conformational change which opens up the complex substantially.

Hazzard and Tollin have measured electron transfer rate constants from free flavin semiquinones to the cytochrome c–cytochrome-c peroxidase complex.[275] The rate constant for reaction with cytochrome c is reduced 73–90% compared with that of the free protein, while the rate constants for the cytochrome-c peroxidase is reduced by about 45%. This indicates that electron transfer to both proteins is partially blocked by the complexation. They found that the two-electron oxidized form of cytochrome-c peroxidase (compound I) underwent intramolecular electron transfer from ferric cytochrome c with a rate constant of 750 s^{-1}. The rate constant from cytochrome c to the one-electron oxidized form of cytochrome c peroxidase was 450 s^{-1}. Recent work of Erman et al. comes to a similar conclusion; the rate constant is greater than 600 s^{-1}.[272] However, Cokic and Erman have found that complex formation between cytochrome c and cytochrome c peroxidase does not block electron transfer from the small neutral reductants tetramethyl-p-phenylenediamine and N-methylphenazonium methosulfate.[271] The recent crystal structure of cytochrome c peroxidase compound I,[276] as well as mutagenesis experiments on cytochrome c peroxidase,[135–137] provide a basis for further work in this area.

Bhattacharyya et al. have investigated electron transfer within complexes of ferredoxin-NADP$^+$ reductase and cytochromes c_2 from *Paracoccus denitrificans* and *R. rubrum*.[277] The rate constants for electron transfer from lumiflavin semiquinone to the cytochromes both free in solution and bound in a ferredoxin–NADP$^+$ reductase–cytochrome c_2 complex were the same, indicating that heme access was not blocked by the reductase. Intramolecular electron transfer rate constants from the reductase free radical to the cytochromes were approximately 700 s^{-1} and 400 s^{-1} for the *Paracoccus* and *Rhodospirillum* proteins, respectively. The authors proposed that the reductase binds to the side of the cytochrome away from the heme edge (redox centers 9–11 Å apart). At high ionic strength, electron transfer from the reductase free radical to the oxidized cytochrome c_2 became second order, as expected because the complexation between the two proteins decreases as the ionic strength increases.

Hazzard et al. have looked at the reduction of the cytochrome c–flavodoxin complex by flavin semiquinones and rubredoxin.[232] Neutral flavin semiquinones reacted two to three times slower with the complexed cytochrome than with the free cytochrome, indicating some blocking of the electron transfer. The ionic strength dependence of the reaction of negatively charged semiquinones indicated a repulsive potential, whereas the reaction with the free cytochrome c indicated a positive potential, showing that local electrostatic interactions were important.

Another effort to alter the electron transfer characteristics of cytochrome c is found in studies by Kuo, Davies, and Smith on the complexation of monoclonal antibodies to cytochromes c with their respective cytochromes.[278,279] In work on human cytochrome c, a monoclonal antibody was found that binds in the area around isoleucine-58. This is not near the heme edge, but in a region that shows a conformational change upon redox change. Addition of the antibody stimulated electron transfer to the reductase and inhibited electron transfer to the oxidase. Kuo et al. proposed that conformational changes may be involved as a control mechanism in electron transfer reactions involving cytochrome c.

6. MISCELLANEOUS TOPICS

A. Multiheme Cytochromes

Study of multiheme cytochromes presents another opportunity to measure electron transfer rate constants in systems of fixed geometry. Those proteins which are monomeric do not suffer any mechanistic ambiguity from the possibility of dissociation of the protein on the time scale of electron transfer. Crystal structures are necessary if a molecular interpretation of the data is to be achieved, however. In addition, these heme proteins have many oxidation states, which can make data analysis difficult.

The cytochromes c_3 are tetraheme proteins isolated for the anaerobic bacteria of the genus *Desulfovibrio*. Crystal structures are available for the proteins from *Desulfovibrio desulfuricans* (Hildenborough)[280,281] and *Desulfovibrio vulgaris* (Miyazaki).[282] A molecule with 4 distinguishable redox centers, each of which can assume 2 oxidation states, has a total of 16 possible states. The greater the spread in the oxidation potentials of the individual redox centers, the fewer of these states that will be occupied significantly. However, even states which have very small populations can have a substantial effect on the observed NMR spectrum if the electron transfer rate constant is fast.

Moura et al. have investigated the electron transfer in the cytochrome c_3 from *Desulfovibrio vulgaris*.[283] They performed a series of saturation transfer experiments at different oxidation states to correlate heme methyl resonances between oxidations states. Intermolecular electron transfer was slow on the NMR time scale [$<5 \times 10^5 \ M^{-1} \ s^{-1}$ (2 mM, 25°C)]. However, some peaks belonging to intermediate oxidation states showed small broadening when the protein concentration or temperature was increased, indicating that some intermolecular transfer occurred with rate constants near this value. Intramolecular electron transfer was observed, however. The rate constant for at least two of the hemes was greater than $5 \times 10^4 \ M^{-1} \ s^{-1}$. This protein is difficult to study because it is necessary to take into account many of the 16 oxidation states to analyze the data.

The cytochrome c_3 from *Desulfovibrio gigas* resembles that from *D. vulgaris* in that the intramolecular electron exchange rate constants are fast on the NMR time scale ($>10^5 \ s^{-1}$).[284] The intermolecular electron transfer rate constants are, however slow on the NMR time scale (273 K, 3 mM protein).

Moura et al. have reported NMR and electron exchange properties of the diheme cytochrome c_{552} from *Pseudomonas perfectomarinus* (M =

25,800).[285] The two hemes in this protein have a large difference in their redox potentials (+174 mV for heme II and −180 mV for heme I), and it is therefore possible to prepare cleanly a mixed redox state with heme II oxidized and heme I reduced. When the fully reduced protein was reoxidized by the addition of aliquots of oxygen, the first step was characterized by shifting and broadening of the lines of heme I, indicating that these resonances were in slow to intermediate exchange. A rate constant for intermolecular electron self-exchange of $2.7 \times 10^5 \ M^{-1} \ s^{-1}$ was calculated. The second oxidation step occurred with no change in either the chemical shift or the line width of the peaks, indicating that electron transfer was slow on the NMR time scale ($<2 \times 10^4 \ M^{-1} \ s^{-1}$ for 2 mM protein at 295 K). A molecular interpretation is not yet possible because there is no X-ray structure of this protein.

Cytochrome c_3 from *Desulfovibrio desulfuricans* Norway has been characterized by Guerlesquin et al.[286] Reduction of heme I by even 0.5% produces shifting and broadening of the resonances. This indicates that intermolecular electron transfer between oxidized and reduced heme I is fast on the NMR time scale (probably $>10^6 \ M^{-1} \ s^{-1}$). Reduction of heme I does not affect the resonances of the other hemes, however, indicating that intramolecular electron transfer is slow in the NMR time scale.

Further progress on multiheme cytochromes will be welcome. Diheme proteins will be easier to characterize than those with more hemes. For example, the X-ray structure of diheme cytochrome c_5 from *Azotobacter* has recently been published.[287] This protein is a dimer of identical 12,000 Da monomers, each containing a single heme. The shortest heme–heme distance is 6.3 Å (Fe to Fe 16.4 Å). Progress will also be aided by improved NMR techniques. Two-dimensional NMR has been used to correlate resonances of oxidized and reduced hemes in the various oxidation states of *D. gigas* cytochrome c_3.[140] When cross-peak volumes can be integrated accurately, two-dimensional NMR can also be used to determine electron transfer rate constants.[129] For the present, it appears that one-dimensional techniques are still those of choice for rate constant measurements, however.

The orientation of the two hemes of *Ps. aeruginosa* cytochrome cd_1 has been investigated by Makinen et al.[288] Polarized single-crystal absorption studies indicated that the hemes were approximately perpendicular. The rate constant for electron transfer is approximately 0.3 s^{-1}. Makinen et al. propose that this slow electron transfer is due to the perpendicular orientation of the two hemes.

B. *Saccharomyces cerevisiae* Missense Mutations

Sherman and co-workers[32,33] have determined the amino acid replacements for a large series of cycle 1 missense mutations of *Saccharomyces cerevisiae* iso-1-cytochrome c. These reduced function or nonfunctional cytochromes appear to have substitutions which alter interaction of the protein with the heme or prevent proper folding of the protein. This work provides a good basis for deciding on future amino acid replacements via site-directed mutagenesis. A few amino acids are especially worthy of note. Trp-64 is located near the floor of the heme crevice and is hydrogen bonded to the propionate at the 7 position on the heme. It can be replaced by phenylalanine, tyrosine, or leucine, however. The phenylalanine mutant has activity nearly that of the native protein, and the leucine mutant also has good activity (growth measured on lactate).[33] Lys-84, on the surface of the protein with its side chain exposed to solvent, hydrogen bonds with Ser-52, stabilizing the heme crevice. However, the Ser-52 is replaced in other species by residues which cannot form a hydrogen bond. The importance of hydrogen bonding in controlling electron transfer is not yet clear. Even the extent of hydrogen bonding can be difficult to ascertain. For example, the crystal structure shows that Tyr-53 is hydrogen bonded to heme propionate 7, but an NMR study by Williams et al.[289] has indicated that this bond may be due to crystal packing. Finally, replacement of the axial methionine with an isoleucine gives a protein which is said to be "slightly functional at a reduced temperature."[33] A more detailed characterization of this protein will be welcome. Louie et al.[290] have completed the X-ray structure of yeast iso-1-cytochrome c. This work provides a basis for continued studies using protein derived from site-directed mutagenesis techniques.[19–22]

ACKNOWLEDGMENTS

I thank the National Institutes of Health for support of our work (DK38826) and the National Science Foundation for a Career Advancement Award (CHE 8707447). The contributions of M. Barbush, X. Hong, S. Woehler, and L. Amis are much appreciated, as are collaborative efforts with R. Timkovich, A. G. Mauk, and F. S. Millett. I also thank S. A. Allison, D. Beratan, D. Edmondson, A. English, J. Erman, H. B. Gray, D. Holten, F. M. Hawkridge, B. Hoffman, S. S. Isied, C. Kirmaier, S. Mathews, A. G. Mauk, G. McLendon, S. Northrup, S. Sligar, and G. Tollin for fruitful discussions and preprints. S. Mathews and R. Cave very kindly provided figures.

REFERENCES

1. Govindjee, Ed. "Photosynthesis: Energy Conversion by Plants and Bacteria"; Academic Press: New York, 1982.
2. Michel-Beyerle, M. E., Ed. "Antennas and Reaction Centers of Photosynthetic Bacteria"; Springer-Verlag: Berlin, 1985.

3. Kirmaier, C.; Holten, D. *Photosynth. Res.,* in press.
4. Hatefi, Y. *Annu. Rev. Biochem.* **1985,** *54,* 1015–1069.
5. Dixit, B. P. S. N.; Vanderkooi, J. M. *Curr. Top. Bioenerg.* **1984,** *13,* 159–202.
6. Poulos, T. L.; Finzel, B. C. In "Peptide and Protein Reviews", Vol. IV; Marcel Dekker: New York, 1984, pp. 115–171.
7. Murray, R. I.; Fisher, M. T.; Debrunner, P. G.; Sligar, S. G. In "Metalloproteins", Part I; Harrison, P. M., Ed.; Verlag Chemie: Deerfield Beach, FL, 1985; pp. 157–206.
8. Dawson, J. H.; Eble, K. S. *Adv. Inorg. Bioinorg. Mech.* **1986,** *4,* 1–64.
9. Ortiz de Montellano, P. R. *Acc. Chem. Res.* **1987,** *20,* 289–294.
10. Calcaterra, L. T.; Closs, G. L.; Miller, J. R. *J. Am. Chem. Soc.* **1983,** *105,* 670–671.
11. Mayo, S. L.; Ellis, W. R.; Crutchley, R. J.; Gray, H. B. *Science* **1986,** *233,* 948–952.
12. Guarr, T.; McLendon, G. *Coord. Chem. Rev.* **1985,** *68,* 1–52.
13. Marcus, R. A.; Sutin, N. *Biochim. Biophys. Acta* **1985,** *811,* 265–322.
14. Mikkelsen, K. V.; Ratner, M. A. *Chem. Rev.* **1987,** *87,* 113–153.
15. Kuki, A.; Wolynes, P. G. *Science* **1987,** *236,* 1647–1652.
16. Liang, N.; Pielak, G. J.; Mauk, A. G.; Smith, M.; Hoffman, B. M. *Proc. Natl. Acad. Sci. U.S.A.* **1987,** *84,* 1249–1252.
17. Beratan, D. N.; Onuchic, J. N.; Hopfield, J. J. *J. Chem. Phys.* **1987,** *86,* 4488–4498.
18. Larsson, S. *J. Chem. Soc., Faraday Trans. 2* **1983,** *79,* 1375–1388.
19. Hampsey, D. M.; Das, G.; Sherman, F. *FEBS Lett.* **1988,** *231,* 275–283.
20. Smith, M. *Annu. Rev. Genet.* **1985,** *19,* 423–462.
21. Pielak, G. J.; Mauk, A. G.; Smith, M. *Nature* **1985,** *313,* 152–154.
22. Holzschu, D.; Principio, L.; Conklin, K. T.; Hickey, D. R.; Short, J.; Rao, R.; McLendon, G.; Sherman, F. *J. Biol. Chem.* **1987,** *262,* 7125–7131.
23. von Bodman, S. B.; Schuler, M. A.; Jollie, D. R.; Sligar, S. G. *Proc. Natl. Acad. Sci. U.S.A.* **1986,** *83,* 9443–9447.
24. (a) Shirazi, A.; Barbush, M.; Ghosh, S. B.; Dixon, D. W. *Inorg. Chem.* **1985,** *24,* 2495–2502. (b) Dixon, D. W.; Barbush, M.; Shirazi, A. *J. Am. Chem. Soc.* **1984,** *106,* 4638–4639.
25. Williams, G.; Moore, G. R.; Williams, R. J. P. *Comm. Inorg. Chem.* **1985,** *4,* 55–98.
26. Moore, G. R.; Eley, C. G. S.; Williams, G. *Adv. Inorg. Bioinorg. Reaction Mech.* **1984,** *3,* 1–96.
27. Meyer, T. E., Kamen, M. D. *Adv. Prot. Chem.* **1982,** *35,* 105–212.
28. Osheroff, N.; Speck, S. H.; Margoliash, E.; Veerman, E. C. I.; Wilms, J.; Kronig, B. W.; Muijsers, A. O. *J. Biol. Chem.* **1983,** *258,* 5731–5738.
29. Yamanaka, T.; Fukumori, Y.; Kamita, Y.; Fujii, K. In "Molecular Evolution, Protein Polymorphism and the Neutral Theory"; Kimura, M., Ed.; Japan Scientific Press: Tokyo, 1982, pp. 315–329.
30. Ferguson-Miller, S.; Brautigan, D. L.; Margoliash, E. In "The Porphyrins", Vol. 7; Dolphin, D., Ed.; Academic Press: New York, 1979, pp 149–240.
31. Wherland, S.; Gray, H. In "Biological Aspects of Inorganic Chemistry", Symp.; Addison, A. W.; Cullen, W. R.; Dolphin D.; and James, B. R., Eds.; Wiley: New York, 1978, pp. 289–368.
32. Ernst, J. F.; Hampsey, D. M.; Stewart, J. W.; Rackovsky, S.; Goldstein, D.; Sherman, F. *J. Biol. Chem.* **1985,** *260,* 13225–13236.
33. Hampsey, D. M.; Das, G.; Sherman, F. *J. Biol. Chem.* **1986,** *261,* 3259–3271; **1987,** *262,* 1926.
34. (a) Ramdas, L.; Sherman, F.; Nall, B. T. *Biochemistry* **1986,** *25,* 6952–6958. (b) Ramdas, L.; Nall, B. T. *Biochemistry,* **1986,** *25,* 6959–6964.
35. Proudfoot, A. E. I.; Wallace, C. J. A.; Harris, D. E.; Offord, R. E. *Biochem. J.* **1986,** *239,* 333–337.
36. Poerio, E.; Parr, G. R.; Taniuchi, H. *J. Biol. Chem.* **1986,** *261,* 10976–10989.
37. Juillerat, M. A.; Taniuchi, H. *J. Biol. Chem.* **1986,** *260,* 2697–2711.
38. Wallace; C. J. A.; Courthésy, B. E. *Prot. Eng.* **1986,** *1,* 23–27.

39. ten Kortenaar, P. B. W.; Adams, P. J. H. M.; Tesser, G. I. *Proc. Natl. Acad. Sci. U.S.A.* **1985,** *82,* 8279–8283.
40. Barstow, L. E.; Young, R. S.; Yakali, E.; Sharp, J. J.; O'Brien, J. C.; Berman, P. W.; Harbury, H. A. *Proc. Natl. Acad. Sci. U.S.A.* **1977,** *74,* 4248–4250.
41. Koul, A. K.; Wasserman, G. F.; Warme, P. K. *Biochem. Biophys. Res. Commun.* **1979,** *89,* 1253–1259.
42. Armstrong, G. D.; Chambers, J. A.; Sykes, A. G. *J. Chem. Soc., Dalton Trans.* **1986,** 755–758.
43. Armstrong, G. D.; Chapman, S. K.; Sisley, M. J.; Sykes, A. G.; Aitken, A.; Osheroff, N.; Margoliash, E. *Biochemistry* **1986,** *25,* 6947–6951.
44. Bosshard, H. R.; Davidson, M. W.; Knaff, D. B.; Millett, F. *J. Biol. Chem.* **1986,** *261,* 190–193.
45. (a) Stonehuerner, J.; O'Brien, P.; Geren, L.; Millett, F.; Steidl, J.; Yu, L.; Yu, C.-A. *J. Biol. Chem.* **1985,** *260,* 5392–5398. (b) Smith, H. T.; Staudenmeyer, N.; Millett, F. *Biochemistry* **1977,** *16,* 4971–4974. (c) Geren, L. M.; Millett, F. *J. Biol. Chem.* **1981,** *256,* 4851–4855. (d) Ahmed, A. J.; Smith, H. T.; Smith, M. B.; Millett, F. *Biochemistry* **1978,** *17,* 2479–2481. (e) Webb, M.; Stonehuerner, J.; Millett, F. *Biochim. Biophys. Acta* **1980,** *593,* 290–298.
46. Domingue, R. P.; Fayer, M. D. *J. Phys. Chem.* **1986,** *90,* 5141–5146.
47. Cave, R. J.; Siders, P.; Marcus, R. A. *J. Phys. Chem.* **1986,** *90,* 1436–1444.
48. Closs, G. L.; Calcaterra, L. T.; Green, N. J.; Penfield, K. W.; Miller, J. R. *J. Phys. Chem.* **1986,** *90,* 3673.
49. Penfield, K. W.; Miller, J. R.; Paddon-Row, M. N.; Costaris, E.; Oliver, A. M.; Hush, N. S. *J. Am. Chem. Soc.* **1987,** *109,* 5061–5065.
50. Oevering, H.; Paddon-Row, M. N.; Heppener, M.; Oliver, A. M.; Costaris, E.; Verhoeven, J. W.; Hush, N. S. *J. Am. Chem. Soc.* **1987,** *109,* 3258–3269.
51. Cowan, J. A.; Sanders, J. K. M.; Beddard,G. S.; Harrison, R. J. *J. Chem. Soc., Chem. Commun.* **1987,** 55–58.
52. Sessler, J. L.; Johnson, M. R. *Angew. Chem., Int. Ed. Engl.,* **1987,** *26,* 678–680.
53. Heiler, D.; McLendon, G.; Rogalskyj, P. *J. Am. Chem. Soc.* **1987,** *109,* 604–606.
54. Gust, D.; Moore, T. A.; Makings, L. R.; Liddell, P. A.; Nemeth, G. A.; Moore, A. L. *J. Am. Chem. Soc.* **1986,** *108,* 8028–8031.
55. Leland, B. A.; Joran, A. D.; Felker, P. M.; Hopfield, J. J.; Zewail, A. H.; Dervan, P. B. *J. Phys. Chem.* **1985,** *89,* 5571–5573.
56. Bolton, J. R.; Ho, T. F.; Liauw, S.; Siemiarczuk, A.; Wan, C. S. K.; Weedon, A. C. *J. Chem. Soc., Chem. Commun.* **1985,** 559–560.
57. Wasielewski, M. R.; Niemczyk, M. P.; Svec, W. A.; Pewitt, E. B. *J. Am. Chem. Soc.* **1985,** *107,* 5562–5563.
58. Lindsey, J. S.; Mauzerall, D. C.; Linschitz, H. *J. Am. Chem. Soc.* **1983,** *105,* 6528–6529.
59. Pielak, G. J.; Concar, D. W.; Moore, G. R.; Williams, R. J. P. *Prot. Eng.* **1987,** *1,* 83–88.
60. Bechtold, R.; Kuehn, C.; Lepre, C.; Isied, S. S. *Nature (London)* **1986,** *322,* 286–288.
61. Lieber, C. M.; Karas, J. L.; Gray, H. B. *J. Am. Chem. Soc.* **1987,** *109,* 3778–3779.
62. Williams, R. J. P.; Concar, D. *Nature (London)* **1986,** *322,* 213–214.
63. Hoffman, B. M.; Ratner, M. A. *J. Am. Chem. Soc.,* **1987,** *109,* 6237–6243.
64. Alger, J. R.; Shulman, R. G. *Q. Rev. Biophys.* **1984,** *17,* 83–124.
65. Senn, H.; Wüthrich, K. *Q. Rev. Biophys.* **1985,** *18,* 111–134.
66. (a) Satterlee, J. D. *Metal Ions Biol. Syst.* **1987,** *21,* 121–185. (b) Satterlee, J. D. *Annu. Rev. NMR Spec.* **1986,** *17,* 79–178.
67. Bertini, I.; Luchinat, C. "NMR of Paramagnetic Molecules in Biological Systems"; Benjamin/Cummings: Menlo Park CA, 1986.
68. Wüthrich, K. "NMR of Proteins and Nucleic Acids"; Wiley: New York, 1986.
69. Keller, R. M.; Wüthrich, K. In "Biological Magnetic Resonance", Vol. 3; Berliner, L. J.; and Reuben, J., Eds.; Vol. 3; Plenum Press: New York, 1981, pp. 1–52.
70. Margoliash, E.; Bosshard, H. R. *Trends Biochem. Sci.* **1983,** *8,* 316–320.

71. Tam, S.-C.; Williams, R. J. P. *Struct. Bond.* **1985,** *63,* 103–151.
72. (a) Matthew, J. B.; Gurd, F. N. R. *Methods Enzymol.* **1986,** *130,* 437–453. (b) Matthew, J. B.; Gurd, F. N. R. *Methods Enzymol.* **1986,** *130,* 413–436.
73. (a) Matthew, J. B. *Annu. Rev. Biophys. Chem.* **1985,** *14,* 387–418. (b) Matthew, J. B.; Gurd, F. R. N.; Garcia-Moreno, B. E.; Flanagan, M. A.; March, K. L.; Shire, S. J. *CRC Crit. Rev. Biochem.* **1985,** *18,* 91–197.
74. (a) Mauk, M. R.; Mauk, A. G.; Weber, P. C.; Matthew, J. B. *Biochemistry* **1986,** *25,* 7085–7091; ibid. **1987,** *26,* 974. (b) Mauk, M. R.; Reid, L. S.; Mauk, A. G. *Biochemistry* **1982,** *21,* 1843–1846.
75. Sinclair-Day, J. D.; Sykes, A. G. *J. Chem. Soc., Dalton Trans.* **1986,** 2069–2073.
76. Brunschwig, B. S.; DeLaive, P. J.; English, A. M.; Goldberg, M.; Gray, H. B.; Mayo, S. L.; Sutin, N. *Inorg. Chem.* **1985,** *24,* 3743–3749.
77. Eley, C. G. S.; Moore, G. R.; Williams, G.; Williams, R. J. P. *Eur. J. Biochem.* **1982,** *124,* 295–303.
78. Mathews, F. S. *Prog. Biophys. Mol. Biol.* **1985,** *45,* 1–56.
79. Mathews, F. S.; Czerwinski, E. W. in "Enzymes Biol. Membr. (2nd Ed.)", Vol. 4; Martonosi, A. N. Ed.; Plenum: New York, 1986, pp. 235–300.
80. Peterson, J. A.; Prough, R. A. in "Cytochrome P-450: Structure, Mechanism and Biochemistry"; Ortiz de Montellano, P. R., Ed.; Plenum Press: New York, 1986; pp. 89–117.
81. Greenwood, C. In "Metalloproteins", Part I; Harrison, P. M., Ed.; Verlag Chemie: Deerfield Beach, FL, 1985, pp. 43–78.
82. (a) Gray, H. B. *Chem. Soc. Rev.* **1986,** *15,* 17–30. (b) Adman, E. T. in "Metalloproteins", Part I; Harrison, P. M., Ed.; Verlag Chemie: Deerfield Beach, FL, 1985, pp. 1–42. (c) Sykes, A. G. *Chem. Soc. Rev.* **1985,** *14,* 283–315.
83. Scott, R. A.; Mauk, A. G.; Gray, H. B. *J. Chem. Educ.* **1985,** *62,* 932–938.
84. Bertrand, P. *Biochimie* **1986,** *68,* 619–628.
85. Dreyer, J. L. *Experientia* **1984,** *40,* 653–675.
86. DeVault, D. "Quantum-Mechanical Tunneling in Biological Systems"; Cambridge University Press: New York, 1984.
87. Jortner, J.; Pullman, B. "Tunneling"; D. Reidel: Dortecht, Holland, 1986.
88. Newton, M. D.; Sutin, N. *Annu. Rev. Phys. Chem.* **1984,** *35,* 437–480.
89. Sutin, N. *Prog. Inorg. Chem.* **1983,** *30,* 441–498.
90. Creutz, C.; Linck, R. G.; Sutin, N. In "Inorganic Reactions and Methods. Volume 15. Electron-Transfer and Electrochemical Reactions. Photochemical and Other Energized Reactions"; Zuckerman, J. J., Ed.; VCH Publ: Deerfield Beach, FL, 1986.
91. Isied, S. S. *Prog. Inorg. Chem.* **1984,** *32,* 443–517.
92. Bowden, E. F.; Hawkridge, F. M.; Blount, H. N. In "Comprehensive Treatise of Electrochemistry", Vol. 10; Srinivasan, S.; Chizmadzhev, Y. A.; Bockris, J. O.; Conway, B. E. and Yeager, E. Eds.; Plenum Press: New York, 1985; pp. 297–346.
93. Hush, N. S. *Coord. Chem. Rev.* **1985,** *64,* 135–157.
94. Haim, A. *Comm. Inorg. Chem.* **1985,** *4,* 113–149.
95. Tollin, G.; Meyer, T. E.; Cusanovich, M. A. *Biochim. Biophys. Acta* **1986,** *853,* 29–41.
96. Peterson-Kennedy, S. E.; McGourty, J. L.; Ho, P. S.; Sutoris, C.; Liang, N.; Zemel, H.; Blough, N. V.; Margoliash, J. L.; Hoffman, B. M. *Coord. Chem. Rev.* **1985,** *64,* 125–133.
97. Cannon, R. D. "Electron Transfer Reactions"; Butterworth: London, 1980.
98. Chance, B.; DeVault, D. C.; Frauenfelder, H.; Marcus, R. A.; Schrieffer, J. R.; Sutin, N., Eds. "Tunneling in Biological Systems"; Academic Press: New York, 1979.
99. Lever, A. B. P.; Gray, H. B., Eds. "Iron Porphyrins"; Addison-Wesley: New York, 1983.
100. Dolphin, D., Ed. "The Porphyrins"; Academic Press: New York, 1979.
101. (a) Takano, T.; Dickerson, R. E. *J. Mol. Biol.* **1981,** *153,* 79–94. (b) Takano, T.; Dickerson, R. E. *J. Mol. Biol.* **1981,** *153,* 95–115.
102. Stellwagen, E. *Nature (London)* **1978,** *275,* 73–74.
103. Mathews, F. S., Czerwinski, E. W., Argos, P. in "The Porphyrins", Vol. 7; Dolphin, D., Ed., Academic Press: New York, 1979, pp. 107–147.

104. Brown, G. M.; Sutin, N. *J. Am. Chem. Soc.* **1979,** *101,* 883–892.
105. Marcus, R. A. *Annu. Rev. Phys. Chem.* **1964,** *15,* 155–196.
106. Miller, J. R.; Beitz, J. V.; Huddleston, R. K. *J. Am. Chem. Soc.* **1984,** *106,* 5057–5068.
107. Beratan, D. N. *J. Am. Chem. Soc.* **1986,** *108,* 4321–4326.
108. Kosower, E. M. *J. Am. Chem. Soc.* **1985,** *107,* 1114–1118.
109. (a) Hupp, J. T.; Weaver, M. J. *J. Phys. Chem.* **1985,** *89,* 1601–1608. (b) Hupp, J. T.; Weaver, M. J. *J. Phys. Chem.* **1985,** *89,* 2795–2804.
110. Miller, J. R.; Calcaterra, L. T.; Closs, G. L. *J. Am. Chem. Soc.* **1984,** *106,* 3047–3049.
111. McGuire, M.; McLendon, G. *J. Phys. Chem.* **1986,** *90,* 2549–2551.
112. Hupp, J. T.; Meyer, T. J. *J. Phys. Chem.* **1987,** *90,* 1001–1003.
113. Calef, D. F.; Wolynes, P. G. *J. Chem. Phys.* **1983,** *78,* 470–482.
114. Sha'afi, R.; Fernandez, S. M., Eds. "Fast Methods in Physical Biochemistry and Cell Biology"; Elsevier: New York, 1983.
115. Gouterman, M. In "The Porphyrins", Vol. 3; Dolphin, D., Ed.; Academic Press: New York, 1979, pp. 1–165.
116. Dixon, D. W.; Kirmaier, C.; Holten, D. *J. Am. Chem. Soc.* **1985,** *107,* 808–813.
117. Gouterman, M.; Holten, D. *Photochem. Photobiol.* **1977,** *25,* 85–92.
118. Ho, P. S.; Sutoris, C.; Liang, N.; Margoliash, E.; Hoffman, B. M. *J. Am. Chem. Soc.* **1985,** *107,* 1070–1071.
119. Liang, N.; Kang, C. H.; Ho, P. S.; Margoliash, E.; Hoffman, B. M. *J. Am. Chem. Soc.* **1986,** *108,* 4665–4666.
120. Conklin, K. T.; McLendon, G. *Inorg. Chem.* **1986,** *25,* 4804–4906.
121. Cheung, E.; Taylor, K.; Kornblatt, J. A.; English, A. M.; McLendon, G.; Miller, J. R. *Proc. Natl. Acad. Sci. U.S.A.* **1986,** *83,* 1330–1333.
122. McLendon, G.; Miller, J. R. *J. Am. Chem. Soc.* **1985,** *107,* 7811–7816.
123. Bechtold, R.; Gardineer, M. B.; Kazmi, A.; van Hemelryck, B.; Isied, S. S. *J. Phys. Chem.* **1986,** *90,* 3800–3804.
124. Crutchley, R. J.; Ellis, W. R.; Gray, H. B. *J. Am. Chem. Soc.* **1985,** *107,* 5002–5004.
125. McLendon, G. L.; Winkler, J. R.; Nocera, D. G.; Mauk, M. R.; Mauk, A. G.; Gray, H.B. *J. Am. Chem. Soc.* **1985,** *107,* 739–740.
126. Espenson, J. H. "Chemical Kinetics and Reaction Mechanisms"; McGraw-Hill: New York, 1981, pp. 66.
127. Hiromi, K. "Kinetics of Fast Enzyme Reactions: Theory and Practice"; Halsted Press: New York, 1979.
128. Chance, B.; Eisenhardt, R.; Gibson, Q., Lundber Holm, K., Eds; "Rapid Mixing and Sampling in Biochemistry"; Academic Press: New York, 1964.
129. Turner, D. L. *J. Magn. Reson.* **1985,** *61,* 28–51.
130. Moore, G. R.; Harris, D. E.; Leitch, F. A.; Pettigrew, G. W. *Biochim. Biophys. Acta* **1984,** *764,* 331–342.
131. (a) Wand, A. J.; Englander, S. W. *Biochemistry* **1985,** *24,* 5290–5294. (b) Wand, A. J.; Roder, H.; Englander, S. W. *Biochemistry* **1986,** *25,* 1107–1114.
132. Hirs, C. H. W.; Timasheff, S. N., Eds. "Methods in Enzymology", Vol. 131; Academic Press: New York, 1986.
133. Englander, S. W.; Kallenbach, N. R. *Q. Rev. Biophys.* **1984,** *16,* 521–655.
134. Gupta, R. K.; Koenig, S. H.; Redfield, A. G. *J. Magn. Reson.* **1972,** *7,* 67–73.
135. Fishel, L. A.; Villafranca, J. E.; Mauro, K. M.; Kraut, J. *Biochemistry* **1987,** *26,* 351–360.
136. Goodin, D. B.; Mauk, A. G.; Smith, M. *Proc. Natl. Acad. Sci. U.S.A.* **1986,** *83,* 1295–1299.
137. Goodin, D. B.; Mauk, A. G.; Smith, M. *J. Biol. Chem.* **1987,** *262,* 7719–7724.
138. Eley, C. G. S.; Ragg, E.; Moore, G. R. *J. Inorg. Biochem.* **1984,** *21,* 295–310.
139. Minton, A. P. *Mol. Cell. Biochem.* **1983,** *55,* 119–140.
140. (a) Santos, H.; Turner, D. L.; Xavier, A. V.; Le Gall, J. *J. Magn. Reson.* **1984,** *59,* 177–180. (b) Boyd, J.; Moore, G. R.; Williams, G. *J. Magn. Reson.* **1984,** *58,* 511–516.
141. Johnston, E. R.; Grant, D. M. *J. Magn. Reson.* **1982,** *47,* 282–291.

142. McCammon, J. A.; Harvey, S. C. "Dynamics of Proteins and Nucleic Acids"; Cambridge University Press: Cambridge, 1987.
143. Isied, S. S.; Vassilian, A.; Magnuson, R. H.; Schwarz, H. A. *J. Am. Chem. Soc.* **1985,** *107,* 7432–7438.
144. Guarr, T.; McGuire, M. E.; McLendon, G. *J. Am. Chem. Soc.* **1985,** *107,* 5104–5111.
145. Moore, G. R.; Pettigrew, G. W.; Rogers, N. K. *Proc. Natl. Acad. Sci. U.S.A.* **1986,** *83,* 4998–4999.
146. Marchon, J.-C.; Mashiko, T.; Reed, C. A. In "Electron Transport and Oxygen Utilization"; Ho, C., Ed; Elsevier North-Holland: New York, 1982, pp. 67–72.
147. Matsuura, Y.; Takano, T.; Dickerson, R. E. *J. Mol. Biol.* **1982,** *156,* 389–409.
148. Korszun, Z. R.; Moffat, K.; Frank, K.; Cusanovich, M. A. *Biochemistry* **1982,** *21,* 2253–2258.
149. Kassner, R. J. *J. Am. Chem. Soc.* **1973,** *95,* 2674–2677.
150. Churg, A. K.; Warshel, A. *J. Biol. Chem.* **1986,** *261,* 1675–1681.
151. Moore, G. R. *FEBS Lett.* **1983,** *161,* 171–175.
152. Leitch, F. A.; Moore, G. R.; Pettigrew, G. W. *Biochemistry* **1984,** *23,* 1831–1838.
153. Rogers, N. K.; Moore, G. R.; Sternberg, M. J. E. *J. Mol. Biol.* **1985,** *182,* 613–616.
154. Rees, D. C.; *Proc. Natl. Acad. Sci. U.S.A.* **1985,** *82,* 3082–3085.
155. Schejter, A.; Aviram, I.; Goldkorn, T. In "Electron Transport and Oxygen Utilization"; Ho, C., Ed; Elsevier North-Holland: New York, 1982, pp. 95–99.
156. Schaap, A. P.; Gagnon, S. D. *J. Am. Chem. Soc.* **1982,** *104,* 3502–3506.
157. Neyhart, G. A.; Meyer, T. J. *Inorg. Chem.* **1986,** *25,* 4808–4810.
158. Nicholls, P. *Biochim. Biophys. Acta* **1974,** *346,* 261–310.
159. Vanderkooi, J.; Erecinska, M. *Arch. Biochem. Biophys.* **1974,** *162,* 385–391.
160. King, G. C.; Binstead, R. A.; Wright, P. E. *Biochim. Biophys. Acta* **1985,** *806,* 262–271.
161. Huang, Y.-Y.; Kimura, T. *Biochemistry* **1984,** *23,* 2231–2236.
162. Koller, K. B.; Hawkridge, F. M. *J. Am. Chem. Soc.* **1985,** *107,* 7412–7417.
163. (a) Kang, C. H.; Ferguson-Miller, S.; Margoliash, E. *J. Biol. Chem.* **1977,** *252,* 919–926. (b) Osheroff, N.; Brautigan, D. L.; Margoliash, E. *Proc. Natl. Acad. Sci. U.S.A.* **1980,** *77,* 4439–4443.
164. Reid, L. S.; Mauk, M. R.; Mauk, A. G. *J. Am. Chem. Soc.* **1984,** *106,* 2182–2185.
165. Sugiyama, T.; Miki, N.; Miura, R.; Miyake, Y.; Yamano, T. *Biochim. Biophys. Acta* **1982,** *706,* 42–49.
166. Salerno, J. C.; Yoshida, S.; King, T. E. *J. Biol. Chem.* **1986,** *261,* 12, 5480–5486.
167. Conroy, C. W.; Tyma, P.; Daum, P. H.; Erman, J. E. *Biochim. Biophys. Acta* **1978,** *537,* 62–69.
168. Nicholls, P.; Mochan, E. *Biochem. J.* **1971,** *121,* 55–67.
169. (a) Vitello, L. B.; Erman, J. E. Unpublished. (b) Moench, S. J.; Satterlee, J. D.; Erman, J. E. *Biochemistry* **1987,** *26,* 3821–3826. (c) Satterlee, J. D.; Moench, S. J.; Erman, J. E. *Biochim. Biophys. Acta* **1987,** *912,* 87–97.
170. Warshel, A.; Russell, S. T. *Q. Rev. Biophys.* **1984,** *17,* 283–422.
171. Dixon, D. W.; Barbush, M.; Shirazi, A. *Inorg. Chem.* **1985,** *24,* 1081–1087.
172. Warshel, A.; Churg, A. K. *J. Mol. Biol.* **1983,** *168,* 693–697.
173. Churg, A. K.; Weiss, R. M.; Warshel, A.; Takano, T. *J. Phys. Chem.* **1983,** *87,* 1683–1694.
174. Gupta, R. K. *Biochim. Biophys. Acta* **1973,** *292,* 291–295.
175. Dixon, D. W.; Barbush, M., unpublished.
176. Nocera, D. G.; Winkler, J. R.; Yocom, K. M.; Bordignon, E.; Gray, H. B. *J. Am. Chem. Soc.* **1984,** *106,* 5145–5150.
177. Isied, S. S.; Kuehn, C.; Worosila, G. *J. Am. Chem. Soc.* **1984,** *106,* 1722–1726.
178. Peterson-Kennedy, S. E.; McGourty, J. L.; Kalweit, J. A.; Hoffman, B. M. *J. Am. Chem. Soc.* **1986,** *108,* 1739–1746.
179. Cho, K. C.; Che, C. M.; Ng, K. M.; Choy, C. L. *J. Am. Chem. Soc.* **1986,** *108,* 2814–2818.
180. Joran, A. D.; Leland, B. A.; Geller, G. G.; Hopfield, J. J.; Dervan, P. B. *J. Am. Chem. Soc.* **1984,** *106,* 6090–6092.

181. Onuchic, J. N.; Beratan, D. N. *J. Am. Chem. Soc.* **1987**, *109*, 6771–6778.
182. Kadish, K. M.; Su, C. H. *J. Am. Chem. Soc.* **1983**, *105*, 177–180.
183. Fisher, M. T.; Sligar, S. G. *J. Am. Chem. Soc.* **1985**, *107*, 5018–5019.
184. (a) Fisher, M. T.; Sligar, S. G. *Biochemistry* **1987**, *26*, 4797–4803. (b) Huang, Y.-Y.; Hara, T.; Sligar, S.; Coon, M. J.; Kimura, T. *Biochemistry* **1986**, *25*, 1390–1394.
185. Fisher, M. T.; Sligar, S. G. *Biochemistry* **1985**, *24*, 6696–6701.
186. Lambeth, J. D.; Kriengsiri, S. *J. Biol. Chem.* **1985**, *260*, 8810–8816.
187. Poulos, T. L.; Finzel, B. C.; Howard, A. J. *Biochemistry* **1986**, *25*, 5314–5322.
188. Peterson-Kennedy, S. E.; McGourty, J. L.; Hoffman, B. M. *J. Am. Chem. Soc.* **1984**, *106*, 5010–5012.
189. McGourty, J. L.; Blough, N. V.; Hoffman, B. M. *J. Am. Chem. Soc.* **1983**, *105*, 4470–4472.
190. Simolo, K. P.; McLendon, G. L.; Mauk, M. R.; Mauk, A. G. *J. Am. Chem. Soc.* **1984**, *106*, 5012–5013.
191. Strittmatter, P. In "Rapid Mixing and Sampling in Biochemistry"; Chance, B.; Eisenhardt, R.; Gibson, Q., and Lundber Holm, K.; Eds; Academic Press: New York, 1964, pp. 71–84.
192. Ng, S.; Smith, M. B.; Smith, H.T.; Millett, F. *Biochemistry* **1977**, *16*, 4975–4978.
193. Stonehuerner, J.; Williams, J. B.; Millett, F. *Biochemistry* **1979**, *18*, 5422–5427.
194. Erman, J. E.; Vitello, L. B. *J. Biol. Chem.* **1980**, *255*, 6224–6227.
195. Leonard, J. J.; Yonetani, T. *Biochemistry* **1973**, *13*, 1465–1468.
196. Kang, C. H.; Brautigan, D. L.; Osheroff, N.; Margoliash, E. *J. Biol. Chem.* **1978**, *253*, 6502–6510.
197. Gupta, R. K.; Yonetani, T. *Biochim. Biophys. Acta* **1973**, *292*, 502–508.
198. Bhattacharyya, A. K.; Meyer, T. E.; Tollin, G. *Biochemistry* **1986**, *25*, 4655–4661.
199. Keller, R.; Groudinsky, O.; Wüthrich, K. *Biochim. Biophys. Acta* **1976**, *427*, 497–511.
200. Keller, R. M.; Wüthrich, K. *Biochim. Biophys. Acta* **1980**, *621*, 204–217.
201. La Mar, G. N.; Burns, P. D.; Jackson, J. T.; Smith, K. M.; Langry, K. C.; Strittmatter, P. *J. Biol. Chem.* **1981**, *256*, 6075–6079.
202. McLachlan, S. N.; La Mar, G. N.; Burns, P. D.; Smith, K. M.; Langry, K. C. *Biochim. Biophys. Acta* **1986**, *874*, 274–284.
203. Jue, T.; Krishnamoorthi, R.; La Mar, G. N. *J. Am. Chem. Soc.* **1983**, *105*, 5701–5703.
204. Levy, M. J.; La Mar, G. N.; Jue, T.; Smith, K. M.; Pandey, R. K.; Smith, W. S.; Livingstone, D. J.; Brown, W. D. *J. Biol. Chem.* **1985**, *260*, 13694–13698.
205. Yamamoto, Y.; La Mar, G. N. *Biochemistry* **1986**, *25*, 5288–5297.
206. La Mar, G. N.; Toi, H.; Krishnamoorthi, R. *J. Am. Chem. Soc.* **1984**, *106*, 6395–6401.
207. Livingston, D. J.; Davis, N. L.; La Mar, G. N.; Brown, W. D. *J. Am. Chem. Soc.* **1984**, *106*, 3025–3026.
208. Light, W. R.; Rohlfs, R. J.; Palmer, G.; Olson, J. S. *J. Biol. Chem.* **1987**, *262*, 46–52.
209. Burns, P. D.; La Mar, G. N. *J. Biol. Chem.* **1981**, *256*, 4934–4939.
210. Northrup, S. H.; Pear, M. R.; Morgan, J. D.; McCammon, J. A.; Karplus, M. *J. Mol. Biol.*, **1981**, *153*, 1087–1109.
211. Henry, E. R.; Eaton, W. A.; Hochstrasser, R. M. *Proc. Natl. Acad. Sci. U.S.A.* **1986**, *83*, 8982–8986.
212. Sutin, N.; Yandell, J. K. *J. Biol. Chem.* **1972**, *247*, 6932–6936.
213. Creutz, C.; Sutin, N. *Proc. Natl. Acad. Sci. U.S.A.* **1973**, *70*, 1701–1703.
214. Creutz, C.; Sutin, N. *J. Biol. Chem.* **1974**, *249*, 6788–6795.
215. Meyer, T. E.; Watkins, J. A.; Przysiecki, C. T.; Tollin, G.; Cusanovich, M. A. *Biochemistry* **1984**, *23*, 4761–4767.
216. Ragg, E.; Moore, G. R. *J. Inorg. Biochem.* **1984**, *21*, 253–261.
217. Senn, H.; Guerlesquin, F.; Bruschi, M.; Wüthrich, K. *Biochim. Biophys. Acta* **1983**, *748*, 194–204.
218. Feinberg, B. A.; Bedore, J. E., Jr.; Ferguson-Miller, S. *Biochim. Biophys. Acta* **1986**, *851*, 157–165.

219. Dyer, C.; Shubert, A.; Timkovich, R.; Feinberg, B. A. *Biochim. Biophys. Acta* **1979**, *579*, 253–268.
220. Ivanetich, K. M.; Bradshaw, J. J.; Fazakerley, G. V. *Biochem. Biophys. Res. Commun.* **1976**, *72*, 433–439.
221. Walker, F. A.; Lo, M.-W.; Ree, M. T. *J. Am. Chem. Soc.* **1976**, *98*, 5552–5560.
222. Satterlee, J. D.; La Mar, G. N.; Frye, J. S. *J. Am. Chem. Soc.* **1976**, *98*, 7275.
223. Doeff, M. M.; Sweigart, D. A. *Inorg. Chem.* **1982**, *21*, 3699.
224. Pasternack, R. F.; Gillies, B. S.; Stahlbush, J. R. *J. Am. Chem. Soc.* **1978**, *100*, 2613–2619.
225. Hwang, Y. C.; Dixon, D. W. *Inorg. Chem.* **1986**, *25*, 3716–3718.
226. Goff, H.; La Mar, G. N.; Reed, C. A. *J. Am. Chem. Soc.* **1977**, *99*, 3641–3646.
227. Parmely, R. C.; Goff, H. M. *J. Inorg. Biochem.* **1980**, *12*, 269–280.
228. Osterhout, J. J., Jr.; Kamalam, M.; Nall, B. T. *Biochemistry* **1985**, *24*, 6680–6684.
229. Gadsby, P. M. A.; Peterson, J.; Foote, N.; Greenwood, C.; Thomson, A. J. *Biochem. J.* **1987**, *246*, 43–54.
230. Gupta, R. K.; Koenig, S. H. *Biochem. Biophys. Res. Commun.* **1971**, *45*, 1134–1143.
231. English, A. M. Personal communication.
232. Hazzard, J. T.; Cusanovich, M. A.; Tainer, J. A.; Getzoff, E. D.; Tollin, G. *Biochemistry* **1986**, *25*, 3318–3328.
233. Beoku-Betts, D.; Sykes, A. G. *Inorg. Chem.* **1985**, *24*, 1142–1147.
234. Koppenol, W. H. *Biophys. J.* **1980**, *29*, 493–508.
235. Van Leeuwen, J. W. *Biochim. Biophys. Acta* **1983**, *743*, 408–421.
236. Koppenol, W. H.; Margoliash, E. *J. Biol. Chem.* **1982**, *257*, 4426–4437.
237. Capaldi, R. A.; Darley-Usmar, V.; Fuller, S.; Millett, F. *FEBS Lett.* **1982**, *138*, 1–7.
238. Augustin, M. A.; Chapman, S. K.; Davies, D. M.; Sykes, A. G.; Speck, S. H.; Margoliash, E. *J. Biol. Chem.* **1983**, *258*, 6405–6409.
239. Concar, D. W.; Hill, H. A. O.; Moore, G. R.; Whitford, D.; Williams, R. J. P. *FEBS. Lett.* **1986**, *206*, 15–19.
240. Waldmeyer, B.; Bosshard, H. R. *J. Biol. Chem.* **1985**, *260*, 5184–5190.
241. Rieder, R.; Wiemken, V.; Bachofen, R.; Bosshard, H. R. *Biochem. Biophys. Res. Commun.* **1985**, *128*, 120–126.
242. Salemme, F. R. *J. Mol. Biol.* **1976**, *102*, 563–568.
243. Poulos, T. L.; Kraut, J. *J. Biol. Chem.* **1980**, *255*, 10322–10330.
244. Matthew, J. B.; Weber, P. C.; Salemme, F. R.; Richards, F. M. *Nature (London)* **1983**, *301*, 169–171.
245. Weber, P. C.; Tollin, G. *J. Biol. Chem.* **1985**, *260*, 5568–5573.
246. Poulos, T. L.; Mauk, A. G. *J. Biol. Chem.* **1983**, *258*, 7369–7373.
247. Berg, O. G.; von Hippel, P. H. *Annu. Rev. Biophys. Biophys. Chem.* **1985**, *14*, 131–160.
248. Berg, O. G. *Biophys. J.* **1985**, *47*, 1–14.
249. Hess, S.; Monchick, L. *J. Chem. Phys.* **1986**, *84*, 1385–1390.
250. Pritchin, I. A.; Salikhov, K. M. *J. Phys. Chem.* **1985**, *89*, 5212–5217.
251. McCammon, J. A.; Northrup, S. H.; Allison, S. A. *J. Phys. Chem.* **1986**, *90*, 3901–3905.
252. Burshtein, A. I.; Doktorov, A. B.; Morozov, V. A. *Chem. Phys.* **1986**, *104*, 1–18.
253. Temkin, S. I.; Yakobson, B. I. *J. Phys. Chem.* **1984**, *88*, 2679–2682.
254. Margoliash, E.; Schejter, A. *Adv. Prot. Chem.* **1966**, *21*, 113–286.
255. Eley, C. G. S.; Moore, G. R. *Biochem. J.* **1983**, *215*, 11–21.
256. Sutin, N. *Chem. Br.* **1972**, *8*, 148–151.
257. Keller, R. M.; Wüthrich, K.; Pecht, I. *FEBS Lett.* **1976**, *70*, 180–184.
258. Timkovich, R.; Dixon, D. W. Unpublished
259. Hasinoff, B. B. *Biochim. Biophys. Acta* **1985**, *829*, 1–5.
260. Tollin, G.; Hanson, L. K.; Caffrey, M.; Meyer, T. E.; Cusanovich, M. A. *Proc. Natl. Acad. Sci. U.S.A.* **1986**, *83*, 3693–3697.
261. Falk, K.-E.; Angstrom, J. *Biochim. Biophys. Acta* **1983**, *722*, 291–296.

262. Livingston, D. J.; McLachlan, S. J.; La Mar, G. N.; Brown, W. D. *J. Biol. Chem.* **1985,** *260,* 15699–15707.
263. Hazzard, J. T.; Tollin, G. *Biochem. Biophys. Res. Commun.* **1985,** *130,* 1281–1286.
264. Tollin, G.; Brown, K.; De Francesco, R.; Edmondson, D. *Biochemistry* **1987,** *26,* 5042–5048.
265. Timkovich, R. *Biochemistry* **1986,** *25,* 1089–1093.
266. Timkovich, R.; Cork, M. S.; Taylor, P. V. *Arch. Biochem. Biophys.* **1985,** *240,* 689–697.
267. Dixon, D. W.; Hong, H. L.; Woehler, S. Unpublished.
268. Williams, G.; Moore, G. R. *J. Inorg. Biochem.* **1984,** *22,* 1–10.
269. Northrup, S. H.; Boles, J. O.; Reynolds, J. C. L. *J. Phys. Chem.* **1987,** *91,* 5991–5998.
270. Northrup, S. H.; Reynolds, J. C. L.; Miller, C. M.; Forrest, K. J.; Boles, J. O. *J. Am. Chem. Soc.* **1986,** *108,* 8162–8170; ibid. **1987,** *109,* 3176.
271. Cokic, P.; Erman, J. E. *Biochim. Biophys. Acta* **1987,** *913,* 257–271.
272. Erman, J. E.; Kim, K. L.; Vitello, L. B.; Moench, S. J.; Satterlee, J. D. *Biochim. Biophys. Acta* **1987,** *911,* 1–10.
273. Tiede, D. M. *Biochemistry* **1987,** *26,* 397–410.
274. Deisenhofer, J.; Epp, O.; Miki, K.; Huber, R.; Michel, H. *Nature (London)* **1985,** *318,* 618–624.
275. Hazzard, J. T.; Tollin, G. *Biochem. J.* **1986,** *49,* 540a.
276. Edwards, S. L.; Xuong, N. H.; Hamlin, R. C.; Kraut, J. *Biochemistry* **1987,** *26,* 1503–1511.
277. Bhattacharyya, A. K.; Meyer, T. E.; Cusanovich, M. A.; Tollin, G. *Biochemistry* **1987,** *26,* 758–764.
278. Kuo, L.-M.; Davies, H. C.; Smith, L. *Biochim. Biophys. Acta* **1985,** *809,* 388–395.
279. Kuo, L.-M.; Davies, H. C.; Smith, L. *Biochim. Biophys. Acta* **1986,** *848,* 247–255.
280. Haser, R.; Pierrot, M.; Frey, M.; Payan, F.; Astier, J.-P. Bruschi, M.; Le Gall, J. *Nature (London)* **1979,** *282,* 806–810.
281. Pierrot, M.; Haser, R.; Frey, M.; Payan, F.; Astier, J.-P. *J. Biol. Chem.* **1982,** *257,* 14341–14348.
282. Higuchi, Y.; Bando, S.; Kusunoki, M.; Matsuura, Y.; Yasuoka, N.; Kakudo, M.; Yamanaka, T.; Yagi, T.; Inokuchi, H. *J. Biochem. (Tokyo)* **1981,** *89,* 1659–1662.
283. Moura, J. J. G.; Santos, H.; Moura, I.; LeGall, J.; Moore, G. R.; Williams, R. J. P.; Xavier, A. V. *Eur. J. Biochem.* **1982,** *127,* 151–155.
284. Santos, H.; Moura, J. J. G.; Moura, I.; LeGall, J.; Xavier, A. V. *Eur. J. Biochem.* **1984,** *141,* 283–296.
285. Moura, I.; Liu, M. C.; LeGall, J.; Peck, H. D.; Payne, W. J.; Xavier, A. V.; Moura, J. J. G. *Eur. J. Biochem.* **1984,** *141,* 297–303.
286. Guerlesquin, F.; Bruschi, M.; Wüthrich, K. *Biochim. Biophys. Acta* **1985,** *830,* 296–303.
287. Carter, D. C.; Melis, K. A.; O'Donnell, S. E.; Burgess, B. K.; Furey, W. F.; Wang, B.-C.; Stout, C. D. *J. Mol. Biol.* **1985,** *184,* 279–295.
288. Makinen, M. W.; Schichman, S. A.; Hill, S. C.; Gray, H. B. *Science* **1983,** *222,* 929–931.
289. Williams, G.; Clayden, N. J.; Moore, G. R.; Williams, R. J. P. *J. Mol. Biol.* **1985,** *183,* 447–460.
290. Louie, G. V.; Hutcheon, W. L. B.; Brayer, G. D. *J. Mol. Biol.* **1988,** *199,* 295–314.

CHAPTER 6

Chemical Studies Related to Iron Protoporphyrin-IX Mixed-Function Oxidases

Thomas C. Bruice

University of California, Santa Barbara, California

CONTENTS

1. The Enzymes. 228
2. Chemical Preparation of Higher Valent Iron-Oxo Porphyrin
   Species. 230
3. Electrochemical Generation of Higher Valent Iron-Oxo
   Porphyrin Species. 233
4. Higher Valent Manganese-oxo Species 234
5. Mechanisms of Oxygen Atom Transfer from Percarboxylic
   Acids and Alkyl Hydroperoxides to Metal(III) Porphyrins 236
6. A Model for the Catalase Reaction: The Mechanism for the
   Formation of Oxygen on Reaction of an Iron(III) Porphyrin
   with Hydrogen Peroxide 245
7. The Mechanism of Reaction of Hydrogen Peroxide with
   Manganese(III) Porphyrin. 248
8. The Rebound Mechanism for Oxygen Insertion into
   Carbon–Hydrogen Bonds 249
9. Mechanisms of Dealkylation Reactions 253
10. Reaction of N,N-Dimethylaniline N-Oxides with Metal(III)
    Porphyrins. 255
11. The Mechanism of the Epoxidation of Alkenes 262

Acknowledgments   273
References   273

1. THE ENZYMES

The prosthetic group of the enzymes peroxidase, catalase, and cytochrome
P-450 is iron(III) protoporphyrin IX. In the case of the peroxidases, the axial
ligand which is distal to the reactive face of the iron(III) porphyrin is an
imidazole of a histidine residue. The distal axial ligand function for catalases
is a tyrosine hydroxyl group,[1] while for cytochrome P-450 enzymes it is a
cysteine thiolate. The mechanism of oxygen activation by horseradish (HR)
peroxidase and catalase enzymes are quite similar. The iron(III) porphyrin
moiety of HR peroxidase undergoes two-electron oxidation upon reaction
with hydrogen peroxide, alkyl hydroperoxides,[2] and percarboxylic acids[3a] to
yield an intermediate known as compound I.[3b] On the basis of visible,[4]
Mossbauer,[5] electron spin resonance (ESR),[6a] and electron nuclear double
resonance (ENDOR)[6b] spectral data, compound I has been assigned the
structure of a low-spin oxo-ligated iron(IV) protoporphyrin IX π-cation radical. An example of a porphyrin π-cation radical is provided in Figure 6-1.
One-electron reduction of compound I results in the formation of an
iron(IV)-oxo protoporphyrin IX species termed compound II.[8] Much the
same type of evidence supports identical intermediates formed on reaction
of catalase with hydrogen peroxide.[9] Almost all oxidations mediated by HR
peroxidase involve (outer-sphere[10]) $1e^-$ or H· transfer from substrate (reaction 6-1). In the reaction, P is

$$(P)Fe^{III} + YOOH \rightarrow (^+\!\cdot P)Fe^{IV}O + YOH$$

Compound I

$$(^+\!\cdot P)Fe^{IV}O + SH_2 \rightarrow (P)Fe^{IV}O + SH\cdot + H^+ \qquad (6\text{-}1)$$

Compound II

$$(P)Fe^{IV}O + SH\cdot \rightarrow (P)Fe^{III} + S_{ox} + H^+$$

the porphyrin, $^+\!\cdot P$ is the porphyrin π-cation moiety, and YOOH represents
H_2O_2, alkyl hydroperoxides, or percarboxylic acids. The catalase enzymes
are charged by hydrogen peroxide and employ hydrogen peroxide as normal
substrate (reaction 6-2). The compound I and II species of the catalases are
capable of carrying out peroxidase chemistry.

$$(P)Fe^{III} + H_2O_2 \rightarrow (^+\!\cdot P)Fe^{IV}O + H_2O$$

$$(^+\!\cdot P)Fe^{IV}O + H_2O_2 \rightarrow (P)Fe^{III}O + H_2O + O_2$$

(6-2)

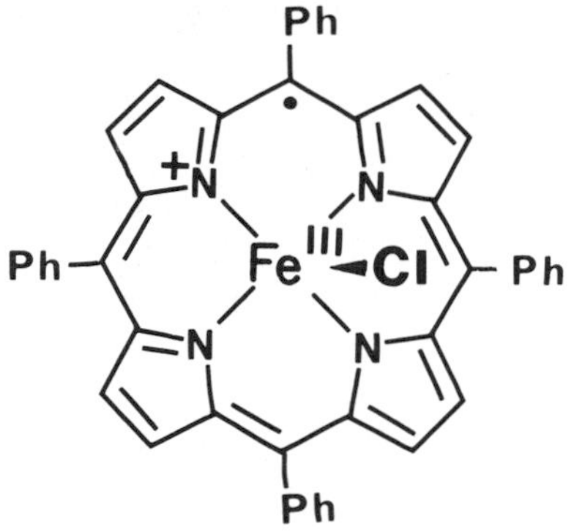

Ph

Ph—Fe^{III}◄Cl—Ph

Ph

Figure 6-1. A synthesizable chloro-ligated iron(III)porphyrin π-cation radical (5,10,15,20-tetraphenylporphinato iron(III) π-cation radical[7]).

It has been proposed that the oxidizing species of cytochrome P-450 is an oxo-ligated iron(IV) porphyrin π-cation radical resembling compound I of the peroxidases.[11,12] Such an assignment of structure is by inference, since a higher valent iron-oxo species of a cytochrome P-450 enzyme has never been prepared for examination. Biochemical synthesis of the P-450 oxidizing species and its reaction with a substrate involve the stepwise reactions of Scheme I. Beginning with the porphyrin iron(III) state, substrate binding is followed by 1e$^-$ reduction and subsequent reaction with dioxygen. These steps have been observed and are well characterized. The second 1e$^-$ reduction is rate determining and the commitment step in the cycle. For this reason, neither the chemistry for the formation of the oxidizing intermediate

CYTOCHROME P-450

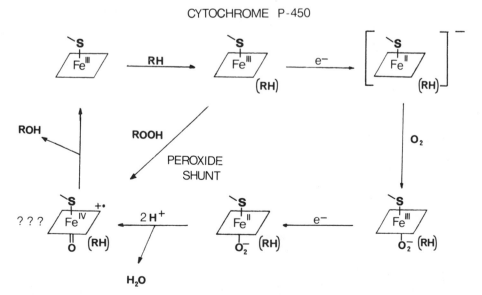

Scheme I

nor its structure can be directly determined. A principle difference between the active oxidizing agents of the peroxidases and cytochrome *P*-450 is their distal axial ligands. With the peroxidases, the imidazole axial ligand allows the formation of an iron(IV)-oxo porphyrin π-cation radical species, while it is questionable whether the thiolate axial ligand of the cytochrome *P*-450 enzymes could. Thus, one must be concerned with the probability of the existence of a thiol anion ligated to a high-valent iron. When R—S⁻ species are ligated to metal ions of higher oxidation state (M^{n+1}), they undergo $1e^-$ oxidation, and a covalent bond is formed between RS· and M^n.[13] It is safe to assume that the structure of the "porphyrin-iron-oxene" species of cytochrome *P*-450 reflects this feature so that the second $1e^-$ deficiency resides with the thiol ligand rather than in the porphyrin ring system as a porphyrin π-cation radical. Other possible structures are shown in Scheme II. The

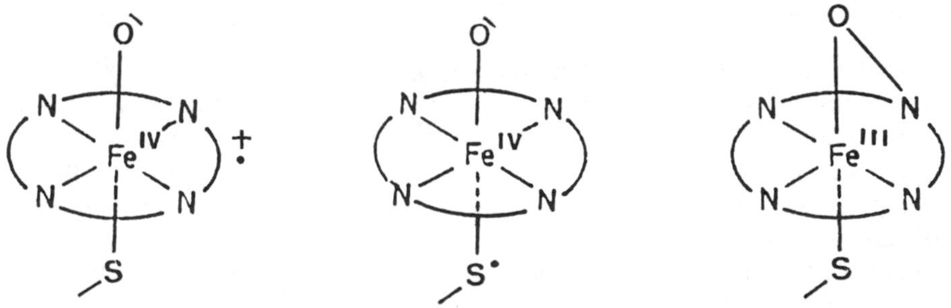

Scheme II

bridging iron-oxo-pyrrole nitrogen structure has been suggested by several investigators.[14] The formation of oxidizing species on reaction of cytochrome *P*-450 with hydroperoxides, percarboxylic acids, etc. (the "peroxide shunt" of Scheme I) resembles the formation of such species with peroxidases and catalases.

The $1e^-$ oxidized products formed on reaction of compound I of HR peroxidase with various substrates may arise by electron transfer to the porphyrin π-cation radical,[10] but oxidations involving catalase and cytochrome *P*-450 enzymes involve reaction at the iron bound oxygen. With cytochrome *P*-450, the most common product-forming reactions involve oxygen transfer in the formation of epoxides, oxygen insertion into H—C bonds to form alcohols, hydroxylation of aromatic rings, and dealkylation of ethers and amines.

2. CHEMICAL PREPARATION OF HIGHER VALENT IRON-OXO PORPHYRIN SPECIES

To protect the meso-positions of the porphyrin ring from oxidation, most investigations involve the use of metallo *meso*-tetrakis(phenyl) porphyrins. The structures of the symmetrically substituted 5, 10, 15, 20,-tetraphenyl-

porphyrins which are referred to, as well as abbreviations for these structures, are provided in Scheme III.

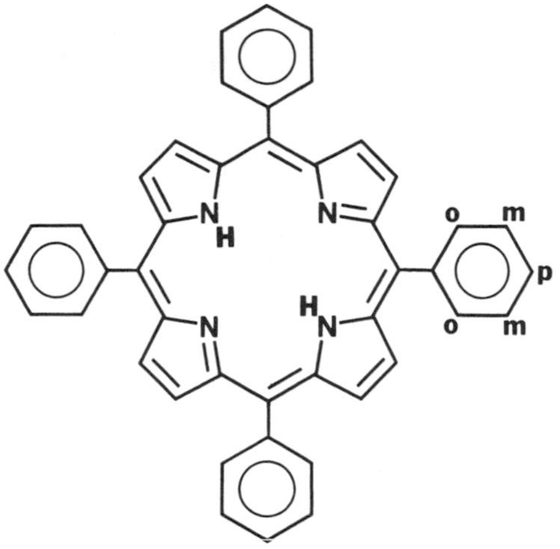

Substituents on Phenyl rings

Ortho		Meta		Para	Abbreviation
H	H	H	H	H	TPP
Me	Me	H	H	H	Me_8TPP
Me	Me	H	H	Me	$Me_{12}TPP$
H	H	Me	H	H	mMe_4TPP
H	H	H	H	Me	pMe_4TPP
Cl	Cl	H	H	H	Cl_8TPP
F	F	H	H	H	F_8TPP
F	F	F	F	F	$F_{20}TPP$
Me	Me	SO_3^-	H	H	Me_8STPP

Scheme III

A number of procedures have been reported for low-temperature preparation of porphyrin iron(IV)-oxo species. The intermediate peroxo species
$[(mMe_4TPP)Fe^{III}\text{-}O\text{-}O\text{-}Fe^{III}(mMe_4TPP)]$ in the autoxidation of the *meso*-
tetra-*m*-tolylporphinato iron(II) $[(mMe_4TPP)Fe^{II}]$ has been thoroughly characterized.[15] The peroxide $[(mMe_4TPP)Fe^{II}]_2O_2$, on reaction with the tertiary
amines N-methylimidazole, pyridine, and piperidine, provides the tertiary
amine (B:) ligated porphinato iron(IV)-oxo species $[(mMe_4TPP)(B)Fe^{IV}O]$.[16]
The magnetic susceptibility and electronic spectra of the latter resemble
closely those properties displayed by compound II of HR peroxidase.

The order of the addition of reagents was found to be critical in determining the nature of the products (reactions 6-3 and 4).[17]

$$B: + [mMe_4TPP)Fe^{II} \rightarrow (mMe_4TPP)]Fe^{II}B \xrightarrow{O_2} (mMe_4TPP)(B)Fe^{II}O_2 \quad (6\text{-}3)$$

$$2(mMe_4TPP)Fe^{II} + O_2 \rightarrow [(mMe_4TPP)Fe^{II}]_2O_2 \xrightarrow{2B:} 2(mMe_4TPP)(B)Fe^{IV}O$$

$$(6\text{-}4)$$

The porphyrin iron(IV)-oxo species was shown to transfer oxygen to triphenylphosphine (reaction 6-5).

$$(mMe_4TPP)(B)Fe^{IV}O + Ph_3P \rightarrow (mMe_4TPP)Fe^{II}(B) + Ph_3PO \quad (6\text{-}5)$$

The peroxide of *meso*-tetrakis(mesityl)porphinato iron(II), which is sterically hindered, undergoes thermal decomposition at $-30°C$ to provide a porphyrin iron(IV)-oxo species which is not ligated at the second axial position (reaction 6-6).[16a] Oxidation ($-70°C$) of $(Me_{12}TPP)Fe^{IV}O$ by Br_2, Cl_2, or the π-cation radical of the porphyrin iron(III) perchlorate generates (reaction 6-7), the iron(IV)-oxo porphyrin π-cation radical species.[18] The same species is obtained by oxidation of $(Me_{12}TPP)Fe^{III}Cl$ with *m*-chloroperbenzoic acid (reaction 6-8).[19]

$$(Me_{12}TPP)Fe^{III}\text{-O-O-}Fe^{III}(Me_{12}TPP) \rightarrow 2(Me_{12}TPP)Fe^{IV}O \qquad (6\text{-}6)$$
$$\text{(Red)}$$

$$(Me_{12}TPP)Fe^{IV}O + (^{+\cdot}Me_{12}TPP)Fe^{III}ClO_4 \rightarrow$$
$$(^{+\cdot}Me_{12}TPP)Fe^{IV}O + (Me_{12}TPP)Fe^{III}ClO_4 \quad (6\text{-}7)$$

$$(mMe_4TPP)Fe^{III}Cl + m\text{-}ClC_6H_4CO_3H \rightarrow$$
$$(^{+\cdot}mMe_4TPP)^{IV}O(Cl) + m\text{-}ClC_6H_4CO_2H \quad (6\text{-}8)$$
$$\text{(Green)}$$

The sequence of reactions shown in reactions 6-9 and 6-10 has been established when using the picket-fence porphyrin tetrakis(pivaloylphenyl)-porphinato iron(II) ($(PP)Fe^{II}$) and iron(III).[20]

$$(PP)Fe^{II} + O_2 \xrightarrow[\text{THF}]{-40°C} (PP)Fe^{II}O_2 \xrightarrow[(RO)_2AlH]{-40°C}$$

$$(PP)Fe^{III}OOH \xrightarrow[CO_2]{-70°C} (^{+\cdot}PP)Fe^{IV}O \quad (6\text{-}9)$$

$$(PP)Fe^{II} + O_2^{\cdot-} + CO_2 \rightarrow (^{+\cdot}PP)Fe^{IV}O \xrightarrow{N\text{-MeIm}}$$

$$(^{+\cdot}PP)(N\text{-MeIm})Fe^{IV}O \quad (6\text{-}10)$$

In these reactions, the peroxide-ligated iron(III) porphyrin is converted to the iron(IV)-oxo porphyrin π-cation radical by using CO_2 to create a better leaving group (reaction 6-11). In much the same fashion acetic anhydride has been employed to create a good leaving group (toluene, $-70°C$; reaction 6-12).[21] The bismethoxy-ligated iron(IV) porphyrin and porphyrin

π-cation radical species have been prepared as shown in reactions 6-13 and 6-14.[22]

$$(PP)Fe^{III}\text{-O-OH} + CO_2 \rightarrow (PP)Fe^{III}\text{-O-O-}CO_2H \rightarrow (^{+\cdot}PP)Fe^{IV}O + HCO_3^-$$
$$(6\text{-}11)$$

$$(Me_{12}TPP)Fe^{III}\text{-O-OH} + (Ac)_2O \rightarrow (Me_{12}TPP)Fe^{III}\text{-O-O-Ac} + AcOH$$
$$(6\text{-}12)$$

$$(Me_{12}TPP)Fe^{III}\text{-O-O-Ac} \rightarrow (^{+\cdot}Me_{12}TPP)Fe^{IV}O + AcO^-$$

$$(Me_{12}TPP)Fe^{III}ClO_4 \xrightarrow{Fe^{III}(ClO_4)_3} (^{+\cdot}Me_{12}TPP)Fe^{III}ClO_4$$
$$(6\text{-}13)$$

$$(^{+\cdot}Me_{12}TPP)Fe^{III}ClO_4 + 2(MeO^-) \rightarrow (Me_{12}TPP)Fe^{IV}(OMe)_2$$

$$(Me_{12}TPP)Fe^{III}ClO_4 \xrightarrow[MeOH]{ArIO} (^{+\cdot}Me_{12}TPP)Fe^{IV}(OMe)_2 \qquad (6\text{-}14)$$

3. ELECTROCHEMICAL GENERATION OF HIGHER VALENT IRON-OXO PORPHYRIN SPECIES

In the cyclic voltammograms of Cl^-, ClO_4^-, etc. ligated iron(III) porphyrins [as *meso*-tetraphenylporphinato iron(III) chloride, $(TPP)Fe^{III}Cl$], when scanning in the anodic direction beyond the porphyrin iron(II)/porphyrin iron(III) couple, two reversible $1e^-$ oxidation waves are seen. It was originally proposed that these $1e^-$ oxidations pertained to the porphyrin iron(III)/ porphyrin iron(IV) and porphyrin iron(IV)/iron(IV) porphyrin π-cation radical couples.[23,24] This proved to be incorrect as shown by the observation that the first oxidation was not iron centered, since the measured potential was independent of the anion employed as the axial ligand bound to the iron atom[7a,25] The physical properties of the products associated with the two $1e^-$ oxidations are consistent with the oxidations of the porphyrin to the π-cation radical and then to its dication (reaction 6-15).

$$(TPP)Fe^{III}Cl \xrightarrow{-e^-} (^{+\cdot}TPP)Fe^{III}Cl \xrightarrow{-e^-} (^{2+}TPP)Fe^{III}Cl \qquad (6\text{-}15)$$

When the strongly basic HO^- and MeO^- ligands replace the weakly basic ligands such as Cl^-, ClO_4^-, etc., the initial $1e^-$ oxidation becomes iron centered.[26,27] It is found that the hydroxy-ligated tetraphenylporphyrin iron(III) salts, $(Me_{12}TPP)Fe^{III}OH$, $(Me_{12}TPP)Fe^{III}OMe$, $(Me_8TPP)Fe^{III}OH$, $(Me_8TPP)Fe^{III}OMe$, $(Cl_8TPP)Fe^{III}OH$, $(Cl_8TPP)Fe^{III}OMe$, $(F_8TPP)Fe^{III}OH$, and $(F_8TPP)Fe^{III}OMe$, electrochemically undergo the sequential $1e^-$ oxidations of reaction 6-16.

$$(Me_{12}TPP)Fe^{III}OH \xrightarrow{-e^-, -H^+} (Me_{12}TPP)Fe^{IV}O$$
$$(Me_{12}TPP)Fe^{IV}O + ClO_4^- \longrightarrow [(Me_{12}TPP)Fe^{IV}(O)ClO_4]^-$$

$$[(Me_{12}TPP)Fe^{IV}(O)ClO_4]^- \xrightarrow{-1e^-} (^{+\cdot}Me_{12}TPP)Fe^{IV}(O)ClO_4$$
$$(^{+\cdot}Me_{12}TPP)Fe^{IV}(O)ClO_4 + ClO_4^- \longrightarrow [(^{+\cdot}Me_{12}TPP)Fe^{IV}(O)(ClO_4)_2]^-$$

$$[(^{+\cdot}Me_{12}TPP)Fe^{IV}(O)(ClO_4)_2]^- \xrightarrow{-1e^-} (^{2+}Me_{12}TPP)Fe^{IV}(O)(ClO_4)_2$$

$$(6\text{-}16)$$

Reaction of electrochemically generated $(Me_{12}TPP)Fe^{IV}O$ with one equivalent of 1-methylimidazole (MeIm) provided a product $((Me_{12}TPP)(MeIm)$-$Fe^{IV}O)$ with the spectral characteristics previously ascribed to $(Me_{12}TPP)$-$(B)Fe^{IV}O$ where B represents a tertiary amine (reaction 6-4), and addition of one equivalent of MeO^- to electrochemically generated $(Me_{12}TPP)Fe^{IV}O$ provided a product with essentially the same spectrum as the proposed product $(Me_{12}TPP)Fe^{IV}(OMe)_2$ of reaction 6-13. The change from porphyrin-centered to metal-centered oxidation is due to the fact that higher oxidation states of metals are stabilized by coordination of the metal to a strongly basic oxyanion.

4. HIGHER VALENT MANGANESE-OXO SPECIES

The study of manganese porphyrins has played an important role in the chemical approach to the mechanism of the cytochrome P-450 enzymes. When Mn(III) is substituted for Fe(III) in cytochrome P-450$_{cam}$, the resultant modified enzyme is capable of the epoxidation of alkenes when charged with PhIO.[28] Furthermore, the reaction was observed to proceed via a spectrally detectable intermediate which closely resembled that of a manganese(V)-oxo porphyrin species.

When plausible, aqueous solutions are best for the investigation of the mechanisms of any reaction which involves acidic and basic species, since the use of the glass electrode allows the determination of all acid–base equilibria and the activity of the hydronium ion, the apparent hydroxide concentration, and the calculation of the concentrations of all acidic and basic species. Oxidations of manganese porphyrins have been investigated in water. Evidence was presented as early as 1963 stating that the addition of one equivalent of hypochlorite to Mn(III) porphyrin (water, pH 13) provides an Mn(IV) porphyrin species.[29] More recently, the oxidation of a number of water-soluble Mn(III) porphyrins have been investigated (pH 14) employing such diverse oxidants as peroxodisulphate, hypochlorite, bromate, hydrogen peroxide, permanganate, lead dioxide, and ferricyanide.[30] Since the oxidation was found to be completely reversible when using the ferricyanide/ferrocyanide couple, it was possible to calculate the pH dependence of a midpoint potential and also the rate constants for the oxidation to the Mn(IV) porphyrin state. Though the rate constants for the oxidation of the Mn(III) porphyrins were dependent upon the structure (formal charge on substituents) of the porphyrin, the E_m values were not. The latter observation suggests the oxidation to be metal rather than porphyrin centered. The finding of a low magnetic moment was ascribed to the existence of the Mn(IV) state as a μ-oxo dimer $[(TPP)Mn^{IV})_2O]$ (reaction 6-17).

$$[(TPP)Mn^{III}]_2O \rightarrow [(TPP)Mn^{IV}]_2O + 2e^- \qquad (6\text{-}17)$$

[When prepared in methanol, the Mn^{IV} porphyrin exists[31] as the monomeric species $(TPP)Mn^{IV}(OMe)_2$.] Further, $1e^-$ oxidation of Mn(IV) porphyrin is easily accomplished at pH 14 by hypochlorite. The actual structure of the product at the Mn(V) oxidation state must be considered to be unknown. It is remarkably stable at high pH in water.[30] From extended X-ray absorption fine structure (EXAFS) spectra of the solid at the Mn(V) oxidation state, prepared by oxidation of Mn(III) porphyrin at high pH by hypochlorite using as solvent a H_2O/CH_2Cl_2 biphase, the structure has been proposed to represent a porphyrin manganese(IV) oxygen radical $[(Porph)Mn^{IV}O\cdot]$ rather than a back-bonded manganese(V)-oxo porphyrin $[(Porph)Mn^V{=}O]$ or manganese(IV)-oxo porphyrin π-cation radical $[(^+Porph)Mn^{IV}{=}O]$.[32] The species formed from manganese(III) tetraphenylporphyrin and ClO^- at high pH, which was assigned a manganese(V)-oxo tetraphenylporphyrin structure, is reported to not react with manganese(III) tetraphenylporphyrin to form the μ-oxo dimer $[(TPP)Mn^{IV})_2O]$ (reaction 6-18).[30]

$$(TPP)Mn^V(OH)_2 + (TPP)Mn^{III}OH + HO^- \not\rightarrow$$
$$[(TPP)Mn^{IV}(OH)]_2O + H_2O \quad (6\text{-}18)$$

This observation is most unexpected and suggests that the structure assignment of the higher valent manganese-oxo porphyrin is not correct.

In CH_2Cl_2 the electrochemical $1e^-$ oxidation of $(TPP)Mn^{III}Cl$ provides the π-cation radical $(^+TPP)Mn^{III}Cl$, whereas $1e^-$ oxidation of $(Me_{12}TPP)$-$Mn^{III}OH$ yields the Mn(IV)-oxo porphyrin species.[33a] Chlorotetraphenylporphinato manganese(III) hexachloroantimonate, upon $1e^-$ oxidation in CH_2Cl_2, has been established by X-ray and other physical measurements to provide the π-cation radical $(^+TPP)Mn^{III}(Cl)(SbCl_6)$.[33b] Treatment with basic methanol results in the formation of $(TPP)Mn^{IV}(MeO)_2$. Thus, as in the case of iron porphyrins $2e^-$ oxidized above the iron(III) state, weakly basic ligands favor an electron deficiency in the porphyrin ring while strongly basic oxo ligands favor metal-centered oxidation.

Addition of two equivalents of hydroxide ion, in aprotic solvent, to the non-μ-oxo dimer forming $(Me_{12}TPP)Mn^{III}X$ (Reference 33a) provides both manganese(IV) and manganese(II) porphyrin species. The mechanism of this reaction is not understood.

The oxidation of $(Me_{12}TPP)Mn^{III}Cl$ to an Mn(V) porphyrin species in methylene chloride via the sequence of reactions 6-19 and 6-20 (where P is $Me_{12}TPP$) has been reported.[34]

$$(P)Mn^{III}X + ArCO_3H \rightarrow (P)Mn^{III}(X)-OOC(O)Ar$$

$$\xrightarrow{\quad HO^- \quad ArCO_2^- \quad} (P)Mn^VO(X) \quad (6-19)$$

$$(P)Mn^{II} + O_2^{\cdot -} \rightarrow (P)Mn^{III}O_2^- \xrightarrow{ArCOCl}$$

$$(P)Mn^{III}-OOC(O)Ar \rightarrow (P)Mn^VO(X) + ArCO_2^- \quad (6-20)$$

Oxidation of $(TPP)Mn^{III}X$ ($X = Cl^-$, Br^-, N_3^-, and OCN^-) by PhIO in organic solvents has been shown to result in the formation of two species.[35] When $X = Cl^-$ or Br^- the product was determined to be an Mn(IV) porphyrin μ-oxo dimer wherein each Mn(IV) species is complexed to a PhIO moiety $\{[(TPP)Mn^{IV}(PhIO)X]_2O\}$, whereas when $X = N_3^-$ or OCN^- there was obtained the simple Mn^{IV} μ-oxo dimer $(((TPP)Mn^{IV})_2O)$.

5. MECHANISMS OF OXYGEN ATOM TRANSFER FROM PERCARBOXYLIC ACIDS AND ALKYL HYDROPEROXIDES TO METAL(III) PORPHYRINS

In the charging of peroxidase enzymes with hydroperoxides and catalase with hydrogen peroxide, and in the peroxide shunt pathway for cytochrome *P*-450 (Scheme I), oxygen transfer to the iron(III) porphyrin might be considered to occur by either heterolytic or homolytic O—O bond scission (Scheme IV). If O—O bond scission were rate determining, then the dependence of the second-order rate constants for oxidative oxygen transfer upon the structure of the oxygen transfer agent could lead to a means of judging the mechanism of the reaction. To this end, studies have been carried out with the *meso*-tetraphenylporphinato chromium(III) chloride, manganese(III) chloride, iron(III) chloride, and cobalt(III) chloride salts [$(TPP)Cr^{III}Cl$, $(TPP)Mn^{III}Cl$, $(TPP)Fe^{III}Cl$, and $(TPP)Co^{III}Cl$, respectively]. The following two observations pertain to all these reactions: (1) When $PhCH_2CO_3H$ serves as the YOOH oxygen donor, $PhCH_2CO_2H$ can be recovered (as the methyl ester) in high yield, and (2) the rates of reaction (k_{YOOH}) are insensitive to bulky substituents on YOOH. Since it is known that the phenylacetoxyl radical rapidly decomposes to CO_2 and benzyl radical,[36] we may conclude that the reaction of peroxycarboxylic acids with the tetraphenylporphinato metal(III) salts investigated involves heterolytic YO—OH bond scission. The slopes $(-\beta_{1g})$ of linear free energy plots of log k_{YOOH} vs the pK_a of the YOH leaving groups are appreciable when YOOH represents percarboxylic acid or acidic alkyl hydroperoxide. In certain cases, a break occurs in the linear free energy plots at pK_a of YOH between 9 and 11. In these instances the values of $-\beta_{1g}$ are quite small for the less-acidic alkyl hydroperoxides. Such changes in $-\beta_{1g}$ are attributed either to a change of rate-limiting step from YO—OH bond scission to complexation of YOOH (reaction 6-21) or to a change in mechanism of YO—OH bond scission from heterolytic to homolytic (Scheme IV).

$$(TPP)Cr^{III}X + YOOH \underset{k_{-1}}{\overset{k_1}{\rightleftarrows}} (TPP)(X)Cr^{III}(HOOY) \overset{k_2}{\longrightarrow}$$

$$(TPP)(X)Cr^VO + YOH \quad (6\text{-}21)$$

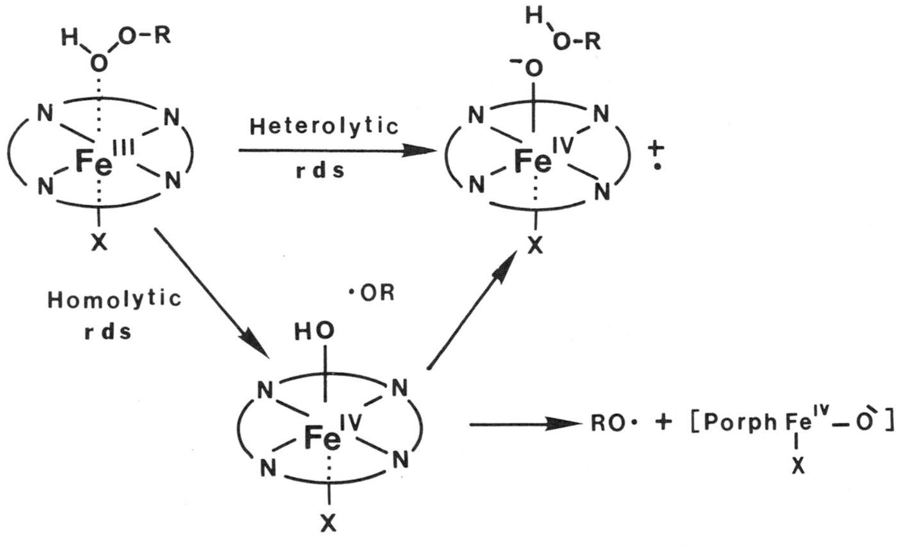

Scheme IV

Higher valent chromium-oxo tetraphenylporphyrins are rather stable. The pentacoordinant chromium(IV)-oxo-complex may be prepared by oxidation of porphinato chromium(III) salts by a number of oxidants.[37] The tetraphenylporphinato chromium(IV)-oxo species [eg, $(TPP)Cr^{IV}O$] are sufficiently stable to bottle and store. Oxidation to the Cr(V) state is most conveniently carried out by electrochemical means.[38] Oxo-chromium(V) porphyrin complexes are hexacoordinate [eg, $(TPP)Cr^VO(Cl)$] and sufficiently stable to allow their synthesis and use.

By employing PhIO as oxygen donor to $(TPP)Cr^{III}Cl$, it was shown that $(TPP)Cr^{III}Cl$ serves as a trap for the generated $(TPP)(Cl)Cr^VO$ species (reaction 6-22).

$$(TPP)Cr^{III}Cl + PhIO \rightarrow (TPP)(Cl)Cr^VO + PhI$$

$$(6\text{-}22)$$

$$(TPP)Cr^{III}Cl + (TPP)(Cl)Cr^VO + H_2O \rightarrow 2((TPP)Cr^{IV}O) + 2(HCl)$$

The $(TPP)Cr^{III}Cl$ salt was employed as both oxygen acceptor and trap (as in reaction 6-22) in order to determine the second-order rate constants (k_{YOOH}) for oxygen transfer from peroxycarboxylic acids and hydroperoxides (reaction 6-21).[39] In Figure 6-2, the values of log k_{YOOH} are plotted vs the pK_a of the YOH leaving groups. Examination of Figure 6-2 shows that (1) the free energies of reaction are a linear function of the standard free energies for proton dissociation of YOH and (2) both peroxycarboxylic acids and alkyl hydroperoxides follow the same linear free energy relationship. The correlation line of Figure 6-2 is generated by reaction 6-23 where $-\beta_{lg} = 0.34$.

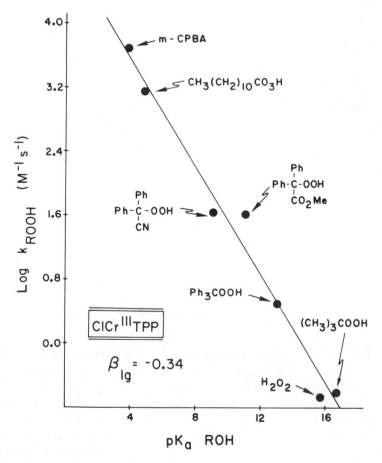

Figure 6-2. A plot of the log of the second-order rate constants (k_{YOOH}) for the reaction $(TPP)Cr^{III}(Cl)$ with percarboxylic acids and hydroperoxides vs the pK_a of the carboxylic acid and alcohol leaving groups (data from Reference 39).

$$\log k_{YOOH} = -\beta_{1g}pK_{YOOH} + C \qquad (6\text{-}23)$$

Because both peroxycarboxylic acids and alkyl hydroperoxides follow the same linear free energy relationship (reaction 6-23), it is reasonable to assume that both share the same heterolytic mechanism of oxygen transfer. The linear free energy plot of Figure 6-2 may be compared to that of Figure 5 of Chapter 8 for oxygen transfer from YOOH species which accompanies nucleophilic attack upon the distal oxygen of YOOH by the sulfide sulfur of thioxane. Much the same linear free energy relationship holds for the reaction of I⁻ and tertiary amines with YOOH compounds. For these nucleophiles, as with (TPP)Cr^{III}Cl, both alkyl hydroperoxides and percarboxylic acids fall on a single linear free energy plot with a given nucleophilic

species. For the oxygen transfers to :S⟨, :N— and I⁻, the value of $-\beta_{1g}$ of 0.6 indicates that $\Delta G\ddagger$ is about twice as sensitive to the leaving tendency of YOH as in the reactions of (TPP)Cr^{III}Cl with YOOH compounds.

The second-order rate constants for reaction of YOOH compounds with manganese(III) tetraphenylporphyrin were determined by trapping the higher valent manganese-oxo species with 2,4,6-tri-*tert*-butylphenol (TBPH) and by monitoring the formation of the TBP· radical species (eg, reaction 6-24).[40] Alkyl hydroperoxides proved to be unreactive. Figure 6-3 shows a plot of log k_{YOOH} for the reaction of a series of peroxycarboxylic acids vs the pK_a of the corresponding carboxylic acids.[40] The lack of reactivity of hydroperoxides becomes understandable from the very large value of $-\beta_{1g}$, which is 1.25.

$$YOOH + (TPP)Mn^{III}Cl \xrightarrow{k_{YOOH}} YOH + (TPP)(Cl)Mn^{V}O$$

$$(6\text{-}24)$$

$$2\ TBPH + (TPP)(Cl)Mn^{V}O \xrightarrow{fast} 2(TBP\cdot) + (TPP)Mn^{III}Cl + H_2O$$

From the equilibrium constants for complexing of (TPP)Mn^{III}Cl with imidazole (ImH) and *N*-methylimidazole (ImMe) and the second-order rate constants for the reaction of manganese(III) porphyrin with YOOH compounds in the presence of various concentrations of ImH and ImMe, there was calculated the second-order rate constants for oxygen transfer to the species (TPP)(ImH)Mn^{III}Cl, (TPP)(ImH)$_2$Mn^{III}Cl and (TPP)(ImMe)Mn^{III}Cl (reaction 6-25 where B represents ImH).

Included in Figure 6-3 are plots of log k_{YOOH} and $k_{YOOH'}$ vs pK_a of YOH for these reactions. There was no observable reaction of YOOH species with (TPP)(ImMe)$_2$Mn^{III}Cl or with the bis-tertiary amine-ligated species when the amine was 3,4-dimethylpyridine. The ability to determine the second-order rate constants for the reactions with alkyl hydroperoxides, when the manganese(III) moiety is ligated to imidazole, is due to both the increase in

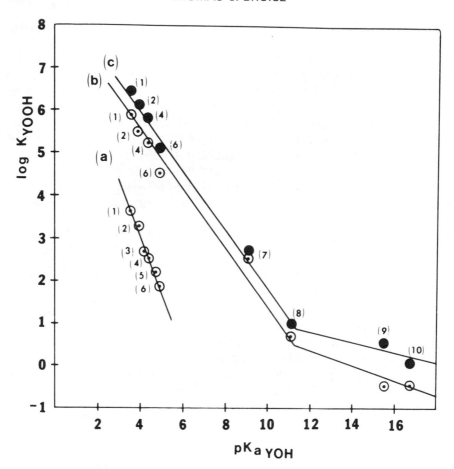

Figure 6-3. Plot of the log of the second-order rate constants (k_{YOOH}) vs the pK_a of the carboxylic acid and alcohol leaving groups. (a) Reactions of (TPP)Mn^{III}Cl with percarboxylic acids in benzonitrile (30°C); (b) reactions of (TPP)Mn^{III}(Cl)(ImH) with percarboxylic acids and hydroperoxides in dichloromethane; (c) the reaction of (TPP)Mn^{III}(Cl)(ImH)$_2$ with percarboxylic acids and hydroperoxides (k_{YOOH}) in dichloromethane. (1) f-Nitroperbenzoic acid, (2) m-chloroperbenzoic acid, (3) 3-chloroperpropionic acid, (4) phenylperacetic acid, (5) 5-chloropervaleric acid, (6) perlauric acid, (7) diphenylhydroperoxyacetonitrile, (8) methyl diphenylhydroperoxyacetate, (9) cumyl hydroperoxide, and (10) t-BuOOH (data from Reference 40b).

$$(TPP)Mn^{III}Cl + B \xrightleftharpoons{K_1} (TPP)(B)Mn^{III}Cl$$

$$(TPP)(B)Mn^{III}Cl + B \xrightleftharpoons{K_2} (TPP)(B)_2Mn^{III}Cl$$

$$(TPP)(B)Mn^{III}Cl + YOOH \xrightarrow{k_{YOOH}} (TPP)(B)Mn^VO(Cl)$$ (6-25)

$$(TPP)(B)_2Mn^VO(Cl) + YOOH \xrightarrow{k_{YOOH'}} (TPP)(B)Mn^VO(Cl) + B$$

rate constants and the decrease in $-\beta_{1g}$ (0.72), as compared to $-\beta_{1g}$ for (TPP)Mn^{III}Cl (1.27). Since both percarboxylic acids and the most acidic alkyl hydroperoxides fit the same linear free energy relationship, the reactions of both must involve heterolytic YO—OH bond scission. However, when the pK_a of YOH for alkyl hydroperoxides becomes greater than approximately 11, the value of $-\beta_{1g}$ decreases to 0.19. This suggests a change of the rate-limiting step from heterolytic YO—OH bond scission to either rate-limiting complexation of YOOH with the porphyrin-bound manganese moiety or a change of the mechanism from heterolytic to homolytic O—O bond scission.

Because YOOH species are excluded from approach to the manganese in (TPP)(ImH)$_2$Mn^{III}Cl, one must conclude that this bis-axial-ligated complex is in equilibrium with one or another reactive isomeric complex or, alternatively, that a second imidazole acts as a catalyst in the reaction of the monoimidazole-ligated species with YOOH. These situations are, of course, kinetically equivalent (Scheme V). Structure **1** is to be eliminated

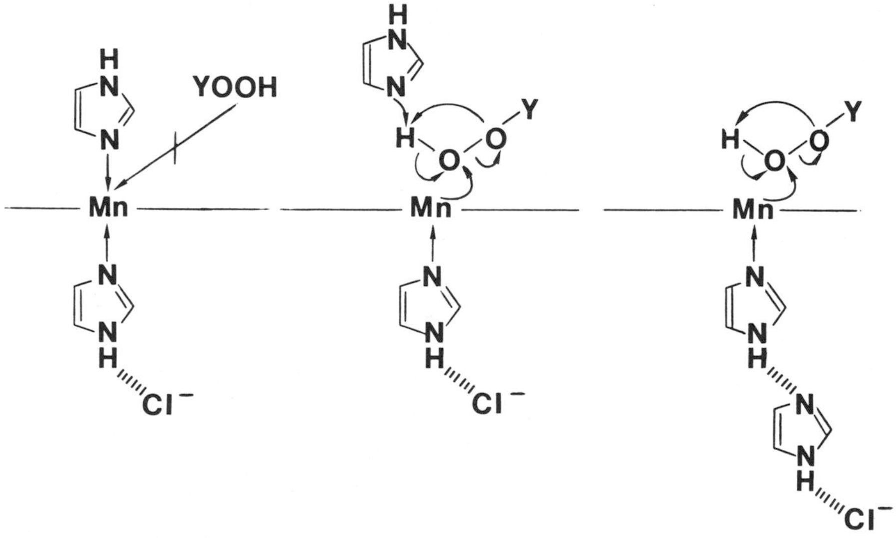

Scheme V

on the basis of steric considerations, while structure **2** must be discounted because there is no detectable reaction of (TPP)(B)$_2$Mn^{III}Cl with YOOH species when B = ImMe or 3,4-lutidine. Structures such as **3** have been proposed to be of importance in other reactions.[41] Since the formation of **3** from **1** should be endergonic, the calculated value of $k_{YOOH'}$ (reaction 6-25) for the reaction of **1** with YOOH is much smaller than the true rate constant for the reaction of **3**.

2,4,6-tri-*tert*-Butylphenol was employed as a trap of the higher valent iron-oxo species in the determination of the second-order rate constants for the reaction of YOOH species with $(TPP)Fe^{III}Cl$, and the recovery of $PhCH_2CO_2H$ from the reaction solution when YOOH = $PhCH_2CO_3H$ showed oxygen transfer from percarboxylic acids to involve heterolytic O—O bond scission. The plot of log k_{YOOH} vs pK_a of YOH is provided in Figure 6-4.[42] The plot is characterized by a sharp break in slope at a pK_a for YOH of about 11 to provide a line with slope ($-\beta_{1g}$) of 0.15. The most acidic alkyl hydroperoxide, as well as the peroxycarboxylic acid rates follows a linear free energy relationsip with slope ($-\beta_{1g}$) equal to that (0.35) seen for heterolytic O—O bond scission with $(TPP)Cr^{III}Cl$.

The question of a change of mechanism in the breaking of the YO—OH bond of peroxycarboxylic acids and alkyl hydroperoxides from heterolysis to homolysis in the reactions of cytochrome *P*-450 has been given considerable attention.[9,18,43–49] A change in the rate-limiting step or mechanism of O—O bond scission is not seen in the reaction of YOOH species with $(EDTA)Fe^{III}$.[50] A single linear free energy plot is obtained with a slope ($-\beta_{1g}$) comparable to that for the reaction of percarboxylic acids and the most acidic alkyl hydroperoxides with $(TPP)Fe^{III}Cl$. The second-order rate con-

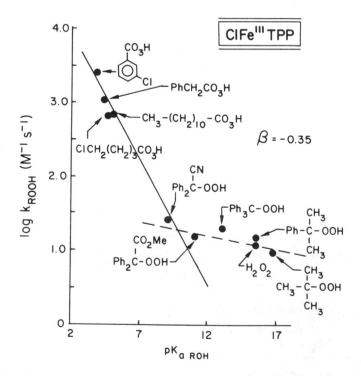

Figure 6-4. Plot of the log of the second-order rate constants (k_{YOOH}) vs the pK_a of the carboxylic acid and alcohol leaving groups for the reaction of $(TPP)Fe^{III}(Cl)$ with percarboxylic acids and hydroperoxides (solvent methanol) (data from Reference 42).

stants associated with O—O bond heterolysis for the reactions with (TPP)-Fe^{III}Cl are about 60-fold greater than for reactions with (EDTA)Fe^{III}. Since the reaction of (EDTA)Fe^{III} with percarboxylic acids has been shown to involve heterolytic YO—OH bond scission and the alkyl hydroperoxides fit the same plot as do the percarboxylic acids, it is probably safe to assume that the rate-limiting step for the reaction of all YOOH species with (EDTA)Fe^{III} involves heterolytic O—O bond scission.

The reaction of YOOH species with (TPP)Co^{III}Cl was followed by stopped-flow spectrophotometric monitoring of the disappearance of the cobalt(III) porphyrin. A plot of log k_{YOOH} vs pK_a of YOH leaving group is shown in Figure 6-5. The linear free energy plot resembles that obtained for (TPP)Fe^{III}Cl but with the break in slope occurring at a pK_a of YOH of about 9.[51] For the heterolytic mechanism of percarboxylic acids and most acidic alkyl hydroperoxide, $-\beta_{lg}$ is equal to 0.72, while for the majority of alkyl hydroperoxides $-\beta_{lg}$ equals 0.15. From a reaction solution when employing

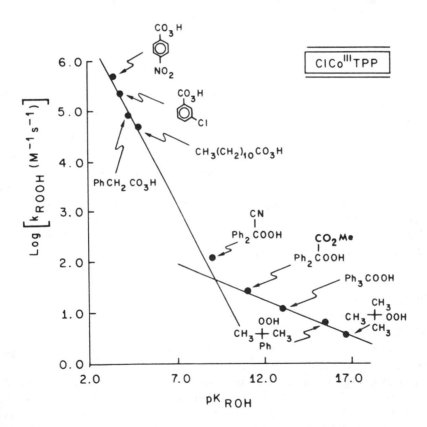

Figure 6-5. Plot of the log of the second-order rate constants (k_{YOOH}) vs the pK_a of the carboxylic acid and alcohol leaving groups for the reaction of (TPP)Co^{III}(Cl) with percarboxylic acids and hydroperoxides (solvent chloroform) (data from Reference 51).

two equivalents of *tert*-BuOOH as oxidant, the following isoporphyrin was isolated:

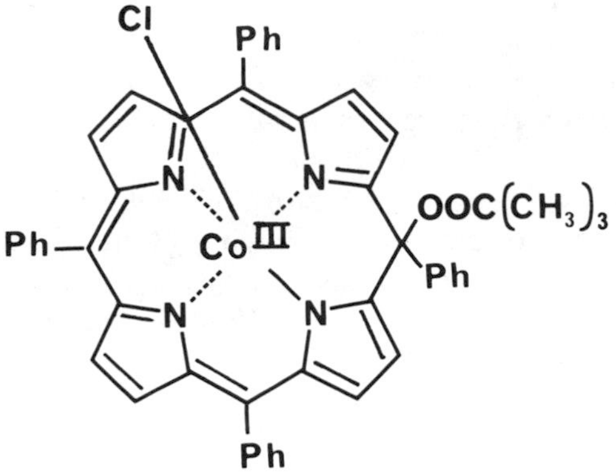

This finding supports the formation of a cobalt(III) porphyrin dication intermediate (reaction 6-26).

$$(TPP)Co^{III}X + YOOH \rightarrow (^{2+}TPP)Co^{III}X + YO^- + H_2O$$

$$(6\text{-}26)$$

$$YO^- + (^{2+}TPP)Co^{III}X \rightarrow (Isoporphyrin)Co^{III}X$$

The experimental results for the reaction of YOOH species with the various tetraphenylporphyrin metal(III) chloride salts and (EDTA)Fe(III) are summarized in Table 6-1. Since (TPP)Zn^{II} does not react with YOOH species at a measurable rate, a requirement for reaction appears to be a d orbital electron vacancy. Transition metal d orbital overlap with both porphyrin π electrons and oxidant serves as a conduit for electron transfer to oxidant. The comparison of one to another of the reactions is made difficult by the inconsistency of the solvent systems employed. However, several observations are allowed. Values of k_{YOOH} and $-\beta_{1g}$ are not dependent on whether oxidation is metal centered or porphyrin centered. Thus, $-\beta_{1g}$ for the reactions of (TPP)(ImH)Mn^{III}Cl and (TPP)Co^{III}Cl are quite similar, with the latter characterized by somewhat larger rate constants. On the other hand, the reaction rates and linear free energy relationship for (TPP)(ImH)Mn^{III}Cl and (TPP)Cr^{III}Cl are quite different, with the former characterized by appreciably larger rate constants. One may also compare (TPP)Fe^{III}Cl and (EDTA)Fe^{III}. The former exhibits larger rate constants, but both show the same sensitivity to the basicity of the leaving group on reaction with percarboxylic acids and the most acidic alkyl hydroperoxides. The break in the linear free energy plots, if attributed to a change of mechanism from heterolytic to homolytic YO—OH bond breaking, can be explained on the basis of the oxidation potentials of the metal porphyrins and a competition between the homolytic and heterolytic mechanisms.[50] Since a homolytic mechanism is not observed with (TPP)Cr^{III}Cl, the intersection of the linear free energy plots for heterolysis and homolysis must occur at a pK_a of YOH > 17 (beyond the limit of pK_a values of alcohol leaving groups). The 2e$^-$ oxidation

TABLE 6-1. Linear Free Energy Relationships for the Reaction of Percarboxylic Acids and Alkyl Hydroperoxides with Tetraphenylporphinato Metal(III) Salts and (EDTA)Fe^{III}

Complex	Solvent	Pertinent observations	Product
(TPP)Cr^{III}Cl	CH$_2$Cl$_2$	Single LFER, $-\beta_{lg} = 0.34$	(TPP)(X)Cr^{V}O
(TPP)Mn^{III}Cl	PhCN	For percarboxylic acids, $-\beta_{lg} = 1.3$	(TPP)(X)Mn^{V}O
(TPP)(ImH)Mn^{III}Cl	CH$_2$Cl$_2$	Break in LFER at pK_a YOH of ~11, $-\beta_{lg} = 0.72$ and 0.19	(TPP)(ImH)Mn^{V}O·X
(TPP)Fe^{III}Cl	CH$_3$OH	Break in LFER at pK_a YOH of ~9.5, $-\beta_{lg} = 0.35$ and 0.15	($^+$TPP)Fe^{IV}O·X
(EDTA)Fe^{III}	CH$_3$OH	Single LFER, $-\beta_{lg} = -0.4$	[(EDTA)(FeO)]$^+$
(TPP)Co^{III}Cl	CHCl$_3$	Break in LFER at pK_a YOH of ~9.5, $-\beta_{lg} = 0.72$ and 0.15	($^{2+}$TPP)Co^{III}X
(TPP)Zn^{II}Cl	CH$_2$Cl$_2$/CH$_3$OH	No reaction	

potentials for the (Me$_{12}$TPP)Fe^{III}OH and (Me$_{12}$TPP)Fe^{III}OMe to give the corresponding iron(IV) porphyrin π-cation radical is 1.14 V [standard calomel electrode (SCE)], and the 2e$^-$ oxidation potential for the (TPP)Fe^{III}OMe is 1.16 V.[26] If one compares this 2e$^-$ oxidation potential with the 2e$^-$ oxidation potential of 0.84 V[52] to generate (TPP)(X)Cr^{V}O from (TPP)Cr^{III}OH, the difference in standard free energies for the 2e$^-$ oxidations is approximately 14 kcal/M. Thus, the oxidation of [(TPP)Cr^{III}]$^+$ by a given hydroperoxide is thermodynamically favored by approximately 14 kcal/M over the oxidation of [(TPP)Fe^{III}]$^+$. For this reason alkyl hydroperoxides can more easily enter into 2e$^-$ oxidation of the Cr(III) porphyrin than with the Fe(III) porphyrin. The 1e$^-$ oxidation of porphinato iron(III) hydroxide salts is thermodynamically favored over the 2e$^-$ oxidation, so it is not surprising that one should see homolytic 1e$^-$ oxidation of (TPP)Fe^{III}Cl by alkyl hydroperoxides. Rather compelling evidence has been presented in support of cumyl—O· as an oxidant when cumyl hydroperoxide is employed to charge cytochrome P-450.[45–48] The radical would be a direct product of homolysis of the alkyl hydroperoxide (Scheme IV). Also, rather good evidence has been presented that the reaction of both (TPP)Fe^{III}Cl and (TPP)Mn^{III}Cl with *tert*-BuOOH provides *tert*-BuO·.[53,54]

6. A MODEL FOR THE CATALASE REACTION: THE MECHANISM FOR THE FORMATION OF OXYGEN ON REACTION OF AN IRON(III) PORPHYRIN WITH HYDROGEN PEROXIDE

The enzyme catalase catalyzes the decomposition of hydrogen peroxide (reaction 6-2). A knowledge of the mechanism of reaction of hydrogen peroxide with iron(III) porphyrins is germane to our understanding of catalase and peroxidase. There have been a number of studies of the dynamics of the reaction of hydrogen peroxide with iron(III) porphyrins,[55–61] but much of this work is incomplete and obscured by the occurrence of metalloporphyrin destruction, aggregation in water, and μ-oxo dimer formation in basic solution. A recent study employed the water-soluble tetrakis(2,6-dimethyl-3-sulfonatophenyl)porphyrinato iron(III) hydrate [(Me$_8$STPP)Fe^{III}OH$_2$] as catalyst (reaction 6-27).[62]

$$(\text{Me}_8\text{STPP})\text{Fe}^{III}\text{OH}_2 \underset{}{\overset{\text{p}K_a\ 7.2}{\rightleftharpoons}} (\text{Me}_8\text{STPP})\text{Fe}^{III}\text{OH} + \text{H}^+ \qquad (6\text{-}27)$$

This iron(III) porphyrin hydrate remains monomeric at all pH values. The reaction of hydrogen peroxide with the iron(III) porphyrin in water is first order in both hydrogen peroxide and iron(III) porphyrin and produces a strong oxidizing agent which is presumably the iron(IV)-oxo porphyrin π-cation radical. 2,2'-Azinobis-(3-methylbenzothiazoline)sulfonic acid (ABTS) was employed as a water-soluble trap to monitor the formation of the higher valent iron-oxo porphyrin species. The 1e$^-$ redox potential of ABTS is not influenced by pH and, on oxidation, it provides the green cation radical ABTS$^{\cdot+}$ (reaction 6-28).[63]

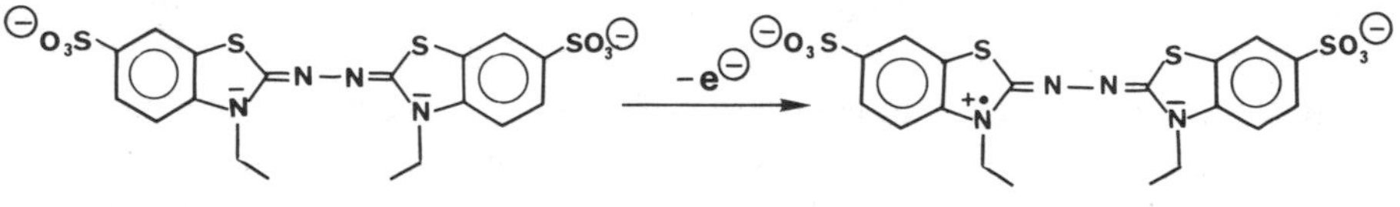

ABTS

ABTS$^{\cdot+}$

660 nm $\qquad$ (6-28)

By adjustment of the ratio of [ABTS] and [H$_2$O$_2$], it was possible to follow the reaction by the appearance of both ABTS$^{\cdot+}$ and O$_2$. Since the rate constants for appearance of both ABTS$^{\cdot+}$ and O$_2$ were identical, within experimental error, oxidation of the iron porphyrin by hydrogen peroxide is rate determining (reaction 6-29, where A is an iron porphyrin species, B is a hydrogen peroxide species, and C is an iron(IV)-oxo porphyrin π-cation species present at the various pH values).

$\qquad$ (a) $\qquad$ A + B $\xrightarrow{\ k_{1y}\ \text{(rate-determining step)}\ }$ C

$\qquad$ (b) $\qquad$ C + 2(ABTS) $\longrightarrow$ A + 2(ABTS$^{\cdot+}$) $\qquad$ (6-29)

$\qquad$ (c) $\qquad$ C + B $\longrightarrow$ A + O$_2$ + H$_2$O or HO$^-$

Values of k_{1y} were determined by buffer dilution experiments from pH 1 to 12. From the log k_{1y} vs pH plot there can be seen the three reactions (30a,b,c; the three plateaus of Figure 6-6).

$$(\text{Me}_8\text{STPP})\text{Fe}^{III}(\text{OH}_2) + \text{H}_2\text{O}_2 \rightarrow (^{+\cdot}\text{Me}_8\text{STPP})(\text{X})\text{Fe}^{IV}\text{O} + \text{H}_2\text{O} + 2\text{H}^+$$
$$(6\text{-}30\text{a})$$

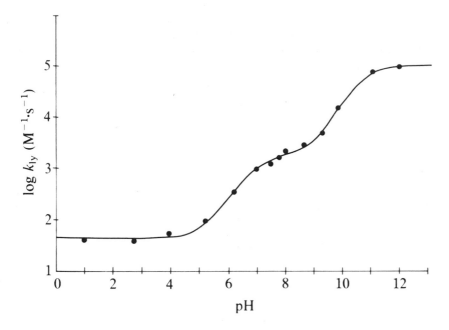

Figure 6-6. Plot of the log of the rate constants for non-buffer-catalyzed (k_{1y}) reaction of $(Me_8STPP)Fe^{III}(H_2O)$ with hydrogen peroxide vs pH (solvent water, 30°C, $\mu = 0.21\ M$) (data from Reference 62).

$$(Me_8STPP)Fe^{III}(OH) + H_2O_2 \rightarrow (^{+\cdot}Me_8STPP)(X)Fe^{IV}O + H_2O + H^+$$
$$(6\text{-}30b)$$

$$(Me_8STPP)Fe^{III}(OH) + HO_2^- \rightarrow (^{+\cdot}Me_8STPP)(X)Fe^{IV}O + H_2O \quad (6\text{-}30c)$$

In reaction 6-30, both a and b, but not c, are catalyzed by 2,4,6-trimethylpy-ridine. From cogent arguments,[38] the most reasonable mechanisms for general catalysis are represented by the transition states of Scheme VI. In TS$_1$

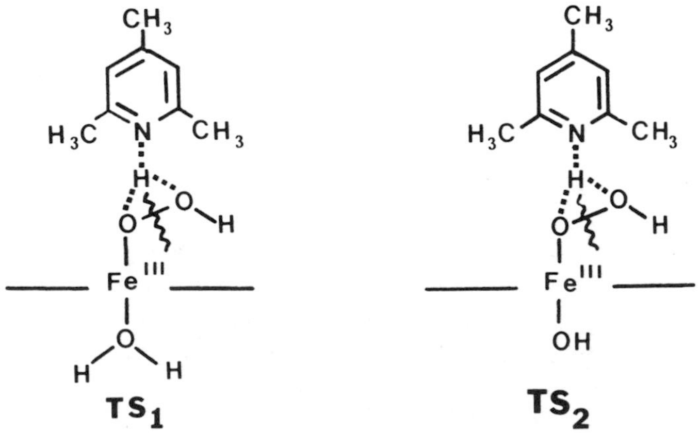

Scheme VI

and TS_2, the base assists the proton migration from the terminal oxygen to the departing oxygen. Since H_2O is the leaving group, these reactions would be required to represent heterolytic HO—OH bond scission (however, see the conclusion of the following section).

7. THE MECHANISM OF REACTION OF HYDROGEN PEROXIDE WITH MANGANESE(III) PORPHYRIN

The solution properties of $(Me_8STPP)Mn^{III}(H_2O)_2$ and the dynamics for the reaction of hydrogen peroxide with this water-soluble manganese(III) porphyrin have also been investigated.[64] As has been found in previous investigations of water-soluble manganese(III) porphyrins (for example, see References 29 and 65), there are two pK_a values for dissociation of lysate ligands (reaction 6-31, P = Me_8STPP).

$$pK_1 = 7 \qquad\qquad\qquad pK_2 = 12$$
$$(P)Mn^{III}(OH_2)_2 \underset{+H^+}{\overset{-H^+}{\rightleftharpoons}} (P)Mn^{III}(OH)(H_2O) \underset{+H^+}{\overset{-H^+}{\rightleftharpoons}} (P)Mn^{III}(OH)_2 \quad (6\text{-}31)$$

The reaction of the manganese(III) porphyrin with hydrogen peroxide is first order in both components at all pH values. Using the technique of competition between ABTS and hydrogen peroxide as employed with $(Me_8STPP)Fe^{III}(H_2O)_2$, it was found that oxygen transfer from hydrogen peroxide is rate limiting and that both $ABTS^{\cdot+}$ and O_2 are formed. As in the case with the iron(III) porphyrin, the yield of these two products at all pH values accounts for the initial concentration of hydrogen peroxide. From the pH dependence of the rate there can be discerned three pathways for oxygen transfer from hydrogen peroxide to Mn(III) porphyrin. The stoichiometry of the three reactions are well defined, though the structures of the hypervalent manganese-oxo porphyrin products may only be suggested (reaction 6-32).

$$(Me_8STPP)Mn^{III}(OH) + H_2O_2 \rightarrow (Me_8STPP)Mn^VO(OH) + H_2O$$

$$(Me_8STPP)Mn^{III}(OH) + HO_2^- \rightarrow (Me_8STPP)Mn^VO(OH) + HO^- \quad (6\text{-}32)$$

$$(Me_8STPP)Mn^{III}(OH)_2 + HO_2^- \rightarrow (Me_8STPP)Mn^VO(OH)_2 + HO^-$$

The addition of imidazole enhances the rate of reaction by five- to sixfold at saturation by this base species. The reactive intermediate is *mono*-imidazole ligated, and the reaction is first order in HOO^-. The structure of the hypervalent manganese-oxo species is not known.

$(Me_8STPP)Mn^{III}(H_2O) + ImH$

$$\underset{}{\overset{K_e}{\rightleftharpoons}} (Me_8STPP)Mn^{III}(ImH)(H_2O) \qquad (6\text{-}33)$$

$(Me_8STTP)Mn^{III}(ImH)(H_2O) + HO_2^-$

$$\xrightarrow{k_r} (Me_8STPP)Mn^VO(ImH)(OH) + H_2O$$

Recall that in the reaction of tetrakis(2,6-dimethyl-3-sulfonatophenyl)-porphinato iron(III) hydrate with hydrogen peroxide, general catalysis of oxygen transfer was not observed with oxygen acids and bases such as $CH_3CO_2H/CH_3CO_2^-$. Catalysis was seen with the single nitrogen buffer 2,4,6-trimethylpyridine/2,4,6-trimethylpyridine·H^+. With tetrakis(2,6-dimethyl-3-sulfonatophenyl)porphinato manganese(III) hydrate there could not be detected catalysis by either oxygen acids and bases or nitrogen acids and bases (the latter including 2,4,6-collidine/2,4,6-collidine·H^+ and imidazole/imidazole·H^+). It is obvious at this time that the role of general catalysis in oxygen transfer to metallo(III) porphyrins is not understood.

8. THE REBOUND MECHANISM FOR OXYGEN INSERTION INTO CARBON–HYDROGEN BONDS

Oxygen insertion into H—C bonds is an important reaction mediated by certain cytochrome P-450 enzymes and in modeling experiments with iron(III) and manganese(III) porphyrins in the presence of suitable oxygen atom donors. The so-called rebound mechanism of reaction 6-34 has been proposed on the basis of stereochemical and isotope distribution studies.[66,67]

$$(^{+}Porph)(L)Fe^{IV}O + H-\overset{\diagup}{\underset{\diagdown}{C}} \rightarrow [(Porph)(L)Fe^{IV}OH + \cdot\overset{\diagup}{\underset{\diagdown}{C}}]$$

$$(6\text{-}34)$$

$$[(Porph)(L)Fe^{IV}OH + \cdot\overset{\diagup}{\underset{\diagdown}{C}}] \rightarrow (Porph)(L)Fe^{III} + HO-\overset{\diagup}{\underset{\diagdown}{C}}$$

Large H—C/D—C isotope effects and some loss of stereochemistry are as seen on oxidation of alkanes by transition metal-oxo species such as chromate and permanganate.[68a,b] For the latter oxidants there was proposed, much as in reaction 6-34, a mechanism of hydrogen atom abstraction to give a caged oxo-metal carbon radical pair which gives way to product with only partial loss of stereochemistry.[68c] The extent of loss of stereochemistry or the formation of alternate products would then be dependent upon the lifetime of the carbon radical within the oxo-metal radical pair and its ability to escape. Cage reactions initiated by single-bond homolysis may proceed with

a high degree of retention of configuration.[68d] What follows is a brief description of the experimental data by which the operation of the mechanism of
reaction 6-34 may be judged.

Since essentially all enzymatic reactions exhibit a high degree of stereospecificity, it is to be expected that this might be so (at least with specific
substrates) for the cytochrome P-450 reactions. In the 5-hydroxylation of d-
camphor by cytochrome P-450$_{cam}$, hydrogen removal occurs from both 5-
exo and 5-endo positions, but oxygen delivery is only to the 5-exo position
(reaction 6-35).[69a]

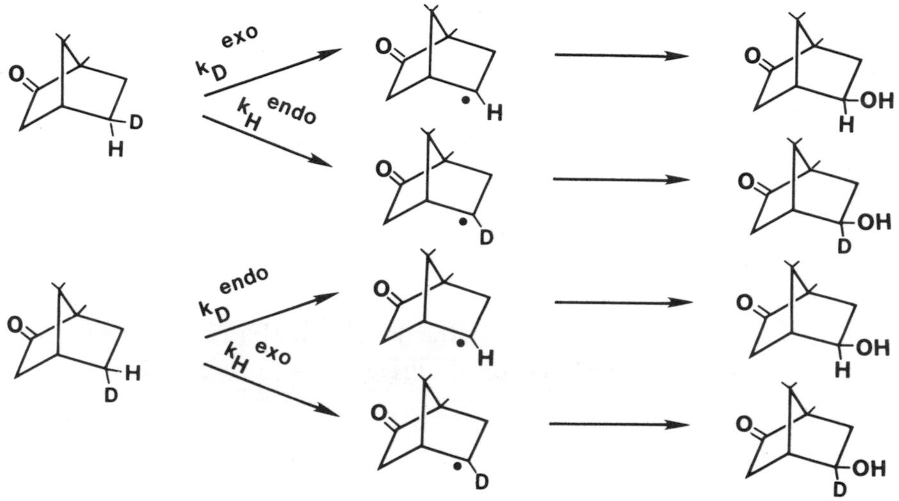

(6-35)

If the 5-exo and 5-endo positions are blocked by fluorine, then hydroxylation
of the methyl carbon at position 9 of the 5,5-difluorocamphor molecule[6-36] is
observed.[69b]

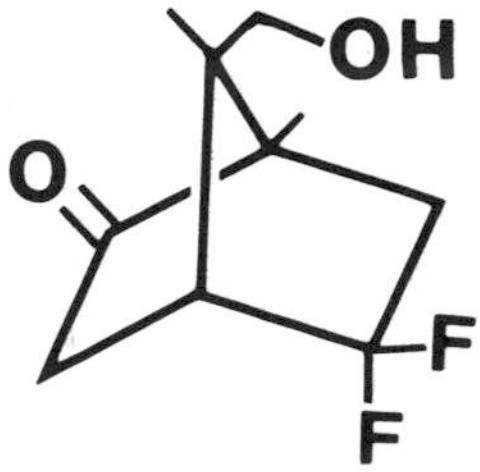

(6-36)

The metabolism of the difluoro analog occurs at a rate one-third that of
camphor. These results show that a change in regioselectivity occurs when
the normal site of reaction is blocked and the reaction of the oxo species
occurs from the exo side of the camphor molecule. Regio- and stereospecificity of cytochrome P-450$_{cam}$ is not absolute, and there would appear to be a

flexibility at the active site. Hydroxylation by the adrenal steroidogenic enzymes[70] and the terminal oxidation of [1-³H,²H,¹H]octane by liver microsomal enzyme[71] are absolutely stereospecific. On the other hand, the hydroxylation of norbornane by cytochrome P-450$_{LM2}$ involves considerable loss of stereochemistry,[72] while hydroxylation of phenylethane by a single isozyme of liver microsomal cytochrome P-450 provides 48% (R)-1-phenylethanol and 52% (S)-1-phenylethanol.[73] One may conclude that the mechanism of hydroxylation by cytochrome P-450 does not per se require retention of stereochemistry and that cases of complete stereospecificity may be ascribed either to orientation of the substrate at the active site by protein[72] or to the very short half-life of an intermediate metal-oxo carbon radical pair (reaction 6-34).

In modeling experiments, it has been found that the hydroxylation of *cis*- and *trans*-decahydronaphthalene by PhIO with both (TPP)Fe^{III}Cl and (F$_{20}$TPP)Fe^{III}Cl occurs with essentially complete retention of stereochemistry at the tertiary carbons. In this system the intermediate carbon radical is of too short a half-life to allow stereochemical equilibration.[74] Oxidation of 1,2-dideuterocyclohexene with PhIO and (TPP)Fe^{III}Cl or (TPP)Cr^{III}Cl provides two isomeric allylic alcohols (reaction 6-37).

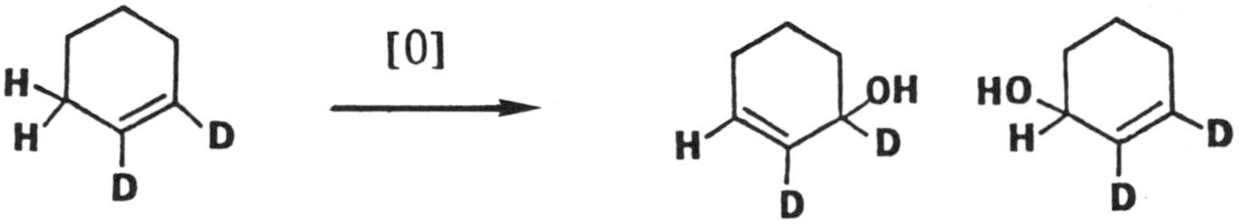

(6-37)

This observation requires the formation of an intermediate radical or carbocation species. Much the same results were obtained with hepatic cytochrome P-450.[67,74] The oxidation of norbornane by cytochrome P-450$_{LM2}$ provides a 3.4:1 mixture of exo and endo alcohols (reaction 6-38).[72]

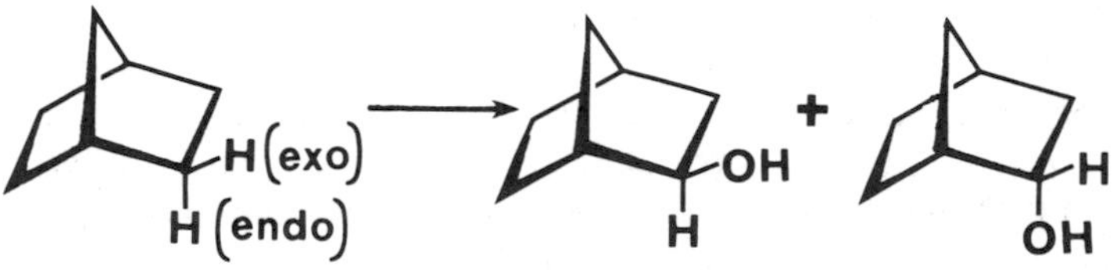

(6-38)

3.4   :   1

When deuterium-labeled norbornane is used, no isotope scrambling in the norbornyl alcohol products is found. This provides rather convincing evidence that formation of a norbornyl cation[75] intermediate does not occur. The required intermediate must, therefore, be a carbon radical species. With all *exo*-[²H]norbornane, the ratio of *exo*-alcohol to *endo*-alcohol products inverts to 0.76:1, reflecting the ^1H—$C/^2H$—C kinetic isotope effect (reaction 6-39).

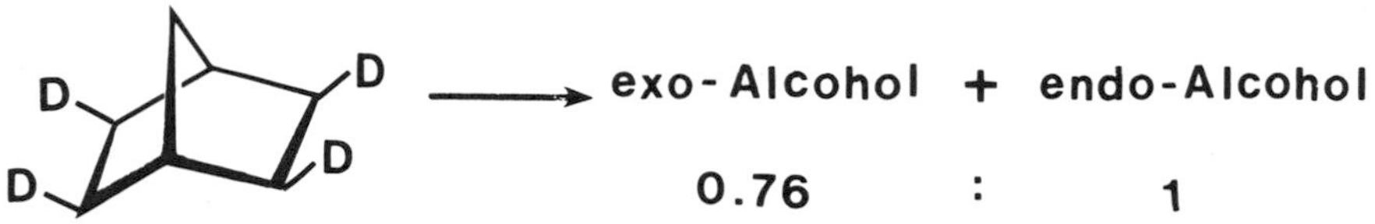

$$\text{exo-Alcohol} \quad + \quad \text{endo-Alcohol}$$
$$0.76 \quad : \quad 1$$

(6-39)

All four deuteriums were found to be present in 25% of the *exo*-alcohols so that an endo hydrogen is abstracted to provide an intermediate that can epimerize prior to oxygenation. These findings have been offered as good evidence for a radical mechanism (reaction 6-40).

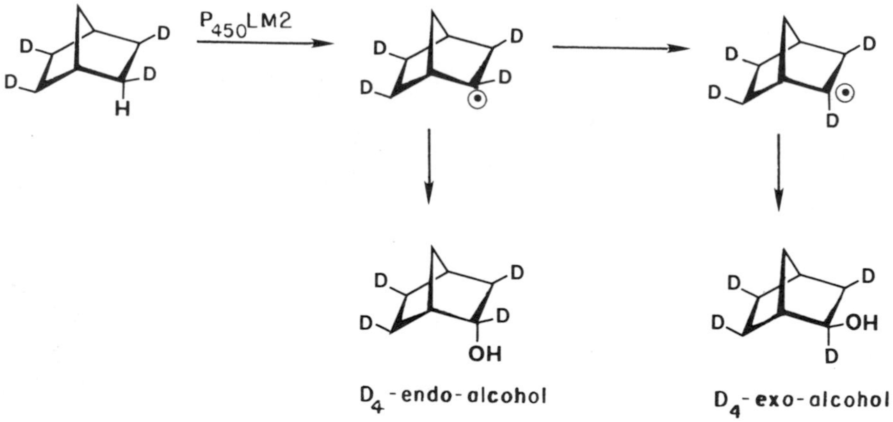

D_4-endo-alcohol D_4-exo-alcohol

(6-40)

The corrected value of ^1H—$C/^2H$—C (11.5) compares favorably to the isotope effects reported for alkane oxidation by higher valent manganese- and chromium-oxo complexes.[68c] Large intramolecular discriminatory ^1H—$C/^2H$—C isotope effects have also been obtained with liver microsomal preparations.[76] In chemical systems, large ^1H—$C/^2H$—C isotope effects (8.0–9.0) have been reported in the hydroxylation of the methyl group of [Me-²H₃]anisole using (TPP)Fe^{III}Cl and (F₂₀TPP)Fe^{III} with PhIO and microsomes.[77] The ^1H—$C/^2H$—C isotope effect for the hydroxylation of [1-³H,²H,¹H,¹⁴C]octane by rat liver microsomes[70] has been reported to be normal. The experimental intramolecular isotope effects for the hydroxylation of camphor analogs containing deuterium at either 5-exo or 5-endo positions by cytochrome P-450$_{cam}$ when using NADH/O₂, PhIO, meta-chloroperbenzoic acid (*m*-CPBA), and H_2O_2 as oxygen-atom-loading sources have been found to be comparable.[68a] This observation suggests that a common mechanism is responsible for hydrogen abstraction in alkane hydroxylation by cytochrome *P*-450$_{cam}$. The determination of the true intramolecular isotope effects from the experimental data is not possible without a knowledge as to whether the geometry of the transition states for both exo and endo hydrogen abstraction is identical. The finding that hydroxylation occurs at only the 5-exo position indicates that the bond lengths in the transition states for endo and exo hydrogen abstraction may not be the same.

A reaction closely related to the rebound mechanism for hydroxylation is the carbon–carbon bond oxidation of the strained hydrocarbon quadricyclane by phenobarbital-induced cytochrome *P*-450 of rat liver microsomes.[78] Cogent arguments support the proposal of a 1e$^-$ oxidation followed by HO· rebound and rearrangement (reaction 6-41).

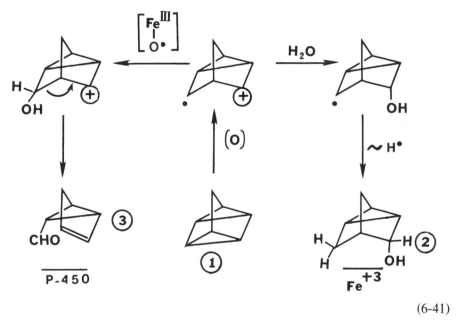

$$(6\text{-}41)$$

The reaction of (TPP)Mn^{III}(X) (X = Cl$^-$, Br$^-$, I$^-$, N$_3^-$) with PhIO and cyclohexane in rigorously dried benzene or chlorocarbon solvents provides, among other products, cyclohexanol, cyclohexyl-X, and a small quantity of dicyclohexyl as well as cyclohexylbenzene. The results suggest the formation of rather long-lived radical species.[79]

9. MECHANISMS OF DEALKYLATION REACTIONS

The oxidative removal of alkyl substituents from nitrogen and oxygen is an important role played by cytochrome *P*-450 in xenobiotic detoxification. Of the possible mechanisms, the two which are to be considered involve initiation by H· or 1e⁻ abstraction. The oxidations proceeding via hydrogen atom abstraction involve α-hydroxylation by the rebound mechansim of reaction 6-35, with the immediate products being either a carbinolamine (reaction 6-42) or a hemiacetal.

$$R_2NCH_2R' \xrightarrow{-H\cdot} R_2N\overset{\cdot}{C}HR \xrightarrow{+HO\cdot} R_2NCH(OH)R' \rightarrow R_2NH + R'CHO \quad (6\text{-}42)$$

Initiation of oxidation by 1e⁻ abstraction is pertinent in the dealkylation of amines. The immediate product is an iminium cation radical, which on loss of a proton converts to a carbon radical. The latter is then combined with HO· to yield carbinolamine. The mechanism of the reaction can be characterized as radical oxygen rebound with intermediate radical reorganization (reaction 6-43).

$$RN(CH_3)_2 \xrightarrow{-e^-} RN(^{+\cdot})(CH_3)_2 \rightarrow RN(CH_3)(CH_2\cdot) \xrightarrow{+HO\cdot}$$
$$RN(CH_3)(CH_2OH) \quad (6\text{-}43)$$

$$RN(CH_3)(CH_2OH) \rightarrow RNH(CH_3) + CH_2O$$

An alternate mechanism would involve H· abstraction and HO· rebound (reaction 6-32). A discussion of these mechanisms and evidence for their involvement follows.

The oxidative dealkylation of Ar—O—alkyl may be conceived as occurring by an addition to the aromatic ring followed by the elimination of an O-alkyl moiety or oxidation of the alkyl substituent which results in the elimination of an O-aryl moiety. Oxidative dealkylation of 4-methoxyacetanilide by rat liver microsomes in the presence of $^{18}O_2$ or $H_2^{18}O$ does not result in the incorporation of label into the product 4-hydroxyacetanilide.[80] Oxidative demethylation of aryldeuteriomethyl ethers proceed with ^1H—C/^2H—C isotope effects of 2–10.[76,81] Oxidative demethylation of [Me-$^2H_3$]anisole by (TPP)Fe^{III}Cl and (F₂₀TPP)Fe^{III}Cl with PhIO are characterized by ^1H—C/^2H—C isotope effects of 9 and 8, respectively, while demethylation with rat liver microsomes supported by NADPH—O₂ provides isotope effects of 11 and, when the oxygen donor is MCPBA, of 8.[77] These results establish that enzymatic and porphyrin-centered chemical dealkylation of aryl—O—CH₃ ethers occur by hydrogen abstraction from the methyl substituent and the breaking of the CH₃—O bond with elimination of an O-aryl moiety. The mechanism of reaction 6-44 would appear to be secure.

$$(^{+\cdot}TPP)(X)Fe^{IV}O + CH_3—OAr \rightarrow (TPP)(X)Fe^{IV}OH + \cdot CH_2—OAr$$

$$(TPP)(X)Fe^{IV}OH + \cdot CH_2—OAr \rightarrow (TPP)Fe^{III}(X) + HOCH_2OAr \quad (6\text{-}44)$$

$$HOCH_2—OAr \rightarrow CH_2O + HOAr$$

There are four recognized mechanisms whereby oxidation of an amine results in dealkylation. The first mechanism involves covalent addition of amine to an electrophile followed by proton abstraction. Examples of this are found in the oxidation of amines by quinones.[82] Hydride transfer to oxidant has been suggested when the oxidant is an arenediazonium ion.[83] Hydrogen atom transfer has been shown for the oxidation of N,N-dimethyl-aniline by photoexcited singlet *trans*-stilbene in hexane.[84] Many examples exist for rate-limiting 1e$^-$ transfer in the 2e$^-$ oxidation of amines.[85,86,87]

Intramolecular ^{1}H—C/^{2}H—C isotope effects have been reported for the N-demethylation of N-methyl-N-(trideuteriomethyl)aniline. The isotope effects for the peroxidases (H$_2$O$_2$, EtO$_2$H), hemoglobin (EtO$_2$H), and myoglobin (EtO$_2$H) demethylation reactions were large (>8.5), while those for the cytochrome P-450 (NADPH + O$_2$ or cumyl hydroperoxide) were low (<3.1).[88] The kinetic isotope effects for direct H· abstraction (permanganate with α-dideuteriobenzylamine,[89] cytochrome P-450 aliphatic hydroxyl-ation,[67,76b] and O-demethylation[76a,81b]) range between 7.0 and 11.5, while the kinetic isotope effects for 1e$^-$ oxidations attributed to anilinium cation radi-cal formation followed by deprotonation range between 1.3 and 3.6 (chlorine dioxide with benzylamines,[90] permanganate with trimethylamine,[91] and oxi-dation of N-methyl-di-N-butylamine by alkaline potassium ferricyanide[92]). These results suggest that demethylation by peroxidase, hemoglobin, and myoglobin (using hydroperoxides as oxygen atom donors) occurs by H· abstraction. The cytochrome P-450 dealkylations must involve proton mo-tion in the transition state and may best be described as 1e$^-$ abstraction with loss of H$^+$ as partially rate determining.

In a model study, relative rate constants (determined by competitive oxi-dation of N,N-dimethylbenzylamine and 3- and 4-substituted N,N-dimethyl-benzylamines) have been employed in the determination of the Hammett ρ values (−0.4 and −0.2, respectively) using (TPP)Fe^{III}Cl and (TPP)Mn^{III}Cl with PhIO as oxidant.[93] It was concluded that the values were consistent with a mechanism of rate-determining electron transfer. This conclusion was confirmed by ^{1}H—C/^{2}H—C isotope effects of about 1.3 determined from the competitive oxidation of N,N-dimethylbenzylamine and its α,α-^{2}H$_2$-labeled analog. For *tert*-BuOOH as oxygen atom donor, a ρ value of zero and isotope effects of about 3 support hydrogen atom abstraction from the α-carbon as rate determining.

Convincing evidence for the 1e$^-$ oxidation of dihydropyridines by cyto-chrome P-450 has been presented.[94] It should be noted that dihydropyridines easily undergo 1e$^-$ oxidation by species such as Fe(CN)$_6^-$ and ferricenium ion.[95]

10. REACTION OF *N,N*-DIMETHYLANILINE *N*-OXIDES WITH METAL(III) PORPHYRINS

In addition to the dealkylation of tertiary amines in hepatic microsomes by cytochrome P-450 (reaction 6-45), there exists a second pathway which involves the conversion of the tertiary amine to the corresponding N-oxide by flavin mixed function oxidase followed by the reaction of the N-oxide with a cytochrome P-445 in the iron(III) state (reaction 6-46).[96]

$$PhN(CH_3)_2 \xrightarrow{\text{cytochrome } P\text{-}450/O_2/NADH} PhNH(CH_3) + CH_2O \quad (6\text{-}45)$$

$$PhN(CH_3)_2 \xrightarrow{\text{Enz-FlH}_2/O_2} \underset{\overset{|}{O_-}}{PhN^+(CH_3)_2} \xrightarrow{\text{cytochrome } P\text{-}445} PhNH(CH_3) + CH_2O$$
$$(6\text{-}46)$$

The reaction of N,N-dimethylaniline N-oxide with (TPP)Fe^{III}(CH$_3$OH) in methanol provides PhN(CH$_3$)$_2$, PhNH(CH$_3$), CH$_2$O, and some PhNH$_2$.[97] The iron(III) porphyrin was found to act as a catalyst and not to be destroyed by the N-oxide. No intermediate higher valent iron-oxo porphyrin species could be trapped by the addition of reagents such as 2,3-dimethyl-2-butene (TME). In a much more detailed study (CH$_2$Cl$_2$ solvent), it was found that (TPP)Fe^{III}Cl is a catalyst for the conversion of p-cyano-N,N-dimethylaniline N-oxide (DNO) to the products DA, MA, FA, H, MD, and A (Chart I).[98] In the presence of TME, DA and TME-epoxide were obtained in 100% yields. When 2,4,6-tri-*tert*-butylphenol (TBPH) was employed as a trap, DA and the phenoxyl radical TBP· were obtained in 100% yields. The rate constants for the formation of DA, TBP·, and TME-epoxide were found to be much the same and comparable to that for the turnover of (TPP)Fe^{III}Cl with DNO in the absence of the TME and TBPH trapping agents. These findings support rate-determining oxygen atom transfer from N-oxide to iron(III) porphyrin (reaction 6-47).

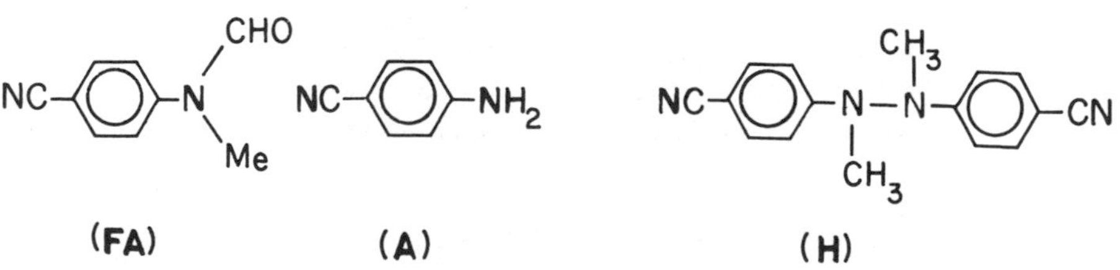

(DA)   (MA)   (MD)

(FA)   (A)   (H)

Chart I

$$(TPP)Fe^{III}Cl + NCPhN^+(CH_3)_2 \rightleftharpoons Complex \rightarrow$$

$$\underset{\overset{|}{O_-}}{}$$

$$(^+\cdot TPP)Fe^{IV}O(Cl) + NCPhN(CH_3)_2 \quad (6\text{-}47)$$

$$(^+\cdot TPP)Fe^{IV}O(Cl) + TME \xrightarrow{\text{fast}} (TPP)Fe^{III}Cl + TME\text{-epoxide}$$

Oxygen atom transfer from the N-oxide to $(TPP)Fe^{III}Cl$ was also shown to be rate determining in the demethylation of DA to form MA by comparison of inter- and intramolecular isotope effects. Thus, an isotope effect of unity was obtained from the comparison of the rate constants for reactions of p-$NCPhN^+(CH_3)_2O^-$ and p-$NCPhN^+(CD_3)_2O^-$. A discriminatory intramolecular isotope effect of 4.5 was observed when p-$NCPhN^+(CH_3)(CD_3)O^-$ was used, and the formation of p-$NCPhNH(CH_3)$ and p-$NCPhNH(CD_3)$ was monitored. The intramolecular isotope effect of 4.5 would be in accord with rate-determining H· abstraction or $1e^-$ oxidation with the loss of H^+ partially rate determining. When the formation of the products of Chart I were monitored with time, their appearance could be computer simulated to the reaction sequence of Scheme VII.

(a) $(TPP)Fe^{III}Cl + DNO \rightleftharpoons (TPP)(Cl)Fe^{III}DNO$

(b) $(TPP)(Cl)Fe^{III}(DNO) \rightarrow (^+\cdot TPP)(Cl)Fe^{IV}O + DA$

(c) $(^+\cdot TPP)(Cl)Fe^{IV}O + DA \rightarrow (TPP)(Cl)Fe^{IV}OH + DA\cdot$

(d) $(TPP)(Cl)Fe^{IV}OH + DA\cdot \rightarrow (TPP)Fe^{III}Cl + CAl$

(e) $Cal \rightleftharpoons MA + CH_2O$

(f) $(^+\cdot TPP)(Cl)Fe^{IV}O + MA \rightarrow (TPP)(Cl)Fe^{IV}O + MA\cdot$

(g) $2\ MA\cdot \rightarrow H$

(h) $2\ MA\cdot \rightarrow MD$

(j) $(^+\cdot TPP)(Cl)Fe^{IV}O + CH_2O \rightarrow (TPP)Fe^{III}Cl + HCOOH$

Scheme VII

Intermolecular and intramolecular isotope effects have been determined for the catalysis of the dealkylation of N,N-dimethylaniline N-oxide and p-cyano-N,N-dimethylaniline N-oxide by cytochrome P-450_{cam} and

TABLE 6-2 Isotope Effects for the Demethylation of N,N-Dimethylaniline (I) and p-Cyano-N,N-Dimethylaniline (II) and Their N-Oxides III and IV, Respectively, by Cytochromes P-450$_{LM2}$ and P-450$_{cam}$

	Exogenous oxidant		$NAD(P)H + O_2$	
	III	*IV*	*I*	*II*
P-450$_{LM2}$				
Intermolecular	1.1	1.0	1.0	1.0
(d_0 vs d_6)				
Intramolecular (d_3)	2.0	3.0	3.9	2.6
P-450$_{cam}$[a]				
Intermolecular	1.7	1.4	—	—
(d_0 vs d_6)				
Intramolecular (d_3)	2.5	4.3	—	—

[a] The specificity of P-450$_{cam}$ precludes its oxidation of the dimethylanilines.

P-450$_{LM2}$.[99] The experimental results ($\pm 8\%$) are shown in Table 6-2. Comparisons of the intermolecular and intramolecular isotope effects for P-450$_{LM2}$ establish that the demethylation of neither the two dimethylanilines nor their N-oxides is rate limiting. The rate-limiting steps must then be oxygen transfer from N-oxides to the cytochrome P-450$_{LM2}$ and the reaction of the enzyme with $NAD(P)H + O_2$, respectively. The intramolecular isotope effects with N-oxides as exogenous oxidants and the dimethylanilines as substrates using $NAD(P)H + O_2$ are comparable. This would be in accordance with the formation of the same oxidizing species at the active site. The magnitude of the intramolecular isotope effects for both cytochromes P-450$_{cam}$ and P-450$_{LM2}$ shows that there is some breaking of the carbon hydrogen bond in the transition state. Since these isotope effects are not large, the mechanism may represent $1e^-$ abstraction with H^+ loss partially rate determining. The intermolecular isotope effects of 1.7 and 1.4 for the reaction of cytochrome P-450$_{cam}$ with the two N-oxides might be interpreted in terms of concerted oxygen transfer and demethylation or the radical mechanism of reaction 6-48 (Reference 100) where the first two steps are rate determining.

$$PhN^+(Me)_2O^- + Enz\text{-}Fe^{III} \rightarrow [PhN(\cdot^+)(CH_3)_2 + Enz\text{-}Fe^{IV}O]$$

$$[PhN(\cdot^+)(CH_3)_2 + Enz\text{-}Fe^{IV}O] \rightarrow$$
$$[PhN(CH_3)(CH_2\cdot) + Enz\text{-}Fe^{IV}OH] \quad (6\text{-}48)$$

$$[PhN(CH_3)(CH_2\cdot) + Enz\text{-}Fe^{IV}OH] \rightarrow PhN(CH_3)(CH_2OH) + Enz\text{-}Fe^{III}$$

In the reaction of the dimethylaniline N-oxides with cytochromes P-450$_{LM2}$ and P-450$_{cam}$[99] as well as P-450$_{PB\text{-}B}$,[99] the immediate product is a dimethylaniline bound at the active site of the enzyme with the enzyme at a higher oxidation state. The rate constant for amine dealkylation within this complex surpasses the rate of dissociation of the dimethylaniline from the complex. This explains the observation[99] that N-oxides are not useful as oxygen transfer agents to cytochrome P-450 for the oxidation of other substrates. In the modeling reactions employing (TPP)Fe^{III}Cl [and other tetraphenylporphinato iron(III) salts] with p-cyano-N,N-dimethylaniline N-oxide and p-chloro-N,N-dimethylaniline N-oxide,[101] iron(IV)-oxo porphyrin π-cation radical species can be generated which are capable of epoxidation, etc.

On addition to oxidizable substrates to a reaction composed of p-cyano-N,N-dimethylaniline N-oxide and (TPP)Fe^{III}Cl a competition for ($\cdot^+$TPP)-Fe^{IV}O by the reactions of Scheme VII (c, d, f, and j) and the oxidation of the added substrate ensues. Employing a series of alkenes as substrate, the percentage yields of epoxides (based upon [DNO]$_i$) are comparable to when PhIO is used as the oxygen donor.[98] The dependence of electronic and steric effects upon amine oxidation vs alkene epoxidation has been examined using (Me$_8$TPP)Fe^{III}Cl,[102] (TPP)Fe^{III}Cl,[98] (Cl$_8$TPP)Fe^{III}Cl,[103] and (F$_{20}$TPP)Fe^{III}Cl (reference 104) as catalysts. With (Me$_8$TPP)Fe^{III}Cl and alkenes at 1 M, the percentage yields of epoxide were 100% (from 2,3-dimethyl-2-butene), 100% (cyclohexene), 80% (norbornene), 100% (styrene), 85% (cyclopentene), and 80% (cis-cylooctene). In the case of (TPP)Fe^{III}Cl, the yields of epoxides were less; for (Cl$_8$TPP)Fe^{III}Cl, epoxidation of 2,3-dimethyl-2-butene and norbornene occurred in but 25% and 34%, respectively. Using (F$_{20}$TPP)-Fe^{III}Cl with cis-cyclooctene, the epoxide was obtained in 6% yield. These results establish that electron removal from the porphyrin ring increases the rate of tertiary amine oxidation in preference to the rate of epoxidation. In part, the preference for amine oxidation over epoxidation has ben shown to be due to the oxidation of a portion of the tertiary amine within a solvent caged pair (Scheme VIII). In Scheme VIII the percentage yields relate to the partitioning of the solvent caged pair. About 30% of demethylation is due to the bimolecular reaction of ($\cdot^+$F$_{20}$TPP)Fe^{IV}O with DA after their diffusion.

$$(F_{20}TPP)Fe^{III}Cl + DNO \rightleftarrows (F_{20}TPP)Fe^{III}(DNO)Cl$$

$$\downarrow \text{r.d.s.}$$

$$[(^{\cdot+}F_{20}TPP)Fe^{IV}O\cdot DA]Cl^-$$

$$(^{\cdot+}F_{20}TPP)Fe^{IV}O + DA \qquad (F_{20}TPP)Fe^{III}Cl + MA \qquad (F_{20}TPP)Fe^{III}Cl + DMP$$

$$+CH_2O$$

$$(67\%) \qquad\qquad (17\%) \qquad\qquad (17\%)$$

Scheme VIII

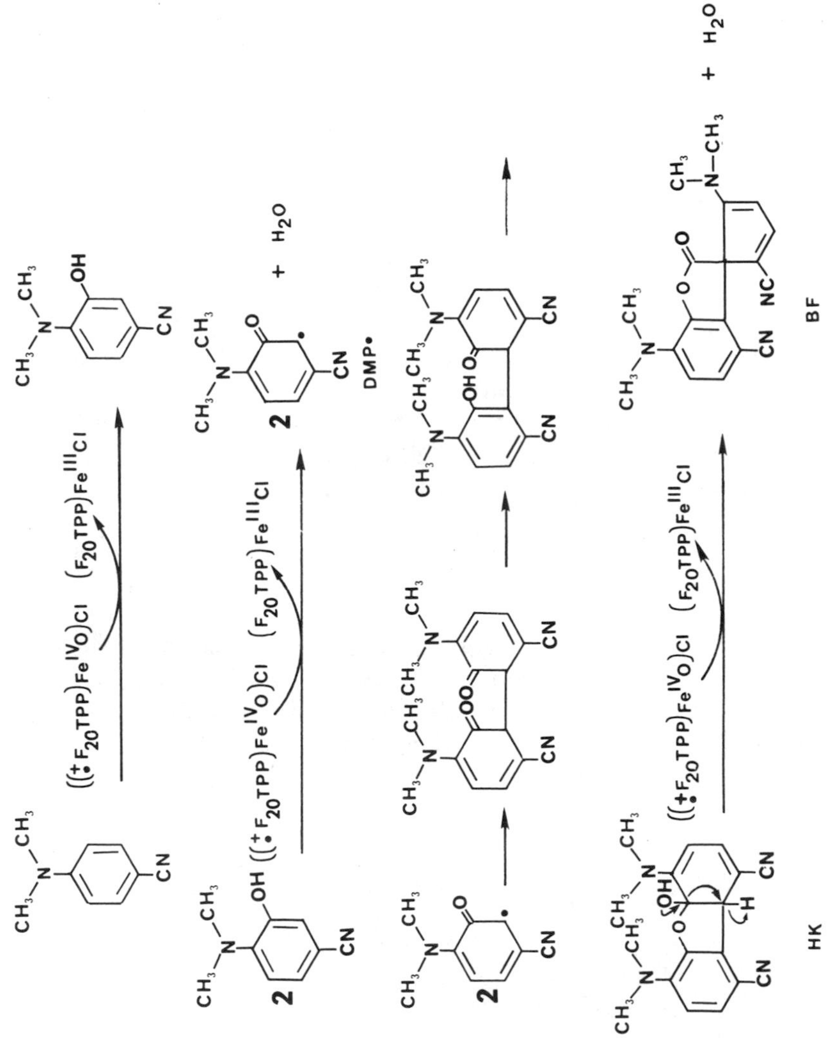

Scheme IX

Within the solvent caged pair, demethylation of DA competes equally with its ring hydroxylation to provide DMP, which is further oxidized to the spiro compound BF (Scheme IX).[104]

The products N,N'-dimethyl-N,N'-bis(p-cyanophenyl)hydrazine (H) and N,N'-bis(p-cyanophenyl)-N-methylenediamine (MD) arise from the reaction of p-cyano-N,N-dimethylaniline N-oxide with iron(III) tetraphenylporphyrins (Chart I) by the coupling of radical species of p-cyano-N-methylaniline (MA) (Scheme VII). With $(F_{20}TPP)Fe^{III}Cl$, hydroxylation of the aromatic ring leads to the formation of 4-cyano-7-dimethylamino-2-benzofuranon-3-spiro-2'-cyano-5'-dimethylaminocyclopentadiene (BF) (Scheme IX).[104] When using the SbF_6^- salt of the iron(III) C_2-capped porphyrin of Baldwin[105] in reactions with p-cyano-N,N-dimethylaniline N-oxide in the presence of alkenes, there the cycloaddition of DA· and alkene with formation of tetrahydroquinolines is observed (Scheme X).[101,102]

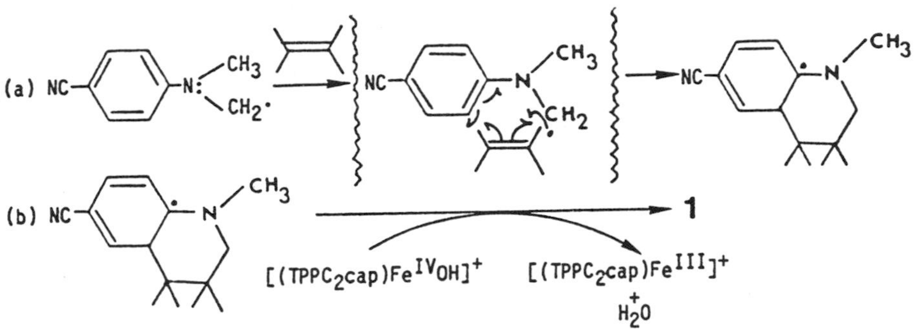

Scheme X

The reactions of p-cyano-N,N-dimethylaniline N-oxide with $(TPP)Mn^{III}X$ (X = F^-, Cl^-, Br^-, I^-, OCN^-) in $PhCN$[106] and CH_2Cl_2 solvents and with $(Me_8TPP)Mn^{III}(ImH)Cl$ (ImH = imidazole) in CH_2Cl_2 solvent[107] have been investigated. The following conclusions were arrived at in the studies with $(TPP)Mn^{III}X$ in PhCN solvent. (1) Oxygen transfer from DNO occurs through the reversible formation of the hexacoordinated species, $(TPP)Mn^{III}(NO)X$, which subsequently decomposes to the intimate pair $[[(TPP)(X)(MnO)]^+ + DA]$. (2) The intimate pair undergoes the competing reactions of internal oxidative demethylation of DA and diffusion. (3) In the presence of alkenes, epoxide formation occurs by reaction of alkene with $[(TPP)(X)(MnO)]^+$ after the latter's diffusion from association with DA. The rates of product formation were shown to be dependent upon the nature of the axial ligand X^-. The pseudo-first-order rate constants for turnover, when $X^- = Cl^-$, Br^-, I^-, and OCN^-, were found to have a first-order dependence on $(TPP)Mn^{III}X$ and to be independent of the concentration of DNO (ie, first order in both species). When X^- is F^-, however, $(TPP)Mn^{III}F$ catalyst was found to be saturated with DNO so that the appearance of products were zero order. In comparing the rate and equilibrium constants for the reaction of (TPP)Mn^{III}F and (TPP)Mn^{III}Cl, it was found that the equilibrium constant for the formation of (TPP)Mn(X)(DNO) species is approximately 700 times larger when X$^-$ = F$^-$, whereas the rate constant for DA demethylation is approximately 70 times smaller. Thus, the nature of the axial ligand has a profound influence upon the overall rate and kinetic behavior in the formation of the higher valent manganese-oxo porphyrin species and also on the rate constants for the reaction of this species with substrates. In CH$_2$Cl$_2$ solvent, the rate of formation of the higher valent manganese-oxo species with (Me$_8$TPP)Mn^{III}(ImH) was found to be 105-fold greater than when using (Me$_8$TPP)Mn^{III}Cl. Also, the higher valent manganese-oxo species with imidazole ligation epoxidized the alkenes 2,3-dimethyl-2-butene, *cis*-cyclooctene, cyclohexene, norbornene, and cyclopentene in approximately 80% yield. Without imidazole ligation, the higher valent manganese-oxo porphyrin did not provide epoxides in the presence of high concentrations of alkenes. Thus, in CH$_2$Cl$_2$ solvent nitrogen base ligation is required for epoxidation, while in PhCN solvent nitrogen base ligation is not required. The requirement of a nitrogen base ligand for the epoxidation of alkenes by manganese(III) porphyrins was first recorded in other studies.[108,109]

11. THE MECHANISM OF THE EPOXIDATION OF ALKENES

In the modeling of cytochrome *P*-450 enzyme alkene epoxidation reactions with metal(III) tetraphenylporphyrins, the following reagents have been used as oxygen atom donors: iodosylbenzene,[11,35c,110,111,112] pentafluoroiodosylbenzene,[113,114] hypochlorite,[114,115,116,117] potassium hydrogen persulfate,[108,118] hydrogen peroxide,[119] alkyl hydroperoxides,[28,115,119a,120] *N,N*-dimethylaniline *N*-oxides[98,102,103,104,107,121] and an oxaziridine.[122] Modeling of cytochrome *P*-450 systems has also been carried out with iron(II) porphyrin generating systems in the presence of O$_2$.[123] Porphyrin destruction, the formation of alcohols (which may be further oxidized to ketones), functional group migrations, etc., compete to an extent, dependent upon the system, with epoxidation. The bridged metallo-oxo porphyrin species (such as seen in Scheme II) formed from the iron(IV)-oxo porphyrin π-cation radical has been favored by theoretical calculations as the actual epoxidizing agent.[124] Such a species has been reported[125] to be formed on reaction of (Me$_{12}$TPP)Fe^{III}OH with *m*-chloroperbenzoic acid. However, there is no evidence that this species is capable of oxygen transfer to an alkene.

Though many valuable investigations have been carried out with iodosylbenzene, the reagent has its limitations. Iodosylbenzene is a polymer,[126] so it cannot be meaningfully employed in kinetic studies. The porphyrin ring of ($^+$·TPP)Fe^{IV}O(X) undergoes rather rapid decomposition at ambient temperatures. The lifetime of the metalloporphyrin catalyst may be prolonged by use of a sufficient excess of oxidizable substrate over iodosylbenzene.[97] Recently introduced are tetraphenylporphinato metal(III) catalysts which are more resistant to oxidation due to the substitution of the phenyl rings by halogens [eg, $(Cl_8TPP)Fe^{III}Cl$ (Reference 113b) and $(F_{20}TPP)Fe^{III}Cl$ (Reference 127)]. Pentafluoroiodosylbenzene (which, like iodosylbenzene, is insoluble in chlorinated solvents) has been recommended for dynamic studies when using $(Cl_8TPP)Fe^{III}X$ in a CH_2Cl_2/CH_3OH (or $CF_3CH_2OH)/H_2O$ (80 : 18 : 2) solvent.[113b] The solvent composition is quite crucial, and it has been proposed that the reactive oxygen transfer agent is $C_6F_5I(OH)(OCH_3)$. Yields of epoxides exceed those obtained with PhIO. Studies with hypochlorite have been carried out in a toluene/H_2O biphase with rapid stirring. When the aqueous phase is at a pH above approximately 10, a phase transfer agent is required to carry the ClO^- to the organic phase in order for it to react with metalloporphyrin. At pH values below 10, a phase transfer agent need not be employed.[128] Perhaps at lower pH the oxygen transfer agent is, along with HClO, the nonionic Cl_2O species.[129] A tertiary amine ligand, such as N-alkyl imidazoles or pyridines, must be present for epoxidation to occur. Though an inexpensive means of preparing epoxides in high yields, the method is useless for mechanism-directed kinetic studies (vide infra). Potassium hydrogen persulfate has been used for epoxidations in a similar biphase.

Hydrogen peroxide and, to a lesser extent, alkyl hydrogen peroxides serve as oxygen transfer agents for epoxidation with manganese(III) porphyrins in the presence of a suitable tertiary nitrogen base ligand. From kinetic studies (CH_2Cl_2 solvent), using $(TPP)Mn^{III}Cl$ and *tert*-BuOOH as oxygen donor, it has been determined that (1) $(TPP)Mn^{III}Cl$ does not react with *tert*-BuOOH in the absence of imidazole and (2) in the presence of varying concentrations of imidazole the pseudo-first-order rate constants for the disappearance of *tert*-BuOOH is always at least two times that for the appearance of epoxide.[40a,120] A kinetically competent explanation for these results involves reversible N-oxidation of imidazole by $[(TPP)MnO(ImH)]^+$ being competitive with epoxidation. Tertiary amines are not required as ligating agents in the epoxidation of alkenes when either 2-phenylsulfonyl-3-(p-nitrophenyl)oxaziridine or p-cyano-N,N-dimethylaniline N-oxide are used as oxygen transfer agents with $(TPP)Mn^{III}Cl$ as catalyst in aprotic solvents.[106,122] p-Cyano-N,N-dimethylaniline N-oxide is soluble in all but hydrocarbon solvents and lends itself to kinetic studies because the appearance of p-cyano-N,N-dimethylaniline (reaction 6-47) can be followed spectrophotometrically.

Epoxidation of certain terminal olefins by P-450 cytochromes results in the suicide inhibition of the enzyme.[130,131,132] Since P-450 cytochromes are not destroyed by the epoxides prepared from these olefins, it was concluded that suicide inhibition was due to an intermediate in the epoxidation reaction. Some of the alkenes which have been investigated are polyhalogenated, and this feature per se could give rise to inhibition, since polyhalogenated alkanes are suicide inhibitors through the formation of porphyrin iron(II) carbenes.[133] However, ethylene and other simple alkenes were established to be suicide inhibitors while their epoxides are not.[134] Suicide inhibition has been shown to be due to the N-alkylation of a pyrrole nitrogen of the porphyrin.[131e,134] Two kinds of *N*-alkylporphyrins can be isolated from 1-alkene oxidation by PhIO catalyzed by iron porphyrins.[135] The N—CH$_2$CHROH porphyrins have been isolated upon oxidation of 3-methyl-1-butene, 4,4-dimethyl-1-pentene, and 1-decene by (Cl$_8$TPP)Fe^{III}Cl (References 113a and 136) and the N-CHRCH$_2$OH porphyrins found upon oxidation of 1-butene, 1-hexene, and 4-methyl-1-pentene by (TPP)Fe^{III}Cl and *meso*-tetrakis(4-chlorophenyl)porphinato iron(III) chloride.[135] Further work is required to understand the mechanisms of these *N*-alkylation reactions. The ratio of enzyme turnover number to a suicide event, using allylisopropylacetamide as substrate with both hepatic tissue and pure cytochrome *P*-450 from phenobarbital-treated rats, has been reported to be approximately 200.[131f,137] It is not known whether the suicide event is due to interception of an intermediate on the path to epoxide formation or to a spurious side reaction of the alkene which occurs once in each 200 turnovers. Likewise, in the evaluation of experiments with chemical systems designed to understand a mechanism by the identification of isolated products, it is most important to understand that the various products may not only arise from intermediates leading to the product of interest but that products may arise from parallel reaction paths. Applied to epoxidation, one may have epoxide arising in a single-step reaction without intermediates and other products arising from competing reactions or, alternatively, products other than epoxide may be formed by side reactions from intermediates along the reaction path to epoxidation.

During the epoxidation of norbornene in a soluble system (CH$_2$Cl$_2$/CF$_3$CH$_2$OH/H$_2$O) or heterogeneous system (CH$_2$Cl$_2$) with C$_6$F$_5$IO and (Cl$_8$TPP)Fe^{III}Cl, with oxidant at high concentration, the hemin is completely converted to an *N*-alkyl hemin of the type of structure shown as 6-49. As the oxidant disappears, the *N*-alkyl hemin reverts to the iron(III) pophyrin. The *N*-alkyl hemin is not an intermediate in the epoxidation reaction but a catalyst comparable to the starting (Cl$_8$TPP)Fe^{III}Cl. Thus, at high oxidant concentration, the *N*-alkylhemin is the catalyst for most of the reaction.[113d]

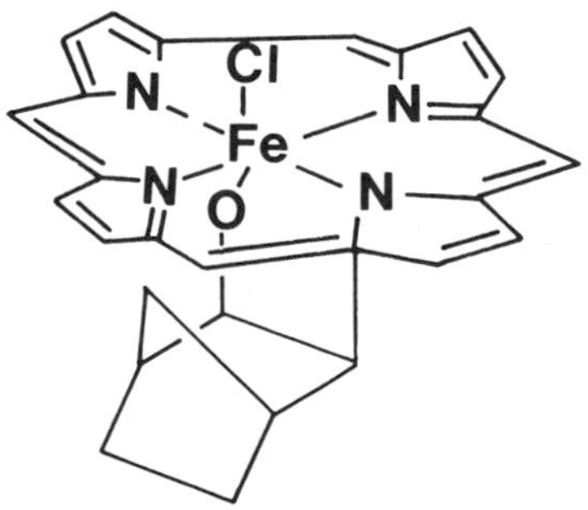

(6-49)

Four intermediates have been proposed in alkene epoxidation by hypervalent iron-oxo and manganese-oxo porphyrins. These are provided in Chart II for the iron species and may be discussed in terms of reactions 6-50 and 6-51. If k_2 is sufficiently greater than k_1, the intermediate carbocation radical pair[111,112,113] would be short lived and the species iron(IV)-oxo carbon radical and iron(III)-oxo carbocation could be looked upon as being formed in reactions with transition states resembling the carbocation radical pair.

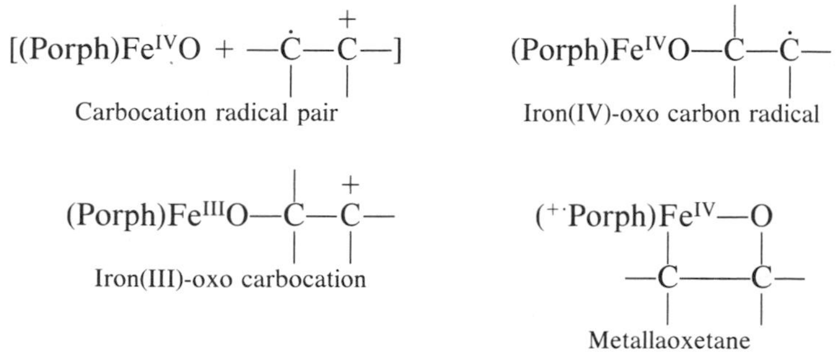

$$[(Porph)Fe^{IV}O \; + \; -\overset{+}{\underset{|}{C}}-\overset{\cdot}{\underset{|}{C}}-]$$

Carbocation radical pair

$$(Porph)Fe^{IV}O-\overset{|}{\underset{|}{C}}-\overset{\cdot}{\underset{|}{C}}-$$

Iron(IV)-oxo carbon radical

$$(Porph)Fe^{III}O-\overset{|}{\underset{|}{C}}-\overset{+}{\underset{|}{C}}-$$

Iron(III)-oxo carbocation

$$(^{+\cdot}Porph)Fe^{IV}-O$$
$$-\overset{|}{\underset{|}{C}}\underline{\qquad}\overset{|}{\underset{|}{C}}-$$

Metallaoxetane

Chart II

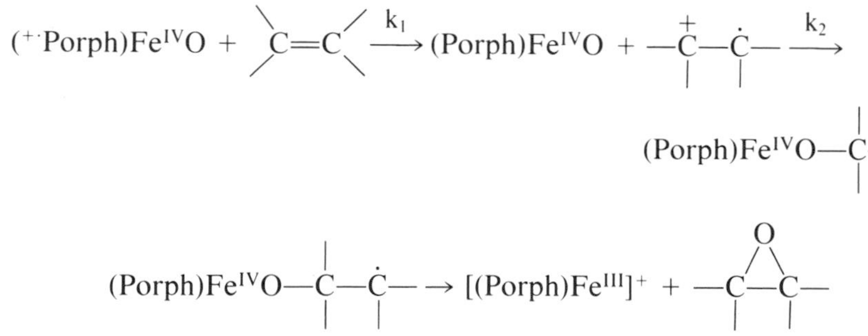

$$(^{+\cdot}Porph)Fe^{IV}O \; + \; \overset{\backslash}{\underset{/}{C}}{=}\overset{/}{\underset{\backslash}{C}} \; \overset{k_1}{\longrightarrow} (Porph)Fe^{IV}O \; + \; -\overset{+}{\underset{|}{C}}-\overset{\cdot}{\underset{|}{C}}- \; \overset{k_2}{\longrightarrow}$$

$$(Porph)Fe^{IV}O-\overset{|}{\underset{|}{C}}-\overset{\cdot}{\underset{|}{C}}-$$

$$(Porph)Fe^{IV}O-\overset{|}{\underset{|}{C}}-\overset{\cdot}{\underset{|}{C}}- \rightarrow [(Porph)Fe^{III}]^+ \; + \; -\overset{O}{\overset{/\backslash}{\underset{|}{C}}-\overset{}{\underset{|}{C}}}- \qquad (6\text{-}50)$$

$$(^{+\cdot}Porph)Fe^{IV}O \; + \; \overset{\backslash}{\underset{/}{C}}{=}\overset{/}{\underset{\backslash}{C}} \; \overset{k_1}{\longrightarrow} (Porph)Fe^{IV}O \; + \; -\overset{+}{\underset{|}{C}}-\overset{\cdot}{\underset{|}{C}}- \; \overset{k_2}{\longrightarrow}$$

$$(Porph)Fe^{III}O-\overset{|}{\underset{|}{C}}-\overset{+}{\underset{|}{C}}-$$

$$(Porph)Fe^{III}O-\overset{|}{\underset{|}{C}}-\overset{+}{\underset{|}{C}}- \rightarrow [(Porph)Fe^{III}]^+ \; + \; -\overset{O}{\overset{/\backslash}{\underset{|}{C}}-\overset{}{\underset{|}{C}}}- \qquad (6\text{-}51)$$

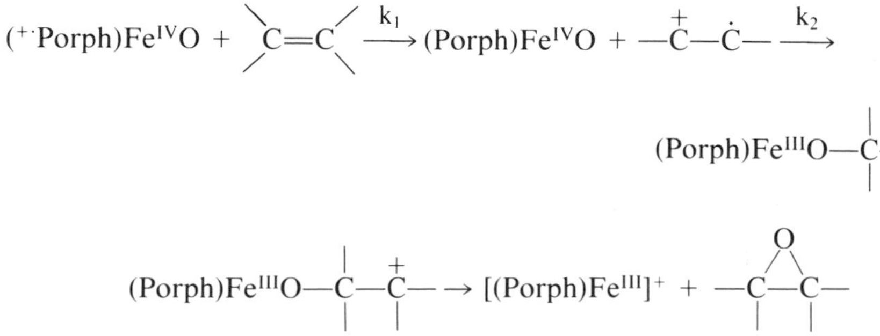

The metallaoxetane structure has been proposed to be formed in a 2a + 2s concerted cycloaddition[114,116] after like proposals for olefin oxidation by oxo transition metal species.[138] Formation of epoxide from metallaoxetane is proposed to occur by a concerted reductive elimination. Treatment of $(^{+\cdot}Me_{12}TPP)Fe^{IV}O$ [generated at $-42°C$ in CH_2Cl_2 by reaction of the iron(III) species with m-chlorobenzoic acid] with cyclooctene provides a color change from green to dark-green. This has been attributed to the formation of either a charge transfer complex or metallaoxetane.[139] The metallaoxetane pathway may be questioned on the basis that *trans*-stilbene oxide is sometimes the major product in the epoxidation of *cis*-stilbene with manganese(III) porphyrins as catalysts.[111,117,118c,d] On the basis that the epoxidation of *cis*-β-methylstyrene by manganese(V)-oxo porphyrin yields three-fold less *trans*-epoxide than does epoxidation with manganese(IV)-oxo porphyrin, it has been proposed that epoxidation by the former is a stereoretentive process while the latter is a nonstereoretentive process.[140] The stereoretentive reaction involving manganese(V)-oxo species may then involve a metallaoxetane, while the reaction with manganese(IV)-oxo was proposed to involve an electron transfer. Electron transfer from alkene to manganese(IV)-oxo species would provide a manganese(III) species, and epoxidation would provide a manganese(II) species. It is likely that additional investigations are required.

The iron(III)-oxo carbocation intermediate of eq 52 would appear to be a likely candidate as an intermediate in the oxidation of styrenes to phenylacetaldehydes (eq 53).[110e]

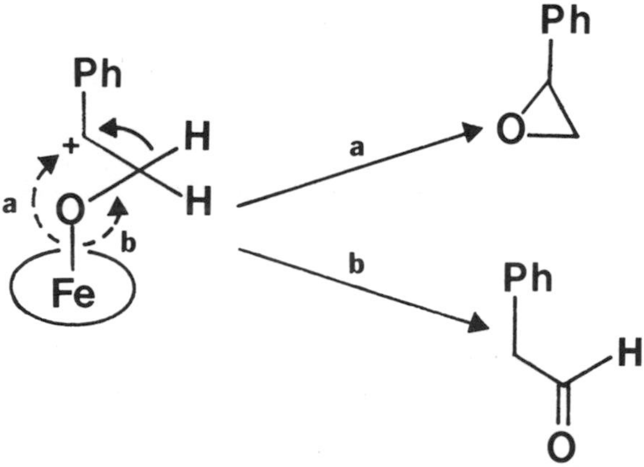

(6-52)

This reaction competes with styrene epoxidation. In oxidations of *cis*- and *trans*-β-^{2}H-styrenes by (1) ClO^- with $(TPP)Mn^{III}Cl$ and $(Me_{12}TPP)Mn^{III}Cl$ in the presence of tertiary amine ligand and (2) C_6F_5IO and $(F_{20}TPP)Fe^{III}Cl$ catalyst, H(D)-migration is intramolecular with a preference for *cis*-H(D)-migration. These results have been considered[116d] in terms of a metallaoxetane intermediate which gives way to an iron(III)-oxo carbocation or iron(IV)-oxo carbon radical intermediate, but this explanation was dismissed on the basis of a lack of sufficient sensitivity of rate to the nature of the alkene; however, these kinetic investigations employed the analysis of competitive rates of epoxidation of an alkene and styrene. The use of styrene for this purpose may have been unfortunate. Thus, it has been reported that in the presence of styrene the rates of epoxidations of other added alkenes are enhanced [using the system ClO$^-$ with manganese(III) porphyrin and pyridine ligand].[117c] The observed association of the time course for the epoxidations with lag phases were suggested to arise from an autocatalysis brought about by formation of a catalytic oxidation product of styrene. Since the addition of benzaldehyde in place of styrene increases the rates of alkene epoxidation, it has been suggested that the promoter formed from styrene is phenylacetaldehyde. From these considerations it would appear that the metal(III)-oxo carbocation is a viable intermediate in the epoxidation of styrene.

It has been observed that both *cis*- and *trans*-stilbene oxide are obtained in the epoxidation of *cis*-stilbene by higher valent manganese-oxo porphyrins [eg, PhIO with (TPP)Mn^{III}Cl (Reference 110b) or (TPP)Mn^{IV}(PhIOAc)$_2$,[35a] and (TPP)Mn^{III}Cl with 2-phenylsulfonyl-3-(*p*-nitrophenyl)oxaziridine[122]]. However, *trans*-stilbene gives only *trans*-stilbene oxide. *trans*-Stilbene reacts slower than *cis*-stilbene because of steric hindrance of its approach to the oxidant but yields the most stable *trans*-epoxide. The direct formation of *trans*-stilbene oxide from *cis*-stilbene would require an intermediate. The carbocation radical pair or the manganese(IV)-oxo carbon radical (Scheme XI[122]) may be considered as intermediates. Diphenylacetaldehyde and deoxybenzoin were not recorded as products. Their absence would indicate that sufficiently stable manganese(III)-oxo carbocation or manganese(IV)-oxo carbon radical were not formed (Scheme XI).

It has recently been found[141a] that cis- to trans-isomerization of *cis*-stilbene occurs when using an iron(III) porphyrin as catalysts. The epoxidation of

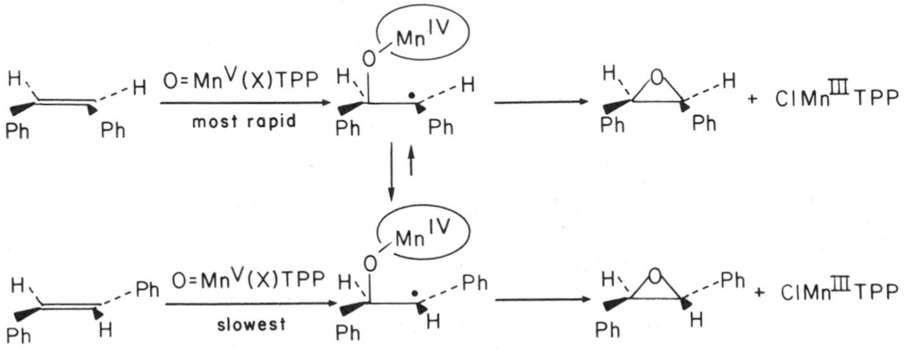

Scheme XI

cis- and *trans-*stilbene has been investigated under an inert atmosphere using C_6F_5IO with $(F_{20}TPP)Fe^{III}Cl$ (dry CH_2Cl_2), $(Cl_8TPP)Fe^{III}Cl$ (dry CH_2Cl_2), or $(Cl_8TPP)Fe^{III}OH$ (CH_2Cl_2/MeOH/H_2O (80 : 18 : 2)). Aside from *cis-*stilbene oxide, careful product analysis showed the presence of *trans-*stilbene, *trans-*stilbene oxide, diphenylacetaldehyde, and deoxybenzoin. The various products were shown not to arise from further reaction of *cis-*stilbene oxide. The formation of diphenylacetaldehyde and deoxybenzoin is best explained as occurring from a metallo(III)-oxo carbocation intermediate by phenyl and hydrogen rearrangements, respectively, while *trans-*stilbene formation may occur by reversible $1e^-$ oxidation of *cis-*stilbene to provide a carbocation radical. In addition to the isomerization of *cis-* to *trans-*stilbene, the observation of radical rearrangements (as shown in reaction 6-53) has been offered as evidence for the intermediacy of a carbocation radical pair (Chart II) in the epoxidation reaction.[113e]

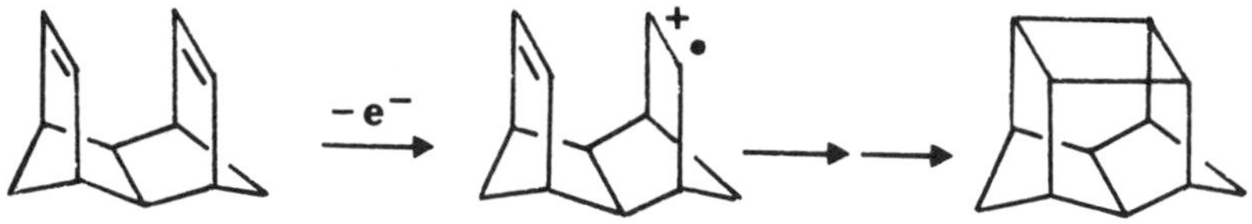

(6-53)

Hypervalent metallo-oxo porphyrins are good $1e^-$ oxidants, and isomerization and rearrangements of alkenes through their carbocation radicals is reasonable; however, the carbocation radical may or may not be on the reaction path to epoxidation.

The observation that the epoxidation of norbornene provides ratios of *exo-* to *endo-*norbornene epoxides of between 5 to 58 along with norcamphor (1–3%) and cyclohexene-4-carboxaldehyde (2–5%) has been offered in support of the formation of intermediates. The active oxidants in these studies were generated from the reaction of both C_6F_5IO and an iodosylxylene with a number of substituted tetraphenylporphyrin and tetrapyridylporphyrin iron(III) salts.[113c] Peracid epoxidation, electrophilic and radical addition as well as electrocyclic additions and metal complex formation generally occur on the exo-side of norbornene. Barring lack of selectivity due to a high degree of activity, electrophilic addition of $(^+Porph)Fe^{IV}O$ to norbornene would be expected to be exo. A general mechanism of epoxidation was proposed to involve an initial $1e^-$ transfer from alkene to $(^+Porph)Fe^{IV}O$ followed by collapse of the intimate carbocation radical pair. In the case of norbornene, collapse of the intimate carbocation radical pair has been proposed to yield both *exo-* and *endo-*norbornene epoxide. The *endo-*norbornene epoxide would arise after rotation of the norbornene carbocation within the intimate pair. Only the intermediacy of the iron(III)-oxo carbocation (as shown in reaction 6-51) can explain the formation of the small yields of norcamphor and cyclohexene-4-carboxaldehyde (reaction 6-54).[113c]

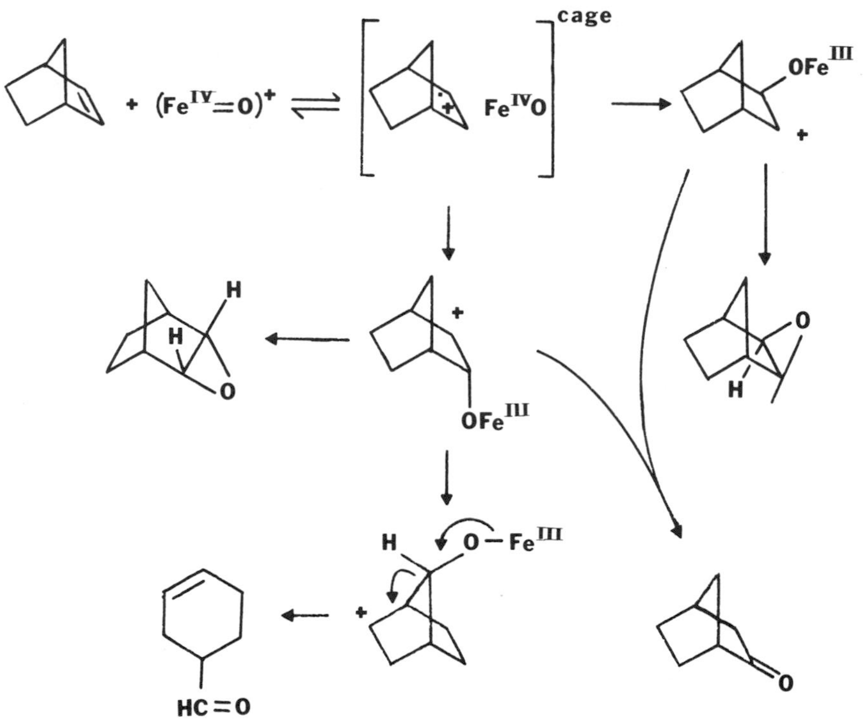

(6-54)

In summary, the presence of at least two species—metal(III)-oxo carbocation and the carbocation radical pair (Chart II)—are required to explain products when using norbornene, styrene, and *cis*-stilbene as substrates. A case may be made that the carbocation radical pair is on the reaction path to endo-epoxidation of norbornene. Otherwise, the two intermediate species could arise in competitive parallel reactions.

Analysis of kinetic results for the epoxidation of norbornene and *cis*-cyclooctene using ClO^- with imidazole ligated $(TPP)Mn^{III}X$ and $(Me_{12}TPP)Mn^{III}X$ in the CH_2Cl_2/H_2O biphase with added phase transfer agent has provided much of the experimental support for the presence of a stable metallaoxetane intermediate.[114,116] It has been pointed out, however, that the system is far too complex to employ the results of initial rates, and alternate explanations have been provided for the reported kinetic results.[141b] Evidence for rate-determining alkene epoxidation with an iron(III) porphyrin has been offered in support of the reversible formation of a stable metallaoxetane intermediate.[114] In these experiments, C_6F_5IO was used with $(Cl_8TPP)Fe^{III}Cl$ and CH_2Cl_2 served as the solvent. This is not, however, an ideal combination for the study of the dynamics of the epoxidation reaction because C_6F_5IO is insoluble in CH_2Cl_2.

Though the kinetic evidence offered in support of the formation of a *stable*

metallaoxetane intermediate in the epoxidation reaction is highly questionable, the formation of an unstable metallaoxetane is a reasonable proposal.
A cyclic intermediate 6-55 is required to explain the stereoelectronic control
observed for preferred migration of cis-H(D) in the reaction of trans- and cis-
β-^{2}H-styrene with ClO$^-$ and (TPP)Mn^{III}Cl.

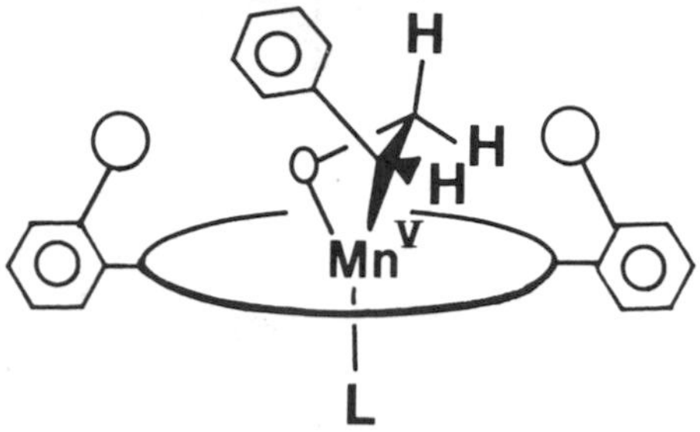

(6-55)

The metallaoxetane 6-55 was proposed as intermediate in the reaction.[116d] In
the epoxidation of a number of alkenes by cationic salen (N,N'-ethylenebis-
(salicylideneiminate)), complete stereoelectronic control was attributed to
the intermediacy of a quasi-closed species (reaction 6-56).[142]

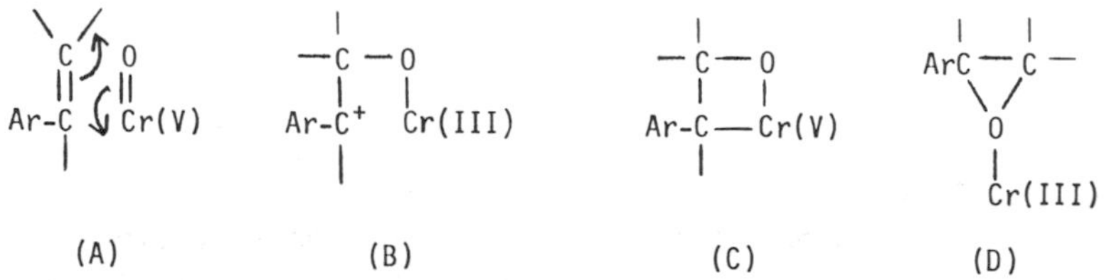

(6-56)

(A)   (B)   (C)   (D)

The authors conclude that their experimental results do not offer a compelling case for a single discrete structure to describe the intermediate, but if
forced to choose a single structure, the metallaoxetane (C) along with (D)
would be the closest representatives. Structure (D) would be closely related
to the activated complex based on the concept of least motion.[143] The choice
of a metallaoxetane intermediate would require the breaking of a porphyrin
nitrogen–chromium bond so that a hypervalent 7-coordinate chromium center is not present. Figure 6-7 shows a means by which a metallo(III)-oxo
carbocation could be stabilized in quasi-closed conformation by a porphyrin
pyrole nitrogen lone pair.[141b] As illustrated in Figure 6-7, H_c and H_t migrations as well as epoxide formation, porphyrin N-alkylation and metallaoxetane formation could derive from such a structure.

We have considered to this point the intermediacy (from Chart II) of the
carbocation radical pair, the metallo(IV)-oxo carbon radical, iron(III)-oxo
carbocation, metallaoxetane, and (from reaction 6-56 and Figure 6-7) quasi-
closed metallo(IV)-oxo carbon radical as well as a metal(III) complex of the

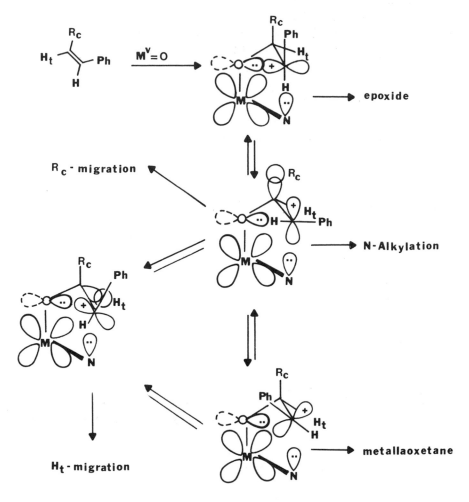

Figure 6-7. Stereoorbital diagram depicting acyclic species as possible intermediates in epoxidation, rearrangement to aldehydes from preferential substituent migration, porphyrin N-alkylation, and metallaoxetane formation (Reference 141a).

epoxide derived by a direct "oxene" insertion into the C=C bond [(D) in reaction 6-56].

As a means of evaluating the possibility of the formation of the radical intermediates the epoxidation of Z-1,2-bis(2^t,3^t-diphenylcyclopropyl) ethene

$$(6\text{-}57)$$

has been investigated.[144] The Z-epoxide was the only oxidized product formed (reaction 6-57). The rate constant for the cyclopropylcarbinyl → allyl carbinyl radical rearrangement of Eq 58 has been determined at 30°C to be $2.7 \times 10^8/s$ and $1.6 \times 10^8/s$ (References 145a and 145b, respectively).

$$(6\text{-}58)$$

This rearrangement involves the formation of a primary radical from a primary radical. The rate constant for the conversion of a secondary cyclopropylcarbinyl radical to a secondary allyl carbinyl radical is essentially the same ($2.5 \times 10^8/s$ at 30°C). The cyclopropylcarbinyl radical rearrangement of reaction 6-59 would provide a benzylic radical as product.

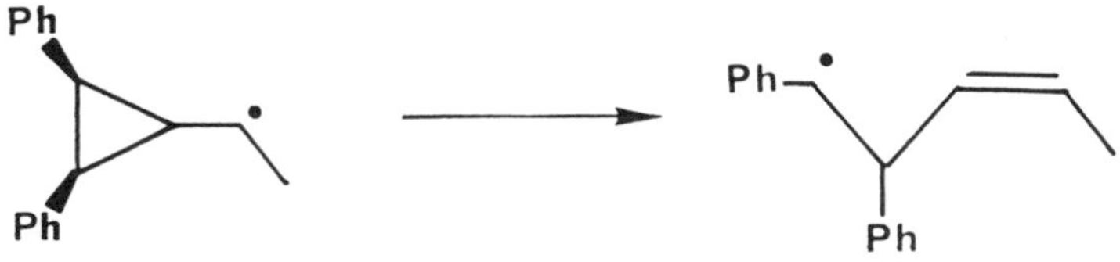

$$(6\text{-}59)$$

One would anticipate a minimum rate enhancement of 100-fold based on the finding that $1e^-$ reduction of 1-(2-phenylcyclopropyl) ketones provide, as the sole product, the benzyl radical.[146] In the intramolecular competitive trapping of the radical intermediate of Equation 6-61 only the $2^t,3^t$-diphenylcyclopropyl ring underwent ring opening.[144]

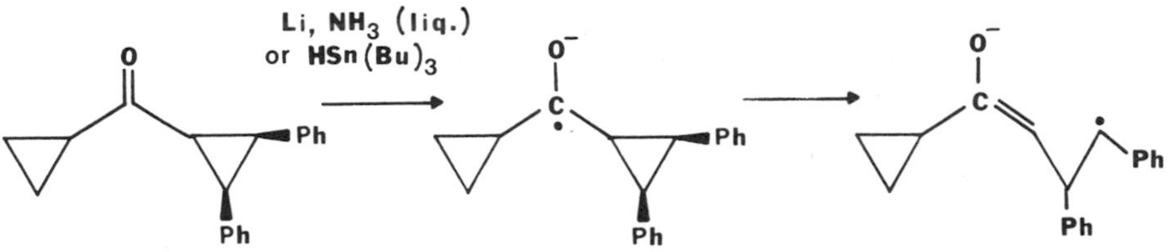

$$(6\text{-}60)$$

The inability to detect cleavage other than of the $2^t,3^t$-diphenylcyclopropyl ring suggests a rate constant for opening of the $2^t,3^t$-diphenylcyclopropyl ring of greater than or equal to $10^{10}/s$. Since the epoxidation reaction of Equation 6-58 is accompanied by less than 0.5% $2^t,3^t$-diphenylcyclopropyl ring opening (detection limit), it may be concluded that the plausible radical intermediate to epoxidation would possess a lifetime no greater than $10^{12}/s$.

ACKNOWLEDGMENTS

This work was supported by the National Institutes of Health and the National Science Foundation.

REFERENCES

1. (a) Fita, I.; Rossmann, M. G. J. *Mol. Biol.* **1986,** *188,* 49. (b) Fita, I.; Rossman, M. G. J. *Mol. Biol.* **1985,** *185,* 21. (b) Reid, T. J. III; Murthy, M. R. N.; Sicignano, A.; Tanaka, N.; Musick, W. D. L.; Rossmann, M. G. *Proc. Natl. Acad. Sci. U.S.A.* **1981,** *78,* 4767.
2. Chance, B. *Arch. Biochem. Biophys.* **1952,** *37,* 235.
3. (a) Schonbaum, G. R.; Lo, S. J. *J. Biol. Chem.* **1972,** *247,* 3353. (b) Theorell, H. *Enzymologia* **1941,** *10,* 250.
4. Dolphin, D.; Forman, A.; Borg, D. C.; Fajer, J.; Felton, R. H. *Proc. Natl. Acad. Sci. U.S.A.* **1971,** *68,* 614.
5. Moss, T. H.; Ehrenberg, A.; Beardon, A. J. *Biochemistry* **1969,** *8,* 4159.
6. (a) Schulz, C. E.; Devaney, P. W.; Winkler, H.; Debrunner, P. G.; Doan, N.; Chiang, R.; Rutler, R.; Hager, L. P. *FEBS Lett.* **1979,** *103,* 102. (b) Roberts, J. E.; Hoffman, B. M.; Rutter, R.; Hager, L. P. *J. Biol. Chem.* **1981,** *256,* 2118.
7. (a) Phillippi, M. A.; Goff, H. M. *J. Am. Chem. Soc.* **1982,** *104,* 6026. (b) Gans, P.; Buisson, G.; Duee, E.; Marchon, J.-C.; Erier, B. S.; Scholz, W. F.; Reed, C. A. *J. Am. Chem. Soc.* **1986,** *108,* 1223.
8. Keilin, D.; Mann, J. *Proc. R. Soc. London Ser. B* **1937,** *122,* 199.
9. Chang, C. K.; Dolphin, D. In "Bioorganic Chemistry"; van Tamelen, E. E., Ed.; Academic Press: New York, **1978,** Vol. IV, 37.
10. Traylor, T. G.; Lee, W. A.; Stynes, D. V. *Tetrahedron* **1984,** *40,* 553.
11. Groves, J. T. *Adv. Inorg. Chem.* **1979,** 119.
12. Guengerich, F. P.; MacDonald, T. L. *Acc. Chem. Res.* **1984,** *17,* 9.
13. Sawyer, D. T.; Srivatsa, G. S.; Bodini, M. E.; Schaefer, W. P. *J. Am. Chem. Soc.* **1986,** *108,* 936.
14. (a) Latos-Grazynski, L.; Cheng, R.-J.; La Mar, G. N.; Balch, A. L. *J. Am. Chem. Soc.* **1981,** *103,* 4270. (b) Chevrier, B.; Weiss, R.; Lange, M.; Chottard, J.-C.; Mansuy, D. *J. Am. Chem. Soc.* **1981,** *103,* 2899. (c) Olmstead, M. M.; Cheng, R.-J.; Balch, A. L. *Inorg. Chem.* **1982,** *21,* 4143. (d) Mansuy, D.; Battioni, J.-P.; Dupre, D.; Sartori, E. *J. Am. Chem. Soc.* **1982,** *104,* 6159. (e) Tatsumi, K.; Hoffman, R. *Inorg. Chem.* **1981,** *20,* 3771.
15. Chin, D.-H.; Gaudio, J. D.; La Mar, G. N.; Balch, A. L. *J. Am. Chem. Soc.* **1977,** *99,* 5486.
16. (a) Chin, D. H.; Balch, A. L.; La Mar, G. N. *J. Am. Chem. Soc.* **1980,** *102,* 1446. (b) Balch, A. L.; Chan, Y.-W.; Cheng, R.-J.; La Mar, G. N.; Latos-Grazynski, L.; Renner, M. W. *J. Am. Chem. Soc.* **1984,** *106,* 7779.
17. Chin, D.-H.; La Mar, G. N.; Balch, A. L. *J. Am. Chem. Soc.* **1980,** *102,* 5945.
18. Balch, A. L.; Latos-Grazynski, L.; Renner, M. W. *J. Am. Chem. Soc.* **1985,** *107,* 2983.
19. Groves, J. T.; Haushalter, R. C.; Nakamura, M.; Nemo, T. E.; Evans, B. J. *J. Am. Chem. Soc.* **1982,** *103,* 2884.
20. Groves, J. T.; Quinn, R.; McMurry, T. S.; Nakamura, M.; Lang, G.; Boso, B. *J. Am. Chem. Soc.* **1985,** *107,* 354.
21. Schappacher, M.; Weiss, R.; Montiel-Montoya, R.; Trautwein, A. *J. Am. Chem. Soc.* **1985,** *107,* 3736.
22. Khenkin, A. M.; Shteinman, A. A. *J. Chem. Soc., Chem. Commun.* **1984,** 1219.
23. Felton, R. H.; Owen, G. S.; Dolphin, D.; Fajer, J. *J. Am. Chem. Soc.* **1971,** *93,* 6332.

24. Felton, R. H.; Owen, G. S.; Dolphin, D.; Forman, A.; Borg, D. C.; Fajer, J. *Ann. N.Y. Acad. Sci.* **1973,** *206,* 504.
25. Phillippi, M. A.; Shimomura, E. T.; Goff, H. M. *Inorg. Chem.* **1981,** *20,* 1322.
26. (a) Lee, W. A.; Calderwood, T. S.; Bruice, T. C. *Proc. Natl. Acad. Sci. U.S.A.* **1985,** *82,* 4301.
27. Calderwood, T. S.; Lee, W. A.; Bruice, T. C. *J. Am. Chem. Soc.* **1985,** *107,* 7198.
28. Gelb, M. H.; Toscano, W. A., Jr.; Sligar, S. G. *Proc. Natl. Acad. Sci. U.S.A.* **1982,** *79,* 5758.
29. Loach, P. A.; Calvin, M. *Biochemistry* **1963,** *2,* 361.
30. (a) Carnieri, N.; Harriman, A.; Porter, G. *J. Chem. Soc., Dalton Trans.* **1982,** 931. (b) Carnieri, N.; Harriman, A.; Porter, G. ibid. 1231.
31. Camenzind, M. J.; Hollander, F. J.; Hill, C. L. *Inorg. Chem.* **1982,** *21,* 4301.
32. Bortolini, O.; Ricci, M.; Meunier, B.; Friant, P.; Ascone, I.; Goulon, J. *Nouv. J. Chim.* **1986,** *10,* 39.
33. (a) Nakagaki, P. C.; Calderwood, T. S.; Bruice, T. C. (b) Spreer, L.; Maliyackel, A. C.; Holbrook, S.; Otvos, J. W.; Calvin, M. *J. Am. Chem. Soc.* **1986,** *108,* 1949.
34. Groves, J. T.; Watanabe, Y.; McMurry, T. J. *J. Am. Chem. Soc.* **1983,** *105,* 4489.
35. (a) Smegal, J. A.; Hill, C. L. *J. Am. Chem. Soc.* **1983,** *105,* 2920. (b) Smegal, J. A.; Schardt, B. C.; Hill, C. L. ibid. **1983,** *105,* 3510. (c) Smegal, J. A.; Hill, C. L. ibid. **1983,** *105,* 3515.
36. Bartlett, P. D.; Ruchardt, C. *J. Am. Chem. Soc.* **1960,** *82,* 1756.
37. Creager, Stephen E.; Murray, R. W. *Inorg. Chem.* **1985,** *24,* 3844 and references cited.
38. (a) Yuan, L.-C.; Calderwood, T. S.; Bruice, T. C. *J. Am. Chem. Soc.* **1985,** *107,* 8273. (b) Creger, S. E.; Murray, R. W. *Inorg. Chem.* **1984,** *24,* 3824.
39. Yuan, L.-C.; Bruice, T. C. *J. Am. Chem. Soc.* **1985,** *107,* 512.
40. (a) Yuan, L.-C.; Bruice, T. C. *Inorg. Chem.* **1985,** *24,* 986. (b) Yuan, L.-C.; Bruice, T. C. *J. Am. Chem. Soc.* **1986,** *108,* 1643.
41. Quinn, R.; Nappa, M.; Valentine, J. S. *J. Am. Chem. Soc.* **1982,** *104,* 2588.
42. Lee, W. A.; Bruice, T. C. *J. Am. Chem. Soc.* **1985,** *107,* 513.
43. Kadlubar, F. F.; Morton, K. C.; Ziegler, D. M. *Biochem. Biophys. Res. Commun.* **1973,** *54,* 1255.
44. Norblum, G. D.; White, R. E.; Coon, M. J. *Arch. Biochem. Biophys.* **1976,** *175,* 523.
45. Blake, R. C.; Coon, M. J. *J. Biol. Chem.* **1980,** *255,* 4100.
46. Blake, R. C.; Coon, M. J. ibid. **1981,** *256,* 12127.
47. McCarthy, M. B.; White, R. E. ibid. **1983,** *258,* 9153.
48. McCarthy, M. B.; White, R. E. ibid. **1983,** *258,* 11610.
49. (a) Mansuy, D.; Bartoli, J.-F.; Chortland, J.-C.; Lang, M. *Angew. Chem.* **1980,** *19,* 909. (b) Mansuy, D.; Bartoli, J. F.; Momenteau, *Tetrahedron Lett.* **1982,** *27,* 2781.
50. Balasubramanian, P. N.; Bruice, T. C. *Proc. Natl. Acad. Sci. U.S.A.* **1987,** *84,* 1734.
51. Lee, W. A.; Bruice, T. C. *Inorg. Chem.* **1986,** *25,* 131.
52. Groves, J. T.; Kruper, W. J., Jr.; Haushalter, R. C.; Butler, W. M. *Inorg. Chem.* **1982,** *21,* 1363. (b) Creager, S. E.; Murray, R. W. *Inorg. Chem.* **1985,** *24,* 3824.
53. Lindsay-Smith, J. R.; Mortimer, D. N. *J. Chem. Soc., Chem. Commun.* **1985,** 64.
54. Mansuy, D.; Bartoli, J.-F.; Momenteau, M. *Tetrahedron Lett.* **1982,** *23,* 2731.
55. von Euler, H.; Josephson, K. *J. Justus Liebigs Ann. Chem.* **1927,** *456,* 111.
56. (a) Kremer, M. L. *Trans. Farad. Soc.* **1965,** *61,* 1453. (b) Kremer, M. L. ibid. **1967,** *63,* 1208. (c) Gatt, R.; Kremer, M. L. ibid. **1968,** *64,* 721.
57. Brown, S. B.; Dean, T. C.; Jones, P. *Biochem. J.* **1970,** *117,* 741.
58. Portsmouth, D.; Beal, A. E. *Eur. J. Biochem.* **1971,** *19,* 479.
59. (a) Jones, P.; Robson, T.; Brown, S. B. *Biochem. J.* **1973,** *135,* 353. (b) Jones, P.; Prudhoe, K.; Robson, T.; Kelly, H. C. *Biochemistry* **1974,** *13,* 4279. (c) Kelly, H. C.; Davies, D. M.; King, M. J.; Jones, P. *Biochemistry* **1977,** *16,* 3543. (d) Jones, P.; Mantle, D.; Davies, D. M.; Kelly, H. C. *Biochemistry* **1971,** *16,* 3974. (e) Jones, P.; Mantle, D.; Wilson, I. *J. Chem. Soc., Dalton Trans.* **1983,** 161.

60. Hatzikonstantinou, H.; Brown, S. B. *Biochem. J.* **1978**, *174*, 893.
61. Traylor, T. G.; Lee, W. A.; Stynes, D. V. *J. Am. Chem. Soc.* **1984**, *106*, 755.
62. (a) Bruice, T. C.; Zipplies, M. F.; Lee, W. A. *Proc. Natl. Acad. Sci. U.S.A.* **1986**, *83*, 4646. (b) Zipplies, M. F.; Lee, W. A.; Bruice, T. C. *J. Am. Chem. Soc.* **1986**, *108*, 4433.
63. (a) Puetter, J.; Becker, R. In "Methods of Enzymatic Analysis", Vol. 3, 3rd ed.; Bergmeyer, H. U., Ed.; Verlag Chemie: Weinheim, **1983**, p. 289.
64. Balasubramanian, P. N.; Bruice, T. C. *J. Am. Chem. Soc.*, **1987**, *109*, 7865.
65. Harriman, A. *J. Chem. Soc., Dalton Trans.* **1984**, 141.
66. Groves, J. T.; McClusky, G. A. *J. Am. Chem. Soc.* **1976**, *98*, 859.
67. Groves, J. T.; Akinbot, O. F.; Avaria, G. E. In "Miscrosomes, Drug Oxidation and Chemical Carcinogenesis"; Coon, M. J.; Conney, A. H.; Estabrook, R. W.; Gelboin, H. V.; Gillette, J. R.; and O'Brien, P. J., Eds.; Academic Press: New York, **1980**, p. 253.
68. (a) Wiberg, K. B.; Foster, G. *J. Am. Chem. Soc.* **1961**, *83*, 423. (b) Brauman, J. I.; Pandell, A. J. ibid. **1970**, *92*, 329. (c) Wiberg, K. B. In "Oxidation in Organic Chemistry"; Wiberg, K. B., Ed; Academic Press: New York, **1965**, p. 69. (d) Lorland, J. P. *J. Am. Chem. Soc.* **1974**, *96*, 2867.
69. (a) Gelb, M. H.; Heimbrook, D. C.; Malkonen, P.; Sligar, S. G. *Biochemistry* **1982**, *21*, 370. (b) Ebie, K. S.; Dawson, J. H. *J. Biol. Chem.* **1984**, *259*, 14389.
70. (a) Orme-Johnson, N. R.; Light, D. R.; White-Stevens, R. W.; Orme-Johnson, W. H. *J. Biol. Chem.* **1979**, *254*, 2103. (b) Nakajin, S.; Shinoda, M.; Haniu, M.; Shively, J. E.; Hall, P. F. *J. Biol. Chem.* **1984**, *259*, 3971.
71. Shapiro, S.; Piper, J. U.; Caspi, E. *J. Am. Chem. Soc.* **1982**, *104*, 2301.
72. Groves, J. T.; McClusky, G. A.; White, R. E.; Coon, M. J. *Biochem. Biophys. Res. Commun.* **1978**, *81*, 154.
73. White, R. E.; Miller, J. P.; Favreau, L. V.; Bhattacharya, A. *J. Am. Chem. Soc.* **1986**, *108*, 6024.
74. Lindsay-Smith, J. R.; Sleath, P. R. *J. Chem. Soc., Perkin Trans. 2* **1983**, 1165. Estabrook, R. W.; Gelboin, H. V.; Gillette, J. R.; O'Brien, P. J., Eds; Academic Press: N.Y., **1980**, p. 253.
75. (a) Winstein, S.; Trifan, D. S. *J. Am. Chem. Soc.* **1949**, *71*, 2953. (b) Roberts, J. D.; Lee, C. C. *J. Am. Chem. Soc.* **1951**, *73*, 5009. (c) Roberts, J. D.; Lee, C. C.; Saunders, W. H. Jr.; *J. Am. Chem. Soc.* **1954**, *76*, 4501.
76. (a) Foster, A. B.; Jarman, M.; Stevens, J. D.; Thomas, P. Westwood, J. H. *Chem. Biol. Interact.* **1974**, *9*, 327. (b) Hjelmeland, L. M.; Arnow, L.; Trudell, J. R. *Biochem. Biophys. Res. Commun.* **1977**, *76*, 541.
77. Lindsay-Smith, J. R.; Sleath, P. R. *J. Chem. Soc., Perking Trans. 2* **1983**, 621.
78. Stearns, R. A.; Ortiz de Montellano, P. R. *J. Am. Chem. Soc.* **1985**, *107*, 4081.
79. Hill, C. L.; Schardt, B. C. *J. Am. Chem. Soc.* **1980**, *102*, 6375.
80. Renson, J.; Weissbach, H.; Udenfriend, S. *Mol. Pharmacol.* **1965**, *1*, 145.
81. (a) Mitoma, C.; Yasuda, D. M.; Tagg, J.; Tanabe, M. *Biochim. Biophys. Acta* **1967**, *136*, 566. (b) Watanabe, Y.; Oae, S.; Iyanagi, T. *Bull. Chem. Soc. Jpn.* **1982**, *55*, 188.
82. Eckert, T. S.; Bruice, T. C. *J. Am. Chem. Soc.* **1983**, *105*, 4431.
83. Suschitzky, H.; Sellers, C. F. *Tetrahedron Lett.* **1969**, 1105.
84. Lewis, F. D.; Ho, T.-I.; Simpson, J. T. *J. Am. Chem. Soc.* **1982**, *104*, 1924.
85. Davis, G. T.; Demek, M. M.; Rosenblatt, D. H. *J. Am. Chem. Soc.* **1972**, *94*, 3321.
86. Audeh, C. A.; Smith, J. R. L. *J. Chem. Soc. B* **1971**, 1741.
87. Galliani, G.; Rindone, B.; Beltrame, P. L. *J. Chem. Soc., Perkin Trans. 2* **1978**, 1803.
88. (a) Miwa, G. T.; Walsh, J. S.; Kedderis, G. L.; Hollenberg, P. F. *J. Biol. Chem.* **1983**, *258*, 14445. (b) Hollenberg, P. F.; Miwa, G. T.; Walsh, J. S.; Dwyer, L. A.; Rickert, D. E.; Kedderis, G. L. *Drug Metab. Dispos.* **1985**, *13*, 272.
89. Wei, M. M.; Stewart, R. *J. Am. Chem. Soc.* **1966**, *88*, 1974.
90. Hull, L. A.; Davis, G. T.; Rosenblatt, D. H.; Williams, H. K. R.; Weglein, R. C. *J. Am. Chem. Soc.* **1967**, *89*, 1163.
91. Rosenblatt, D. H.; Davis, G. T.; Hull, L. A.; Forberg, G. D. *J. Org. Chem.* **1968**, *33*, 1649.
92. Lindsay-Smith, J. R.; Mead, L. A. V. *J. Chem. Soc., Perkin Trans. 2* **1973**, 206.

93. Lindsay-Smith, J. R.; Mortimer, D. N. *J. Chem. Soc., Chem. Commun.* **1985,** 64.
94. Ortiz de Montellano, P. R.; Beilan, H. S.; Kunze, K. L. *J. Biol. Chem.* **1982,** *256,* 6708. Augusto, O.; Beilan, H. S.; Ortiz de Montellano, P. R. *J. Biol. Chem.* **1982,** *257,* 11288.
95. (a) Sinha, A.; Bruice, T. C. *J. Am. Chem. Soc.* **1984,** *106,* 7291. (b) Carlson, B. W.; Miller, L. L.; Neta, P.; Grodkowski, J. *J. Am. Chem. Soc.* **1984,** *106,* 7233.
96. Hamill, S.; Cooper, D. Y. *Xenobiotica* **1984,** *14,* 139.
97. Shannon, P.; Bruice, T. C. *J. Am. Chem. Soc.* **1981,** *103,* 4580.
98. Dicken, C. M.; Lu, F.-L.; Nee, M. W.; Bruice, T. C. *J. Am. Chem. Soc.* **1985,** *107,* 5776.
99. Heimbrook, D. C.; Murray, R. I.; Egeberg, K. D.; Sligar, S. G.; Nee, M. W.; Bruice, T. C. *J. Am. Chem. Soc.* **1984,** *16,* 1514.
100. Burka, L. T.; Guengerich, F. P.; Willard, R. J.; Macdonald, T. L. *J. Am. Chem. Soc.* **1985,** *107,* 2549.
101. Dicken, C. M.; Lu, F.-L.; Bruice, T. C. *Tetrahedron Lett.* **1986,** *27,* 5967.
102. Woon, T. C.; Dicken, C. M.; Bruice, T. C. *J. Am. Chem. Soc.* **1986,** *108,* 7990.
103. Dicken, C. M.; Woon, T. C.; Bruice, T. C. ibid. **1986,** *108,* 1636.
104. Ostovic, D.; Knobler, C. B.; Bruice, T. C. ibid. **1987,** *109,* 3444.
105. Almog, J.; Baldwin, J. E.; Crossley, M. J.; Debernardis, J. F.; Dyer, R. L.; Huff, J. R.; Peters, M. K. *Tetrahedron* **1981,** *37,* 3589.
106. Powell, M. F.; Pai, E. F.; Bruice, T. C. *J. Am. Chem. Soc.* **1984,** *106,* 3277.
107. Wong, W.-H.; Ostovic, D.; Bruice, T. C. *J. Am. Chem. Soc.* **1987,** *109,* 3428.
108. De Poorter, B.; Meunier, B.; *Nouv. J. Chim.* **1985,** *9,* 393.
109. Battioni, P.; Renaud, J.-P.; Bartoli, J. F.; Mansuy, D. *J. Chem. Soc., Chem. Commun.* **1986,** 341.
110. (a) Groves, J. T.; Nemo, T. E.; Richard, S. M. *J. Am. Chem. Soc.* **1979,** *101,* 1032. (b) Groves, J. T.; Kruper, W. J., Jr.; Haushalter, R. C. ibid. **1980,** *102,* 6375. (c) Groves, J. T.; Kruper, W. J., Jr.; Nemo, T. F.; Myers, R. S. *J. Mol. Catal.* **1980,** *7,* 169. (d) Groves, J. T.; Nemo, T. E. *J. Am. Chem. Soc.* **1983,** *105,* 5786. (e) Groves, J. T.; Myers, R. S. ibid. **1983,** *105,* 5791.
111. Lindsay-Smith, R. J.; Sleath, P. R. *J. Chem. Soc. Perkins Trans. 2* **1982,** 1009.
112. (a) Fontecave, M.; Mansuy, D. *J. Chem. Soc., Chem. Commun.* **1984,** 879. (b) Mansuy, D.; Leclaire, J.; Fontecave, M.; Dansette, P. *Tetrahedron* **1984,** *40,* 2847.
113. (a) Mashiko, T.; Dolphin, D.; Nakano, T.; Traylor, T. G. *J. Am. Chem. Soc.* **1985,** *107,* 3735. (b) Traylor, T. G.; Marsters, J. C., Jr.; Nakano, T.; Dunlap, B. E. ibid. **1985,** *107,* 5537. (c) Traylor, T. G.; Nakano, T.; Dunlap, B. E.; Traylor, P. S.; Dolphin, D. ibid. **1986,** *108,* 2782. (d) Traylor, T. G.; Nakano, T.; Mikszal, A. R.; Dunlap, B. E. *J. Am. Chem. Soc.* **1987,** *109,* 3625. (e) Traylor, T. G.; Mikszal, A. R. *J. Am. Chem. Soc.* **1987,** *109,* 2770.
114. Collman, J. P.; Kodadek, T.; Raybuck, S. A.; Brauman, J. I.; Papazian, L. M. *J. Am. Chem. Soc.* **1985,** *107,* 4343.
115. (a) Guilmet, E.; Meunier, B. *Tetrahedron Lett.* **1980,** *21,* 4449. (b) Guilmet, E.; Meunier, B. ibid. **1982,** *23,* 2449. (c) Guilmet, E.; Meunier, B. *Nouv. J. Chim.* **1982,** *6,* 511. (d) Guilmet, E.; Meunier, B. *J. Mol. Catal.* **1984,** *23,* 115. (e) Meunier, B.; Guilmet, E.; De Cavalho, M.-E.; Poilblanc, R. *J. Am. Chem. Soc.* **1984,** *106,* 6668.
116. (a) Collman, J. P.; Kodadek, T.; Raybuck, S. A.; Meunier, B. *Proc. Natl. Acad. Sci. U.S.A.* **1983,** *80,* 7039. (b) Collman, J. P.; Brauman, J. I.; Meunier, B.; Raybuck, S. A.; Kodadek, T. ibid. **1984,** *81,* 3245. (c) Collman, J. P.; Brauman, J. P.; Meunier, B.; Hayashi, T.; Kodadek, T.; Raybuck, S. A. *J. Am. Chem. Soc.* **1985,** *107,* 2000. (d) Collman, J. P.; Kodadek, T.; Brauman, J. I. ibid. **1986,** *108,* 2588.
117. (a) Van Der Made, A. W.; Nolte, R. J. M. *J. Mol. Catal.* **1984,** *26,* 333. (b) Razenberg, J. A. S. J.; Nolte, R. J. M.; Drenth, W. *Tetrahedron Lett.* **1984,** *25,* 789. (c) Razenberg, J. A. S. J.; Nolte, R. J. M.; Drenth, W. *J. Chem. Soc., Chem. Commun.* **1986,** 277.
118. (a) De Poorter, B.; Meunier, B. *J. Chem. Soc., Perkin Trans. 2* **1985,** 1735. (b) De Poorter, B.; Ricci, M.; Meunier, B. *Tetrahedron Lett.* **1985,** *26,* 4459, (c) Bortolini, O.; Meunier, B. *J. Chem. Soc., Perkin Trans. 2* **1984,** 1967. (d) Meunier, B.; Guilmet, E.; De Carvalho, M.-E.; Poilblanc, R. *J. Am. Chem. Soc.* **1984,** *106,* 6668.

119. (a) Mansuy, D.; Battioni, P.; Renaud, J.-P. *J. Chem. Soc., Chem. Commun.* **1984,** 1255. (b) Renaud, J.-P.; Battioni, P.; Bartoli, J. F.; Mansuy, D. *J. Chem. Soc., Chem. Commun.* **1985,** 888. (c) Battioni, P.; Renaud, J.-P.; Bartoli, J. F.; Mansuy, D. *J. Chem. Soc., Chem. Commun.* **1986,** 341.

120. Balasubramanian, P. N.; Sinha, A.; Bruice, T. C. *J. Am. Chem. Soc.* **1987,** *109,* 1456.

121. Bruice, T. C.; Dicken, C. M.; Balasubramanian, P. N.; Woon, T. C. and Lu, F.-L. *J. Am. Chem. Soc.* **1987,** *109,* 3436.

122. Yuan, L.-C.; Bruice, T. C. *J. Chem. Soc., Chem. Commun.* **1985,** 868.

123. (a) Tabushi, I.; Yazaki, A. *J. Am. Chem. Soc.* **1981,** *103,* 7371. (b) Khenkin, A. M.; Shteinman, A. A. *J. Chem. Soc., Chem. Commun.* **1984,** 1219. (c) Fauvet, M. P.; Gaudemer, A. *J. Chem. Soc., Chem. Commun.* **1981,** 874.

124. Jorgensen, K. A. *J. Am. Chem. Soc.* **1987,** *109,* 698.

125. Groves, J. T.; Watanabe, Y. *J. Am. Chem. Soc.* **1986,** *108,* 7836.

126. Bell, R.; Morgan, K. J. *J. Chem. Soc.* **1960,** 1209.

127. Chang, C. K.; Evina, F. *J. Chem. Soc., Chem. Commun.* **1981,** 118.

128. Montanari, F.; Penso, M.; Quici, S.; Vigano, P. *J. Org. Chem.* **1985,** 4888.

129. Rodgers, K. R.; Goff, H. M. *J. Am. Chem. Soc.* **1987,** *109,* 611.

130. Guengerich, F. P.; Strickland, T. W. *Mol. Pharmacol.* **1977,** *13,* 993.

131. (a) Ortiz de Montellano, P. R. O.; Mico, B. A.; Yost, G. S. *Biochem. Biophys. Res. Commun.* **1978,** *83,* 132. (b) Ortiz de Montellano, P. R. O.; Yost, G. S.; Mico, B. A.; Dinizo, S. F.; Correia, M. A.; Kumbara, H. *Arch. Biochem. Biophys.* **1979,** *197,* 524. (c) Ortiz de Montellano, P. R. O.; Kunze, K. L.; Mico, B. A. *Mol. Pharmacol.* **1980,** *18,* 602. (d) Ortiz de Montelano, P. R. O.; Beilan, H. S.; Kunze, K. L.; Mico, B. A. *J. Biol. Chem.* **1981,** *256,* 4395. (e) Ortiz de Montelano, P. R. O.; Kunze, K. L.; Beilan, H. S.; Wheeler, C. *Biochemistry* **1982,** *21,* 1331. (f) Ortiz de Montellano, P. R. O.; Mico, B. A. *Arch. Biochem. Biophys.* **1981,** *206,* 4325.

132. Miller, R. E.; Guengerich, F. P. *Biochemistry* **1982,** *21,* 1090.

133. (a) Mansuy, D.; Lange, M.; Chottard, J. C.; Bartoli, J. F.; Chevrier, B.; Weiss, R. *Angew. Chem.* **1978,** *17,* 781. (b) Lando, D.; Manassen, J.; Hodes, G.; Cahen, D. *J. Am. Chem. Soc.* **1979,** *101,* 3971. (c) Pohl, L. R.; George, J. W. *Biochem. Biophys. Res. Commun.* **1983,** *117,* 367.

134. (a) Ortiz de Montellano, P. R. O.; Mico, B. A. *Mol. Pharmacol.* **1980,** *18,* 128. (b) Kunze, K. L.; Mangold, B. L. K.; Wheeler, C.; Beilan, H. S.; Ortiz de Montellano, P. R. O. *J. Biol. Chem.* **1983,** *258,* 4202. (c) Kunze, K. L.; Mangold, B. L. K.; Wheeler, C.; Beilan, H. S.; Ortiz de Montellano, P. R. O. ibid. 4208.

135. Artaud, I.; Devocelle, L.; Battioni, J.-P.; Girault, J.-P.; Mansuy, D. *J. Am. Chem. Soc.* **1987,** *109,* 3782.

136. Collman, J. P.; Hampton, P. D.; Brauman, J. I. *J. Am. Chem. Soc.* **1986,** *108,* 7861.

137. Loosemore, M. J.; Wogan, G. N.; Walsh, C. *J. Biol. Chem.* **1981,** *256,* 8705.

138. Sharpless, K. B.; Teranishi, A. Y.; Backvall, J.-E. *J. Am. Chem. Soc.* **1977,** *99,* 3120.

139. Groves, J. T.; Watanabe, Y. *J. Am. Chem. Soc.* **1985,** *108,* 507.

140. Groves, T. G.; Stern, M. K. *J. Am. Chem. Soc.* **1987,** *109,* 3812.

141. (a) Castellino, A. J.; Bruice, T. C. *J. Am. Chem. Soc.* **1988,** *110,* 158 (b) Nakagaki, P. C.; Lee, R. W.; Balasubramanian, P. N.; Bruice, T. C. *Proc. Natl. Acad. Sci. U.S.A.* **1988,** *85,* 641.

142. Samsel, E. G.; Scrinivasan, K.; Kochi, J. K. *J. Am. Chem. Soc.* **1985,** *107,* 7606.

143. (a) Rice, F. O.; Teller, E. *J. Chem. Phys.* **1938,** *6,* 489. (b) Hine, J. *Adv. Phys. Org. Chem.* **1977,** *15,* 1.

144. Castellino, A. J.; Bruice, T. C. *J. Am. Chem. Soc.* **1988,** *110,* 1313.

145. (a) Mathew, L.; Warkentin, J. *J. Am. Chem. Soc.* **1986,** *108,* 7981. (b) Maillard, B.; Forrest, D.; Ingold, K. U. *J. Am. Chem. Soc.* **1976,** *98,* 7024.

146. (a) Zimmerman, H. E.; Rieke, R. D.; Scheffer, J. R. *J. Am. Chem. Soc.* **1967,** *89,* 2033. (b) Zimmerman, H. E.; Hancock, K. G.; Licke, G. C. *J. Am. Chem. Soc.* **1968,** *90,* 4892. (c) Godet, J.-Y.; Pereyre, M. *J. Organomet. Chem.* **1972,** *40,* C23. (d) Castaing, M.; Pereyre, M.; Tatier, M.; Blum, P. M.; Davies, A. G. *J. Chem. Soc., Perkins Trans. 2* **1979,** 589.

CHAPTER 7

Porphyrin Metalation Reactions in Biochemistry

David K. Lavallee

City University of New York, New York, New York

CONTENTS

1. Introduction: Biological Porphyrin Metabolism 279
2. Metalation of Porphyrins in Vitro 280
3. Formation of Metal Complexes of Macrocycles in Vitro 286
4. Ferrochelatase Inhibition and Chlorophyll Biosynthesis 307
5. Pathological Conditions Associated with Ferrochelatase. 309
6. Conclusion. 311
Acknowledgments 311
References 311

1. INTRODUCTION: BIOLOGICAL PORPHYRIN METALATION

The iron and magnesium complexes of porphyrins and closely related chlorin ligands are ubiquitous in animals and plants. Among their many functions are oxygen transport, catalysis of oxygen incorporation in organic molecules, electron transport, and photosynthesis. In the biosynthesis of hemes, the protoporphyrin IX ring is synthesized and the metal subsequently added.

279

Formation of hemes such as the c-type cytochromes and heme a occurs from protoheme. Synthesis of chlorophyll in plants occurs if magnesium is added to protoporphyrin rather than iron. In the case of the iron porphyrins, there is considerable evidence for essential participation of an enzyme called ferrochelatase, but no corresponding enzyme has been characterized for the introduction of magnesium into the chlorins and bacteriochlorins to form chlorophylls and bacteriochlorophylls. A third type of macrocyclic ring complex is that of the corrinoids, as found in vitamin B_{12}. In the biosynthesis of these compounds, which proceeds from uroporphyrinogen III as in the cases of hemes and chlorophylls, the cobalt atom is inserted at some stage after initial ring formation and perhaps not until ring contraction has occurred. It is not known whether an enzyme is involved. Therefore, it may be expected that some aspects of porphyrin metalation reactions observed in vitro would relate directly to biochemical metalation, but in the case of heme formation via ferrochelatase there may be interesting differences.

Substances that interfere with the metalation step of heme synthesis (including lead and numerous substances that destroy cytochrome P-450 in the liver) result in pathological symptoms that may be clinically evident because of the accumulation of uncomplexed protoporphyrin IX or its derivatives (lead intoxication, porphyrias, and malfunction of cytochromes may be detected by zinc protoporphyrin, free porphyrins, and "green pigments," respectively). This chapter summarizes general aspects of the mechanism of porphyrin metalation in vitro and in vivo. Emphasis is given to reactions of ferrochelatase, including its inhibition and role in certain pathologies.

2. METALATION OF PORPHYRINS IN VITRO[1]

A. General Features

One of the first concerns in attempting to evaluate biochemical porphyrin metalation reactions is the rate-determining step(s) of the mechanism and any special aspects of the reaction medium that promote or inhibit metalation reactions. A general mechanism for porphyrin metalation based on that first proposed by Hambright and Chock[2] and further supported by work of the groups of Hambright,[3] Lavallee,[4] and Tanaka[5] is given in Scheme I (H_2P represents a general porphyrin).

For nonenzymatic or enzymatic reactions, important considerations are the reactivities of different oxidation states of the metal atom [Fe(II) or Fe(III), Co(II) or Co(III)], the dependence of reactivity on the ligands about the metal atom (for aqua complexes and complexes with water and other ligands in the coordination sphere, then, pH is important), the reversibility of the reaction, and the dependence of the overall rate of the concentrations of metal ion and macrocycle. Important features of this mechanism for a consideration of enzymatic activity involve ways in which the typical rate-

Deformation
$$H_2P \rightleftharpoons H_2P^*$$

Outer-sphere complexation
$$ML_6 + H_2P^* \rightleftharpoons [L_6M, H_2P^*]^{2+}$$

Ligand dissociation and first metal–nitrogen bond formation
$$[ML_6, H_2P^*]^{2+} \rightleftharpoons [ML_5\text{---}L\text{---}H_2P]^{2+} \rightleftharpoons L_5M\text{---}H_2P^{2+} + L$$

Second metal–nitrogen bond formation
$$L_5M\text{---}H_2P^{2+} \rightleftharpoons L_nM\text{=\!=}H_2P^{2+} + (5\text{-}n)L$$

Metalloporphyrin formation (proton release)
$$L_nM\text{=\!=}H_2P^{2+} \rightleftharpoons MP + nL + 2H^+$$

Scheme I

determining step(s) can be accelerated: distortion of the planar porphyrin ring, alteration of ligand dissociation from the metal atom, enhancement of outer-sphere complex formation, and/or promotion of loss of protons from the porphyrin ring in the final step.

a. Porphyrin Ring Distortion

There are several experiments that bear on the question of porphyrin ring distortion. The most direct experiment was that of Pasternack, Sutin, and Turner, who used temperature-jump spectroscopy to determine the lifetime of a thermally excited form of a water-soluble porphyrin with a low tendency to self-associate [tetrakis(N-methyl-4-pyridyl)porphyrin].[6] They concluded that the absorbance change induced by the temperature jump was due to distortion of the porphyrin ring. From kinetic data for the formation and decay of the distorted form, the equilibrium constant is 0.04, representing a free energy difference of about 8 kJ/mol (2 kcal/mol) between the planar and distorted states. This is the only experimental value for a distortion energy available and, of course, the extent or nature of the ring distortion is not revealed by this experiment.

Porphyrin ring distortions are also found in some crystal structures.[7] The extent of the distortion, a twisting of the pyrrole rings about the methine bridge carbon, is most evident for synthetic *meso*-tetraaryl porphyrins. (See Figure 7-1 for numbering and nomenclature of porphyrins.) This distortion probably allows the molecules to pack together more efficiently and allows *meso*-aryl groups to avoid unfavorable contacts. It is difficult to confidently estimate the energies involved.

Less direct, but nonetheless interesting, information about the effect of ring distortion comes from studies of N-alkylporphyrins. These porphyrins have a permanently distorted ring structure that exposes nitrogen lone pair electrons to an incoming metal ion. The comparisons of rates of metalation

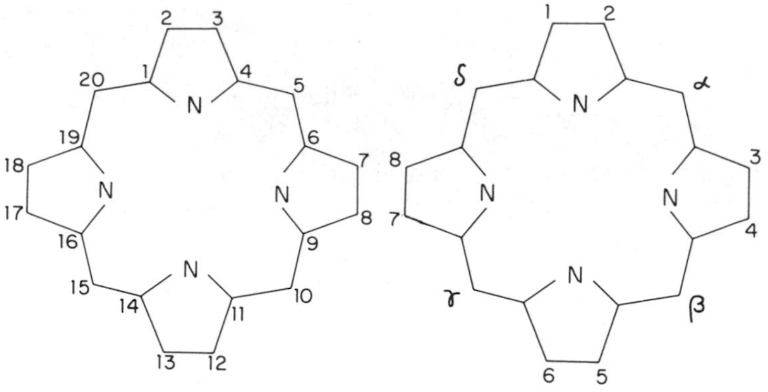

Figure 7-1. Common numbering schemes for porphyrins. The numbering of the ring on the left is consistent with IUPAC convention and is becoming common. It is most often used for synthetic porphyrins, such as those with aryl substituents at positions 5, 10, 15, and 20—the methine bridge or meso positions. The numbering of the ring at the right is very common for naturally derived porphyrins which have substituents on the pyrrolenine rings (the numbered positions) and hydrogen atoms at the positions with Greek designations. A possible source of confusion is that atropoisomers of porphyrins with nonsymmetric substituents which may be on the same or opposite sides of the porphyrin plane have Greek designations to distinguish their positions (α for "up" and β for "down").

of N-alkylporphyrins and corresponding planar porphyrins have shown that the distorted porphyrins bind metal ions 10^3 to 10^5 times more rapidly.[4,5a,5b,8] It is evident, then, that an enzyme could increase the rate of metalation by bending the porphyrin ring to expose the nitrogen lone pairs. The resonance energy of the porphyrin ring system is remarkably large—about 1 MJ (5 times that of benzene) for both natural and synthetic porphyrins.[9] What is not known is the degree to which this resonance energy would be sacrificed by bending a porphyrin to the degree that makes the N-alkylporphyrins so reactive. Certainly, these porphyrins are still highly aromatic (they have absorption maxima very near those of their planar analogs, but generally red shifted on the order of 10 nm; their extinction coefficients are as high or higher and the bond lengths throughout the porphyrin ring systems are similar for both meso- and non-meso-substituted N-alkylporphyrins).[10] Unfortunately, there are not yet any calculations of the resonance energy of the porphine ring system as a function of distortions from planarity to ascertain the probability of distortion by thermal energy.

b. Outer-Sphere Complexation

The parameters that determine the magnitude of the outer-sphere equilibrium constant (a multiplicative factor in the overall rate law and, therefore, an important rate determinant) include charge, size, and the polarity of the medium.[1] These factors are useful predictors for homogeneous reactions in solution, but are not as generally applicable to enzymatic reactions. A common feature of enzymatic reactions is enhanced outer-sphere association due to substrate binding at proximal sites. The enzyme catalyzes the reaction involving two substrates by accumulating them in close proximity to increase the probability of reaction. An example of this in vitro is the study of Buckingham and co-workers that showed marked rate enhancement for reactions of 5,10,15,20-$\alpha,\alpha,\alpha,\alpha$-tetrakis(o-N-maleeimidoanilio)porphine complexation of copper(II), zinc(II), cobalt(II), and nickel(II) relative to tetraphenylporphine. In that porphyrin, there are carboxylate groups all poised above one side of the coordination site (the designation α). The equilibrium constant for binding of copper(II) to the carboxylate site is 1.5 $\times$ 10^3 and rate constants for the reaction of the aqua complexes bound to the porphyrin (ie, the movement of the metal ion into the coordination site of the porphyrin from its position above that site where it is bound to the carboxylates) are 5.6/s for copper(II), 1.5/s for zinc(II), 0.1/s for cobalt(II), and 3 $\times$ 10^{-4}/s for nickel(II), where the reactions were run in 1:1 DMF/H_2O (DMF, dimethylformamide).[13] These rates are all much faster than those for tetraphenylporphyrin in DMF with the metal ions copper(II) (0.03 $M^{-1}s^{-1}$ at 25°C, cobalt(II) (4.8 $\times$ 10^{-4}/M/s at 44°C), and nickel(II) (3.6 $\times$ 10^{-3} $M^{-1}s^{-1}$ at 74°C).[14] For Buckingham's porphyrin, the preequilibrium in which the metal ion binds to the carboxylate groups replaces the typical outer-sphere complexation step: the more favorable equilibrium constant of this step increases the overall rate of reaction.

c. Metal–Ligand Dissociation

The third step shown in the general mechanism is the loss of a ligand from the coordination sphere of the metal ion. This step is not thought to be totally dissociative, but rather an assisted dissociation in which some bond formation with the incoming ligand takes place in the activated complex. The rates of many ligand replacement reactions are closely related to rates of the loss of the ligands in the coordination sphere and typically show only a slight dependence on the nature of the incoming ligand.[11]

Since it appears that ligand dissociation and porphyrin deformation generally require more time than that during which an outer-sphere complex between the initial metal complex and the porphyrin remains intact,[1a] one can view the overall process in homogeneous solution (ie, in vitro) as follows: A complex in which a ligand has already begun to detach and a porphyrin that is already substantially deformed collide and form an outer-sphere complex. If the geometry is correct for a lone pair of electrons from the porphyrin to occupy the vacant site provided by ligand dissociation, an initial metal–nitrogen bond can be formed. At this stage, complexation is reversible (as shown by extensive kinetics data for acid hydrolysis of metalloporphyrins indicating that a complex with only one metal–nitrogen bond dissociates readily).[1,12] If a second bond is formed by the combination of ligand dissociation and porphyrin deformation (the porphyrin deformation is

probably much easier at this point and may specifically involve the pyrrolenine ring trans to the one forming the initial bond), the reaction proceeds to
completion.

Enzymatic porphyrin metalation can be expected to proceed somewhat
differently. In this case, the close association of the metal atom and the
porphyrin in the active site can be long lived with respect to the time needed
for ligand dissociation. Thus, a process analogous to outer-sphere complexation occurs without requiring prior deformation or any ligand dissociation,
unlike the process that occurs in a homogeneous reaction.

Rapid ligand dissociation is favored by lower oxidation states (ferrous and
cobaltous rather than ferric and cobaltic), high-spin states (relevant to reactions of iron), high pH (coordinated hydroxide groups dissociate much faster
than water), and geometries that are not octahedral (all tetrahedral complexes with nonchelating ligands exchange ligands much more rapidly than
octahedral complexes with the same type of ligand).

d. The Final Step: Proton Loss for the Porphyrin Nitrogen Atoms

There has been only one study of the kinetic isotope effect on porphyrin
metalation in which effects due to the solvent have been ruled out.[15] The
results showed that the final step in the general mechanism is not rate determining because there was no difference in rate for a porphyrin with protons
or deuterons bound to the pyrroleninic nitrogen atoms ($H_2TPPS_4^{4-}$ and
$D_2TPPS_4^{4-}$). Certainly, additional studies of this type would be useful in
affirming the conclusion. The lack of an effect of proton dissociation on the
overall rate infers that general base catalysis by an enzyme directed toward
removal of the protons from the porphyrin coordination site would not be
useful in increasing the complexation rate. The presence of basic residues in
the active site could prevent the reverse reaction, acidic demetalation, however. This would be most important for metalloporphyrins that are easily
demetalated, such as the magnesium and zinc complexes.

e. Activation Parameters of the Overall Reaction

An interesting and relatively straightforward activation parameter to interpret mechanistically is the volume of activation. As long as the rate-determining step is the same for the series of reactions, a positive volume of
activation indicates a reaction in which the process of formation of the
activated complex is dissociative while a negative activation volume is characteristic of an associative process. Tanaka has found that the volume of
activation is positive for metalloporphyrin formation for a series of metal
ions (Table 7-1) and has attributed this result to the ligand dissociation from
the metal.[5c] Although these results would also be consistent with proton
dissociation in the final step, in that case the values would be expected to be
more similar. Proton dissociation as the rate-determining step would also
conflict with studies which show either that more basic porphyrins (which

TABLE 7-1. Activation Parameters for Metalloporphyrin Formation

Metal ion	$\Delta H^{\ddagger}$ (kJ/mol)	$\Delta S^{\ddagger}$ (J/K·mol)	$\Delta G^{\ddagger}$ (298 K) (kJ/mol)	$\Delta V^{\ddagger}$ cm³/mol
Tetraphenylporphine[a]				
Cu(II)	68	−41	85	
Zn(II)	64	−76	90	8.9[b]
Co(II)	104	21	96	8.0[b]
Mn(II)	—	—	—	12.9[b]
Ni(II)	97	2.5	97	7.0[b]
Tetrakis(4-N-methylpyridyl)porphine[c]				
Cu(II)	64	−32	74	
Zn(II)	64	−62	83	
Co(II)	85	−15	90	
Ni(II)	51	−180	105	
N-methyltetraphenylporphine[d]				
Cu(II)	70	40	58	
Zn(II)	59	−28	68	
Co(II)	85	36	74	
Mn(II)	90	19	85	
Ni(II)	90	−13	94	

[a] As the chloride salt, in dimethylformamide (giving mixed, chloro/DMF precursor complexes; Reference 14).

[b] As the nitrate salts in dimethylformamide (giving mixed complexes or, in some cases, hexadimethylformamide species; Reference 5a).

[c] In aqueous buffer at pH 4.0 except for the copper(II), pH 2.1 (Reference 2).

[d] As perchlorate salts of the hexakis(dimethylformamide) complexes, in dimethylformamide (Reference 4).

might be expected to hold their protons more firmly) react faster[16] or that basicity has little effect.[1b]

It is typical that the relative rates of metalation for a series of metals, such as those which have been reported for reactions of *meso*-tetraphenylporphyrin[14] and N-methyl-*meso*-tetraphenylporphyrin[4,5a] in dimethylformamide, and *meso*-tetrakis(4-methylpyridyl)porphine in aqueous solution[2] parallel those of the solvent exchange reactions of the metal ions. Some care must be exercised in interpreting both relative rates and activation parameters because there is some tendency for them to show an isokinetic effect—compensation of more unfavorable enthalpies of activation by more favorable entropies of activation.[1,17] The values of the activation free energies for reactions of different metals with the nonplanar N-methyl-*meso*-tetraphenylporphyrin relative to planar *meso*-tetraphenylporphyrin in dimethylformamide are about 20 kJ/mol more favorable for those metal ions that exchange ligands rapidly, but the values are similar for slowly exchanging nickel(II). There are no data for comparison of reactions of iron(II), but the similarity of its solvent exchange rate to that of cobalt(II) [the rate of water exchange at 25°C is about 3 times faster for iron(II) than cobalt(II)][18] could indicate that the ring distortion would be a distinct advantage with regard to the activation free energy. Distortion of the ring, which leads to an increase in rate of 10^3 to 10^5 at 25°C,[4,5a,5b] is analogous to a change in a rapid preequilibrium step rather than a lowering of the activation energy of the slowest step in reaction. Because the overall reaction is affected by equilibria and rates of several steps, the overall activation enthalpies and entropies actually reflect temperature effects on all of these processes. The most useful general result with respect to these activation parameters is their consistency.

B. Conclusions from Studies of Metalation in Vitro

From these general results, then, the means that an enzyme could use to facilitate porphyrin metalation would be to (1) deform the porphyrin ring, (2) bind the metal with residues that can dissociate readily, (3) increase the degree of formation of the outer-sphere complex, and (4) keep the complex in which initial bond formation has occurred intact so that formation of the second (and subsequent) bonds can take place. Additional aspects that bear on both enzymatic and nonenzymatic metalation are that (1) overall rates will be sensitive to the dissociation rate of the ligands about the metal, which, in turn, generally depend on the oxidation state of the metal ion [being much slower for Co(III) than Co(II) and considerably slower for either high-spin or low-spin complexes of Fe(III) than for high-spin complexes of Fe(II)]; (2) the overall rates depend on metal ion and porphyrin concentration (in straightforward cases where the metal ion exists as a single, monometallic complex and the porphyrin is not self-associated, the overall reaction is first order in each concentration); and (3) the overall reaction is reversible is two ways—other ligands can compete at the stage where only a single metal–porphyrinic nitrogen bond has been formed, and acid can cause dissociation of an intact metalloporphyrin. In addition, acid can prevent the entire reaction from the start because only the free base form of the porphyrin, with just two of the four pyrroleninic nitrogen atoms protonated, is reactive. Thus, the pH is limited to the region where the porphyrin is in the free base form (above 5) and where the metal ion forms soluble complexes.

3. FORMATION OF METAL COMPLEXES OF MACROCYCLES IN VIVO

A. Chlorophylls

No strong case has been made for requirement of an enzyme in chlorophyll metalation. In fact, there is a significant body of evidence that the magnesium atom is introduced not into a chlorin ring but into protoporphyrin IX, and that it is this magnesium complex that undergoes several modifications of the porphyrin ring to form chlorophylls.[19] The ligand exchange reactions of magnesium are relatively fast.[18] Since magnesium is a nontransition element, without ligand field stabilization energy, complexation in the strong ligand field of the constraining, planar protoporphyrin IX ring system does not impart the great stability to the magnesium complex that it does to transition metal complexes such as those of iron. Magnesium is readily removed from the porphyrins by traces of acid or by competing ligands, as chlorophylls readily lose magnesium to form pheophytins when chloroplasts are ruptured. At low acid concentrations, the protoporphyrin IX complex loses magnesium more slowly than the chlorophylls and they, in turn, are more stable than the bacteriochlorophylls.[1,12] Since concentrations of magnesium in plants are relatively high and its ligand dissociation reaction is fast, there would be little to gain by intermediacy of an enzyme. It is most important that the magnesium complex, once formed, be protected from an acidic medium or good competing ligands, however.

B. Cobalamins

Although the metal ions nickel(II), palladium(II), and zinc(II) were used as templates for the laboratory synthesis of vitamin B_{12},[20] there is no evidence that cobalt acts as a template in vivo. In fact, although the path for metalation in vivo is not clear, there is strong circumstantial evidence to suggest that several modifications of aromatic porphyrinlike rings (isobacteriochlorins) occur before cobalt is incorporated, and it is even possible that ring contraction to form a corrin ring occurs before the cobalt atoms is introduced.[21] (This work has been based on the isolation of macrocyclic pigment intermediates in the vitamin B_{12} biosynthetic pathway obtained from *Propionibacterium shermanii* grown in the absence of cobalt.[21a]) The only organisms that synthesize the corrinoids are bacteria. Toohey and co-workers have analyzed a number of strains of the photosynthetic bacteria *Chromatium* which synthesize metal-free corrins.[22] Toohey attempted to incorporate cobalt(II), nickel(II), iron(II), and manganese(II) into the five metal-free corrinoids he had isolated, but only cobalt(II) reacted at a reasonable rate (10 h at 22°C, apparently for completion of the reaction—no kinetic data are available) under his reaction conditions (0.01 M metal salt, 0.2 M NH_4OH).[22b] The basic conditions were probably necessary because the pK_a of the metal-free corrinoids at about 9 is due to conversion of a positively charged corrinoid ring (which has protons on two of the pyrrolinenic nitrogen atoms) to the neutral form (which has one protonated pyrrolenine ring; the structural assignments have been made from the characteristics of the changes in the visible absorption spectrum that occur[23]). Since it has been found for metalation reactions of porphyrins that cationic, protonated forms are unreactive, it is reasonable that alkaline conditions would be necessary for corrinoid metalation in vitro. The unreactivity of ferric ion under alkaline conditions is understandable because of its insolubility. Manganese(II) and nickel(II) react with porphyrins more slowly than cobalt(II)[1] (Table 7-1), and their reactivity is paralleled, then, for the metal-free corrinoids. From the porphyrin work, copper(II) and zinc(II) would be expected to react more rapidly, but copper(II) forms an insoluble ammine complex under Toohey's conditions and he did not report an attempt to incorporate zinc(II).

Koppenhagen and co-workers and Rubison have formed complexes between zinc, copper, and rhodium with metal-free corrinoids obtained from *Chromatium* cultured in the presence of added 5,6-dimethylbenziimidazole.[23] Zinc complexes are obtained by briefly boiling a solution of the metal-free corrinoid with zinc acetate,[23b,c] while rhodium insertion requires the use of a low-valent carbonyl complex in an organic solvent.[23d] Rubison showed that copper(II) is easily inserted—after only a few minutes in aqueous solution at 60°C.[23e] A zinc(II) complex of a synthetic corrinoid was formed in a template reaction, and the resistance of the resulting complexes to demetalation has been studied.[24] Only zinc(II) has been removed from a corrinoid without destroying the macrocyclic ring system (it is removed under relatively mild conditions, trifluoroacetic acid in acetonitrile[24b]). The high pH required for cobalt(II) incorporation in vitro and the relatively slow rate [if the reaction is first order in cobalt(II), as expected for a macrocycle metalation, it would be exceedingly slow at the concentrations of cobalt(II) present in the *Chromatium* medium, 10 μg/L or 0.2 μM][22a] would appear to make unassisted cobalt incorporation into corrinoids in bacterial cytoplasm unlikely. From corresponding reactions of porphyrins, zinc ion at a particular concentration would be expected to react faster with metal-free corrinoids than cobalt (the relative rates cannot be inferred directly from the experiments of by Koppenhagen and co-workers[23b,c]). Considering the fact that the concentration of zinc(II) in cytoplasm is greater than that of cobalt, it would be expected that Zn(II) would handily "out compete" Co(II) for incorporation into a metal-free corrin if there is no specially mediated metalation mechanism. Replacement of Zn(II), once incorporated into the macrocyclic ring, by Co(II) would be exceedingly slow if the Zn(II) had to first dissociate. The only possibility, which hardly seems feasible, is one in which complexation of Zn(II) actually accelerates incorporation of a second metal ion [the second metal ion presumably attacks from the side of the porphyrin plane opposite the zinc(II) ion and displaces it], in analogy to the studies in which Hambright and Tanaka and their co-workers have found, in reactions of cadmium(II), mercury(II), and lead(II), complexes of synthetic porphyrins.[25] No experiments of this type have been reported using natural metal-free corrins. The isolation of metal-free corrins has been performed by so few investigators that experiments with metal-free corrin have been very limited.

C. Heme Formation in Vivo

a. Investigations That Implicated Enzymatic Metalation

Extensive information about the role of ferrochelatase (protoheme ferro-lyase, EC 4.99.1.1) in heme formation has been obtained from experiments involving membrane extracts from duck, chicken, cow, pig, sheep, human, yeasts, nonphotosynthetic and photosynthetic bacteria, barley, and spinach.[26] In eukaryotes, the enzyme is bound to the inner mitochrondrial membrane,[27,28] while in prokaryotes it is bound to the cytoplasmic membrane.[26j] The enzyme has been purified to homogeneity from chicken, rat, cow, mouse, and a photosynthetic bacterium,[29] allowing investigators to study both structural and kinetic aspects.

In the enzyme preparations, only ferrous iron leads to heme formation, which has led to the hypothesis that an iron-reducing site is present on the inner mitochrondrial membrane.[30] However, Tokunaga and co-workers have shown that if there is a suitable reducing agent present, such as NAD(P)H and FMN or FAD, ferric chloride and mesoporphyrin are converted to mesoheme at the same rate as when ferrous ion is used.[31] They have also shown that conversion of ferric ion to heme occurs with ferrochelatase and NADH dehydrogenase from complex I of bovine heart mitochondria when NADH is added.[28] Thus, a soluble component of the inner mitochondrial compartment could maintain sufficient ferrous ion in equilibrium with ferric iron to permit heme formation via ferrochelatase without a spatially anchored proximal reducing site. From experiments with rat liver mitochondria, Tangeras has proposed that there is a pool of available iron in the inner mitochrondrial compartment (1 nmol/mg of protein) that can be used by ferrochelatase for heme formation.[26e] Using both exogenous deuteroporphyrin and protoporphyrin (100 μM; iron from the available pool, 10 μM; protein at 10 mg/mL), he found a rate of 0.3 nmol of heme formation per hour, an amount about 5 times that necessary for the turnover of hemoproteins in hepatocytes.[32]

b. Comparisons with the Nonenzymatic Reaction

One of the troublesome difficulties with in vitro comparisons of heme formation that attempt to mimic the conditions in vitro, but without the use of the enzyme ferrochelatase, is the low solubility of protoporphyrin IX in neutral, buffered aqueous solution. Second, attempts to determine whether iron(III) instead of iron(II) can bind at a rate comparable to biological reactions are hampered because simple complexes of iron(III) are insoluble (or form highly aggregated complexes that may be colloidal) at neutral pH. Unfortunately, protoporphyrin IX is unreactive toward metalation when it is in the diprotonated form—the form present in the acidic pH range where iron(III) is soluble. Attempts to use ligands that solubilize iron(III) are flawed because the overall rate strongly depends on the ligand dissociation rate and, hence, on the identity of the ligands about the metal atom—legitimate comparisons should involve similar ligands.

The results from four studies have been used to cast doubt on the need for catalytic assistance for the incorporation of iron into protoporphyrin in vivo. In a study by Granick and Mauzerall on porphyrin biosynthesis in erythrocytes,[33] the following statement was made: "However, under conditions whereby PROTO is not colloidal at neutral pH and 38°C, ferrous iron will coordinate to form heme in the absence of enzymes." No experimental details or date were presented to support the statement. Kassner and Wang reported the reaction of iron(II) with several porphyrins quite similar to protoporphyrin IX but more soluble in water: hematoporphyrin, uroporphyrin, and coproporphyrin.[34] They found that the rates (obtained in 10% aqueous pyridine solutions) increased with the number of pendant carboxylate groups, perhaps because these groups could themselves coordinate iron and create a greater local concentration of iron. Of the three reactions, the rate constant of the reaction of hematoporphyrin IX, $1.5/M^{-1}s^{-1}$ at 40°C should be closest to that for protoporphyrin IX. The activation free energies were the same (100 kJ/mol) within normal experimental uncertainty for such determinations, showing the typical isokinetic compensation of enthalpy and entropy.

In a related study, Kassner and Walchak also investigated the reaction of iron(II) and naturally derived porphyrins in aqueous solution and stated that the rate of protoheme formation achieved, 19 nmol/h at 40°C [using 0.05 mM protoporphyrin IX, 0.10 mM iron(II), and 5 mM glutathione], was equivalent to rates claimed for enzyme preparations. They showed that the reactions without any cellular material exhibit a pH maximum (not unexpected based on other studies of porphyrin metalation in vitro[3b]) near that of the purported enzyme preparations,[35] casting doubt on the existence of ferrochelatase. A significant difference between the cell-material-free reactions and the reports of ferrochelatase activity by other authors is that porphyrins such as coproporphyrin and uroporphyrin are more reactive than protoporphyrin IX in the former system but they are inactive in the enzyme assay. The results of Kassner and Walchak are also inconsistent with the general finding that heating deactivates the purported enzyme: If the reaction occurs without an enzyme, preheating should have no effect. In a later study by Tokunaga and Sano,[36] the statement was made that "It was observed that Fe^{2+} was well incorporated into the protoporphyrin ring in the enzymic system, but that rate and yield of protoheme formation were not higher than that of the nonenzymic system." In this work, the nonenzymatic reaction took place in a colloidal suspension using fifty times more iron(II) than was used for the enzymatic reaction. Although the authors did not perform experiments as a function of metal ion concentration, as would be expected in a kinetic study to determine the effect of metal ion and enzyme concentrations, one would assume that this reaction would, like other metalloporphyrin formation reactions, be dependent on iron(II) concentration. Thus, a valid comparison with the iron in the same concentration for both enzymatic and nonenzymatic reactions would probably show a considerable difference in rate.

Another consideration with this study is the extent to which pure ferrochelatase had been obtained. From the data that 67 nmol of heme was formed after 2 h of incubation with 5.7 mg of protein at 37°C [with 200 nmol of iron(II)],[36] the formation rate of heme would be 0.01 nmol/min/mg of enzyme. In more recent papers by Tokunaga and co-workers, the purification procedure for ferrochelatase appears to be markedly improved.[29b] In an assay for ferrochelatase activity they have used 0.5 μmol of a ferrous salt that is added under a nitrogen atmosphere to 1 mL of a solution that was 0.05 mM in mesoporphyrin, 1 mM in dithiothreitol, and 0.1 M in Tris HCl buffer at pH 8.0. With these concentrations, the heme synthesis rate at 37°C reported for ferrochelatase isolated from rat liver mitochondria is 58 nmol/min/mg of ferrochelatase. Using this value and the concentrations of iron(II) employed in the enzymatic assay (0.5 mM) and the nonenzymatic reaction previously reported by Tokunaga and Sano[36] (1.2 mM), the differences in rate constants for the enzymatic reaction with the more highly purified enzyme and the nonenzymatic reaction is a factor of 1.4×10^4. [It should be noted that some current studies employ a different assay reaction for ferrochelatase, using the rate of the reaction of cobalt(II) with mesoporphyrin IX.[37] Neither of these substrates is present at appreciable concentrations in tissue avoiding interference by endogenous materials.] Purification of ferrochelatase by a factor of 1000- to about 2000-fold (using the assay rate as the criterion) are typical.[29]

Strong evidence for the role of ferrochelatase is the fact that heat-denatured mitochrondrial preparations are unable to use iron(II) to form heme at 37°C[26e] and that the conversion of free base porphyrins (protoporphyrin, deuteroporphyrin, or mesoporphyrin) to metalloporphyrins [of iron(II), cobalt(II), or nickel(II)] is linearly dependent on the amount of the purified protein that is proposed to be the enzyme. In addition, antibodies to the enzyme bovine ferrochelatase have been shown to significantly inhibit the metal incorporation reaction in vitro.[38] Dailey and Lascelles have also reported isolation of a bacterial mutant that lacks ferrochelatase activity and has an absolute requirement for heme.[26j] Inhibition of heme formation in the presence of N-alkylporphyrins (vide infra) also demonstrates enzyme activity because N-alkylporphyrins added to solutions of protoporphyrin do not affect the rate of metalloporphyrin formation in the absence of the protein that is proposed to be ferrochelatase.[39]

There is no doubt whatsoever that ferrochelatase acts as a powerful catalyst for formation of metalloporphyrins from cobalt(II), since this ion is very soluble and well behaved in aqueous solution and many studies have shown that metalloporphyrin formation reactions of cobalt(II) are much slower in the absence of the enzyme. The case with iron has been less clear because the few relevant studies of the nonenzymatic reaction and the Kassner's studies both indicate that iron(II) reacts at with porphyrins in the absence of enzyme at rates that are similar to those found in some systems that have been proposed to contain an enzyme. The role of the enzyme in the case of iron(II) may involve, in fact, the utilization of the ferric ion pool in a way that is not feasible in the absence of the enzyme. This possibility will be explored further when the mechanistic propositions based on the kinetic studies of Dailey and Flemming[29b] and Camardo and co-workers[26g] are discussed.

c. Properties of Ferrochelatase

In early attempts to isolate ferrochelatase, little catalytic activity was retained. Since ferrochelatase is tightly bound to the inner mitochondrial membrane in eukaryotes and to the cytoplasmic membrane of bacteria, detergent is required. When the detergent was removed in later stages of the procedures, the activity of the enzyme was lost. The first procedure that substantially improved the separation of ferrochelatase from other mitochondrial membrane proteins was reported by Mailer et al. in 1980.[40] They purified ferrochelatase from rat liver using 2-vinyl-4-[3′-(N-3″-aminopropyl)]acrylamidodeuteroporphyrin on an affinity column and found that 2-mercaptoethanol was required to preserve activity (and, at that, for only a 12 h half-life at 4°C).[40] Using Blue Sepharose CL-6B chromatography in the isolation procedure, Tokunaga and co-workers obtained rat liver ferrochelatase with a molecular weight (about 40,000) that appears to be more consistent with values obtained for other purified ferrochelatase enzymes than Mailer's value (about 60,000).[29b] Blue Sepharose CL-6B (with tandem Red Sepharose CL-6) has also been used to purify the enzyme from bovine liver.[29c] To purify the enzyme from the bacterium *Rhodopseudomonas sphaeroides,* separation by DEAE-Sephacrel was followed by absorption on an Amicon Blue B column.[29e] The purified rat liver enzyme contains considerable amounts of fatty acids and is enhanced markedly by addition of exogenous lipid, whereas the purified bovine liver enzyme contains neither fatty acids nor phospholipids and is enhanced much less by addition of lipid.[29b] Ferrochelatase does not appear to require cofactors. Thus, if care is taken to use the same amount and type of lipid in the test medium, results from different preparations of ferrochelatase should produce comparable results. The molecular weights for all but the *P. sphaeroides* are about 40,000 to 50,000, with the bacterial enzyme at about 115,000.[29e]

d. Implications for the Mechanism of Heme Formation by Ferrochelatase from Kinetics Studies

Independent Binding of Iron and Protoporphyrin IX. Binding constants (as K_m values) for protoporphyrin and iron, respectively, obtained from kinetics measurements for a ferrochelatase from a variety of sources are shown in Table 7-2. The binding constants are independent for the two substrates.

TABLE 7-2. Michaelis Constants for Ferrochelatase Preparations

Source	K_m, Iron (μM)	K_m, Protoporphyrin IX (μM)	Reference
Bovine liver[a]	11	80	29c
Chicken erythrocytes[b]	166	83	29a
Duck erythrocytes[c]	70	80	26a
Human liver[d]	0.5	0.35	26h
Ovine liver[e]	0.8		26g
Spirrillum itersonii[a]	47	20	26j
Rhodopseudomonas sphaeroides[2]	18	50	29d

[a] Conditions: 0.05 mM Protoporphyrin IX, 0.05 μCi, 0.02 mM [^{59}Fe] ferric citrate, 0.5 mM sodium succinate, 35 mM Tris HCl, pH 7.6.

[b] Conditions: 0.2 mM Ferrous ammonium citrate, 0.1 mM deuteroporphyrin, 0.050 M Tris HCl, pH 8.1, 5 mM dithiothreitol, 0.2% Triton X-100.

[c] Conditions: 0.6 mM Protoporphyrin IX, 0.6 mM ferric chloride, 1.2 mM cysteine, 60 mM Tris buffer (acid unspecified), pH 8.0.

[d] Conditions: 0.020 mM Protoporphyrin IX, 0.02 mM ferrous sulphate, 40 mM glutathione, 0.10 M Tris acetate (pH unspecified).

[e] Conditions: according to the original description of this procedure (R. J. Porra and O. T. G. Jones, *Biochem. J.* **1963,** 181), there would be 50 mM protoporphyrin IX and 100 mM cobaltous chloride in the assay mixture. More likely, the amounts were nanomoles rather than micromoles, giving 0.050 mM porphyrin and 0.10 mM cobalt chloride. Other constituents: 10 mM glutathione, 100 mM Tris HCl pH 7.2, 0.1% (v/v) Tween 80.

Porphyrins other than protoporphyrin IX and metals other than iron have been studied. Variations in the K_m value for sheep liver ferrochelatase as a function of the substituents on the periphery of the porphyrin are given in Table 7-3, while those with purified bovine liver ferrochelatase are given in Table 7-4 and those for purified murine hepatic ferrochelatase are given in

TABLE 7-3. Michaelis Constants For Several Porphyrins with Sheep Liver Ferrochelatase[a]

Porphyrin	K_m (μM)	V_{max} (nmol/min/mg)
Deuteroporphyrin IX	4.0	1.4
Mesoporphyrin IX	1.9	1.0
Protoporphyrin IX	0.8	0.11
Protoporphyrin XIII	1.4	0.25
5,6-Dimethyl-6,7-dipropionylporphine	5.0	2.8
6,8-Dimethyl-5,7-dipropionylporphine	0.3	0.4
6,7-Dimethyl-5,8-dipropionylporphine	—	0
Mesoporphyrin I	—	0
Protoporphyrin I	—	0
Rhodoporphyrin XV	—	0
G-Phylloporphyrin XV	—	0
1,4,5-Triethyl-2,3,6,8-tetramethyl-7-propionylporphine	—	0

[a] Reference 26f. Note that the nomenclature is that used in the original reference. It does not conform to IUPAC conventions. Conditions: 0.100 M Tris HCl, pH 7.2, 0.1% (v/v) Tween 80, 37°C. For these assays, Co^{2+} was used instead of Fe^{2+}

DAVID K. LAVALLEE

TABLE 7-4. Kinetic Parameters For Several Porphyrins with Bovine Liver Ferrochelatase

Porphyrin[a]	K_m, Iron(II)[b] (μM)	K_m, Porphyrin[b] (μM)	Specific activity[b] (nmol heme/mg/min)
Protoporphyrin IX	46	64	88
Mesoporphyrin IX	44	46	413
Deuteroporphyrin IX		36	224
Hematoporphyrin		55	99
2,4-Diacetyldeuteroporphyrin		479	128
2,4-Diformyldeuteroporphyrin		—	0
Porphyrin c		—	3
Coproporphyrin		—	0

Porphyrin	A and B ring substituents (positions 2 and 4)	K_m, Porphyrin[c] (μM)	K_i, Porphyrin[c] (μM)
Protoporphyrin IX	$CH\!=\!CH_2$	11	—
Mesoporphyrin IX	CH_2CH_3	34	—
Deuteroporphyrin IX	H	47	—
Hematoporphyrin IX	$CH(OH)CH_3$	22	—
Monohydroxyethylmonovinyl-deuteroporphyrin	$CH\!=\!CH_2, CH(OH)CH_3$	23	—
O,O'-Diacetylhematoporphyrin	$CH(OAc)CH_3$	—	11
2,4-Bis(acetal)deuteroporphyrin	$CH_2CH(OCH_3)_2$	—	13
2,4-Bis(glycol)deuteroporphyrin	$CH(OH)CH_2OH$	—	67
2,4-Disulphonicdeuteroporphyrin	SO_3^-	—	70

[a] The nomenclature is that used in the original references and does not conform to IUPAC conventions. For example, propionyl rather than 2-carboxyethyl is used for $-CH_2CH_2COOH$ and acetal rather than 2,2-dimethoxyethyl is used for $-CH_2CH(OCH_3)_2$.

[b] Reference 29b. Conditions: 0.10 mM ferric citrate, 0.050 mM porphyrin, 0.100 M Tris HCl, pH 8.0, 1 mM dithiothreitol, 0.5% Triton X-100.

[c] Reference 26m. Conditions: 0.2 mM ferrous ammonium citrate, 0.1 mM porphyrin, 0.050 M Tris HCl, pH 8.1, 5 mM dithiothreitol, 0.2% Triton X-100.

Table 7-5. A variation in iron binding for different workers can be expected because there is a competitive equilibrium involving different media (for example, in the case of the relatively poor binding of iron for the chicken erythrocyte enzyme, the iron source was the relative strongly ligated iron citrate[29a]). In contrast, the Michaelis constants for protoporphyrin IX in Table 7-2 are very consistent with the exception of the human hepatic enzyme. The sensitivity of the enzyme to the substituents on the porphyrin periphery, shown for ovine, bovine, and murine enzymes in Tables 7-3 to 7-5, show that the rings of protoporphyrin IX that bear the vinyl groups (at positions 2 and 4 of the A and B rings) can be substituted with only relatively small alkyl groups and retain strong binding. The data from the sheep enzyme (Table 7-3) demonstrate that it is important that the propionic side chains be on adjacent rings in the proper positions (designated 6 and 7 by conventional nomenclature), since protoporphyrin I and mesoporphyrin I

TABLE 7-5.   Michaelis and Inhibition Constants for Several Porphyrins and
Metalloporphyrins with Murine Liver Ferrochelatase[a]

Porphyrin	K_m (μM)	Metalloporphyrin	K_i (μM)
Protoporphyrin IX	9	Iron (II) protoporphyrin IX	2.0
Hematoporphyrin IX	9	Cobalt(II) protoporphyrin IX	3.0
Monovinyl, monohydroxyethyl-		Zinc(II) protoporphyrin	4.5
deuteroporphyrin IX	10	Tin(IV) protoporphyrin IX	13
Monovinyl, monohydroxymethyl-			
deuteroporphyrin IX	62		
Mesoporphyrin	156		
Deuteroporphyrin IX	247		

[a] H. A. Dailey, private communication; Dailey, H. A., Jones, C., and Karr, S. W., manuscript in preparation. Conditions (as in 29d): 0.1 mM porphyrin, 0.050 M Tris acetate, pH 8.1, 0.2 mM ferrous ammonium sulfate (as ^{59}Fe for some assays), 5 mM dithiothreitol, 0.2% Triton X-100.

are not substrates. The murine enzyme (Table 7-5) is especially sensitive to the nature of the substituents on the A and B rings, and there appears to be some preference exhibited for the vinyl groups. It is interesting to note how strongly the metal complexes of protoporphyrin IX inhibit this enzyme.

Inhibition by Other Metal Ions. The ability of ferrochelatase enzymes to incorporate metals other than iron into porphyrins and the inhibition of heme formation by metals other than iron have been studied by several workers. The results (Table 7-6) demonstrate (1) that zinc, cobalt, and nickel can be incorporated, (2) that manganese, magnesium, and calcium as well as lead and mercury are competitive inhibitors but are not readily incorporated, and (3) that copper is neither incorporated nor inhibitory. Those metals which bind well (iron, zinc, and cobalt) all have a tendency to form tetrahedral complexes, and tetrahedral complexes are typically very labile,[11] which would allow for rapid heme formation. Magnesium, calcium, and manganese may inhibit by inducing structural changes in the protein that are not due to binding of the metal in the active site. Lead and mercury protoporphyrin complexes are weak. They do tend to bind strongly to sulfhydryl groups, which may be the reason for their inhibitory properties. It is important to emphasize that the inhibitory effect of a metal, like its Michaelis constant, can depend strongly on the medium employed for the assay.

The Michaelis constants previously discussed indicate that one role of ferrochelatase, possibly the most important one, is its ability to effectively concentrate the two substrates in close proximity. The outer-sphere complexation constants for divalent metal ion complexes and porphyrins approach a value of about 3 at the high ionic strength of the cytoplasm.[1c] With K_m values in the μM range for both iron(II) and protoporphyrin, the effective outer-sphere complexation constants is much higher. If we consider the

TABLE 7-6.   Inhibition of Ferrochelatase-Assisted Heme Formation
by Metals Other Than Iron

Source of enzyme	Metal ion	Concentration (μM)	Relative activity (%)
Bovine liver[a]	Nickel(II)	10	100
	Copper(II)	10	104
	Magnesium(II)	10	96
	Cadmium(II)	10	70
	Manganese(II)	10	44
	Lead(II)	10	37
	Mercury(II)	10	15
Human liver[b]	Zinc(II)	0.58	50
Rat liver[c]	Tin(II)	600	100
	Magnesium(II)	600	77
	Calcium(II)	600	73
	Cadmium(II)	600	60
	Nickel(II)	600	52
	Lead(II)	300	58
	Copper(II)	200	50
	Manganese(II)	50	47
	Zinc(II)	50	40
	Mercury(II)	50	37
Murine erythroleukemia cells[d]	Cobalt(II)	10	100
	Cobalt(II)	50	48
	Manganese(II)	15	56
	Cadmium(II)	8.6	64

[a] Purified enzyme, Reference 29c. Note that the nitrate salt of lead was used. Chloride ion in vivo could reduce the effect of lead. Conditions: 0.1 mM deuteroporphyrin IX, 5 mM dithiothreitol, 0.2% Triton X-100, 50 mM Tris acetate, pH 8.1, 37°C.

[b] Mitochondrial extract, Reference 26h. Conditions: 25 μM protoporphyrin IX, succinate (concentration after anerobiosis unspecified) 0.2–2 mg protein/mL, 0.1 M potassium phosphate, pH 7.6, 30°C.

[c] Mitochondrial extract, Reference 26e. Conditions: 0.02 mM protoporphyrin IX, 100 mM Tris, 40 mM ascorbic acid or glutathione, 38°C.

[d] Cells in culture media, Reference 41. Conditions as in footnote a above.

situation where the enzyme is saturated with one of the two substrates (which appear to exhibit K_m values independent of each other), the equilibrium constant for the second substrate would be in the range of 10^4 to 10^5 for bovine ferrochelatase [based on K_m values of 11 and 80 μM for iron (II) and porphyrin, respectively[29c]] and in the range of 10^6 to 10^7 for human ferrochelatase (based on K_m values of 0.5 and 0.35 μM[26g]) rather than 3.

The kinetic studies of Dailey and Fleming using highly purified bovine liver ferrochelatase[29b] and of Camardo, Ibraham, and Levere using human liver mitochrondrial preparations,[26h] and the pattern of inhibition of other metal ions strongly support an independent, bi-bi mechanism[29b] as shown in Scheme II (with the addition of the ferrous iron scavenging equilibrium).

$$\begin{array}{ccccc}
Fe^{3+} \rightleftharpoons Fe^{2+} & & PP\ IX & & heme \\
\updownarrow & & \updownarrow & & \updownarrow \\
E \rightleftharpoons E:Fe^{2+} & \rightleftharpoons & \left[\begin{array}{c} (E:Fe^{2+}:PP) \\ \updownarrow \\ (E:heme:2H^+) \end{array}\right] & \rightleftharpoons & E:2H^+ \rightleftharpoons E + 2H^+
\end{array}$$

Scheme II

In this mechanism, the role of ferrochelatase in promoting the outer-sphere complexation constant may be effective in the following way. If we consider that the primary form of iron in the cytoplasm is ferric but that there are labile oxidoreductant systems available (for example, NADH dehydrogenase and cofactors), the small amount of ferrous iron present at the effective electrical potential of the cytoplasm will be rapidly restored as it is depleted. By binding the available ferrous iron, ferrochelatase causes the reduction of additional ferric iron and a continual supply of ferrous iron is available. This interpretation is consistent with the kinetic studies cited previously and with the study of heme synthesis from ferric iron in the presence of the NADH dehydrogenase-rich fraction of rat liver mitochondria.[31]

e. Structural Studies of the Active Site of Ferrochelatase

Spectroscopic Studies. Since there have been no reports of crystallographic structural determinations for ferrochelatases, the most definitive information available is spectroscopic. Unfortunately, there are few spectroscopic data useful for defining the coordination geometry, but some interesting results from spectrofluorometric experiments on bovine ferrochelatase, reported by Dailey,[42] as well as results caused by chemical modifications of the enzyme provide at least a partial picture of the active site. Modification of sulfhydryl groups of either eukaryotic[43] or prokaryotic[44] ferrochelatases inactivates the enzyme. Since the presence of iron(II) protects the enzymes from deactivation, Dailey has proposed that the iron binding site contains the sulfhydryl residue. Modification of arginyl residues indicates that arginine is present at the protoporphyrin binding site. Arginine alteration does not affect the K_m for iron(II), so its function should be associated with the porphyrin. The spectrofluorometric results, along with the fact that two propionic acid side chains and relatively small 2,4-substituents are necessary for high porphyrin-binding affinity (Tables 7-3 to 7-5), indicate that the porphyrin is bound in a hydrophobic pocket and surrounded on all sides but one. Because good substrates have small 2,4-substituents, one end of the pocket is probably very restricted.[26b,42] Inhibition studies using N-alkylporphyrins (vide infra) also bear on the amount of space available in the pocket adjacent to the porphyrin site. The luminescence properties of tryptophan[42] in the bovine enzyme indicate that one or more tryptophan residues are present in the pocket and that they are exposed to water in the pocket but not to bulk water. The well-defined maximum in emission spectrum at 347 nm is evidence that no tyrosine was excited. Phenylalanine is not eliminated, however, since it has a considerably different excitation spectrum and much weaker emission intensity.

Inhibition for Ferrochelatase and the Structure of the Active Site of Metalation. Information about the nature of the active site of ferrochelatase has become available from inhibition studies. For about 30 years, it has been known that certain drugs lead to accumulations of iron-free protoporphyrin and that rats treated with these drugs accumulate significant amounts of "green pigments" in the liver. Comparisons of electronic spectra[45] and nuclear magnetic resonance (NMR) spectra[46] with synthetic compounds have demonstrated that these green pigments are *N*-alkylated protoporphyrin IX derivatives. Administration of *N*-alkylporphyrins to animals has been shown to result in the same symptoms as administration of the porphyria-inducing drugs,[47] and addition of *N*-alkylporphyrins to ferrochelatase (either isolated or in microsomal preparations) leads to marked inhibition.

The question is this: "Why are the *N*-alkylporphyrins such potent inhibitors of ferrochelatase?" That is, "Why are certain *N*-alkylprotoporphyrins stabilized more by binding to the protein than is protoporphyrin IX itself?" In the absence of complete structural characterization of the porphyrin binding site of the enzyme, there are at least two reasonable explanations.

The first concerns the basicity of N-alkylated porphyrins. It has been known since the very first report of an *N*-alkylporphyrin (*N*-methyletioporphyrin II, synthesized as a possible monoprotic titration indicator[48]) that they are about nine orders of magnitude stronger bases than the corresponding non-N-alkylated precursor.[49] (It should be noted that this great difference is found for porphyrins with hydrogen atoms at the methine bridge, or meso positions, the category of all naturally derived porphyrins, but not for the synthetic *meso*-tetraarylporphyrins.[50]) Perhaps there are acidic residues at the binding site that can interact favorably with the lone pairs of electrons of the pyrroleninic nitrogen atoms. More likely, the *N*-alkylprotoporphyrin binds to ferrochelatase as a monocation. This possibility is reasonable because the monocation is the major form of *N*-methylprotoporphyrin IX at physiological pH since its pK_3 is much higher than 7. Since the core of the porphyrin would bear a positive charge, the monocation could be stabilized by electron-donating residues near the active site. A residue such as lysine could act as the stabilizing electron donor, effectively sharing the additional proton at the core of the *N*-alkylprotoporphyrin monocation.

Another possibility that could occur with or without the charge stabilization is that the distorted pyrrolenine rings are in a better position to undergo favorable π-stacking interactions with aromatic residues near the active site. An attractive feature of this possibility is that the stabilization of an *N*-alkylporphyrin, in which the pyrrolenine rings are rotated with respect to the plane of the four pyrroleninic nitrogen atoms, mimics the activated complex needed for insertion of a metal atom into a non-N-alkylated porphyrin.

At the end of this section, data from studies of the structural dependence of inhibition will be combined with results from Dailey's spectrofluorometric study[42] to assemble plausible components for the active site of ferrochelatase.

Investigations of the Structural Dependence of Inhibition. The desire to learn how certain drugs functioned and what metabolites they provide led several workers to the conclusion that heme formation was being inhibited by a novel type of porphyrin that had previously only been known to exist as a synthetic product, the N-substituted porphyrins. At first, the major effort was to characterize these products and account for their formation—their overall structure and biological origin. Soon, however, De Matteis and co-workers could draw conclusions about the structure of ferrochelatase based on the differential reactivity of the N-substituted porphyrins derived from the attack of these drugs on cytochrome *P*-450 enzymes. When the first phase of structural tests was reported, the nature of the green pigment produced from DDC (3,5-diethoxycarbonyl-1,4-dihydro-2,4,6-trimethylpyridine) treatment had been deduced to be *N*-methylprotoporphyrin IX,[53] and it was postulated (later verified) that the product from treatment with ethylene gas was a two-carbon fragment.[54] De Matteis and co-workers investigated the inhibitory properties of *N*-methylmesoporphyrin IX in order to test the activity of a species derived from a good substrate for the enzyme (mesoporphyrin IX) that has a small N-substituent against the activity of *N*-ethylmesoporphyrin IX which has a two-carbon N-substituent (like those derived from inactive pigments produced by ethylene). They also used *N*-methylcoproporphyrin to test a derivative of a porphyrin that is found at elevated levels in porphyria victims.[53] At the same time, they treated rats with DDC and ethylene and extracted the green pigments from their livers for comparison with the synthetically produced porphyrins of established structure.

The major finding evident in Table 7-7 is that *N*-methylmesoporphyrin IX

TABLE 7-7. Inhibition of Ferrochelatase by *N*-Alkylmesoporphyrins and "Green Pigments" from DDC and Ethylene[a,b]

Inhibitor	Inhibitory activity (units/nm)
Green pigment from DDC treatment (*N*-methylprotoporphyrin IX)	20.2
N-Methylmesoporphyrin IX (synthetic, mixture of isomers)	11.4
N-Ethylmesoporphyrin IX (synthetic, mixture of isomers)	0.89
Green pigment from ethylene (*N*-2-hydroxyethylprotoporphyrin IX)	0.15
N-Methylcoproporphyrin	0.15

[a] Reference 53.

[b] nmol of mesoporphyrin converted to the cobalt(II) complex/min/mg of enzyme.

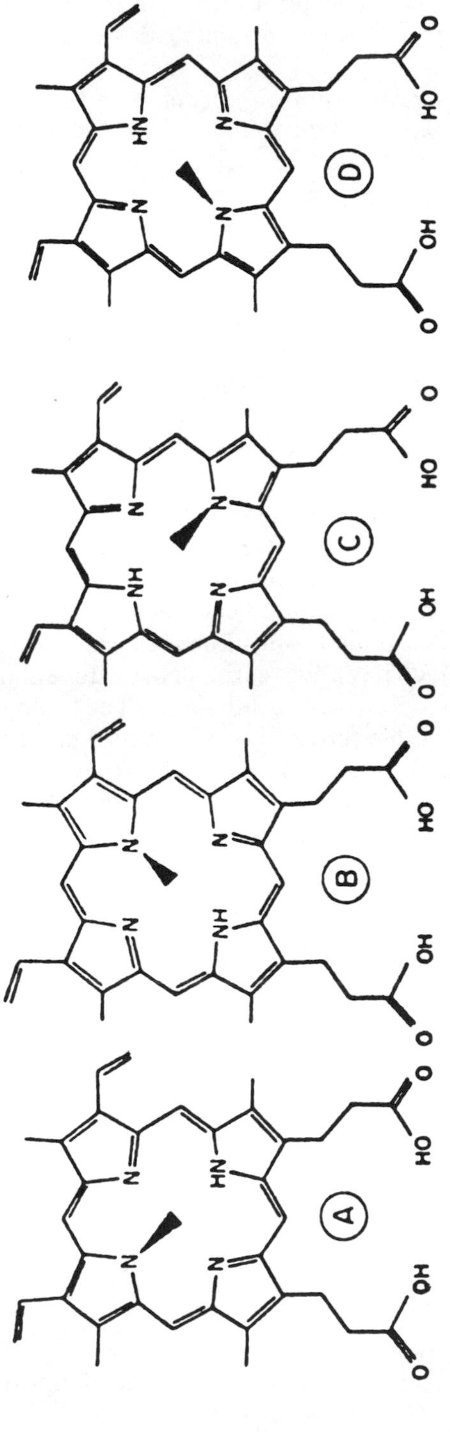

Figure 7-2. The four geometric isomers of N-alkylated protoporphyrin IX.

TABLE 7-8.   Ferrochelatase Inhibition by the Isomers of Free Base and Chlorozinc
N-Methylprotoporphyrin IX[a]

Compound tested	N-Methylated ring	Inhibitory activity (units/nmol)
N-Methylprotoporphyrin	A	10.6
	A and B	12.6
	D	9.6
	C and D	11.7
Chloro-N-methylprotoporphinatozinc(II)	A	12.8
	A and B	12.9
	D	2.1
	C and D	2.0

[a] Reference 59. Pure N_B and N_C isomers were apparently not isolated.

is strongly inhibitory, whereas the green pigment [now known to be N-(CH$_2$CH$_2$OH)protoporphyrin IX] extracted from the livers of animals treated with ethylene as well as N-ethylmesoporphyrin IX and N-methylcoproporphyrin were weaker inhibitors. It should be kept in mind that four structural isomers of N-alkylmesoporphyrin (or N-alkylprotoporphyrin) are produced in nearly equal amounts by in vitro chemical alkylation reactions and each of these exists as a racemic mixture. The isomeric distribution that is produced naturally, however, is instead weighted to particular isomers and enantiomers. The relative contributions of the isomers present in synthetic and naturally produced N-alkylporphyrins should, therefore, be taken into account when considering reported inhibition values.

De Matteis and co-workers separated the geometric isomers of N-methylprotoporphyrin IX and pooled the fractions corresponding to the isomers with the methyl groups on the A and B rings (the N_A and N_B isomers, Figure 7-2) and those of the C and D isomers (the N_C and N_D isomers). They also tested two of the individual isomers (N_A and N_D) and studied the inhibitory properties of the zinc complexes of corresponding isomers, obtaining the results shown in Table 7-8.[54] Interestingly, there was little difference among the isomers of the free base porphyrin, but a distinctly greater inhibition for the zinc complexes of the N_A and N_B isomers in comparison with the N_C and N_D isomers. A comparison of the structures of an N-methylporphyrin free base[55] and its corresponding zinc complex[56] shows difference in the angle of the N-substituted pyrrole with respect to the mean porphyrin plane (Figure 7-3) of 11° (27.7 vs 38.5°). A comparison of a monoprotonated N-alkylporphyrin (N-ethoxycarbonylmethyloctaethylporphyrin)[57] and its zinc complex[58] shows a much greater difference in this angle, 25° (19.1 vs 44.1°).

It is evident that the N-substituted pyrrole ring is titled from the mean porphyrin plane to a greater extent in the zinc complex than in either the free base or monoprotonated species. The greater cant of the pyrrole ring for the zinc complexes would indeed cause the propionic side chains of the N_C and N_D isomers to be separated to a greater extent than those of the N_A or N_B

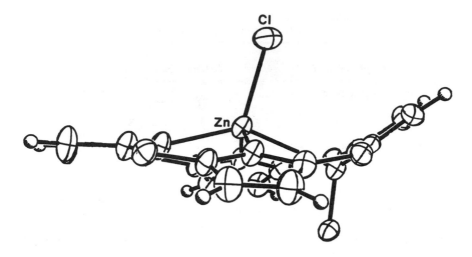

Figure 7-3. The structure of a zinc(II) complex of an *N*-alkylprotoporphyrin. The N-alkylated pyrrorenine ring is highly canted.

isomers, as proposed by De Matteis.[59] On one hand, the flexibility of the propionic acid side chains should allow them to adopt the best position for interaction with the enzyme. On the other, the difference in affinities is about a factor of six, equivalent to a free energy difference of only 5 kJ/mol, so it could result from rather subtle differences of the fit of the porphyrin to the active site of the enzyme. The asymmetric manner in which the zinc(II) atoms binds its axial ligand (Figure 7-3) could give rise to the difference in binding constants, for example. When the alkyl group is on the A or B rings, the best bond of the zinc atom to a residue in the active site would be formed with a residue over the A or B rings, providing a difference between these isomers and the N_C and N_D isomers.

In 1981, Ortiz de Montellano and co-workers reported the separation [by high-performance liquid chromatography (HPLC)] of the four regioisomers of *N*-ethylprotoporphyrin IX and the inhibition of rat liver ferrochelatase in vitro for each of the isomers.[61] They found that the isomers in which the ethyl group is present on pyrrole rings A or B (having vinylic β-substituents) are as strongly inhibitory as the corresponding *N*-methylprotoporphyrin derivatives (a 50% decrease in enzymatic activity as a ratio of 2.4×10^{-6} g of *N*-ethylporphyrin to protein), whereas the isomers substituted on rings C or D (which have propionic acid substituents in the β-position) were 30–100 times less inhibitory ($1-3 \times 10^{-4}$ g of *N*-ethylporphyrin were required to cause a decrease in activity of 50%).

The results of the inhibition studies of De Matteis and co-workers were extended and summarized in a 1985 article.[62] In this study, it was shown that the N_C and N_D isomers of *N*-methylmesoporphyrin IX, *N*-ethylmesoporphyrin IX, *N*-*n*-propylmesoporphyrin IX, and *N*-methylprotoporphyrin IX inhibit ferrochelatase activity to essentially the same degree (Table 7-8). The first fraction of the isomers of N-methyldeuteroporphyrin which elutes chromatographically (probably the N_C and N_D isomers) also shows the same activity. Typically, the N_A and N_B isomers are much less inhibitory, with the pronounced exceptions of N-methyl- and N-ethylprotoporphyrins, which show about equal effects, and N-n-propylprotoporphyrin IX and the zinc complex of N-methylprotoporphyrin IX, in which the N_A and N_B isomers are significantly more inhibitory. On the basis of these results, De Matteis concluded that the interaction of the vinyl groups of protoporphyrin with the enzyme may be quite important and that bulky N-substituents, like the zinc complexes, show greater differences between the N_C and N_D isomers and the N_A and N_B isomers because of greater deviations of the N-substituted pyrrole ring from the mean porphyrin plane.

A comparison of the structures of N-methyl, N-phenyl-, and N-benzylporphyrins, however, shows little difference in the pyrrole ring–porphyrin plane angle, even though the groups differ greatly in size (the angles are 38.5, 42.0, and 32° for the N-methyl, N-phenyl, and N-benzyl derivatives, respectively),[60] so it does not seem reasonable at this time to attribute the difference in affinities of various N-substituted porphyrins to the different distances between the propionic groups on the porphyrin periphery.[60] It seems reasonable to assume that the bulky N-substituents are somewhat weaker inhibitors because they require distortion of the protein in the region near the center of the porphyrin substrate.

Summary of Data concerning the Structure of the Active Site of Ferrochelatase. From Dailey's spectrofluorometric experiments,[42] the kinetics data demonstrating that metal ion binding is independent of protoporphyrin IX binding,[29b] and the dependence of ferrochelatase inhibition on the structure of N-substituted porphyrins discussed in the previous section, a picture of the active site is beginning to take shape. (The studies by Dailey and co-workers have been performed with bovine ferrochelatase while the inhibition studies have mostly involved the rat enzyme, but properties of ferrochelatase are similar for a wide variety of animals.[26–29] so it seems likely that major aspects of the active site for the enzyme from various sources are similar.)

The following assumptions are used in constructing a model of the active site:

1. The propionic acid side chains of bound protoporphyrin IX extend out of the active site toward the cytosol. This is consistent with the structures of many cytochrome active sites that contain the iron complex of protoporphyrin IX and with the dependence of Michaelis constants of ovine, bovine, and murine liver ferrochelatase on porphyrin structure (Tables 7-3 to 7-5).[23f]

2. The inner portion of the porphyrin-binding pocket is narrow. This follows from the order of the Michaelis constants for porphyrins that differ only by substituents on the A and B rings (Tables 7-3 to 7-5). The decrease in inhibition for N-alkylprotoporphyrins and N-alkylmesoporphyrins for large

N-substituents[62] (Table 7-9) is also consistent with a constricted binding site for the porphyrin.

　　3. At least one cysteine binds to the iron atom. This follows from Dailey's findings that reagents that deactivate sulfhydryl groups inhibit the enzyme in the absence of iron(II) but that bound iron(II) protects the enzyme.[42,43] Cysteinyl sulfur binds well to iron(II), stabilizing the 2+ oxidation state, and yet it can release the iron atom rapidly.

　　4. There is an arginine residue that is located on the side of the pocket which binds the porphyrin. Dailey has presented evidence for an arginine residue at the active site and that modification of arginine (by 2,3-butanedione or camphorquinone-10-sulfonate) leads to inactivation of the enzyme but does not alter the K_m for iron(II).[42-44] Such a residue could play one of several roles: as an anchor for one of the propionic acid side chains of protoporphyrin IX, as a ligand for the iron atom, or as recipient for the protons as they are released when the iron inserts to form heme. The lack of any effect of arginine alteration on the K_m for iron(II) indicates that arginine does not serve as a ligand for iron(II). Therefore, it is probably positioned on the side of the pocket that binds the porphyrin. The two other possible roles arginine could serve are as a proton shuttle (accepting a proton released from the porphyrin as iron binds and transferring its proton to another residue) or as an anchor for a propionic acid group of protoporphyrin. The proton shuttle or anchor roles are distinguished by the position of the arginine terminus—either toward the center of the porphyrin ring or toward the

TABLE 7-9.　Ferrochelatase Inhibition of N-Alkylporphyrin Isomers[a]

Porphyrin	N-Alkylated ring	Inhibitory activity
N-Methylprotoporphyrin	A and B	10.8
	C and D	9.24
N-Ethylprotoporphyrin	A and B	5.86
	C and D	6.67
N-n-Propylprotoporphyrin	A and B	0.11
	C and D	6.06
N-Methylmesoporphyrin	A and B	8.01
	C and D	15.0
N-Ethylmesoporphyrin	A and B	8.17
	C and D	0.14
N-n-Propylmesoporphyrin	A and B	7.85
	C and D	<0.05
N-methyldeuteroporphyrin	A and B (?)	8.78
	C and D (?)	0.32
Chloro-N-methylprotoporphyinatozinc(II)	A and B	15.7
	C and D	2.1
Chloro-N-mesoporphinatozinc(II)	A and B	6.96
	C and D	0.5

[a] Negligible inhibition by N-methylated uro- and coproporphyrins; Reference 62.

periphery, respectively. The hydrophilic nature of arginine makes the peripheral position more likely. Although the conclusion is not straightforward because chemical active-site modification can cause unexpected changes in protein structure, the data are most consistent with arginine's role as an anchor for the propionic acid side chains.

5. There is another good Lewis base, perhaps a lysine residue, on the side of the pocket that binds the iron(II) atom. This speculation is based on the inhibition studies using the zinc(II) complexes of N-alkylated porphyrins and the metal-free ligands. The differences in behavior of the zinc(II) complexes of N-alkylated porphyrins with respect to the metal-free ligands are sizeable enough to demonstrate that the zinc(II) atom remains bound when the enzyme–substrate complex is formed (Tables 7-8 and 7-9). Assuming that the zinc(II) atom resides in the active site on the same side of the porphyrin from which the iron would normally enter is reasonable since both zinc(II) and iron(II) would be stabilized in a polar environment while the porphyrin ligand would be stabilized by hydrophobic, especially aromatic, residues. The amine of lysine, for example, would be a good axial ligand for zinc(II) complexes of N-alkylporphyrin. Differences in the position where the zinc(II) atom could bind best to a Lewis base such as lysine [based on the asymmetric binding of zinc(II) complexes of N-alkylporphyrins[56,60]] in the various isomers of its N-methylprotoporphyrin and N-methyldeuteroporphyrin IX complexes can account for their different inhibitions of ferrochelatase activity: The A and B isomers are more inhibitory than the C and D isomers by 6 to 14 times (Tables 7-8 and 7-9).

Another fact consistent with the presence of an amine is the greater Michaelis constant of an N-methylporphyrin than the corresponding unsubstituted porphyrin, 7 nm vs 11 μM in case of N-methylprotoporphyrin IX and protoporphyrin IX, for example.[29b] A difference of over 10^3 in equilibrium constant corresponds to over 17 kJ/mol, too much to be accounted for by a difference in more favorable hydrophobic or aromatic residue stacking interactions brought about by changes in the topology of the porphyrin ring caused by N-substitution. It seems more plausible that the N-alkylporphyrin is protonated and that this species is stabilized by polar residues—of which at least some are the same residues that ordinarily bind the iron atom. A lysine residue is a strong candidate for this interaction, placing it on the same side the iron normally occupies. Although it has been reported that lysyl deactivating reagents do not diminish the activity of bovine hepatic ferrochelatase,[44] it is possible that a lysine deep in the pocket might be too inaccessible to be modified.

6. At least one tryptophan residue is located toward the inner end of the active site and it is (or they are) probably on the side of the pocket where the porphyrin binds. The presence of tryptophan at the active site and the fact that it is (or they are) not exposed to bulk water is indicated from Dailey's spectrofluorometric study. Stacking of the tryptophan ring with the porphyrin would stabilize the enzyme–substrate complex. The most hydrophilic

environment for the iron atom and most hydrophobic environment for the porphyrin would probably be achieved if the tryptophan residue(s) was on the side of the pocket opposite the iron atom; the A and B isomers of zinc(II) complexes of N-alkylprotoporphyrin IX are the most inhibitory, and they are the innermost rings (they have the N-alkyl substituent on the pyrrolenine ring with the vinyl group, not a propionic acid side chain). If the extra stabilization of binding is due, at least in part, to favorable interactions with the canted pyrrolenine ring by tryptophan, the tryptophan would need to be in the inner end of the pocket. Phenylalanine is not ruled out by the fluorescence experiments, and its aromatic ring could also stabilize the porphyrin as it distorts to bind the iron atoms. Hence, it seems quite possible for phenylalanine to be present in the porphyrin–binding region. Heme formation could be promoted by such aromatic residues being oriented to favor a puckered conformation of porphyrin ring or by their being sufficiently flexible to "follow" the pyrrolenine ring as distortion occurs.

In the proposed structure for the active site of bovine liver ferrochelatase shown in Figure 7-4, the opening to the solvent is at the bottom of the figure.

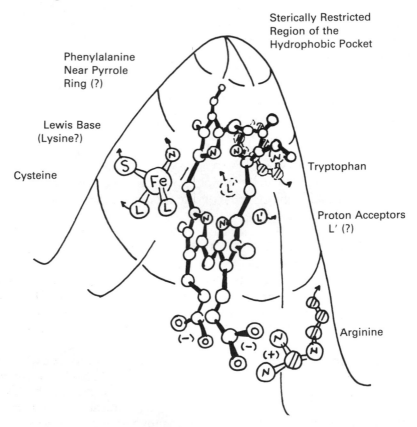

Figure 7-4. A proposed structure for the active site of bovine ferrochelatase.

Features which are totally speculative are denoted by question marks. The positioning of those features which have been identified experimentally is also speculative (vide supra). Based on the arguments discussed above, N-alkylporphyrins are predicted to bond such that the N-alkyl substituent is on the side of the porphyrin plane opposite the metal-binding site. That portion of the pocket is also sterically restricted based on the varied inhibitory activity of isomers of N-n-propylporphyrins (Table 7-9). (One possible reason for the differences among isomers is that the pendant alkyl group causes the porphyrin to be pushed away from favorable contacts with the hydrophobic pocket and the orientation of the porphyrin will depend on the position of the alkyl group.)

4. FERROCHELATASE INHIBITION AND CHLOROPHYLL BIOSYNTHESIS

The biosynthetic pathways to the plant cytochromes and chlorophylls apparently diverges after the formation of protoporphyrin IX. If the introduction of iron is enzymatically controlled in plants, the regulation or inhibition of the enzyme should affect chlorophyll production. Thus, the inhibition of ferrochelatase activity by N-substituted porphyrins has also been used as a means of studying chlorophyll synthesis. In these studies, the effect of arresting iron complexation with protoporphyrin IX to form cytochromes and the consequent capability of magnesium to compete better for protoporphyrin, leading perhaps to enhanced chlorophyll synthesis, is the point under scrutiny.

Hemoproteins are degraded in animals to produce bile pigments which apparently have no function except as a means of eliminating unneeded heme. In plants and blue-green algae, however, the bilins with structures similar to the biliverdin and bilirubin produced in animals play functional roles in photosynthesis and photomorphogenesis. The photosynthetic antennae of plants and some algae are composed of bilins associated with proteins. Troxler, Brown, and their co-workers found that there is a precursor–product relationship between the plant bilins and heme.[64] Since N-methylprotoporphyrin has been shown to inhibit ferrochelatase in mammalian systems, Brown and co-workers reasoned that it could be used to determine the role of ferrochelatase in the plant biosynthesis of bilins.[65] They chose the unicellular rhododyte Cyanidium caldarium because it shows ferrochelatase activity and also because it can be grown aerobically in the dark without the formation of photosynthetic pigments and then induced to form photosynthetic pigments in two quite different ways: (1) by suspension in minimal medium and exposure to light, giving phycocyanin and chlorophyll a without significant cell division, and (2) by resuspension in the dark in a medium containing glucose and δ-aminolevulinate, resulting in excretion of large amounts of porphyrins and phycocyanobilin, the bilin chromophore of phycocyanin.[66] They found that N-methylprotoporphyrin strongly inhibits bilin synthesis both in the light and in the dark if δ-aminolevulinic acid is added to promote the synthesis of phycocyanobilin. This result implies that ferrochelatase (and, hence, heme) is required for algal bilin synthesis.

They also found that N-methylprotoporphyrin has little effect on the synthesis of chlorophyll a in the dark. Even the step in which magnesium is inserted is not inhibited in the dark. However, the synthesis of chlorophyll a in the light is inhibited in a manner parallel to the inhibition of phycocyanin synthesis. The explanation they advanced for these results is that the primary inhibition of chlorophyll a synthesis in the light is actually the inhibition of phycocyanin synthesis and that the synthesis of chlorophyll in the light is strictly coordinated to phycocyanin synthesis. The results also imply that the synthesis of chlorophyll a in this organism occurs by different routes in the dark and light.

Houghton and co-workers studied inhibition of ferrochelatase by N-methylprotoporphyrin dimethyl ester both in isolated membranes and growing cultures of the photosynthetic bacterium *Rhodopseudomonas sphaeroides*.[67] Their objective was to determine the relationship between heme and bacteriochlorophyll synthesis in this organism. Under conditions of high aeration, this organism is able to utilize nutrients in the medium without assembling the photosynthetic apparatus characteristic of the photosynthetic form. In this case, hemes that serve as prosthetic groups of cytochromes are synthesized, but the magnesium tetrapyrroles used for photosynthesis are not. There is evidence that both the hemes and the bacteriochlorophylls originate from protoporphyrin,[68] but under conditions of iron deficiency, coproporphyrin is the final product rather than the magnesium pigments.

N-methylprotoporphyrin dimethyl ester (possibly deesterified to the free acid during the experiment) was found to be a potent inhibitor of ferrochelatase in *R. sphaeroides* membranes (30% inhibition was measured at a concentration of 1.5 nM). A lag time of a few seconds was noted (the time for deesterification, perhaps). In cultures of whole cells of *R. sphaeroides*, N-methylprotoporphyrin dimethyl ester at 8×10^{-7} M causes a decrease in cytochrome b and c content by 40%, and the ferrochelatase activity is reduced by a factor of 10. The inhibitor-induced decrease in the concentration of free heme in the cells is accompanied by the production of magnesium tetrapyrroles. This result indicates that the N-methylprotoporphyrin is specific for enzymatic iron chelation and does not affect magnesium chelation, consistent with the results found by Brown and co-workers for *C. caldarium*,[67] as discussed previously. In cells of *R. sphaeroides* competent to make bacteriochlorophyll, inhibition of ferrochelatase by N-methylprotoporphyrin causes the same effects on pigment production as are found under conditions of iron deficiency. In green plants, similar effects are found under conditions of iron deficiency,[69] implying that the synthesis of chlorophyll in bacteria and higher plants may be quite similar.[67] The point was made that the normal ferrochelatase activity in these cells is far in excess of that needed for the synthesis of cytochromes and, hence, sufficient tetrapyrroles could be made even in the presence of inhibitors.

5. PATHOLOGICAL CONDITIONS ASSOCIATED WITH FERROCHELATASE

A. Natural and Drug-Induced Porphyrias: Inhibition of Protoporphyrin IX Metalation

In healthy animals, only small quantities of the intermediates on the heme synthesis pathway, δ-aminolevulinic acid and porphobilinogen, are excreted. Porphyrias (erythropoietic, arising in the bone marrow, or hepatic, arising in the liver) result in the accumulation of large amounts of these intermediates through increased activity of δ-aminolevulinic acid synthetase. The concomitant production of highly colored porphyrins is sometimes obvious. Porphyrias can be inherited in both humans and other animals, and porphyric cattle have provided a commercial source of uroporphyrins. In variegate porphyria, a distinctive pattern of porphyrin excretion exists (a marked increase in fecal protoporphyrin) suggesting that the defect occurs in the conversion of protoporphyrin IX to heme.[70] Evidence has been provided for the inhibition of ferrochelatase in skin fibroblasts of porphyric patients in South Africa[71] and for protoporphyrin oxidase but not ferrochelatase in American patients with variegate porphyria.[72] The extent to which ferrochelatase activity is inhibited in particular porphyrias and the mechanism by which the inhibition occurs are not well understood in general. A recent study by Bloomer and co-workers on bovine protoporphyria, however, has demonstrated that the hepatic ferrochelatase of porphyric cattle has a defect that is likely a point gene mutation that causes a minor change in enzyme structure leading to a porphyric condition.[73]

Drug-induced pathologies in animals that result in porphyrialike symptoms have been much better understood than the natural disease state. Whether or not ferrochelatase is the principal site at which the regulation of heme synthesis is normally disrupted in the natural porphyrias, it is now evident that drugs which induce the excretion of excessive amounts of porphyrins in animals produce N-substituted porphyrins (green pigments) which inhibit ferrochelatase activity.[74] DeMatteis and Gibbs found that the antifungal drugs griseofulvin and isogriseofulvin act like the drug that had earlier been shown to produce such pigments, DDC, by producing hepatic green pigments that strongly inhibit ferrochelatase.[75] It is now known that a wide variety of compounds that become alkyl group donors on oxidation, as well as many terminal olefins and alkynes, can produce N-alkylporphyrins from the prosthetic heme of cytochrome P-450 enzymes.[10,70]

In 1980, Tephly and co-workers found a correlation between the amount and rate of inhibitor formed by treatment of rats or mice with DDC and the extent and course of the decrease in mitochondrial ferrochelatase activity in their excised livers.[76] The correlation between the porphyrinogenesis produced by DDC, the inhibition of ferrochelatase, and the formation of inhibitor (N-methylprotoporphyrin) established N-methylprotoporphyrin as the substance responsible for experimental animal porphyria produced by DDC. It is reasonable to assume that the inhibition of ferrochelatase caused by this drug and other compounds that produce N-alkylporphyrins in the animals limits the amount of free heme to such low levels that either δ-aminolevulinic acid synthetase or protoporphyrin oxidase is deregulated and excess porphyrin production results.

B. Zinc Protoporphyrin IX Formation by Ferrochelatase and Lead Poisoning

A widespread and relatively straightforward diagnostic method for determining lead intoxication is measurement of fluorescence in blood samples due to the presence of the zinc(II) complex of protoporphyrin IX. Although zinc(II) forms porphyrin complexes relatively rapidly,[1] there does not appear to be any accumulation of the zinc protoporphyrin IX complex in cases of erythropoietic protoporphyria, in which there is ferrochelatase deficiency.[77] In addition, it has been demonstrated that ferrochelatase is a potent catalyst of zinc protoporphyrin formation (Table 7-6). Tokunaga and co-workers have shown that lead(II) does not inhibit heme formation when there is a high level of ferrous ion available (0.5 mM), but it does inhibit heme formation when the only ferrous ion available is that produced from ferric ion in the presence of biological oxidation–reduction systems such as the NADH dehydrogenase-rich fraction of rat liver mitochondria and NAD(P)H (with or without FMN or FAD).[31] Therefore, they postulated that the lead inhibition of heme formation may occur because lead interferes with ferrous iron production rather than by a direct interference with ferrochelatase. Since it seems likely that the formation of heme in vivo uses ferrous ion produced in this manner, this was a highly relevant experiment. In this article, they showed that lead(II) inhibits the formation of zinc(II) protoporphyrin to a smaller extent [50% inhibition of zinc(II) complexation required nearly 100 times more lead(II) under their conditions]. Since iron deficiency often accompanies lead intoxication and zinc complexation is much less inhibited than iron (from a ferric ion source) complexation, then ferrochelatase activity can readily lead to accumulation of zinc protoporphyrin. However, the formation of zinc porphyrins near neutral pH is so rapid even without an enzyme present, it would be surprising if at least some of the complex that accumulates in lead-intoxicated individuals doesn't result from a nonenzymatic reaction.

6. CONCLUSION

Recent work has contributed greatly to our understanding of the enzymatic formation of hemes. The most important features of the enzymatic reaction are (1) the concentration effect resulting from the high binding constants for iron(II) and protoporphyrin IX, (2) the ability of the enzyme to bind iron(II) selectively, thereby shifting the ferric–ferrous equilibrium in the mitochondrion, and (3) the binding of iron(II) in a relatively labile coordination environment (possibly tetrahedral or highly distorted) within the pocket. While there is as yet no definitive (ie, crystallographic) determination of the structure of the active site for any of these related enzymes, spectrofluorometric experiments and determinations of Michaelis constants for a wide variety of porphyrins, including the strongly inhibiting N-alkylporphyrins, have indicated probable components and structural limitations. It appears very likely that the iron atom is bound as ferrous iron by one or more cysteinyl residues and the porphyrin is anchored by an arginine residue that stabilizes a propionic side chain near the opening of the pocket. The hydrophobic side of the pocket that binds protoporphyrin IX is constrained at the innermost end. It is likely that it is also relatively crowded at the center of the porphyrin-binding side opposite the iron-binding site.

Much less is yet known about the metalation reactions in the formation of chlorophylls and corrinoids, but it is likely that enzymatic intervention is not necessary for the chlorophylls but would be reasonable for insertion of cobalt at some point in the biosynthesis of cobalamin.

ACKNOWLEDGMENTS

I appreciate the help of Dr. Harry Dailey, who supplied me with useful preprints as well as suggestions. I am grateful to the National Cancer Institute for generous support of my research in metalloporphyrin chemistry (CA 25427), and for support by the PSC-CUNY grants program of the City University of New York. This work was completed under the auspices of the Center for the Study of Gene Structure and Function, Research Centers in Minority Distributions Program at the NIH.

REFERENCES

1. (a) Lavallee, D. K. *Coord. Chem. Rev.* **1985,** *61,* 55. (b) Hambright, P. In "Porphyrins and Metalloporphyrins"; Smith, K. M., Ed.; Elsevier: Amsterdam, 1975, pp. 233–278. (c) Schneider, W. *Struct. Bond. (Berlin)* **1975,** *23,* 123.
2. Hambright, P.; Chock, P. B. *J. Am. Chem. Soc.* **1974,** *96,* 3123.
3. Turay, J.; Hambright, P. *Inorg. Chem.* **1980,** *19,* 562.
4. Bain-Ackerman, M. J.; Lavallee, D. K. *Inorg. Chem.* **1979,** *18,* 3358.

5. (a) Matsushima, Y.; Sugata, S. *Chem. Pharm. Bull.* **1979**, *27*, 3049. (b) Funahashi, S.; Yamaguchi, Y.; Tanaka, M. *Bull. Chem. Soc. Jpn.* **1984**, *57*, 204. (c) Funahashi, S.; Yamaguchi, Y.; Tanaka, M. *Inorg. Chem.* **1984**, *23*, 2449. (d) Funahashi, S.; Yamaguchi, Y.; Ishihara, K.; Tanaka, M. *J. Chem. Soc., Chem. Commun.* **1982**, 976.

6. Pasternack, R.; Sutin, N.; Turner, D. *J. Am. Chem. Soc.* **1976**, *98*, 1908.

7. Fleischer, E. B. *Acc. Chem. Res.* **1970**, *3*, 105.

8. Shah, B.; Shears, B.; Hambright, P. *Inorg. Chem.* **1971**, *10*, 1818.

9. The value of the resonance energy depends on the choice of reference values for the single bonds between the sp^2 hybridized carbon atoms. In his critical survey of various methods (George, P. *Chem. Rev.* **1975**, *75*, 85), George estimated the range for this value [$E(C_d-C_d)$] to be between a low value of 372.9 kJ/mol (I) and a high value of 396.1 kJ/mol (II). The corresponding values for the resonance energy of benzene given by George are 203.3 kJ/mol (48.58 kcal/mol) (RE_I) and 133.7 kJ/mol (31.94 kcal/mol)(RE_{II}). From thermodynamic data of Stern and Klebs for nine naturally derived porphyrins (Stern, A., Klebs, G. *Justus Liebigs Ann. Chem.* **1933**, *505*, 295) and Lavallee and Hamilton for *meso*-tetraphenyl-porphyrin (Lavallee, D. K.; Hamilton, W. S. *J. Phys. Chem.* **1977**, *80*, 854.), the resonance energies on the two scales are 1017 ± 96 kJ/mol (RE_I) and 653 ± 95 kJ/mol (RE_{II}) and 1022 kJ/mol and 659 kJ/mol, respectively, for the two types of porphyrins. On either scale, the resonance energies of both natural and synthetic porphyrins are about five times that of benzene.

10. Lavallee, D. K. "The Chemistry and Biochemistry of *N*-Substituted Porphyrins"; VCH Publ: New York, 1987.

11. Basolo, F.; Pearson, R. G. In "Mechanisms of Inorganic Reactions," 2nd ed.; Wiley (Interscience): New York, 1967.

12. Small, T.; Lavallee, D. K. "Abstracts of Papers"; 183rd Natl. A.C.S. Meeting, Seattle, March 21–25, 1982; American Chemical Society: Washington, D.C.; Paper No. INOR-138.

13. Buckingham, D. A.; Clark, C. R.; Webley, W. S. *J. Chem. Soc., Chem. Commun.* **1981**, 192.

14. Brown, E. M. Ph.D. Thesis, Drexel University, Philadelphia, 1971.

15. Lavallee, D. K.; Onady, G. M. *Inorg. Chem.* **1981**, *20*, 907.

16. Adeyemo, A. O.; Shamim, A.; Hambright, P.; Williams, R. F. X. *Ind. J. Chem.* **1982**, *21A*, 763.

17. Longo, F. R.; Brown, E. M.; Quimby, D. J.; Adler, A. D.; Meotner, M. *Ann. N.Y. Acad. Sci.* **1973**, *206*, 420.

18. (a) Eigen, M. *Pure Appl. Chem.* **1963**, *6*, 105. (b) Bennett, H. D.; Caldin, B. F. *J. Chem. Soc. A* **1971**, 2198.

19. Marks, G. S. "Heme and Chlorophyll: Chemical, Biochemical and Medical Aspects"; Van Nostrand: London, 1969.

20. Echenmoser, A. *Quart. Rev.* **1970**, *24*, 366.

21. (a) Dolphin, D., Ed. "B_{12}", Vols. 1 and 2; Wiley: New York, 1982. (b) Battersby, A. R. *Acc. Chem. Res.* **1986**, *19*, 147.

22. (a) Toohey, J. I. *Proc. Nat. Acad. Sci., U.S.A.* **1965**, *54*, 934. (b) Toohey, J. I. *Fed. Proc., Fed. Am. Soc. Exp. Biol.* **1966**, *25*, 1628.

23. (a) Koppenhagen, V. B. In "B_{12}"; Dolphin, D., Ed.; Wiley Interscience, New York, 1982, pp. 105–150. (b) Koppenhagen, V. B.; Pfiffner, J. J. *J. Biol. Chem.* **1970**, *245*, 5865. (c) Koppenhagen, V. B.; Pfiffner, J. J. *J. Biol. Chem.* **1971**, *246*, 3075. (d) Koppenhagen, V. B.; Wagner, F.; Pfiffner, J. J. *J. Biol. Chem.* **1973**, *248*, 7999. (e) Rubison, K. A. *J. Am. Chem. Soc.* **1979**, *101*, 6105.

24. (a) Thomson, A. J. *J. Am. Chem. Soc.* **1969**, *91*, 2780. (b) Fischli, A.; Eschenmoser, A. *Angew. Chem.* **1967**, *6*, 866.

25. (a) Tabata, M.; Tanaka, M. *Anal. Lett.* **1980**, *13*, 427. (b) Tabata, M.; Tanaka, J. *J. Chem. Soc., Dalton Trans.* **1983**, *9*, 1955. (c) Tabata, M.; Tanaka, M. *Pure Appl. Chem.* **1983**, *55*, 151. (d) Adeyemo, A. O.; Krisnamurty, M. *Int. J. Chem. Kinet.* **1984**, *16*, 1975.

26. (a) Yoneyama, Y.; Ohyama, M.; Sugita, Y.; Ysohikawa, H. *Biochim. Biophys. Acta* **1962**, *62*, 261. (b) Schwartz, H. C.; Cartwright, E. E.; Smith, E. L.; Wintrobe, M. M. *FEBS Proc.* **1959**, *18*, 545. (c) Porra, R. J.; Ross, B. D. *Biochem. J.* **1965**, *94*, 557. (d) Porra, R. J.; Jones, O. T. G. *Biochem. J.* **1963**, *87*, 181. (e) Labbe, R. F.; Hubbard, N. *Biochim. Biophys. Acta* **1961** *52*, 130. (f) Tangeras, A. *Biochim. Biophys. Acta* **1985**, *843*, 199. (g) Honeybourne, C. L.; Jackson, J. T.; Jones, O. T. G. *FEBS Lett.* **1979**, *98*, 207. (h) Camadro, J.-M.; Ibraham, N. G.; Levere, R. D. *J. Biol. Chem.* **1984**, *259*, 5678. (i) Labbe, P.; Volland, C.; Chain, P. *Biochim. Biophys. Acta* **1968**, *159*, 527. (j) Dailey, H. A., Jr.; Lascelles, J. *Arch. Biochem. Biophys.* **1974**, *160*, 523. (k) Jones, M. S.; Jones, O. T. G. *Biochem. J.* **1970**, *119*, 953. (l) Goldin, B. R.; Little, H. N. *Biochim. Biophys. Acta* **1969**, *171*, 321. (m) Dailey, H. A.; Smith, A. *Biochem. J.* **1984**, *223*, 441.
27. (a) Jones, M. S.; Jones, O. T. G. *Biochem. J.* **1970**, *113*, 507. (b) Harbin, B. M.; Dailey, H. A., Jr. *Biochemistry* **1975**, *24*, 366.
28. Taketani, S.; Tanaka-Yoshioka, A.; Masaki, R.; Tashiro, Y.; Tokunaga, R. *Biochim. Biophys. Acta* **1986**, *886*, 277.
29. (a) Hanson, J. W.; Dailey, H. A. *Biochem. J.* **1984**, *222*, 695. (b) Taketani, S.; Tokunaga, R. *J. Biol. Chem.* **1981**, *256*. 12748. (c) Dailey, H. A.; Fleming, J. E. *J. Biol. Chem.* **1983**, *258*, 11453. (d) Dailey, H. A.; Fleming, J. E.; Harbin, B. M. *Methods Enzymol.* **1986**, *123*, 401. (e) Dailey, H. A. *J. Biol. Chem.* **1982**, *257*, 14714. (f) Dailey, H. A.; Fleming, J. E.; Harbin, B. M. *J. Bacteriol.* **1986**, *165*, 1.
30. Barnes, R.; Connelly, J. L.; Jones, O. T. G. *Biochem. J.* **1972**, *128*, 1043.
31. Taketani, S.; Tanaka, A.; Tokunaga, R. *Arch. Biochem. Biophys.* **1985**, *242*, 291.
32. Granick, S.; Beale, S. I. *Adv. Enzymol.* **1978**, *46*, 33.
33. Granick, S.; Mauzerall, D. *J. Biol. Chem.* **1958**, *232*, 1119.
34. Kassner, R. J.; Wang, J. K. *J. Am. Chem. Soc.* **1966**, *88*, 5170.
35. Kassner, R. J.; Walchak, H. *Biochim. Biophys. Acta* **1973**, *304*, 294.
36. Tokunaga, R.; Sano, S. *Biochim. Biophys. Acta* **1972**, *264*, 263.
37. Tephly, T. R.; Gibbs, A. H.; De Matteis, F. *Biochem. J.* **1979**, *180*, 241.
38. Taketani, S.; Tokunaga, R. *Eur. J. Biochem.* **1982**, *127*, 443.
39. Lavallee, D. K. Unpublished results.
40. Mailer, K.; Poulson, R.; Dolphin, D.; Hamilton, A. D. *Biochem. Biophys. Res. Commun.* **1980**, *96*, 777.
41. Fadigan, A.; Dailey, H. A. *Biochem. J.* **1987**, *243*, 419.
42. Dailey, H. A. *Biochem.* **1985**, *24*, 1287.
43. Dailey, H. A. *J. Biol. Chem.* **1984**, *259*, 2711.
44. Dailey, H. A. *J. Bacteriol.* **1986**, *165*, 1.
45. Tephly, T. R.; Gibbs, A. H.; DeMatteis, F. *Biochem. J.* **1979**, *180*, 241.
46. Kunze, K. L.; Ortiz de Montellano, P. R. *J. Am. Chem. Soc.* **1981**, *103*, 4225.
47. DeMatteis, F. *Pharmacol. Rev.* **1967**, *19*, 523.
48. McEwen, W. K. *J. Am. Chem. Soc.* **1936**, *58*, 1127.
49. Neuberger, A.; Scott, J. J. *Proc. R. Soc. London Ser. B.* **1952**, 307.
50. Lavallee, D. K.; Gebala, A. *Inorg. Chem.* **1974**, *13*, 2004.
51. (a) Ortiz de Montellano, P. R.; Kunze, K. L.; Mico, B. A. *Mol. Pharmacol.* **180**, *18*, 602. (b) DeMatteis, F.; Gibbs, A. H. *Biochem. J.* **1980**, *187*.
52. (a) DeMatteis, F.; Cantoni, L. *Biochem. J.* **1979**, *183*, 99. (b) Ortiz de Montellano, P. R.; Kunze, K. L.; Yost, G. S. *Biochem. Biophys. Res. Comm.* **1978**, *83*, 132.
53. DeMatteis, F.; Gibbs, A. H.; Smith, A. G. *Biochem. J.* **1980**, *189*, 645.
54. DeMatteis, F.; Gibbs, A. H.; Tephly, T. R. *Biochem. J.* **1980**, *188*, 145.
55. Lavallee, D. K.; Anderson, O. P. *J. Am. Chem. Soc.* **1982**, *104*, 4707.
56. Lavallee, D. K.; Anderson, O. P.; Kopelove, A. *J. Am. Chem. Soc.* **1978**, *100*, 3025.
57. McLaughlin, G. M. *J. Chem. Soc., Perkin Trans. 2* **1974**, 136.
58. Goldberg, D. E.; Thomas, K. M. *J. Am. Chem. Soc.* **1976**, *98*, 913.
59. DeMatteis, F.; Jackson, A. H.; Gibbs, A. H.; Rao, K. R. N.; Atton, J.; Weerasinghe, S.; Hollands, C. *FEBS Lett.* **1982**, *142*, 44.

60. Schauer, C.; Anderson, O. P.; Lavallee, D. K.; Battioni, J.-P.; Mansuy, D. *J. Am. Chem. Soc.* In press.
61. Ortiz de Montellano, P. R.; Kunze, K. L.; Cole, S. P. C.; Marks, G. S. *Biochem. Biophys. Res. Commun.* **1981**, *103*, 581.
62. DeMatteis, F.; Gibbs, A. H.; Harvey, C. *Biochem. J.* **1985**, *226*, 537.
63. (a) Bottomley, L. A.; Kadish, K. M. *Inorg. Chem.* **1981**, *20*, 1348. (b) Kadish, K. A. In "Iron Porphyrins", Part 2; Lever, A. B. P.; and Gray H. B., Eds.; Addison-Wesley: Reading, MA, 1983, pp. 161–251 and references therein.
64. (a) Troxler, R. F.; Brown, S. B. *Biol. Bull.* **1980**, *159*, 502. (b) Brown, S. B.; Holroyd, J. A.; Troxler, R. F.; Offner, J. D. *Biochem. J.* **1969**, *6*, 116. (c) Brown, S. B.; Holroyd, J. A.; Troxler, R. F.; Offner, J. D. *Biochem. J.* **1981**, *194*, 137.
65. Brown, S. B.; Holroyd, J. A.; Vernon, D. I.; Troxler, R. F.; Smith, K. M. *Biochem. J.* **1982**, *208*, 487.
66. Troxler, R. F.; L. Bogorad, L. *Plant Physiol.* **1966**, *41*, 491.
67. Houghton, J. D.; Honeybourne, C. L.; Smith, K. M.; Tabba, H. D.; Jones, O. T. G. *Biochem. J.* **1982**, *2208*, 479.
68. Jones, O. T. G. In "The Photosynthetic Bacteria"; Clayton, R. K.; and Sistrom, W. D., eds.; Plenum Press: New York, 1978, pp. 750–777.
69. Spiller, S. C.; Castelfranco, A. M.; Castelfranco, P. A. *Plant Physiol.* **1982**, *69*, 107.
70. Cole, S. P. C.; Marks, G. S. *Mol. Cell. Biochem.* **1984**, *64*, 127.
71. Viljoen, D. J.; Cayanis, E.; Becker, D. M.; Kramer, S.; Dawson, B.; Bernstein, R.; *Am. J. Hematol.* **1979**, 185.
72. Brenner, D. A.; Bloomer, J. R. *N. Engl. J. Med.* **1980**, *302*, 765.
73. Bloomer, J. R.; Hill, H. D.; Morton, K. O.; Anderson-Burnham, L. A.; Straka, J. G. *J. Biol. Chem.* **1987**, *262*, 667.
74. De Matteis, F.; Gibbs, A. H.; Tephly, T. R. *Biochem. J.* **1980**, *188*, 145.
75. DeMatteis, F.; Gibbs, A. H. *Biochem. J.* **1975**, *146*, 285.
76. De Matteis, F.; Gibbs, A. H.; Tephly, T. R. *Biochem. J.* **1980**, *180*, 145.
77. (a) Bonkowsky, H. L.; Bloomer, J. R.; Ebert, P. S.; Mahoney, M. J. *J. Clin. Invest.* **1975**, *56*, 1139. (b) Bottomley, S. S.; Tanaka, M.; Everett, M. A. *J. Lab. Clin. Med.* **1975**, *86*, 126.

CHAPTER 8

Chemical Studies and The Mechanism of Flavin Mixed Function Oxidase Enzymes

Thomas C. Bruice

University of California, Santa Barbara, California

CONTENTS

1. Chemical Studies Related to the Flavoenzyme
   Mixed-Function Oxidases: General Aspects 316
2. Historical Aspects and Postulations of Mechanisms 319
3. The Practical Synthesis of Models of the
   4a-Hydroperoxyflavin and the Postulated
   6-Amino-5-oxo-3*H*,5*H*-uracil Intermediate 321
4. The Reaction of 1,5-Dihydroflavins with Molecular Oxygen ... 326
5. Monooxygen Transfer from 4a,5-Dihydro-4a-hydroperoxy-5-
   alkyllumiflavins 331
6. Comparison of Reactions of Flavoenzyme Mixed-Function
   Oxidase Enzymes and 5-Ethyl-4a,5-Dihydro-4a-
   Hydroperoxylumiflavin which Result in Heteroatom
   Oxygenation 338
7. Reactions Involving Nucleophilic Additions to Carbonyl
   Functions.. 339
8. Hydroxylation of Electron-Rich Aromatic Rings 342
   Acknowledgments 350
   References... 350

1. CHEMICAL STUDIES RELATED TO THE FLAVOENZYME MIXED-FUNCTION OXIDASES: GENERAL ASPECTS

Scheme I depicts the type reaction carried out by the so-called flavin mixed-function oxidases (flavomonooxygenases). Enzyme-bound flavin cofactor is reduced by a dihydropyridine cofactor (a), and the resultant enzyme-bound dihydroflavin reacts with molecular oxygen to yield a reactive intermediate (b). The latter, upon reaction with an appropriate substrate, transfers an oxygen atom or "oxene equivalent" to the substrate molecule (c). In the absence of suitable substrate, the enzyme–dihydroflavin–O_2 intermediate decomposes to yield hydrogen peroxide and flavoenzyme–flavin (d). The

(a) enzyme–flavin + 1,4-dihydropyridine →
$$\text{enzyme–dihydroflavin + pyridinium}$$
(b) enzyme–dihydroflavin + O_2 → enzyme–dihydroflavin–O_2
(c) enzyme–dihydroflavin–O_2 + substrate →
$$\text{enzyme–flavin + substrate–O + H}_2\text{O}$$
(d) enzyme-dihydroflavin–O_2 → enzyme–flavin + H_2O_2

Scheme I

formation of hydrogen peroxide establishes that the O—O bond remains intact in the protein bound intermediate derived from O_2 and dihydroflavin. In the cycle involving substrate monooxygenation, 1,4-dihydropyridine cofactor and oxygen are consumed and water is a product (reaction 8-1).

$$\text{1,4-dihydropyridine} + O_2 + \text{substrate} \xrightarrow{\text{flavoenzyme}}$$
$$\text{pyridinium} + H_2O + \text{substrate—O} \quad (8\text{-}1)$$

The term mixed-function oxidase stems from the observation that these enzymes oxidize both substrate and 1,4-dihydropyridine cofactor by using one of the two oxygen atoms of O_2 for each reaction. The name flavoenzyme monooxygenase follows from the fact that the enzymes require flavin as cofactor and combines one of the oxygen atoms of O_2 with the substrate molecule. Flavins [in the form of flavin mononucleotide (FMN) and flavin dinucleotide (FAD)] are only present free in solution in the cell meluei at very low concentrations. Pyridine (nicotinamide) cofactors are present at much higher concentrations. For this reason, flavoenzymes are constituted by the very tight binding (often covalent[1]) of flavin cofactor and apoenzyme. Regeneration of the reactive oxidation state of the flavoenzyme is often carried out by its reaction with (as in the case of the flavoenzyme mixed-

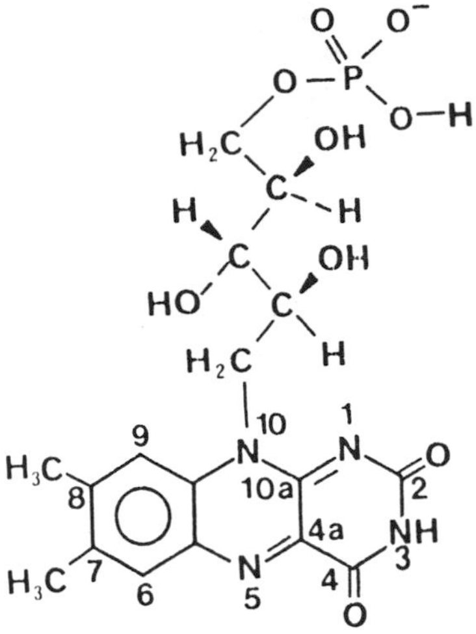

FMN

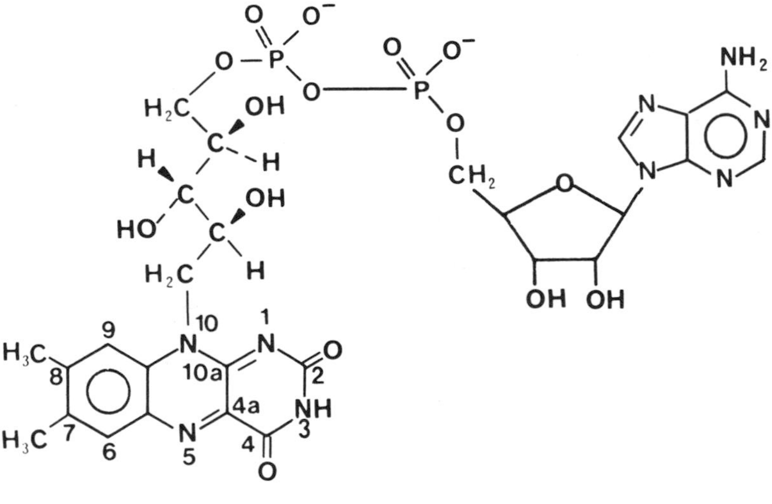

FAD

function oxidases) pyridine nucleotide coenzyme in the appropriate oxidation state. Reduction of flavin (Fl_{ox}) to 1,5-dihydroflavin (FlH_2) by 1,4-dihydronicotinamides has been shown, in model studies, to involve initial complex formation followed by intracomplex hydride transfer.[2] The structure of 1,5-dihydro, semiquinone, and oxidized states of a typical flavin (lumiflavin) and associated acid–base equilibria are provided in Scheme II.[3]

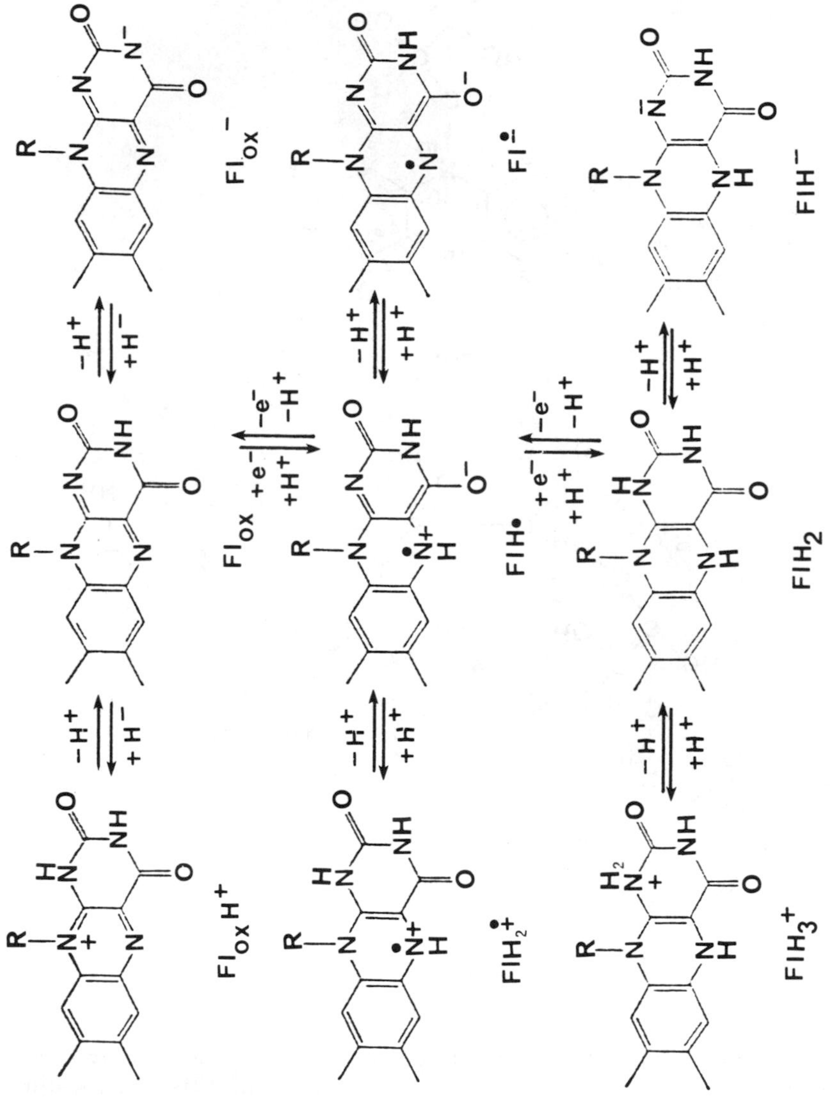

Scheme II

2. HISTORICAL ASPECTS AND POSTULATIONS OF MECHANISMS

An understanding of the mechanisms of the flavin mixed-function oxidases is dependent upon a knowledge of the answers to several questions. First, the mechanism for the reaction of O_2 with 1,5-dihydroflavins and the structure of the product which contains the elements of both O_2 and dihydroflavin must be known. Second, the mechanism for the transfer of a single oxygen from the dihydroflavin–O_2 adduct to substrate must be known. The various flavoenzyme mixed-function reactions may be divided into three types. These are (1) oxygen transfer to nucleophilic amines, organic sulfides, phosphines, etc., to provide N-oxides, sulfoxides, phosphine oxides, etc., (2) chemiluminescent oxidation of aldehydes to carboxylic acids, (3) oxidation of cyclic ketones to provide lactones, and (4) the hydroxylation of electron-rich benzenoid structures.

Most proposals of mechanism have assumed that O_2 reacts with 1,5-dihydroflavin to form a hydroperoxyflavin (FlHOOH, reaction 8-2) which then

$$1,5\text{-FlH}_2 + O_2 \rightarrow \text{FlHOOH} \qquad (8\text{-}2)$$

rearranges to form the monooxygenating species (Scheme III). Of these various proposals those of Hamilton have received particular attention.[8] In Hamilton's mechanism (Scheme III) there is proposed to be formed a 4a,5-dihydro-4a-hydroperoxyflavin (4a-FlHOOH) which undergoes ring scission between the C-4a and N-5 atoms to provide a carbonyl oxide. The latter is proposed to act as monooxygen transfer agent or to rearrange into a vinylogous ozonide which is the actual oxidizing agent. Following oxygen transfer to substrate a 6-amino-5-oxo-3H,5H-uracil is generated, which upon ring closure yields a 4a,5-dihydro-4a-hydroxylumiflavin (4a-FlHOH) (8-3). Loss of the elements of H_2O from 4a-FlHOH provides oxidized flavoenzyme.

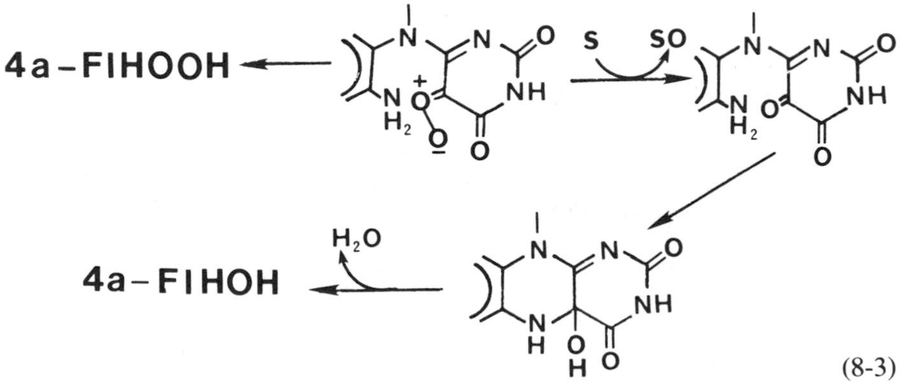

$$(8\text{-}3)$$

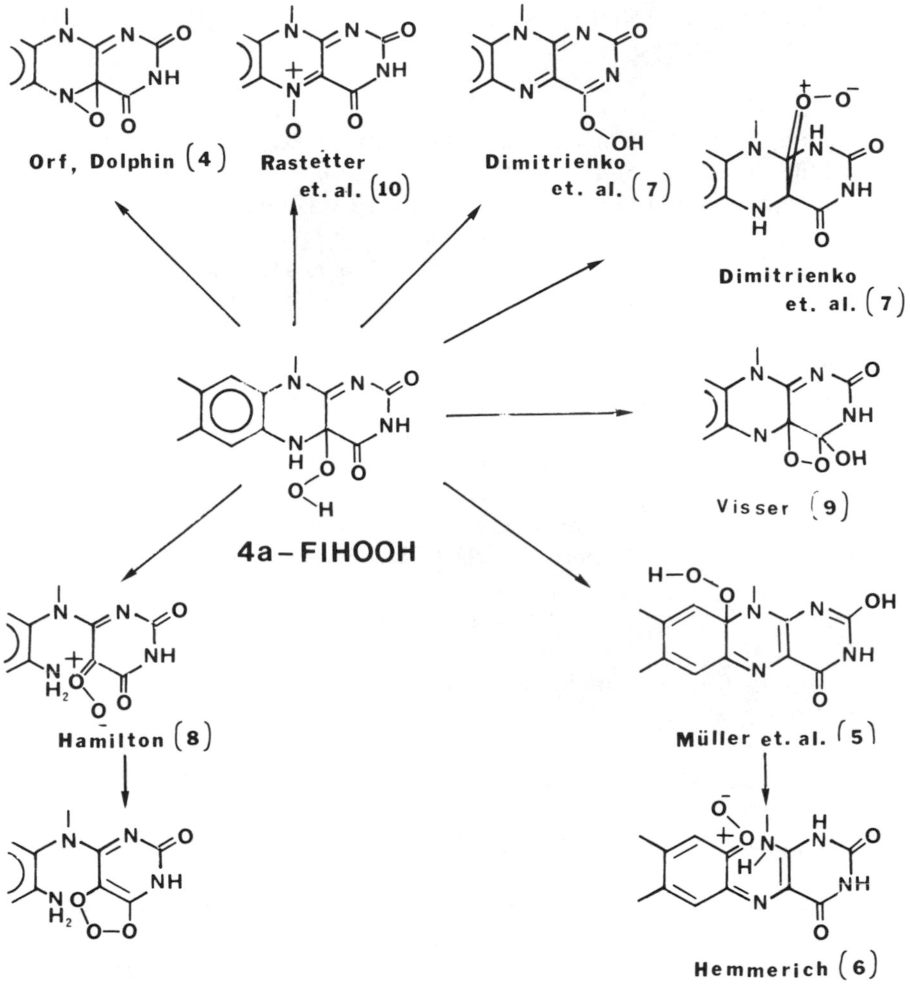

Scheme III

Much the same mechanism was postulated for the pterin- (biopterin)-dependent phenylalanine hydroxylase enzyme reaction (8-4).[8]

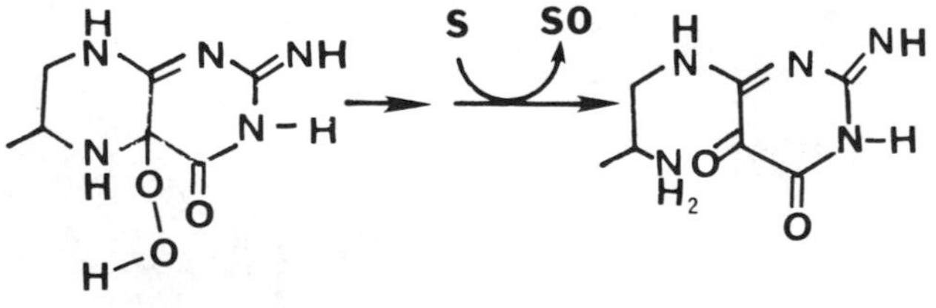

$$(8\text{-}4)$$

3. THE PRACTICAL SYNTHESIS OF MODELS OF THE 4a,5-DIHYDRO-4a-HYDROPEROXYFLAVIN AND THE POSTULATED 6-AMINO-5-OXO-3H,5H-URACIL INTERMEDIATE

Direct chemical approaches to the problem of the mechanism of the flavin mixed-function oxidase enzymes involved the synthesis and investigations of the chemistries of models for the postulated 4a,5-dihydro-4a-hydroperoxyflavin and 6-amino-5-oxo-3H,5H-uracil.

Because of the facile reversibility of the reaction, 4a,5-dihydro-4a-hydroperoxyflavin cannot be prepared by addition of H_2O_2 to oxidized flavin (reaction 8-5). Alkylation of the N-5 position of 4a,5-dihydro-4a-hydroperoxy-

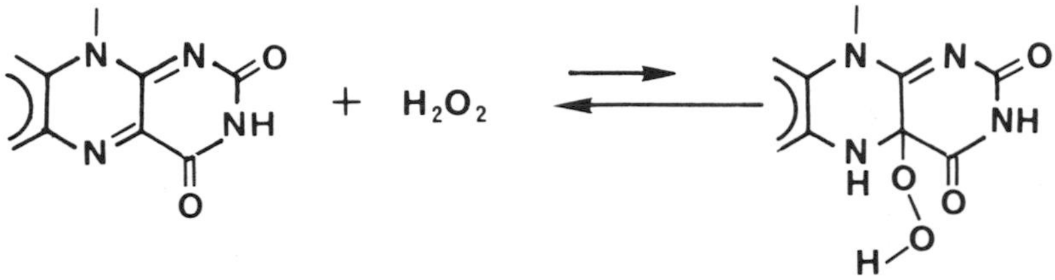

$$(8\text{-}5)$$

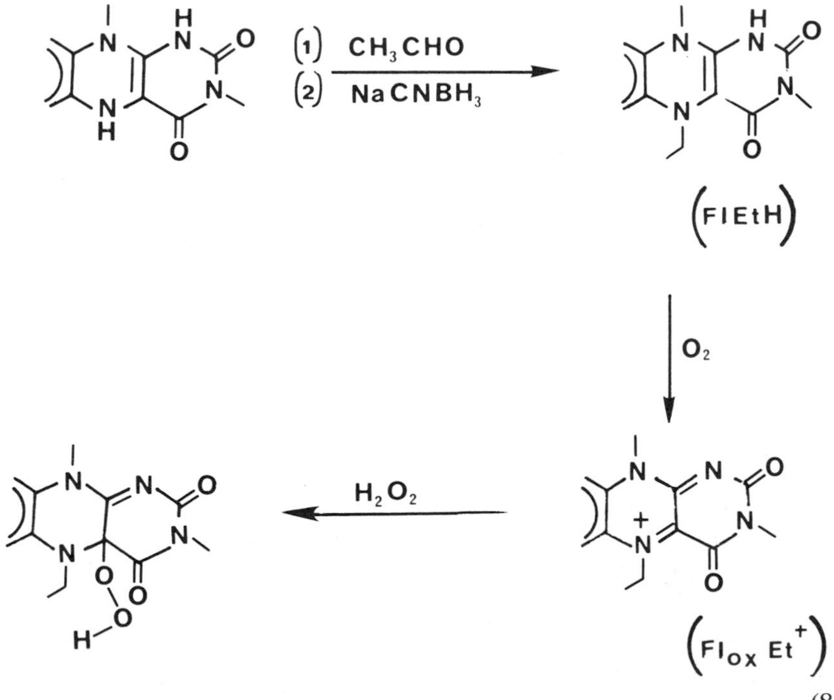

$$(8\text{-}6)$$

flavin species was expected to provide it with greater stability. Such as N^5-alkylated 4a-hydroperoxyflavin (4a-FlROOH) would most easily be realized by addition of H_2O_2 to an N^5-flavinium cation. Such flavinium cations $(Fl_{ox}R^+)$ can be prepared by alkylation of a 1,5-dihydrolumiflavin $(1,5\text{-}FlH_2)$ followed by oxidation. Alkylation of $1,5\text{-}FlH_2$ is best accomplished by its reaction with the appropriate aldehyde and reduction of the resultant N^5-imine (reaction 8-6).[12,28] When $R = CH_3$, the flavinium cation is not stable in buffered aqueous solution due to its susceptibility to general base-catalyzed deprotonation (reaction 8-7) followed by a complex but well-understood

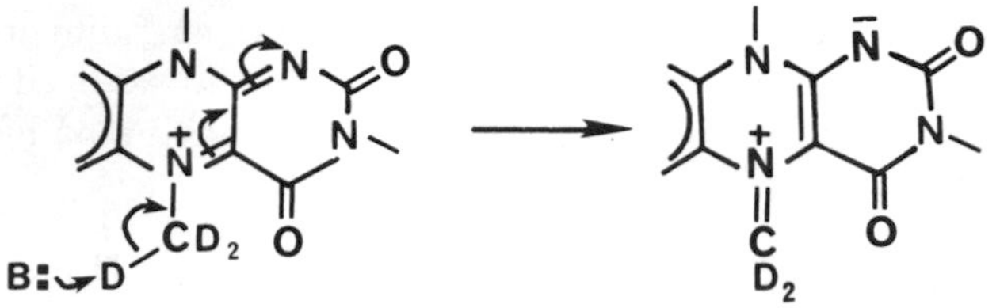

$$(8\text{-}7)$$

series of electron transfer and solvolytic reactions.[11] The kinetic deuterium isotope effect $(Fl_{ox}CH_3^+/Fl_{ox}CD_3^+)$ associated with reaction 8-7 were found to be 10.5 and 14.0 when $B: = H_2O$ and CH_3COO^-, respectively. Due to this large isotope effect, $Fl_{ox}CD_3^+$ was chosen for study as was the rather stable N^5-ethyl flavinium ion $Fl_{ox}Et^+$. The addition of either $Fl_{ox}CD_3^+$ or $Fl_{ox}Et^+$, as their perchlorate salts in acetonitrile, to a chilled, buffered (pH $\sim$ 6) aqueous H_2O_2 solution results in the precipitation of the corresponding 5-alkyl-1,5-dihydro-4a-hydroperoxylumiflavin (4a-FlROOH) as a yellow powder in up to approximately 96 purity (reaction 8-8).[12] When dissolved in H_2O

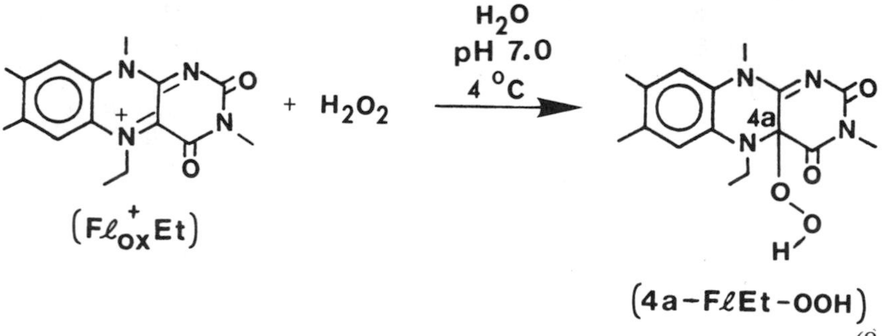

$$(8\text{-}8)$$

or CH_3OH the compounds 4a-FlCD$_3$OOH and 4a-FlEtOOH undergo an exchange reaction as shown in reaction 8-9. Solutions of the 4a,5-dihydro-4a-hydroperoxylumiflavins in dry and oxygen-free $tert$-butyl alcohol are stable

$$4a\text{-}FlEtOOH + CH_3OH \rightarrow 4a\text{-}FlEtOCH_3 + H_2O_2 \qquad (8\text{-}9)$$

for several days, and studies of the dynamics of their reactions have been successfully carried out in this solvent. The 5-alkyl-1,5-dihydro-4a-hydroperoxyflavins decompose immediately on dissolving in acetonitrile. The visible spectra of 4a-FlROOH species are typical of 4a-substituted 4a,5-dihydroflavins.[13]

The synthesis of an appropriate 6-amino-5-oxo-3*H*,5*H*-uracil was required in order to evaluate the proposal that such a compound is formed as an intermediate in flavoenzyme mixed function oxidase reactions. Compounds of this type were unknown. A priori, the simplest approach would involve 2e⁻ oxidation of an appropriate 6-amino-5-hydroxyuracil (reaction 8-10); however, it is known that oxidation of 6-amino-5-hydroxyuracil per se pro-

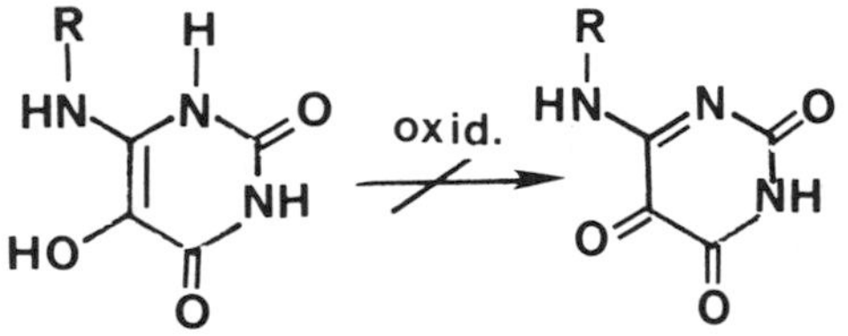

(8-10)

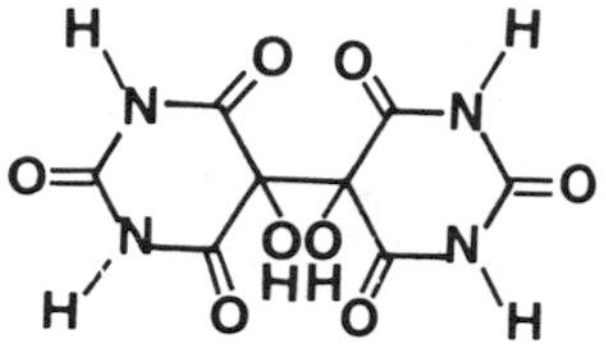

Alloxantin

vides the dimeric product alloxantin.[14] The synthesis of 6-[[2-(dimethyl-amino)-4,5-dimethylpheynl]methylamino]-3-methyl-5-oxo-3*H*,5*H*-uracil (**I**) is shown in Scheme IV.[15] The synthesis of **II**, in low yield, was accomplished

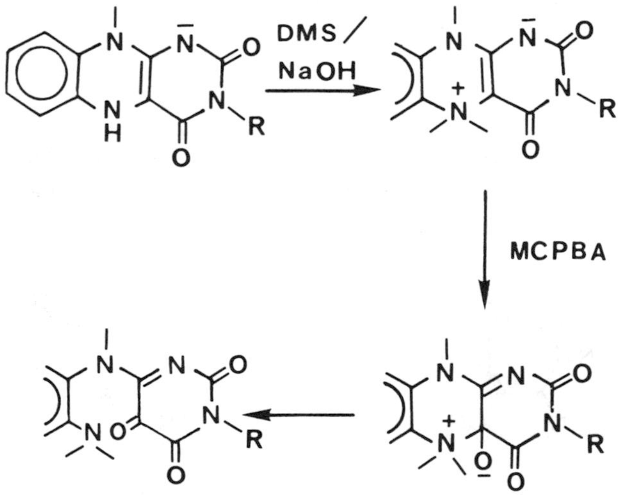

I, R = CH₃
II, R = H

Scheme IV

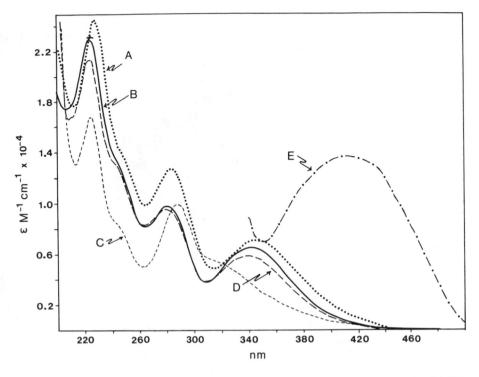

Figure 8-1. Comparison of the UV/visible spectra of **I** [in CH_3CN (line A)] and **II** [in CH_3CN (line B)] in *t*-BuOH in the presence of a 10-fold excess of *t*-BuO$^-$K$^+$ (line C) and in *t*-BuOH (line D) with the spectra of the intermediate observed to arise in time between enzyme bound 4a-FlHOOH and 4a-FlHOH in the reaction of *p*-hydroxybenzoate hydroxylase with the alternative substrate 2,4-dihydroxybenzoate (line E).

in much the same manner.[16] The ultraviolet visible (UV/vis) spectrum of **I** is included in Figure 8-1. The solution properties of **I** are provided in Scheme V. The compound is subject to hydrolysis by H_3O^+ catalysis involving the **I** ·

IIa   **IIa$^-$**

H$^+$ species, spontaneous hydrolysis of the **I** · H$^+$ and **I** species, and HO$^-$ catalyzed hydrolysis of **I** and **I** · HO$^-$ species. The compound is most stable

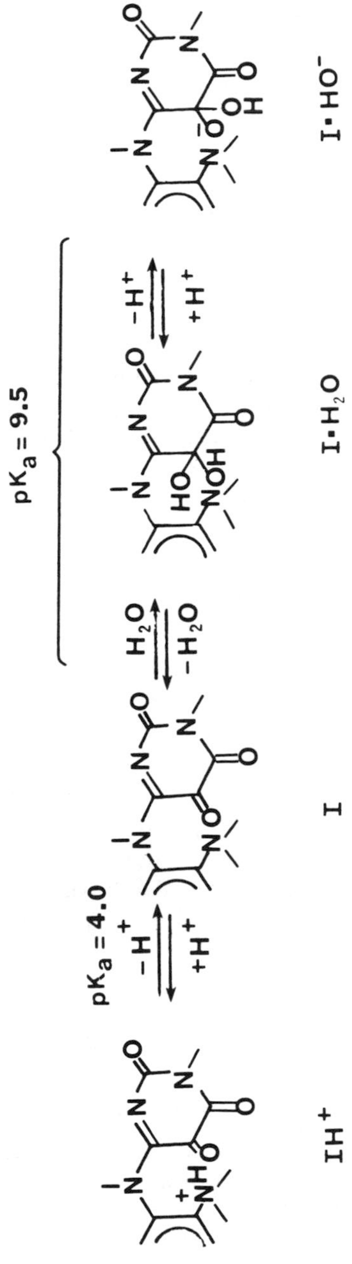

Scheme V

at neutrality. The hydrolytic reaction involves nucleophilic attack of water at C(6) to provide $N, N, N', 4,5$-pentamethyl-o-phenylene diamine as a product. **I** exhibits a quasi-reversible $1e^-$ reduction at -0.42 V which likely involves the formation of a semiquinone species which dimerizes to an alloxatin-like compound.

4. THE REACTION OF 1,5-DIHYDROFLAVINS WITH MOLECULAR OXYGEN

Spectral evidence (UV/vis) for an enzyme-bound dihydroflavin oxygen compound (Scheme 1) has been found for p-hydroxybenzoate hydroxylase and other phenol hydroxylases,[17,18] bacterial luciferase,[19] and microsomal flavin dependent monooxygenase.[20] The assignment of structure to the enzyme-bound dihydroflavin oxygen species is based on (1) our synthesis of the N^5-blocked 4a,5-dihydro-4a-hydroperoxyflavins (4a-FlROOH) and the establishment of the virtual identity of their spectra with those of the enzyme-bound dihydroflavin oxygen species (Table 8-1) and (2) the use of carbon-13 nuclear magnetic resonance (^{13}C-NMR) spectroscopy in the case of bacterial luciferase to show that dioxygen is bound to the C-(4a) position of the dihydroflavin cofactor.[22]

It has been appreciated for some time that the reaction of 1,5-dihydroflavins with oxygen constitutes a radical chain process (reaction 8-11).[23] The direct reaction of oxygen and 1,5-dihydroflavin constitutes an initiation step for the formation of oxidized flavin, and the major portion of 1,5-FlH_2 is consumed by its reacting with Fl_{ox} to provide flavin semiquinone ($FlH \cdot$) which reacts with oxygen. Attempts to elucidate the details of the mechanism for the direct reaction of 1,5-dihydroflavin with oxygen by use of initial rates[23d] have been shown to be fruitless, since this step constitutes only a small percentage of the reaction.[24]

TABLE 8-1. λ_{max} of Enzyme-Bound FlH_2O_2 Species and Comparison to the Spectrum of 4a-FlEtOOH

Enzyme	λ_{max} (nm)	ε (per M/cm)	Ref.
p-Hydroxybenzoate hydroxylase	382	8500	2a,b
and melilotate hydroxylase			2c
Phenol hydroxylase	380		2d
Bacterial luciferase	372		2e
4a-FlEtOOH	372	8200	12

(a) $1,5\text{-FlH}_2 + O_2 \rightarrow Fl_{ox} + H_2O_2$

(b) $Fl_{ox} + 1,5\text{-FlH}_2 \rightarrow 2(FlH \cdot)$

(c) $FlH \cdot + O_2 \rightarrow Fl_{ox} + HOO \cdot$

(d) $FlH \cdot + HOO \cdot \rightarrow Fl_{ox} + H_2O_2$ (8-11)

There are two possible mechanisms for the reduction of molecular oxygen to the hydroperoxide level by 1,5-dihydroflavins.[13] The first would require $2e^-$ transfer from the dihydroflavin to oxygen. Such a process would likely require that $^3O_2$ undergo a spin inversion and reaction of the resultant $^1O_2$ with singlet 1,5-dihydroflavin, or that 1,5-dihydroflavin be excited to its triplet state in order to react with $^3O_2$. The second mechanism involves two consecutive $1e^-$ transfers to O_2 with a radical pair intermediate (reaction 8-12). We have presented experimental evidence which supports the mechanism of reaction 8-12.[13,24]

1,5-dihydroflavin   Flavin radical

$$+ \quad \underset{k_2}{\overset{k_1}{\rightleftharpoons}} \quad + \quad \overset{k_3}{\longrightarrow} Fl_{ox} + H_2O_2 \qquad (8\text{-}12)$$

O_2 superoxide

The standard potentials for $1e^-$ reduction of $^3O_2$ to $O_2{}^{\cdot-}$ have been determined in a number of investigations[25] and may be corrected to any pH by use of the Nernst equation and the pK_a for $HOO \cdot \rightleftharpoons O_2 + H^+$ ($pK_a = 4.6$). We have determined[24] the pH dependence of the potentials for consecutive $1e^-$ reduction of a series of various substituted flavins. Combination of the $1e^-$ reduction potentials of reaction 8-13 at a given pH (E_0 pH) provides an electromotive force (emf) for the $1e^-$ reduction of $^3O_2$ by the 1,5-dihydroflavin species at this pH (reaction 8-14).

$$O_2 + 1e^- \rightarrow \text{superoxide}$$

$$\text{flavin radical} + 1e^- \rightarrow 1,5\text{-dihydroflavin} \qquad (8\text{-}13)$$

$$1,5\text{-dihydroflavin} + {}^3O_2 \rightarrow \text{flavin radical} + \text{superoxide} \qquad (8\text{-}14)$$

The pH dependence of the standard free energy, for the formation of flavin radical and superoxide from reduced flavin plus oxygen ($\Delta G°_{pH}$), was calculated from the pH-dependent emf values. Electrochemical calculations were also used to determine the pH-dependent standard free energies for the overall reaction of O_2 with dihydroflavin to yield oxidized flavin and H_2O_2. This was accomplished by combination of the known E_0 pH values for $1e^-$ reduction of $O_2{}^{\cdot-}$ and the determined values for $1e^-$ reduction of flavin to flavin radical. The pH dependence of the standard free energes of activation ($\Delta G\ddagger_{pH}$) for reaction of 1,5-dihydroflavins with oxygen were determined from kinetic data. From $\Delta G\ddagger_{pH} - \Delta G°_{pH}$ one obtains the pH-dependent free energies of activation for reduction of flavin radical by superoxide.

$$
\begin{array}{lcl}
O_2{}^{\cdot-} + le^- & \rightleftarrows & \text{hydrogen peroxide} \\
\text{flavin} + le^- & \rightleftarrows & \text{flavin radical}
\end{array}
\qquad (8\text{-}15)
$$

$$\text{flavin radical} + O_2{}^{\cdot-} \rightleftarrows \text{flavin} + \text{hydrogen peroxide}$$

By this means, k_1, k_2, and the overall equilibrium constant for reaction 8-12 at various pH values can be calculated. For a series of dihydroflavins it was found, at all pH values, that $\Delta G°_{pH} < \Delta G\ddagger_{pH}$, showing that the radical pair intermediate is always of lower free energy content than the critical transition state at all pH values (the computed values of k_2 vary from 10^{11} to 10^7 s^{-1}). Because the reactions of reduced flavins with O_2 are neither spe- cific- nor general acid catalyzed,[13,24b,26c] there can be no proton in the transi- tion state for the formation of flavin radical and superoxide. Furthermore, the value of k_2 is of such magnitude that diffusional capture of a proton by flavin radical or superoxide is not possible except at very acidic pH values. The only means by which the reactions may proceed to the eventual forma- tion of $Fl_{ox} + H_2O_2$ is by collapse of the radical pair to form a more stable covalent intermediate which may then capture a proton. This was estab- lished in a detailed study[24] of the reaction of O_2 with 5-ethyl-1,10-ethano-1,5- dihydrolumiflavin (**III**). This dihydroflavin does not possess a transferable proton.

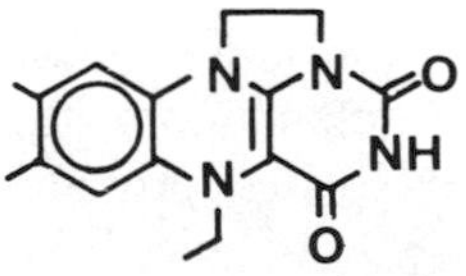

$$\textbf{III} \qquad\qquad \textbf{III}\cdot \qquad\qquad \text{not likely } \textbf{III}^{2+}$$

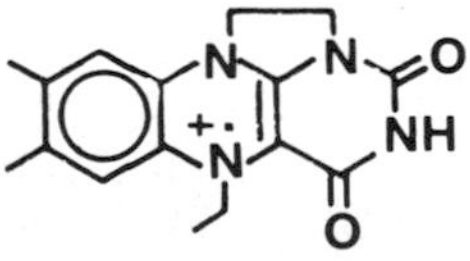

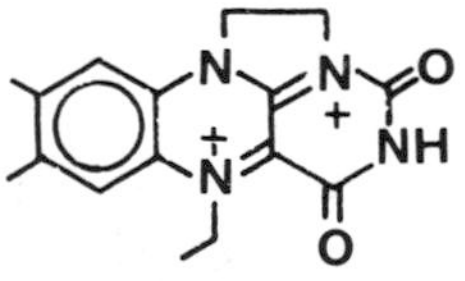

Also, **III** does not undergo an autocatalytic reaction with O_2 (reaction 8-11) due to the instability of $\textbf{III}^{2+}$. The established mechanism for the direct reaction of O_2 with reduced flavin in water to yield the 4,5-dihydro-4a- hydroperoxyflavin can be summarized as in reaction 8-16.

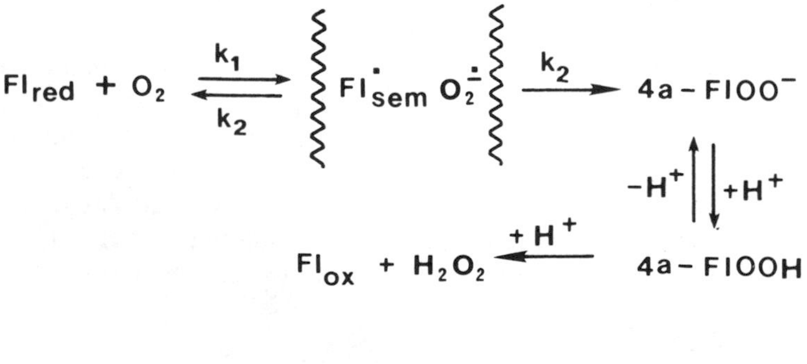

$$(8\text{-}16)$$

With 1,5-dihydro-5-ethyl-3-methyllumiflavin, the pH dependence of the apparent second-order rate constant for reaction with O_2 can be quantitatively explained by reaction 8-17. Plots of free energies of activation ($\Delta G‡$)

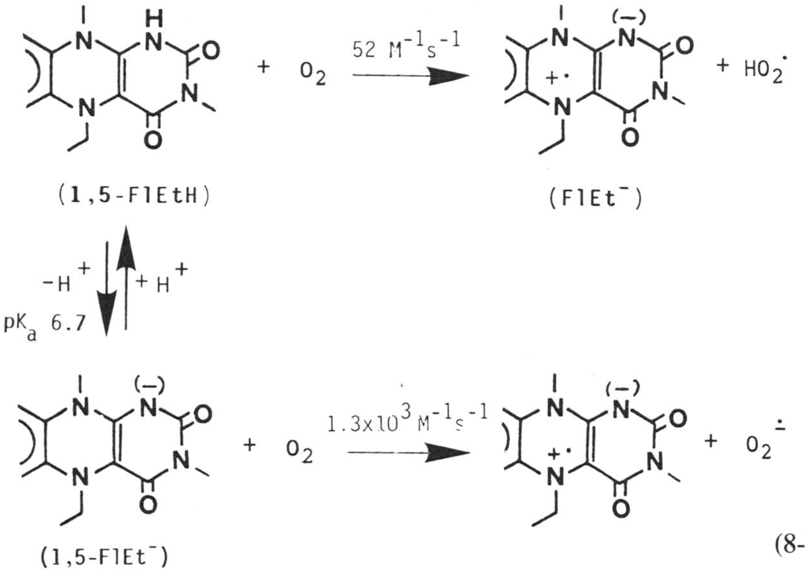

(8-17)

and electrochemically calculated standard free energies of formation of the flavin radical plus superoxide intermediate ($\Delta G°$) vs pH (Figure 8-2) are seen to possess the same shapes with a constancy of $\Delta\Delta G = \Delta G‡ - \Delta G°$. The rate constants for O_2 oxidation of undissociated and dissociated dihydroflavin are balanced so that the mole fraction of $O_2^{\cdot -}/HO_2\cdot$ formed at a given pH is predictable from the pK_a of $HO_2\cdot$. Thus, the radical species are formed in their most thermodynamically stable states. There seems to be little doubt concerning the validity of reaction 8-16. Figure 8-3 presents the calculated reaction coordinate for the O_2 oxidation of FlEtH at pH 4.6

Once formed, the participation of 4a-FlEtOOH in the overall oxidation of the 1,5-dihydroflavin species ($FlHCH_3$) in aqueous solution is dependent upon pH (Scheme VI).[13]

$$FlHCH_3 + O_2(+H^+) \xrightarrow{k_1} Fl_{ox}CH_3 + H_2O_2 \quad \text{(a)}$$

$$FlHCH_3 + Fl_{ox}{}^+CH_3 \xrightarrow{k_2} 2FlCH_3\cdot + H^+ \quad \text{(b)}$$

$$FlCH_3\cdot + O_2 \xrightarrow{k_3} Fl_{ox}{}^+CH_3 + O_2^-\cdot \quad \text{(c)}$$

$$O_2^-\cdot + FlHCH_3(+H^+) \xrightarrow{k_4} FlCH_3\cdot + H_2O_2 \quad \text{(d)}$$

$$Fl_{ox}{}^+CH_3 + {}^-OH \xrightarrow{k_5} 4a\text{-}FlCH_3\text{—}OH \quad \text{(e)}$$

pH > 8: (a), (b), (e); pH 7–8: (a), (b); pH < 7: (a), (b), (c), (d)

Scheme VI

In anhydrous *t*-BuOH and $HCONMe_2$, N^5-alkyl-1,5-dihydroflavins react with O_2 to yield stable solutions of 4a-FlROOH.[13] The same 4a-FlEtOOH is

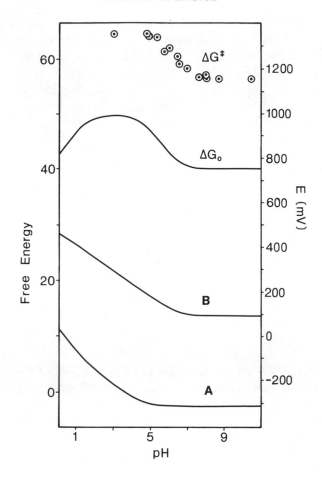

Figure 8-2. Plots of the experimentally determined free energies of activation ($\Delta G^\ddagger$) and calculated standard free energies (ΔG°) for formation of flavin radical plus superoxide on reaction of 1,5-dihydro-5-ethyl-3-methyllumiflavin with O_2 (30°C; H_2O) vs pH. Lines a and b represent the pH dependencies of the E_m for $O_2 + 1e^- \rightarrow O_2^-\cdot$ and FlEt$\cdot$ + 1e$^- \rightarrow$ FlEt$^-$, respectively.

formed on reaction of FlEt$\cdot$ with $O_2^{\cdot-}$ in HCONMe$_2$ solvent[27] and by the addition of H_2O_2 to N^5-alkylflavinium cations.[13,28]

$$\text{(1,5-FlEtH)} \quad + \ O_2 \quad \longrightarrow \quad \text{(4a-FlEt-OOH)} \qquad (8\text{-}18)$$

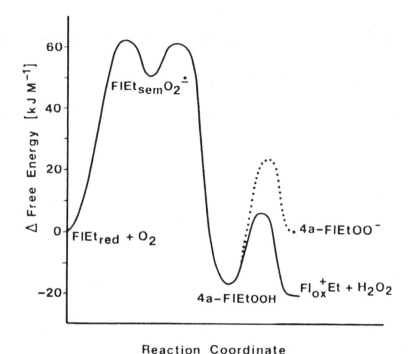

Figure 8-3. Reaction coordinate for the oxidation of N^5-ethyl-1,5-dihydro-3-methyllumiflavin by O_2 at 30°C in H_2O at pH 4.6.

5. MONOOXYGEN TRANSFER FROM 4a,5-DIHYDRO-4a-HYDROPEROXY-5-ALKYLLUMIFLAVINS

In humans the hepatic flavomonooxygenase plays a crucial role in the detoxification of natural and xenobiotic amines and sulfides by N and S oxidations. This is an important route for drug metabolism.[29] 4a-FlEtOOH reacts with hydroxylamines (reaction 8-19), tertiary amines (reaction 8-20), second-

$$4a\text{-}FlEtO_2H + PhCH_2\overset{\overset{\text{H}}{\overset{\text{O}}{\mid}}}{N}CH_3 \longrightarrow 4a\text{-}FlEtOH + PhCH_2\overset{\overset{\text{H}}{\overset{\text{O}}{\mid}}}{N}\rightarrow O$$

$$\underset{\text{CH}_3}{\mid}$$

$$-H_2O$$

$$PhCH=\underset{\overset{\mid}{\text{CH}_3}}{N}\rightarrow O \quad + \quad PhCH_2\underset{\overset{\parallel}{\text{CH}_2}}{N}\rightarrow O$$

ENZYMATIC   3.0 : 1

MODEL REACTION   2.2 : 1   (8-19)

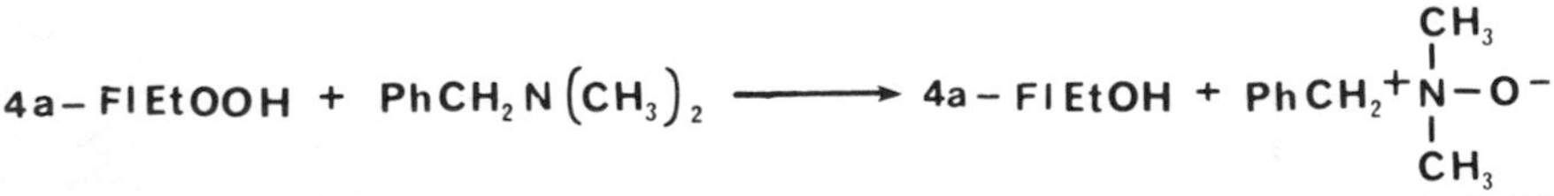

$$4a\text{-FlEtOOH} + PhCH_2N(CH_3)_2 \longrightarrow 4a\text{-FlEtOH} + PhCH_2\overset{CH_3}{\underset{CH_3}{\overset{+}{N}}}-O^-$$

(8-20)

$$4a\text{-FlEtOOH} + PhCH_2NH(CH_3) \longrightarrow 4a\text{-FlEtOH} + PhCH_2-\underset{CH_3}{\overset{}{N}}-OH$$

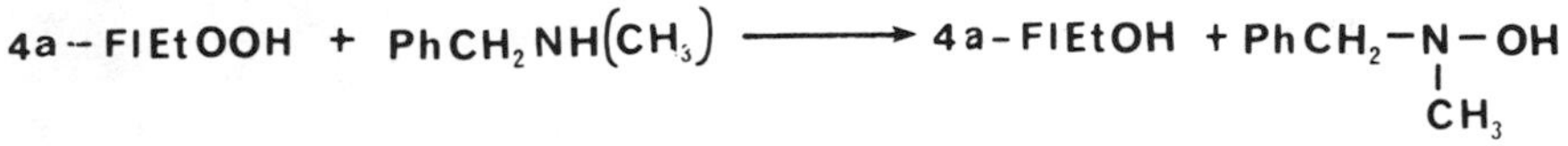

(8-21)

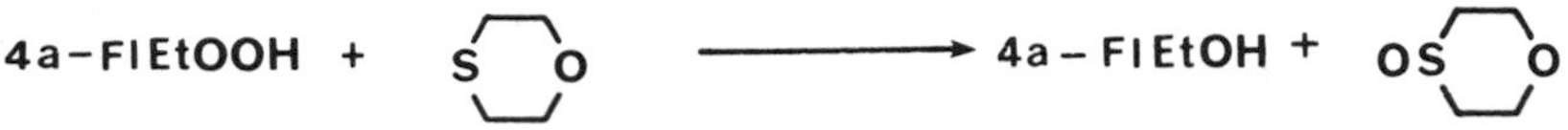

(8-22)

ary amines (reaction 8-21), and alkyl sulfides (reaction 8-22)[26,30a] to give the same products as does the flavoenzyme microsomal mixed-function oxidase.[31] The reactions 8-19 to 8-22 were studied in abs. *tert*-BuOH (30°C) where they are first order in [4a-FlEtOOH] and first order in [substrate].[13,26,32] The monooxygenated substrate and 4a-hydroxyfavin (4a-FlEtOH) are both formed in 100% yield. There is no evidence for the formation of any intermediate in these reactions. Separate studies have established that the N^5-blocked 4a-hydroperoxyflavin does not rearrange[33] to form either of the previously proposed[34] 9a- or 10a-isomers. The common

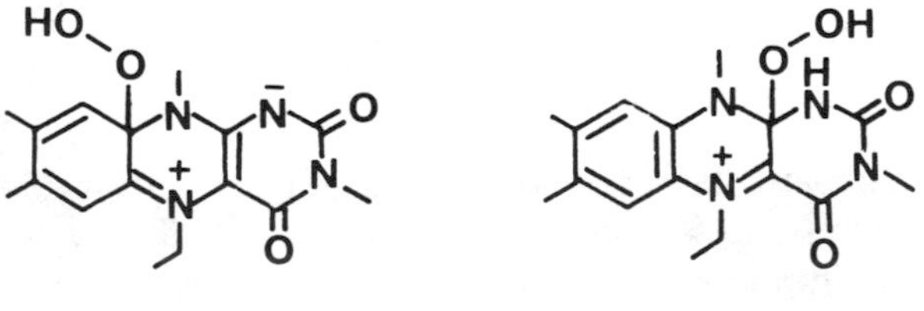

9a–FlEtOOH   10a–FlEtOOH

mechanism of reaction 8-23 suffices.

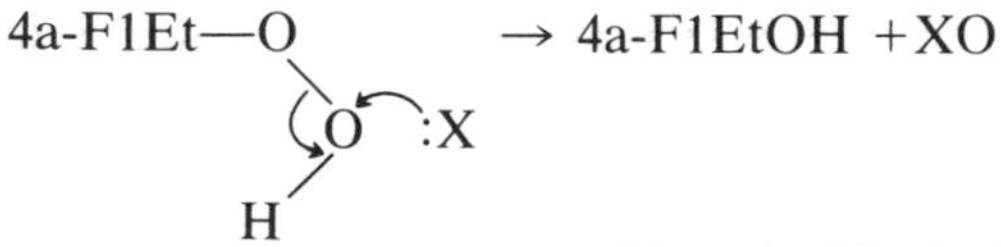

$$\text{X:} = \gg\!N\!:, >\!S, I^-$$

(8-23)

The logarithms of the second-order rate constants for the reaction of 4a-FlEtOOH with hydroxylamines, tertiary amines, and secondary amines (re-

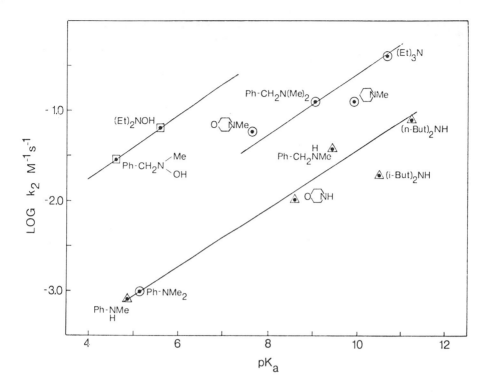

Figure 8-4. Plots of log k_2 for the reactions of 4a-FlEtOOH with amines in t-BuOH (30°C) vs pK_a of the amine in H_2O.

actions 8-19 through 8-21) are plotted vs the pK_a's (in H_2O) of the conjugate acids of the nitrogen bases in Figure 8-4.[26b] The slopes of the Brønsted plots (β_{nuc}) equal approximately 0.2. This establishes that the rates of N-oxidation by 4a-FlEtOOH are rather insensitive to the basicity of the nitrogen base. From Figure 8-4 the order of reactivity of amines with 4a-FlEtOOH is hydroxylamines > $tert$-amines > sec-amines > $prim$-amines. N-Oxidation of primary amines by 4a-FlEtOOH could not be detected.

The relative second-order rate constants for the oxidation of I^- to I_2 (k_I) and for the N- and S-oxygenation of N,N-dimethylbenzylamine (k_N) and thioxane (k_S) by different kinds of hydroperoxides are provided in Table 8-2.[32b] Examination of Table 8-2 establishes that 4a-FlEtOOH is not as good an oxygen donor as the peracid m-ClC$_6$H$_4$CO$_3$H, but that it is a much better oxygen donor than other hydroperoxides which we have investigated. If one plots log k_S vs log k_N, a straight line is obtained with a slope of 1.0. Likewise, if one plots log k_S vs log k_I, a linear plot is obtained of slope 1.1. The conclusion is inescapable; monooxygen transfer to the highly polarizable and negatively charged I^-, the neutral and polarizable S, and the basic but

TABLE 8-2. Second-Order Rate Constants for Reaction of I^- (k_I), Thioxane (k_S) and N,N-Dimethylbenzylamine (k_N) with YOOH Substrates.

Substrate	$k_I{}^a$	$k_S{}^a$	$k_N{}^a$
1	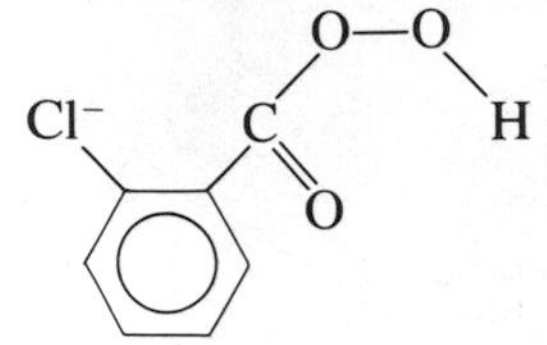 7.6×10^2	7.4×10^2	(2.14×10^3)
2	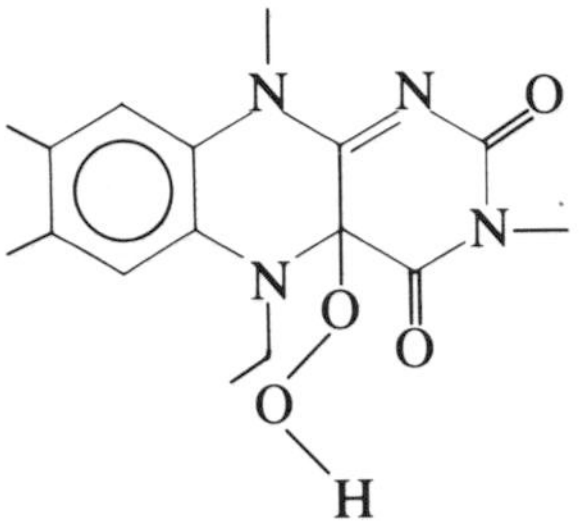 1.0 $(6.0 \text{ M}^{-1}\text{s}^{-1})$	1.0 $(0.12 \text{ M}^{-1}\text{s}^{-1})$	1.0 $(0.12 \text{ M}^{-1}\text{s}^{-1})$
3	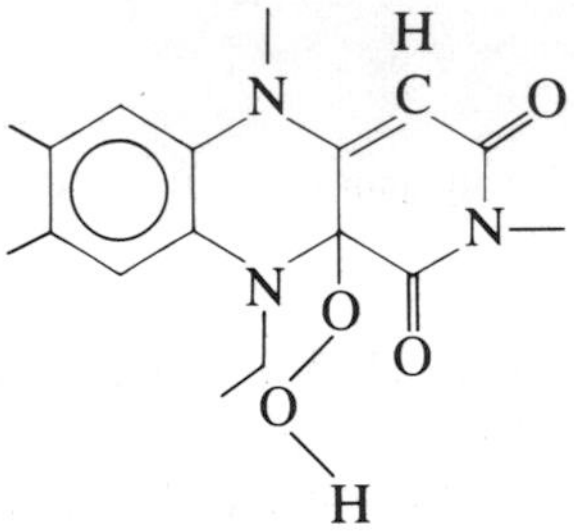 3×10^{-1}	5.7×10^{-2}	8.6×10^{-2}
4	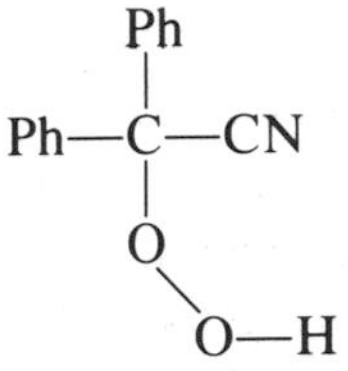 5.6×10^{-2}	8.7×10^{-1}	3.3×10^{-1}
5	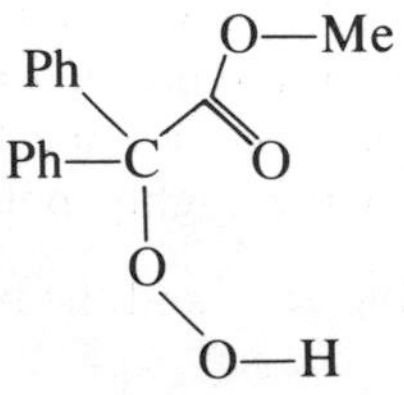 5.5×10^{-2}	3.2×10^{-2}	1.1×10^{-2}

TABLE 8-2. Cont.

Substrate	$k_I{}^a$	$k_S{}^a$	$k_N{}^a$
6 (spiro bicyclic hydroperoxide structure)	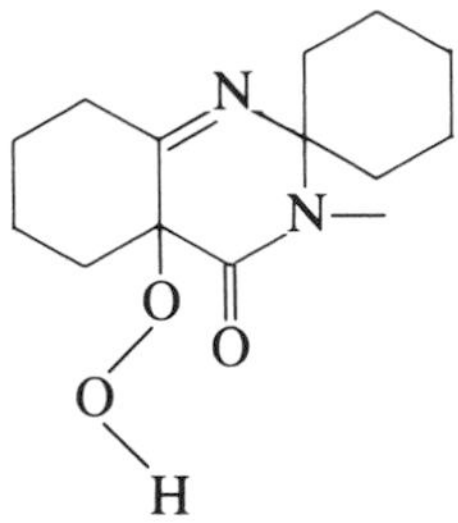 9.0×10^{-3}	9.2×10^{-4}	1.2×10^{-4}
7 $\quad$ H_2O_2	1.0×10^{-3}	1.4×10^{-4}	2.8×10^{-5}
8 $\quad$ $CH_3-\overset{\overset{\displaystyle CH_3}{|}}{\underset{\underset{\displaystyle CH_3}{|}}{C}}-O-OH$	1.0×10^{-3}	5.8×10^{-6}	$>1 \times 10^{-6}$

^{a} k_1, Oxidation of I^- to I_2.

hard $N\!\!<$ is dependent upon the same electronic and steric features of the hydroperoxides. Furthermore, the logarithms of the second-order rate constants for monooxygen transfer from hydroperoxides (YOOH) are directly related to the pK_a of YOH as shown in Figure 8-5.[32] In Figure 8-5 the point for hydroperoxide **6** (not shown) exhibits a large negative deviation due, presumably, to a marked steric effect brought about by its spiro structure. The slope of the line in Figure 8-5 is $\beta_{lg} = -0.6$. The Brønsted β_{nuc} for the

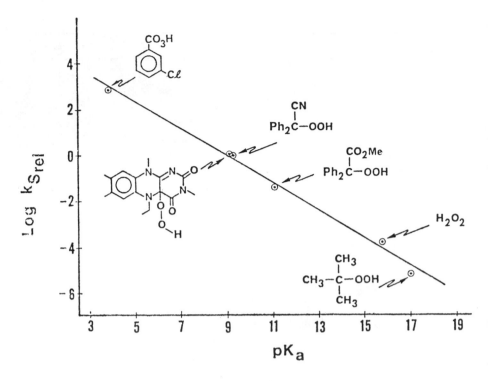

Figure 8-5. Plot of the log of the second-order rate constants relative to the rate constant for 4a-FlEtOOH) for sulfoxidation of thioxane (k_S) by the ROOH species vs the pK_a of ROH species (solvent absolute t-BuOH for the rate constants and H_2O for pK_a values, 30°C).

reaction of hydroxylamines, tertiary amines, and secondary amines with 4a-FlEtOOH is +0.2 (Figure 8-5). We may conclude that (1) from β_{nuc}, the transition state for nucleophilic attack of :S⟨ , :N— ,or I⁻ upon the terminal oxygen of YO—OH is early, and the rate of oxygen transfer is dependent upon the ground state inductive polarization of the O—O bond and the leaving ability of YO⁻ (both determined by β_{lg}); (2) percarboxylic acids, organic hydroperoxides, and hydrogen peroxide comprise a common series of YOOH oxygen donors, the advantage of percarboxylic acids being the greater stability of YO⁻. It may also be concluded that cyclic proton transfer, as proposed by Bartlett[35] for epoxidation by percarboxylic acids, is not of importance in the percarboxylic acid oxidation of I⁻, amines, and organic sulfides. This is shown by the observation that percarboxylic acids and alkyl hydroperoxides must share a common mechanism, and the alkyl hydroperoxides can not undergo cyclic proton transfer as shown in 8-24. The internal proton transfer mechanism proposed by Rebek[36] for the epoxidation of olefins by 1,1-diphenyl-1-hydroperoxyacetate ester and 1,1-diphenyl-1-hydroperoxyacetonitrile is also not important, since these hydroperoxides share in common a mechanism with other alkyl hydroperoxides, which are incapable of internal hydrogen bonding of the hydroperoxide proton.

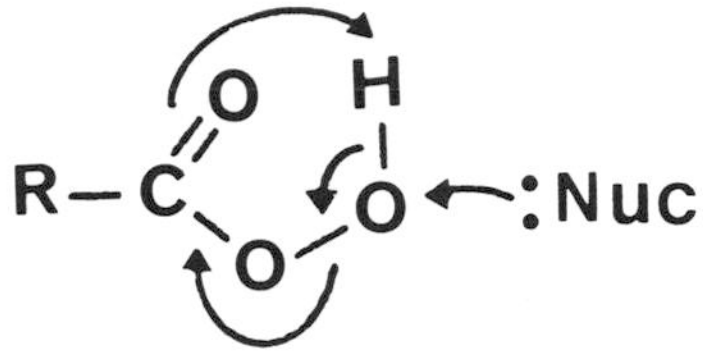

$$R-C \quad O-H \quad :Nuc \tag{8-24}$$

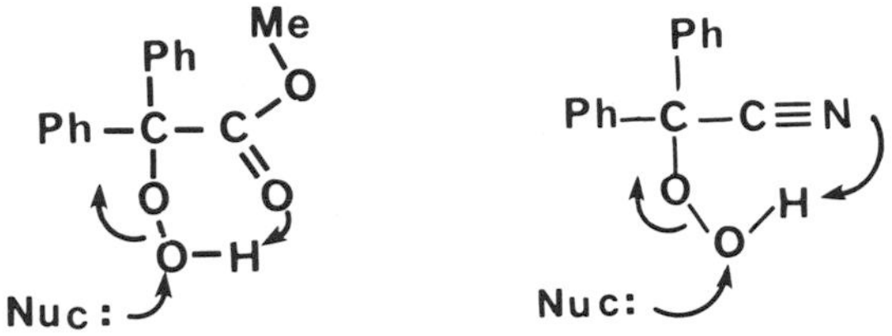

$$\tag{8-25}$$

The pK_a of the alcohol (4a-FlEtOH) obtained upon oxygen transfer from 4a-FlEtOOH to I⁻, S or N— etc., cannot be determined by base titration due to the ring contraction reaction 8-26 (and in the presence of oxygen, base hydrolysis of the flavin C-ring to provide 1,6,7-trimethyl-4-ethylquinoxaline-2,3-dione).[37,38] An apparent pK_a value of 9.9 for proton dissociation of 4a-FlEtOH was obtained from the pH dependence of the rate constants associated with the formation of the 10a-spirohydantoin and quinoxaline dione.[38] An alternate approach to the estimation of the pK_a involved the use of σ_i values[39] and the consideration of 4a-FlEtOH as a substituted methanol.[40] The electronic model chosen to best represent 4a-FlEtOH was [Ph(Et)N-

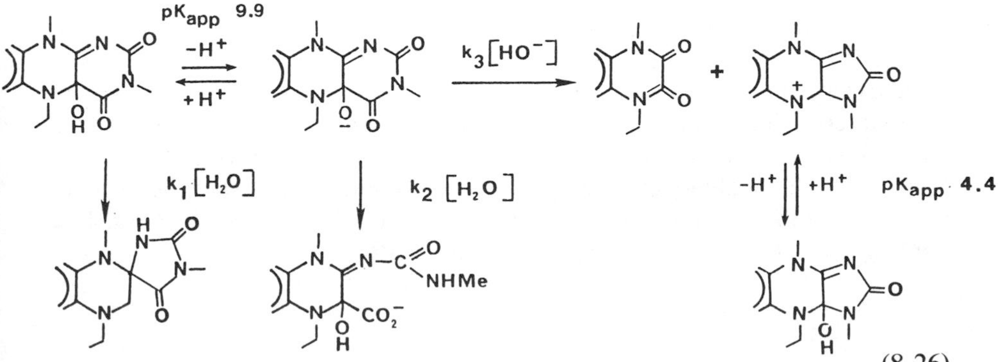

$$\tag{8-26}$$

[H$_2$NCO][HCONHCO]C—OH, and the pK_a obtained was 9.4. Inspection of Figure 8-5 shows that the reactivity of the biologically important 4a-hydroperoxyflavins is due to the electronegativity of the 4a-position so that its derived YOH species possesses a pK_a comparable to phenol. The electrophilicity of the 4a-position is due to the electronegativity of the ring nitrogens at positions 1,5 and 10 and the carbonyl group at position 4.

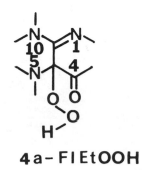

4a–FlEtOOH

6. COMPARISON OF THE REACTIONS OF FLAVOENZYME MIXED-FUNCTION OXIDASE ENZYMES AND 5-ETHYL-4a,5-DIHYDRO-4a-HYDROPEROXYLUMIFLAVIN WHICH RESULT IN HETEROATOM OXYGENATION

The rates of oxygen transfer to the nitrogen of amines by hepatic flavomonooxygenase follows the order hydroxylamines > tertiary amines > secondary amines > primary amines which do not serve as substrates.[29] The very same order of reactivity is found when 4a-FlEtOOH is the oxidant and primary amines are not oxidized at a measurable rate. Oxidation of *N*-benzyl-*N*-methylhydroxylamine by both hepatic flavomonooxygenase and 4a-FlEtOOH provides two isomeric nitrones, and in each case they are found to be formed in the same ratio (see reaction 8-19). Finally, reaction of reduced hepatic flavomonooxygenase with molecular oxygen provides an enzyme-bound flavin derivative that posseses a visible spectrum assignable to a 4a-hydroperoxyflavin, and monooxygenation of substrate yields a second enzyme intermediate whose spectrum is assignable to a 4a-hydroxyflavin. The latter converts to oxidized flavoenzyme (reaction 8-27).[41] Through the investigation of chemical models, we have gained an understanding of the mechanism of mammalian microsomal flavin mixed-function oxidases. No other approach could have been used to gain this insight.

$$\text{Enz·FlH}_2 \xrightarrow{\text{O}_2} \text{Enz·4}\alpha\text{-FlHOOH} \xrightarrow{\text{S} \quad \text{SO}}$$

$$\text{Enz·4}\alpha\text{-FlHOH} \rightarrow \text{Enz·Fl}_{\text{ox}} + \text{H}_2\text{O}$$

(8-27)

Oxidations by all flavoenzyme mixed-function oxidases involve an enzyme bound 4a-hydroperoxyflavin. The bacterial flavoenzyme cyclohexanone oxygenase converts cyclic ketones into ring-expanded lactones (reaction 8-29). The 4a-hydroperoxyflavin species of cyclohexanone oxygenase carries out heteroatom monooxygenation much like the hepatic flavoenzyme mixed function oxidase.[42] A class of bacterial luciferases are known to be flavoenzyme mixed-function oxidases. They catalyze the chemiluminescence oxidation of aldehyde substrates (vida infra). These enzymes also catalyze heteroatom monooxygenation insofar that they convert organic sulfides to sulfoxides.[43]

7. REACTIONS INVOLVING NUCLEOPHILIC ADDITIONS TO CARBONYL FUNCTIONS

The bacterial luciferases are a class of flavin mixed-function oxidases responsible for the bioluminescence of certain fish.[44] Chemiluminescence accompanies the decomposition of an adduct of the 4a-hydroperoxide of flavin mononucleotide and a long chain aldehyde. Rather convincing evidence has been presented in support of the emitter being the enzyme bound 4a-hydroxyflavin[45] which has been proposed to be formed via a Baeyer–Villiger rearrangement (reaction 8-28).[46] The bacterial cyclohexanone oxygenase

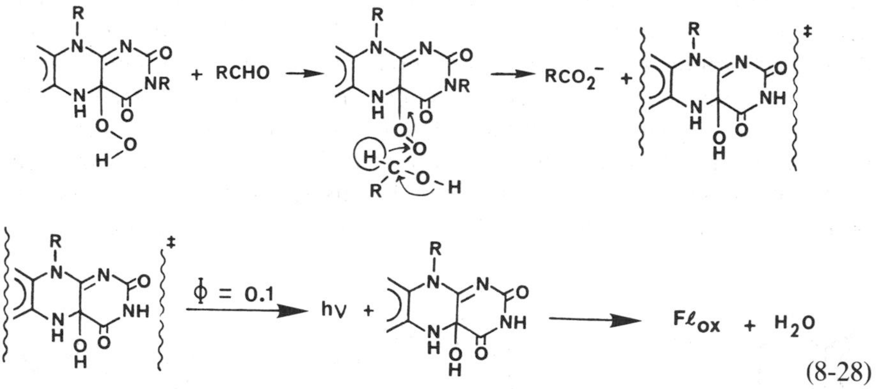

$$(8-28)$$

converts cyclic ketones into ring-expanded lactones. The proposed mechanism involves nucleophilic addition of the peroxide moiety of the 4a-FlHOOH to the ketone carbonyl followed by Baeyer–Villiger rearrangement (reaction 8-29).

Though the mechanism of reaction 8-29 represents a typical hydroperoxide reaction, our attempts to carry out such a cyclic ketone oxidation with 4a-FlEtOOH in *tert*-butanol solvent have been unsuccessful. Even the ring-strained cyclobutanone proved to be unreactive with 4a-FlEtOOH.

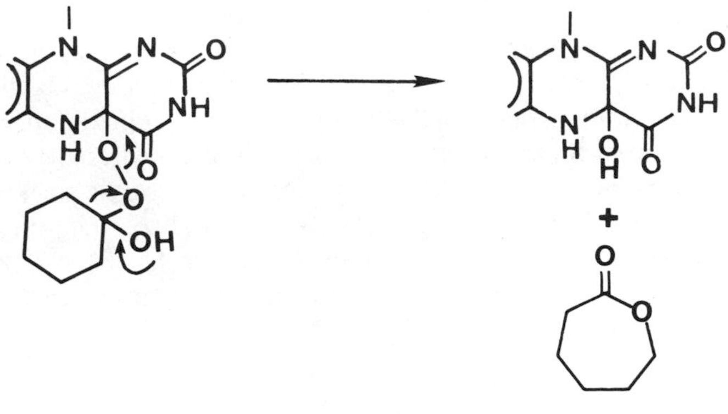

$$(8\text{-}29)$$

Some success has been obtained in the modeling of the bacterial luciferase reaction.[12,28,30a,47] We have found (dry *tert*-BuOH solvent under N_2) that the reaction of 4a-FlEtOOH with excess of an aldehyde follows pseudo-first-order kinetics and that the reaction can be followed both by spectrophotometrically monitoring the formation of oxidized flavin and by photon counting. The pseudo-first-order rate constants determined by both methods are equivalent. Light emission on decomposition of adducts of reactive hyroperoxides with aldehydes is not a general phenomenon. Thus, benzaldehyde adducts of the reactive peroxides **4** and **5** (Table 8-2) do not chemiluminesce whereas the benzaldehyde adduct of 4a-FlEtOOH does. Chemiluminescence accompanies the decomposition of compounds of structure **IV** and **V** [quantum yield (Φ) $= 10^{-3}$ to 10^{-4}], but no chemiluminescence can be detected on decomposition of compounds of structure **VI**. That the light emission on decomposition of **V** compounds is comparable to that for **IV** compounds is not in accordance with the Baeyer–Villiger mechanism suggested (reaction 8-28) for the enzymatic reaction.[47] There can be no doubt, however, that light emission in the enzymatic reaction stems from the generation of excited 4a-hydroxyflavin species.[46] 4a-Hydroxyflavins are highly fluorescent when bound to a protein or immobilized in a glass. Due to this feature and the

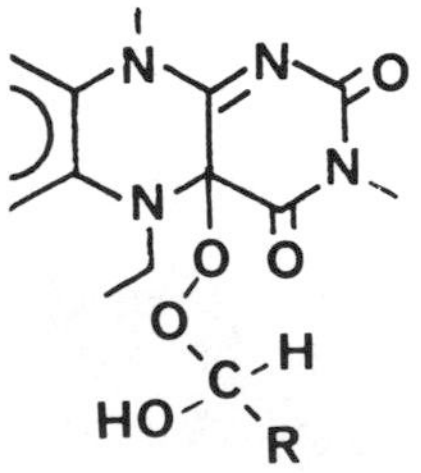

IV 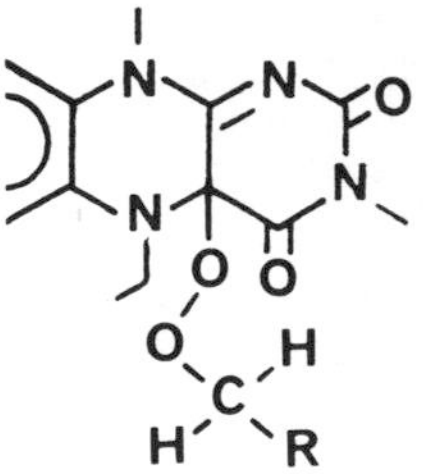   **V** 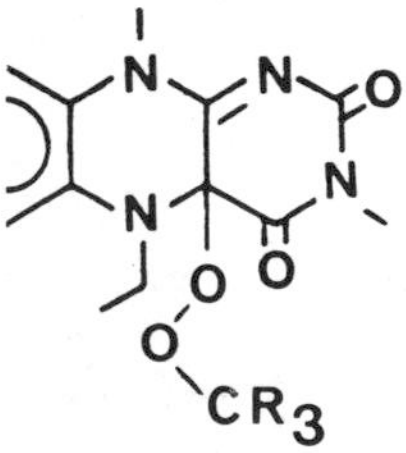   **VI**

quantum yield of the enzymatic reaction ($\Phi = 0.2$), it is possible to obtain the spectrum of the emitted light. It proves to be comparable to that of the fluorescence spectrum of 4a-hydroxy-5-ethylriboflavin when bound to protein.[48] The actual emitter in the chemical systems is not known due to the quantum yield being lower than in the enzymatic reaction as well as the rates of reaction being slower than in the enzymatic reaction. These features result in a low intensity of light emission per given time which has prevented the determination of the chemiluminescent spectra of the excited species. The lower quantum yield for the chemical system may be attributed to the low efficiency of light emission from excited 4a-FlEtOH when in homogeneous solution. It is not unreasonable to tentatively assume that excited 4a-FlEtOH is the emitter in the chemical system.

Substitution of the α-protons of **IV** or **V** by deuterium decreases the quantum yield by twofold but does not influence the rate of reaction. Much the same isotope effect is observed in the enzymatic reaction. This finding, along with the lack of chemiluminescence of **VI**, supports a mechanism which involves C—H/D bond scission in the light-producing reaction. Possible mechanisms which may be considered include a Russell fragmentation reac-

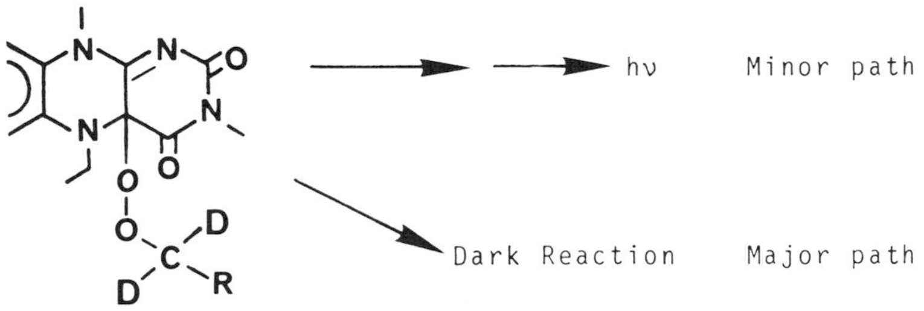

tion (reaction 8-30) which would yield the 6-amino-5-oxo-3*H*,5*H*-uracil of Hamilton as the emitting species.[47a] The synthesis of the 6-amino-5-oxo-3*H*,5*H*-uracils **I** and **II** and our finding that these compounds are nonfluorescent in solution rules out the mechanism of reaction 8-30. The mechanism of the chemiluminescent decay of compounds **IV** and **V** is not yet understood. We propose the free radical chemically initiated electron-exchange lumines-

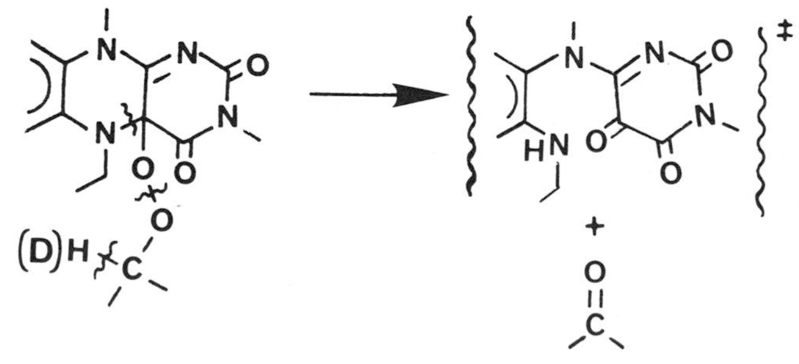

(8-30)

cence (CIEEL)[47b] mechanism of reaction 8-31. Attempts to employ N[5]-meth-

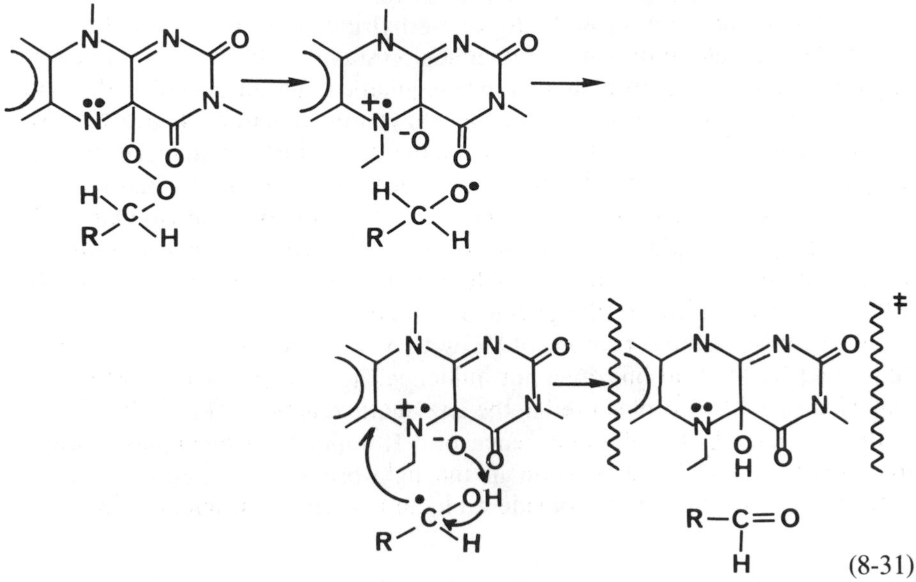

$$(8\text{-}31)$$

ylated FMN in the enzymatic reaction have been thwarted due to the inability of the N[5]-alkylated cofactor to bind to apoenzyme. Apparently there is not bulk tolerance at the active site for such a substitution. This observation may explain the stability of 4a-hydroperoxy FMN at the active site since elimination of the N[5]-proton may be sterically blocked by the protein structure or by hydrogen bonding (8-31).

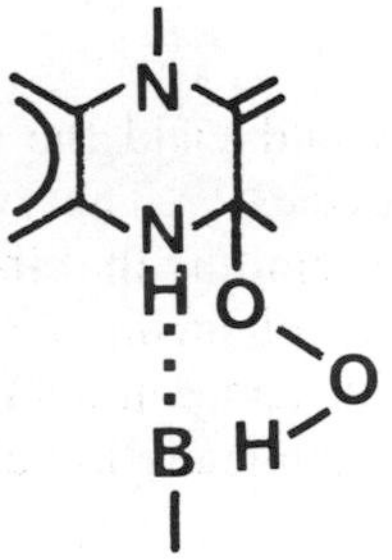

$$(8\text{-}32)$$

8. HYDROXYLATION OF ELECTRON-RICH AROMATIC RINGS

The hydroxylation of electron-rich aromatic ring structures by another class of flavoenzyme mixed-function oxidases is not yet understood. Epoxidation by flavin hydroperoxide to yield an intermediate arene oxide[49] can probably be ruled out, since 4a-FlEtOOH is incapable of the epoxidation of such reactive alkenes as 2,3-dimethyl-2-butene.[32b] Hamilton's proposal[8] that the flavin–oxygen species responsible for aromatic hydroxylation is a carbonyl oxide or vinylogous ozonide (see discussion with reaction 8-3) has received particular attention in the consideration of mechanisms for these enzymes. A carbonyl oxide type intermediate has also been proposed for the biopterin requiring phenylalanine hydroxylase. In this instance rearrangements of a 4a-hydroperoxy pteridine and oxygen transfer to phenylalanine was proposed to yield a 6-amino-5-oxo-3H,5H-uracil as the immediate product.[50] Results of Benkovic and co-workers establish that the 5-oxo-3H,5H-uracil is not an intermediate in the phenylalanine hydroxylase reaction.[51]

Entch, Ballou, and Massey[21b] observed, in the oxidation of certain alternate substrates by p-hydroxybenzoate hydroxylase, that a spectrally observable intermediate (λ_{max} 395–420; ε = 15,000) appears in time between the enzyme bound 4a-FlHOOH and 4a-FlHOH. The mechanism of reaction 8-33 was proposed. The investigation is of considerable importance, since it represents the only instance of the observation with a flavoenzyme mixed-function oxidase of the appearance of an intermediate on conversion of 4a-FlHOOH to 4a-FlHOH with accompanying monooxygenation of substrate. It represents the only support for Hamilton's intermediate. From comparison of the spectrum of the enzyme intermediate to the spectrum of I and II in solvents of various dielectric constants and basicities, one may conclude that the species formed in the enzymatic reaction does not exhibit the spectral characteristics of the 6-amino-5-oxo-3H,5H-uracils (for example, see Figure 8-1).

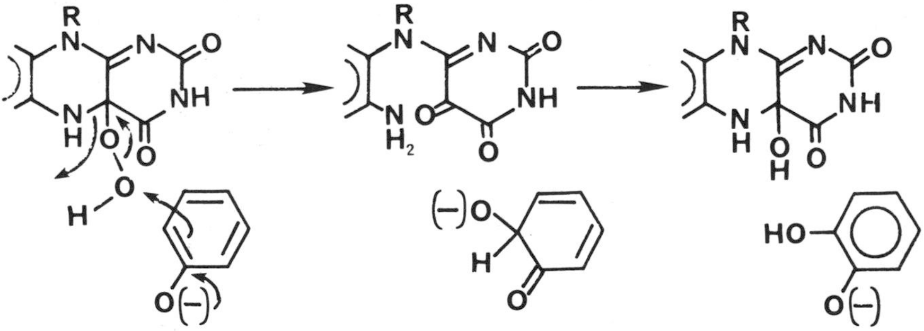

(8-33)

It is known that phenolate serves as the actual substrate for phenol hydroxylase enzymes. The enzymatic hydroxylation of phenolate anions by bacterial flavin mixed-function oxidases may simply involve nucleophilic attack of the ambident substrate upon the terminal oxygen of 4a-FlEtOOH (reaction 8-34). This would represent a conservation of mechanisms with the heteroatom oxygenations. For the operation of this mechanism, one might envision the phenolate oxygen at the active site surrounded by nonpolar functional groups. This would result in the lack of charge solvation and the delocalization of the negative charge to the aromatic ring where it could partake in the nucleophilic attack upon the terminal peroxy oxygen as shown in reaction 8-34. It is not possible to test this mechanism in a chemical

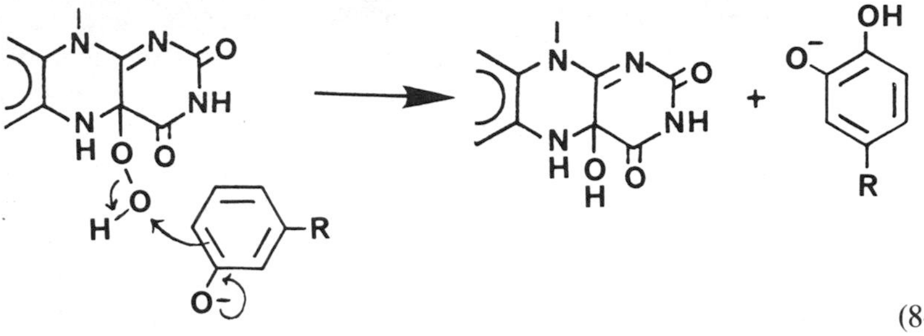

(8-34)

system because the addition of sufficient base to dissociate the phenol proton also dissociates the proton of the flavin hydroperoxide. We were surprised to find that transfer of dioxygen from FlEtOO$^-$ to phenolate ion occurred in tert-BuOH solvent (dry under N$_2$ atmosphere with sufficient tert-BuO$^-$K$^+$ to dissociate reactants).[52]

Dioxygen transfer from FlEtOO$^-$ to phenolate provides an alternate mechanism for hydroxylation (reaction 8-35).[52d] The dioxygen transfer from

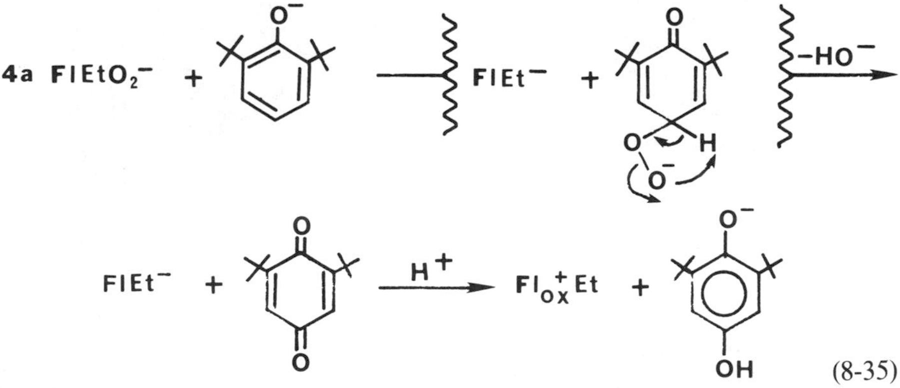

(8-35)

4a-FlEtO$_2$$^-$ to ambident nucleophiles has no precedence in organic peroxide chemistry. Interestingly, oxygen transfer from 4a-FlEtOO$^-$ results in the regeneration of 1,5-dihydro-5-ethylflavin anion (FlEt$^-$). Since FlEt$^-$ reacts with molecular oxygen to provide 4a-FlEtOO$^-$, there is obtained a catalysis, by FlEt$^-$, of the peroxidation of substrate by molecular oxygen (for example, reaction 8-36).

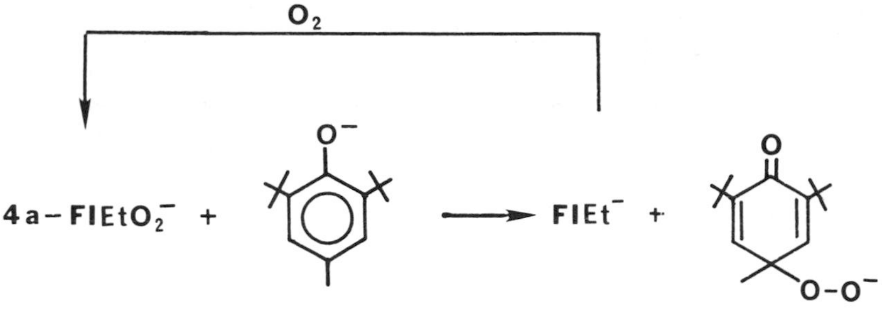

$$O_2$$

$$4a-FlEtO_2^- \ + \quad [\text{structure}] \quad \longrightarrow \quad FlEt^- \ + \quad [\text{structure}]$$

$$> 95\% \qquad > 95\%$$

$$(8\text{-}36)$$

The formation of a peroxide anion from the ambident nucleophile may result in C—C bond scission (reactions 8-37 and 8-38), a characteristic of certain flavoenzyme mixed-function oxidases.[53] The C—C bond scissions result

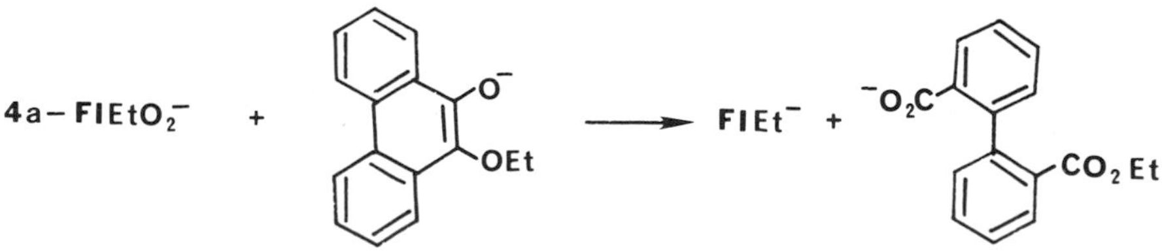

$$4a-FlEtO_2^- \quad + \quad [\text{structure}] \quad \longrightarrow \quad FlEt^- \ + \quad [\text{structure}]$$

$$(8\text{-}37)$$

from a Hock rearrangement of the initially formed hydroperoxide anion (8-39).[55]

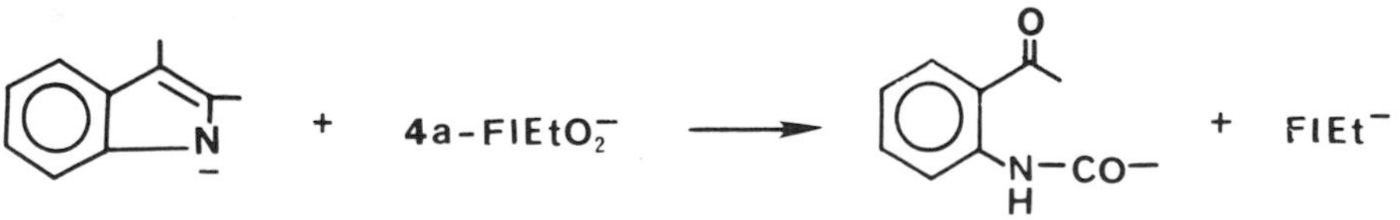

$$[\text{structure}] \quad + \quad 4a-FlEtO_2^- \quad \longrightarrow \quad [\text{structure}] \quad + \quad FlEt^-$$

$$(8\text{-}38)$$

$$[\text{structure}] \quad \longrightarrow \quad [\text{structure}] \quad \longrightarrow \quad [\text{structure}]$$

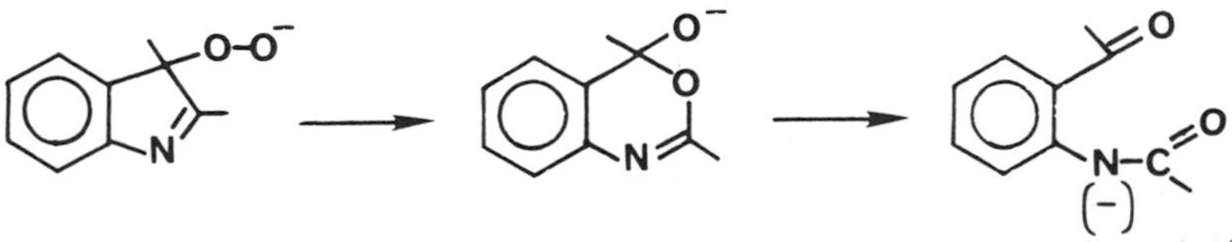

$$(8\text{-}39)$$

The transfer of the dioxygen moiety from $4a\text{-}FlEtO_2^-$ to substrate takes place by the trapping of a species **(VII)** formed from $4a\text{-}FlEtO_2^-$ (reaction 8-40).

THOMAS C. BRUICE

$$4a\text{-}FlEtO_2^- \; \underset{k_2}{\overset{k_1}{\rightleftharpoons}} \; VII \; \xrightarrow{k_3(\text{substrate}^-)} \; FlEt^- + \text{Substrate-}O_2^- \quad (8\text{-}40)$$

Substrate	$k_1(s^{-1})$	k_2/k_3 (M)
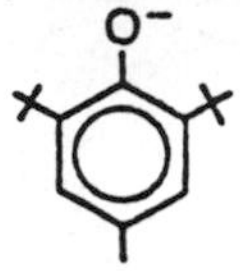 (2,4,6-trisubstituted phenolate, O^-)	0.36	2.2×10^{-4}
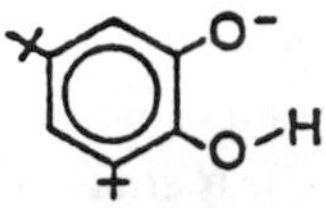 (catechol, O^-, $O\text{-}H$)	0.37	2.8×10^{-4}
 (2,6-disubstituted phenolate, O^-)	0.39	6×10^{-4}
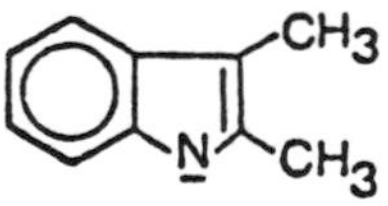 (2,3-dimethylindole, CH_3, CH_3)	0.33	8.3×10^{-3}
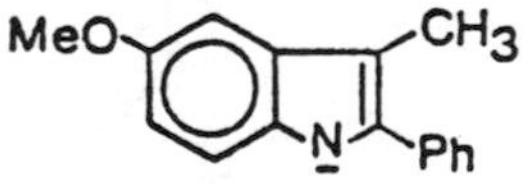 (MeO-substituted indole, CH_3, Ph)	0.37	1.3×10^{-2}
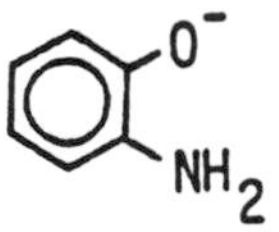 (2-aminophenolate, O^-, NH_2)	0.30	4.8×10^{-4}

(*o*- and *p*-aminophenols provide quinone imines as the immediate product)

The constant k_1 is independent of the nature of the substrate while the partition coefficient (k_2/k_3) is dependent upon the substrate. Thus, the intermediate **VII** must be formed in an endothermic reaction and trapped by substrate. Furthermore, **VII** must represent a species in which the peroxide O—O bond is intact. The species **VII** might be conceived of as solvent-separated flavin and oxygen moieties. Choices for solvent-separated flavin and oxygen species include **IX, X,** or **XI** which can be imagined to be formed upon complete heterolytic or homolytic dissociation of 4a-FlEtO$_2^-$. In what follows it will be described how these species were shown not to represent **VII**.[52]

$$FlEt\cdot + O_2^{-\cdot} \qquad\qquad FlEt^- + {}^3O_2 \qquad\qquad FlEt^- + {}^1O_2$$
$$\textbf{IX} \qquad\qquad\qquad\quad \textbf{X} \qquad\qquad\qquad\quad \textbf{XI}$$

If dioxygen transfer from 4a-FlEtO$_2^-$ was to involve the intermediacy of solvent-separated FlEt$\cdot$ + O$_2^{-\cdot}$, then the mechanism of oxygen transfer would be as shown in reaction 8-41. The homolytic dissociation mechanism of reaction 8-41a is the microscopic reverse of the coupling of FlEt$\cdot$ with

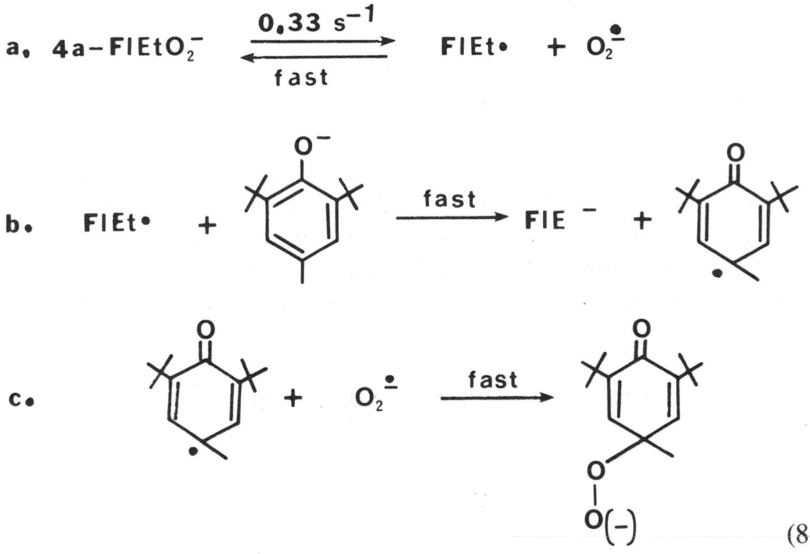

$$\tag{8-41}$$

O$_2^{-\cdot}$ to yield 4a-FlEtOO$^-$.[24,27] All substrates which have been investigated and which accept the dioxygen moiety from 4a-FlEtO$_2^-$ have been shown to reduce FlEt$\cdot$ by one electron to yield substrate radical and FlEt$^-$ (reaction 8-41b). Furthermore, the rate constants for these 1e$^-$ reductions (reaction 8-42) were found to be great enough to allow the kinetic competency of this step in the dioxygen transfer reaction. The mechanism of reaction 8-41 would appear to suffer, however, from reports that O$_2^{-\cdot}$ does not couple with the radical of 2,6-di-t-butyl-4-methylphenol (as required in reaction

$$A^- + FlEt\cdot \xrightarrow{\quad k_r\ M^{-1}\ s^{-1}\quad} A\cdot + FlEt^-$$

$A^- =$

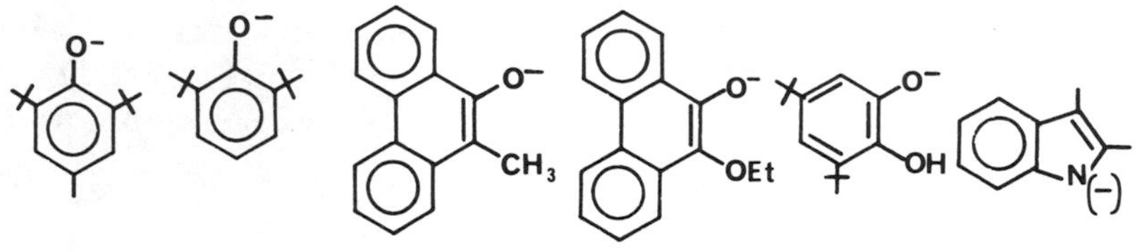

$$k_r = \quad >10^7 \quad 2.0 \times 10^4 \quad 10 \quad\quad 1 \times 10^3 \quad 2 \times 10^4 \quad 2 \times 10^2$$

$A^- =$

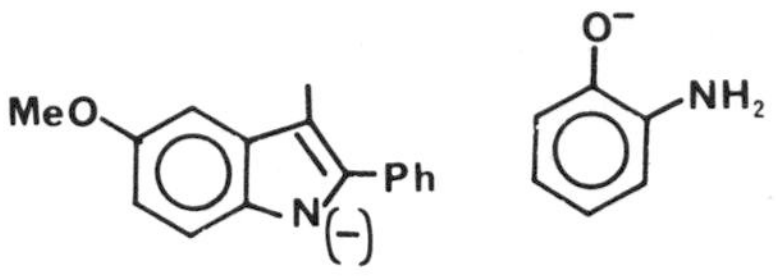

$$k_r = \quad 2.6 \times 10^3 \quad\quad >10^7 \quad\quad\quad\quad\quad\quad\quad (8\text{-}42)$$

8-41c).[54] The mechanism of reaction 8-43 would appear to be unreasonable on the basis that the determined second-order rate constants for the reactions of substrate anions with $^3O_2$ are too small to allow its kinetic competency. The involvement of $^1O_2$ can be ruled out by the finding that $^1O_2$ traps such as 2,3-dimethyl-2-butene and 2,5-dimethylfuran do not serve as substrates (reaction 8-44).

$$4a\text{-}FlEtO_2^- \rightleftharpoons FlEt^- + {}^3O_2$$

$$(8\text{-}43)$$

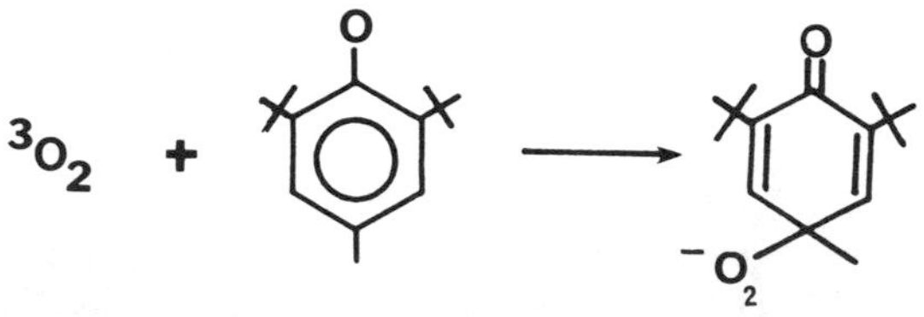

Since **VII** does not represent solvent-separated flavin and oxygen species (reactions 8-41, 8-43, and 8-44), it must represent a species in which the oxygen has not departed from the flavin moiety. The species **VII** could represent the nonsolvent separated pair [FlEt$^-$O$_2$] formed by dissociation of O$_2$ from 4a-FlEtO$_2^-$ in which O$_2$ possesses a heightened reactivity. Other than this possibility, one must conclude that 4a-FlEtOO$^-$ is converted in an

$$4a\text{-}FlEtO_2^- \; \rightleftharpoons \; FlEt^- + {}^1O_2$$

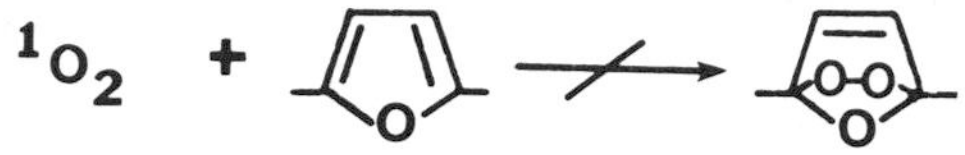

$$(8\text{-}44)$$

endothermic and reversible manner to an isomeric peroxide prior to dioxygen transfer. Possible structures for **VII** would include the dioxetane and *endo*-peroxide of reaction 8-45.

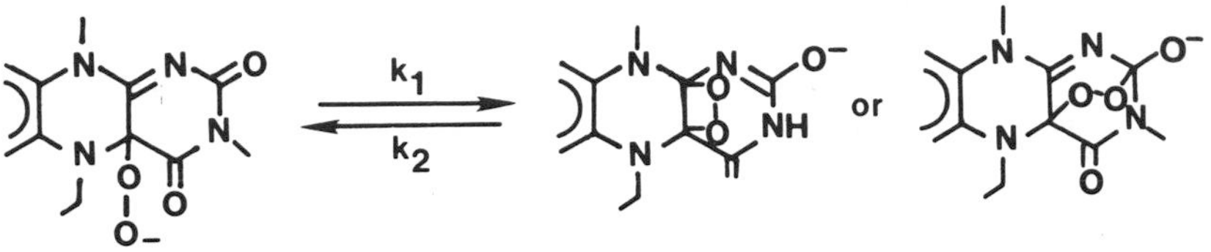

$$(8\text{-}45)$$

In either case the C_4-carbonyl group would act as an electron sink allowing an overall nucleophilic attack upon a peroxy oxygen without O—O bond scission (reaction 8-46).

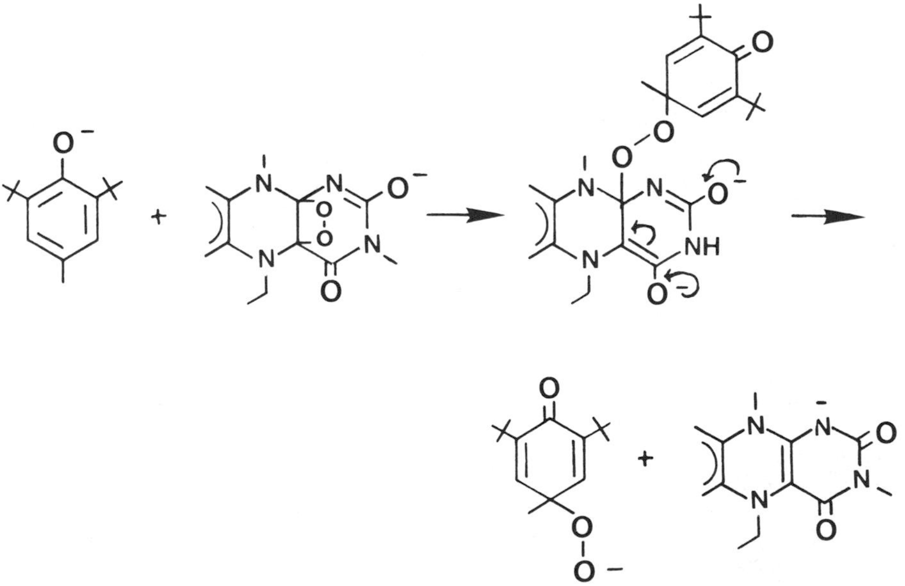

$$(8\text{-}46)$$

ACKNOWLEDGMENTS

This work was supported by the National Institutes of Health and the National Science Foundation.

REFERENCES

1. Contributed reports from p. 237–295 in "Flavins and Flavoproteins," Proceedings of the 6th International Symposium; Yagi, K.; and Yamano, T., Eds.; University Park Press: Baltimore, 1980, pp. 237–295.
2. Bruice, T. C.; Main, L.; Smith, S; Bruice, P. Y. *J. Am. Chem. Soc.* **1971,** *93,* 7327. Blankenhorn, G. *Eur. J. Biochem.* **1975,** *50,* 351. Porter, D. J.; Bright, H. J. *J. Biol. Chem.* **1980,** *255,* 7362. Powell, M. F.; Bruice, T. C. *J. Am. Chem. Soc.* **1983,** *105,* 1014.
3. Bruice, T. C. In "Biomimetic Chemistry"; Dolphin, D.; McKenna, C.; Muratami, Y.; and Tibushi, I., Eds; Advances in Chemistry, Series 191; American Chemical Society: Washington, D.C., 1980, Chapter 6.
4. Orf, W. H.; Dolphin, D. *Proc. Natl. Acad. Sci. U.S.A.* **1974,** *71,* 2646–.
5. Muller, F.; Grande H. J.; Jarbandhau, T. In "Flavins and Flavoproteins"; Singer, J. P., Ed.; Elsevier: New York, 1976, pp. 38
6. Hemmerich, P.; Wessiak, A. In "Flavins and Flavoproteins"; Singer, J. P., Ed.; Elsevier: New York, 1976, pg. 9.
7. Dimitrienko, I.; Snieckus, S.; Viswanatha, T. *Bioorg. Chem.* **1977,** *6,* 421.
8. Hamilton, G. A. In "Progress in Bioorganic Chemistry"; Kaiser, E. T.; and Kezky, F. J., Eds.; Wiley Interscience: New York, 1971.
9. Visser, C. M. *Eur. J. Biochem.* **1983,** *135,* 543.
10. Rastetter, W. H.; Gadek, T. R.; Tane, J. P.; Frost, J. W. *J. Am. Chem. Soc.* **1979,** *101,* 2228.
11. Kemal, C.; Bruice, T. C. *J. Am. Chem. Soc.* **1976,** 98, 3955.
12. Kemal, C.; Bruice, T. C. *Proc. Natl. Acad. Sci. U.S.A.* **1976,** *73,* 995.
13. Kemal, C.; Chan, T. W.; Bruice, T. C. *J. Am. Chem. Soc.* **1977,** *99,* 7272.
14. Bien, S.; Salemnik, G.; Zamir, L.; Rosenblum, M. *J. Chem. Soc. C* **1968,** 496.
15. Wessiak, A.; Bruice, T. C. *J. Am. Chem. Soc.* **1983,** *105,* 4809.
16. Wessiak, A.; Noar, J. B.; Bruice, T. C. *Proc. Natl. Acad. Sci. U.S.A.* **1984,** *81,* 332.
17. Massey, V. In "Biochemical and Clinical Aspects of Oxygen"; Caughey, W. W., Ed.; Academic Press: New York, 1979, pp. 473.
18. (a) Flashnwe, M. S.; Massey, V. In "Molecular Mechanisms of Oxygen Activation"; Hayaishi, O., Ed.; Academic Press: New York, 1974, p. 245. (b) Massey, V.; Hemmerich, P. In "The Enzymes", 3rd ed.; Boyer, P. D., Ed.; Academic Press: New York, 1975, Vol. 12, p. 191. (c) Ballou, D. P. In "Flavins and Flavoproteins", Massey, V; and Williams, C. H., Eds.; Elsevier: Amsterdam, 1982, p. 301.
19. (a) Hastings, J. W. In "Bioluminescence in Action"; Herring, P. J., Ed.; Academic Press: London, 1978, pp. 129. (b) Ziegler, M. M.; Baldwin, T. O. *Curr. Top. Bioenerg.* **1981,** *12,* 65.
20. Poulsen, L. L.; Ziegler, D. M. *J. Biol. Chem.* **1979,** *254,* 6449.
21. (a) Spencer, T.; Massey, V. *J. Biol. Chem.* **1972,** *247,* 5632. (b) Entsch, B.; Ballou, D. P.; Massey V. *J. Biol. Chem.* **1976,** *251,* 2550. (c) Strickland, S.; Massey, V. *J. Biol. Chem.* **1973,** *248,* 2953. (d) Massey, V.; Neujahr, H. Y. In "The Enzymes"; 3rd ed.; Boyer, P. D., Ed.; Academic Press: New York, 1975, Vol. 12, p. 222. (e) Hastings, J. W.; Balny, C.; Le Peuch, C.; Douzou, P. *Proc. Natl. Acad. Sci. U.S.A.* **1973,** *70,* 3468.
22. Ghisla, S.; Hastings, J. W.; Favaudon, V.; Lhoste, J.-M. *Proc. Natl. Acad. Sci. U.S.A.* 1978, *75,* 5860.

23. (a) Gibson, G. H.; Hastings, J. W. *Biochem. J.* **1962,** *83,* 368. (b) Massey, V.; Palmer, G.; Ballou, D. In "Flavins and Flavoproteins"; Kamin, H. Y., Ed.' University Park Press: Baltimore, **1971,** pp. 349. (c) Massey, G.; Palmer, G.; Ballou, D. In "Oxidases and Related Redox Systems"; King, J. E.; Mason, H. S.; and Morrison, M., Eds.; University Park Press: Baltimore, **1973,** Vol. I, pp. 25. (d) Favaudon, V. *Eur. J. Biochem.* **1977,** *78,* 293.

24. (a) Eberlein, G.; Bruice, T. C. *J. Am. Chem. Soc.* **1982,** *104,* 1449. (b) Eberlein, G.; Bruice, T. C. ibid. **1983,** 105, 6684.

25. (a) Rabini, J.; Matheson, M. S. *J. Am. Chem. Soc.* **1964,** *86,* 3175. (b) Koppenol, W. H. *Nature (London)* **1976,** *262,* 420. (c) Koppenol, W. H. *Photochem. Photobiol.* **1978,** *28,* 431. (d) George, P. In "Oxidases and Related Redox Systems"; King, T. D.; Mason, H. S.; and Morrison, M., Eds.; Wiley: New York, 1965, Vol. I, pp. 3.

26. (a) Ball, S. S.; Bruice, T. C. *J. Am. Chem. Soc.* **1979,** *101,* 4017. (b) Ball, S. S.; Bruice, T. C. *J. Am. Chem. Soc.* **1980,** *102,* 6498. (c) Ball S. S.; Bruice, T. C. *J. Am. Chem. Soc.* **1981,** *103,* 5494.

27. Nanni, E. J.; Sawyer, D. T.; Ball, S. S.; Bruice, T. C. *J. Am. Chem. Soc.* **1981,** *103,* 2797

28. Kemal, C.; Bruice, T. C. *Proc. Natl. Acad. Sci. U.S.A.* **1977,** *74,* 405.

29. (a) Poulsen, L. L.; Hyslop, R. M.; Ziegler, D. M. *Biochem. Pharmacol.* **1974,** *23,* 3431. (b) Hajjan, N. P.; Hodgson, E. *Science,* **1980,** *209,* 1134. (c) Prough, R. A.; Ziegler, D. M. *Arch. Biochem. Biophys.* **1977,** *180,* 363. (d) Ziegler, D. M.; McKee, E.; Poulsen, L. P. *Drug Metab. Dispos.* **1973,** *1,* 314. (e) Sofer, S. S.; Ziegler, D. M. *Drug Metab. Dispos.* **1978,** *6,* 232. (f) Pousen, L. L.; Hyslo, R. M.; Ziegler, D. M. *Arch. Biochem. Biophys.* **1979,** *198,* 78. (g) Hajjan, N. P.; Hodgson, E. *Biochem. Pharmacol.* **1982,** *31,* 745.

30. (a) Kemal, C.; Chan, T. C.; Bruice, T. C. *Proc. Natl. Acad. Sci. U.S.A.* **1977,** *74,* 405. (b) Ghisla, S.; Entsch, B.; Massey, V.; Husein, M. *Eur. J. Biochem.* **1977,** *76,* 139.

31. (a) Poulsen, L. L.; Kadlubar, F. F.; Diegler, D. M. *Arch. Biochem. Biophys.* **1972,** *164,* 774. (b) Ziegler, D. M.; Mitchell, C. H.; *Arch. Biochem. Biophys.* **1972,** *150,* 116. (c) Hajjar, N. P.; Hodgson, E.; *Science* **1980,** *209,* 1134.

32. (a) Bruice, T. C. *J. Chem. Soc., Chem. Commun.* **1983,** 14. (b) Bruice, T. C.; Noar, J. B.; Ball, S. S.; Venkataram, U. V. *J. Am. Chem. Soc.* **1983,** *105,* 2452.

33. (a) Miller, A.; Bruice, T. C. *J. Chem. Soc., Chem. Commun.* **1979,** 896. (b) Bruice, T. C.; Miller, A. ibid. **1980,** 693.

34. Hemmerich, *Prog. Chem. Nat. Prod.* **1976,** *33,* 451.

35. Bartlett, P. C. *Rec. Chem. Prog.* **1957,** *31,* 419.

36. Rebek, J.; McCready, R. *J. Am. Chem. Soc.* **1980,** *102,* 5602.

37. Iwata, M.; Bruice, T. C.; Carrell, H. L.; Glusker, J. P. *J. Am. Chem. Soc.* **1980,** *102,* 5036.

38. Venkataram, U. V.; Bruice, T. C. *J. Chem. Soc., Chem. Commun.* **1984,** 899.

39. Charton, M. *J. Org. Chem.* **1964,** *29,* 1222.

40. Fox, J. P.; Jencks, W. P. *J. Am. Chem. Soc.* **1974,** *96,* 1436.

41. Beaty, N. B.; Ballou, D. P. *J. Biol. Chem.* **1980,** *255,* 3817. *J. Biol. Chem.* **1981,** *256,* 4611, 4619.

42. (a) Ryerson, C. C.; Ballou, D. P.; Walsh, C. T. *Biochemistry.* **1982,** *21,* 2644. (b) Branchaud, B. P.; Walsh, C. T. *J. Am. Chem. Soc.* **1985,** *107,* 2153. (c) Latham, J. A. Jr.; Walsh C. T. *J. Chem. Soc., Chem. Commun.* **1986,** 527.

43. McCapra, F.; Hart, R. *J. Chem. Soc., Chem. Commun.* **1976,** 273.

44. Hastings, J. W. *CRC Crit. Rev. Biochem.* **1978,** 163.

45. Kurfurst, S.; Ghisla, S.; Presswood, R.; Hastings, J. W. *Eur. J. Biochem.* **1982,** *123,* 355.

46. Eberhard, A.; Hastings, J. W. *Biochem. Biophys. Res. Commun.* **1972,** *47,* 348.

47. (a) Kemal, C.; Bruice, T. C. *J. Am. Chem. Soc.* **1977,** *99,* 7064. (b) Schuster, G. B. *Acc. Chem. Res.* **1979,** *12,* 366.

48. (a) Kurfurst, M.; Ghisla, S.; Hastings; J. W. *Proc. Natl. Acad. Sci. U.S.A.* **1984,** *81,* 2990. (b) Kurfurst, M.; Ghisla, S.; Presswood, R.; Hastings, J. W. *Eur. J. Biochem.* **1982,** *123,* 355.

49. Bruice, T. C.; Bruice, P. Y. *Acc. Chem. Res.* **1976,** *9,* 378.

50. Bailey, S. W.; Ayling, J. *J. Biol. Chem.* **1980,** *255,* 7774.

51. Lazarus, R. A.; Dietrich, R. F.; Wallick, D. E.; Benkovic, S. J. *Biochemistry* **1981**, *20*, 6834.
52. (a) Kemal, C.; Bruice, T. C. *J. Am. Chem. Soc.* **1979**, *101*, 1635. (b) Muto, S.; Bruice, T. C. ibid. **1980**, *102*, 4472. (c) ibid. **1980**, *102*, 7559. (d) ibid. **1982**, *104*, 2284. (e) Keum, S. R.; Bruice, T. C. Unpublished results.
53. (a) Kishore, G. M.; Snell, E. E. *J. Biol. Chem.* **1981**, *256*, 4228, 4234. (b) Kido, T.; Soda, K.; Asada, K. *J. Biol. Chem.* **1978**, *253*, 226.
54. Muto, S.; Bruice, T. C. *J. Am. Chem. Soc.* **1980**, *102*, 7379.
55. (a). Nishinaga, A.; Itahara, T.; Tomita, H.; Nishizawa, K.; Matsuura, T. *Photochem. Photobiol.* **1978**, *28*, 687.

CHAPTER 9

Glutathione-Dependent Aldehyde Oxidation Reactions

Donald J. Creighton and Tayebeh Pourmotabbed

University of Maryland Baltimore County, Baltimore, Maryland

CONTENTS

1. Introduction.. 353
2. Distribution of Glutathione, Methylglyoxal, and
   Formaldehyde in Cells 356
3. The Glyoxalase Enzyme System......................... 358
4. Kinetic Model for the Conversion of Methylglyoxal to
   D-lactate in Erythrocytes.............................. 369
5. Optimization of Efficiency in the Glyoxalase Pathway........ 374
6. The Formaldehyde Dehydrogenase/S-Formylglutathione
   Hydrolase Enzyme System 378
7. Future Directions 382
Acknowledgments....................................... 382
References 382

1. INTRODUCTION

The aim of this chapter is to review recent developments pertaining to the mechanisms of action and substrate specificities of enzyme systems that catalyze the net, glutathione-dependent oxidation of biogenic aldehydes to

353

the corresponding carboxylic acids. Wherever possible, particular emphasis will be placed on the degree to which the properties of these enzyme systems are adapted to their particular metabolic function within the cell.

There are two known enzyme systems that catalyze the oxidation of biogenic aldehydes to carboxylic acids wherein glutathione (GSH) is a cofactor. The first is the glyoxalase enzyme system, composed of glyoxalases I (Glx I) and II (Glx II), that catalyzes the net conversion of methylglyoxal to D-lactate via S-D-lactoylglutathione (reaction 9-1).

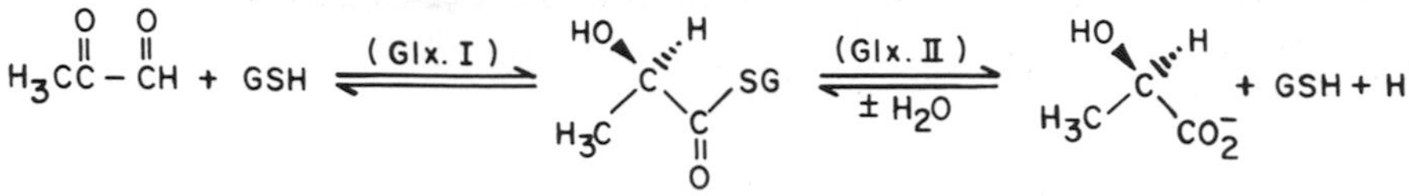

$$(9\text{-}1)$$

The second is the formaldehyde dehydrogenase (FDH)/S-formylglutathione hydrolase (FH) enzyme system that catalyzes the net, NAD(P)$^+$-dependent conversion of formaldehyde to formate via S-formylglutathione (reaction 9-2).

$$H_2CO + NAD(P)^+ + GSH \xrightleftharpoons{(FDH)} GSCHO + (NAD(P)H + H^+) \xrightleftharpoons[\pm H_2O]{(FH)} HCO_2^- + GSH + H^+$$

$$(9\text{-}2)$$

A putative physiological function of these enzyme systems is to remove from cells cytotoxic methylglyoxal and cytotoxic formaldehyde that arise either as normal or as aberrant products of intermediary metabolism. Cells are also exposed to formaldehyde as an external environmental pollutant.

These aldehydes may exist as multiple equilibrium forms in the aqueous environment of cells which normally contain high concentrations of glutathione ($\sim$2 to 12 mM). In the case of methylglyoxal, these forms are methylglyoxal (M), the aldehydrol (M'), and the dihydrate (M'') of methyl-

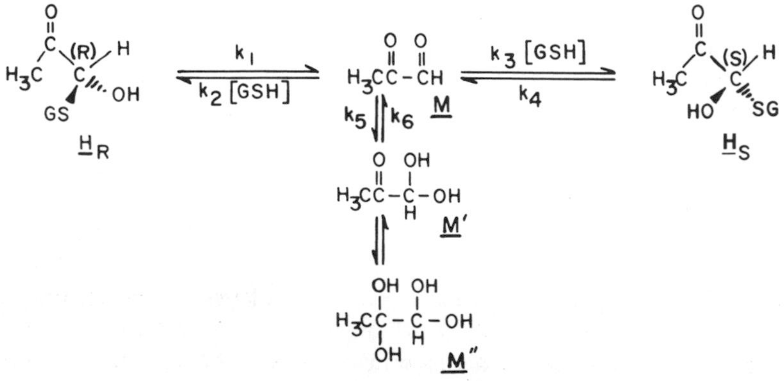

$$(9\text{-}3)$$

glyoxal, and two diastereotopic thiohemiacetals (H_R and H_S) (reaction 9-3). In the case of formaldehyde, these forms are formaldehyde, hydrated formaldehyde, and the thiohemiacetal, S-hydroxymethylglutathione (reaction 9-4).

$$H_2C(OH)_2 \rightleftharpoons H_2CO \xrightarrow{\text{[GSH]}} GSCH_2OH \qquad (9\text{-}4)$$

Therefore, any comprehensive understanding of the metabolism of these aldehydes must include the mechanism by which these various equilibrium forms are accommodated by the enzymes involved in their metabolism.

Over the past several years, a substantial amount of new information has been obtained with respect to the kinetic properties, substrate specificities, and catalytic mechanisms of the above enzyme systems. Moreover, there is enough information available in the literature to estimate with fair accuracy the microscopic rate constants associated with the equilibria of reaction 9-3. These rate constants, together with the estimated concentrations of glyoxalases I and II in mammalian erythrocytes (as a model cell), have permitted the formulation of a quantitative kinetic model for the conversion of methylglyoxal to D-lactate in erythrocytes. On this basis, the following conclusions emerge:

1. Methylglyoxal may arise from both enzymatic and nonenzymatic processes within cells. The apparent physiological function of the glyoxalase pathway is to remove cytotoxic methylglyoxal from cells as D-lactate.

2. The probable rate-determining step for the removal of methylglyoxal from mammalian erythrocytes is the uncatalyzed, bimolecular addition of glutathione to methylglyoxal to form H_R and H_S. The uncommon ability of glyoxalase I to use both H_R and H_S directly as substrates means that the rate of product formation is not limited by the nonenzymatic rate of interconversion of these diasteriomers.

3. With respect to the reaction mechanism of glyoxalase I, the conversion of bound H_R/H_S to bound product by an enediol-proton-transfer mechanism can be rationalized on the basis of the special ability of divalent sulfur to stabilize α-carbanions.

4. In several respects, the kinetic and thermodynamic factors that govern the efficiency of the glyoxalase pathway in erythrocytes recapitulate those that probably govern the efficiencies of individual enzyme molecules.

5. Cells are exposed to formaldehyde from the environment as well as from endogenous processes within cells. The most likely physiological function of the formaldehyde dehydrogenase pathway is to remove cytotoxic formaldehyde from cells as formate, perhaps in conjunction with other degradative pathways. This enzyme system also appears to be specific for the thiohemiacetal, S-hydroxymethylglutathione, as substrate.

2. DISTRIBUTION OF GLUTATHIONE, METHYLGLYOXAL, AND FORMALDEHYDE IN CELLS

These substrate forms are widely distributed in the cells of various organisms.

A. Glutathione

This sulfhydryl-containing tripeptide was first discovered by Hopkins in 1921.[1] The structure of glutathione was initially suggested on the basis of chemical analysis studies[2-4] and later confirmed by independent chemical synthesis.[5]

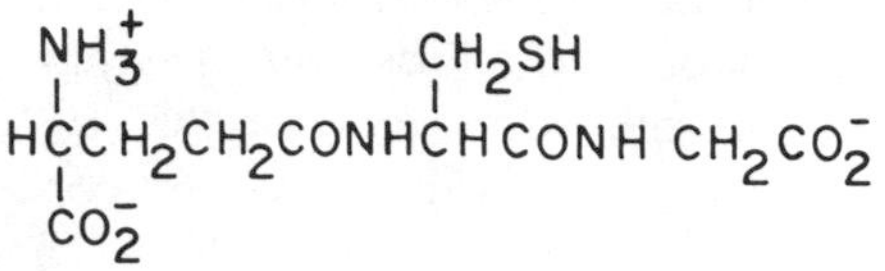

$$\underset{\underset{CO_2^-}{|}}{H\overset{\overset{NH_3^+}{|}}{C}}CH_2CH_2CONH\overset{\overset{CH_2SH}{|}}{C}HCONH\ CH_2CO_2^-$$

Glutathione

Glutathione is now known to be widely distributed in the intracellular space of plants, animals, and microorganisms.[6] In mammalian tissues the concentration of glutathione is generally in the range of 2–12 mM (Table 9-1). A multitude of different biochemical processes require glutathione as a cofactor.[7] Two important general functions are to remove toxic metabolites from the cell and to maintain cellular sulfhydryl groups in their reduced form.

TABLE 9-1. Distribution of Glutathione in Various Mammalian Tissues[a]

Cell or tissue type	Glutathione (mM)
Erythrocytes (human, rat, sheep)	2–3
Leukocytes (human, rabbit: granulocytes, lymphocytes)	3.5–5
Kidney (rat)	2.5
Liver (rat)	4.5–6.5
Liver (mouse)	
Maternal	3.0
Fetal	8.7
Placenta (mouse)	0.85
Cornea (bovine)	
Epithelium	2.7
Endothelium	1.6
Lens (monkey, human, rat, rabbit)	2.6–12.0
Nervous tissue	2–3.4

[a] Reference 6.

B. Methylglyoxal

Opinions about the metabolic origins of methylglyoxal in biological systems have gone through several phases over the last 70 years. Originally, methylglyoxal was thought to be a normal intermediate in glycolysis on the basis of the observation that methylglyoxal is commonly found in the homogenates of mammalian tissues.[8] However, subsequent studies demonstrated that methylglyoxal is unlikely to be an obligatory intermediate in glycolysis.[9]

Recent studies emphasize that methylglyoxal may arise nonenzymatically from the pools of trioses and triose phosphates in tissues. Both D,L-glyceraldehyde and dihydroxyacetone are slowly converted to methylglyoxal at physiological pH, a process catalyzed by polyvalent ions like phosphate and bicarbonate.[10,11] The triose phosphates D,L-glyceraldehyde-3-phosphate (G3P) and dihydroxyacetone phosphate (DHAP) also give rise to methylglyoxal near neutral pH by a mechanism that probably involves an enediolate intermediate (reaction 9-5).[12] This may explain the cell growth inhibitory effect of glycerol on certain mutant strains of *Escherichia coli* and of yeast, since glycerol can be metabolically converted to triose/triose phosphates that may then be nonenzymatically converted to methylglyoxal.[13,14]

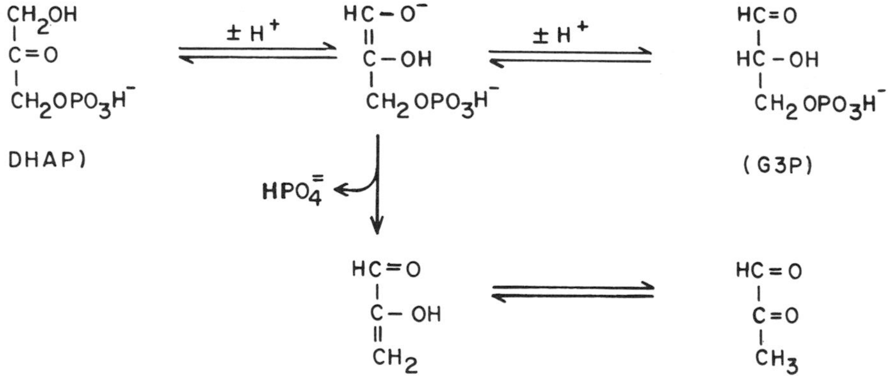

$$
\begin{array}{ccccc}
\underset{\displaystyle CH_2OPO_3H^-}{\overset{\displaystyle \underset{|}{\overset{|}{C}}=O}{\underset{|}{\overset{|}{C}H_2OH}}} & \xrightleftharpoons{\pm H^+} & \underset{\displaystyle CH_2OPO_3H^-}{\overset{\displaystyle \underset{|}{\overset{||}{C}}-OH}{\underset{|}{\overset{||}{H}C-O^-}}} & \xrightleftharpoons{\pm H^+} & \underset{\displaystyle CH_2OPO_3H^-}{\overset{\displaystyle \underset{|}{\overset{|}{H}C}-OH}{\underset{|}{\overset{|}{H}C=O}}}
\end{array}
$$

(DHAP)   (G3P)

$HPO_4^=$ ←

$$
\underset{\displaystyle CH_2}{\overset{\displaystyle \underset{||}{\overset{|}{C}}-OH}{\underset{|}{\overset{||}{H}C=O}}} \quad \rightleftharpoons \quad \underset{\displaystyle CH_3}{\overset{\displaystyle \underset{|}{\overset{|}{C}}=O}{\underset{|}{\overset{|}{H}C=O}}}
$$

(9-5)

Enzyme-catalyzed formation of methylglyoxal is also possible. Convincing evidence has been presented that methylglyoxal is synthesized from dihydroxyacetone phosphate in certain types of bacteria, due to the action of methylglyoxal synthase (reaction 9-6).[15–18]

$$
HOCH_2\overset{\displaystyle \overset{O}{\|}}{C}CH_2OPO_3H^- \longrightarrow OHC\overset{\displaystyle \overset{O}{\|}}{C}CH_3 + H_2PO_4^-
$$

(9-6)

This reaction may serve as a "bypass" mechanism for maintaining minimum levels of monophosphate in the cell in order to sustain normal glycolysis. An analogous methylglyoxal synthase has not, as yet, been isolated and purified from mammalian tissues. Therefore, the significance of the bypass mechanism in mammalian cells is uncertain. Nevertheless, in a recent study methylglyoxal was demonstrated to be formed from dihydroxyacetone phosphate in dialyzed whole-cell homogenates of rat liver at rates that could not easily be explained on the basis of the nonenzymatic decomposition of dihydroxyacetone phosphate.[19] One possibility is that methylglyoxal is an aberrant product of the action of triose-phosphate isomerase on dihydroxyacetone phosphate, perhaps involving a reaction sequence analogous to that shown in reaction 9-5.[20,21] This process has been suggested to be a significant source of methylglyoxal in rat muscle ($\sim$0.25 μmol/h/g of wet tissue).[22]

The biosynthesis of methylglyoxal from aminoacetone during threonine metabolism has been demonstrated in bacterial systems.[23,24] The significance of this process in mammalian systems is unclear.

C. Formaldehyde

Cells are exposed to formaldehyde arising from the environment[25] as well as from various metabolic processes within the cell. Endogenous sources of formaldehyde that have been identified include the oxidative demethylation of xenobiotics by the cytochrome P-450 monooxygenase system[26] and the oxidation of methanol by both alcohol dehydrogenase and catalase.[27] Other endogenous compounds that give rise to formaldehyde upon incubation with rat liver homogenates include choline, N,N-dimethylaminoethanol, N,N-dimethylglycine, and methionine.[27,28]

3. THE GLYOXALASE ENZYME SYSTEM

Several previous reviews have been written covering various aspects of the glyoxalase enzyme system.[9,29–31]

A. Physiological Function

A very tentative hypothesis has been proposed that the glyoxalase system plays a direct role in cell growth in the normal and cancerous states by regulating the intracellular levels of growth-inhibitory α-ketoaldehydes.[32] However, the absence of any clear evidence that the glyoxalase system is under allosteric control suggests that its principle function is to simply detoxify cells of methylglyoxal.[33] In accordance with this general role, the glyoxalase system is widely distributed in bacteria, yeast, and animal tissues.[34,35] Importantly, a mutant strain of the yeast *Saccharomyces cerevisiae*, defective in glyoxalase I, is eventually killed by exposure to glycerol and excretes methylglyoxal into the growth medium.[14]

In contrast to the glyoxalase system, 2-oxoaldehyde dehydrogenase catalyzes the $NAD(P)^+$-dependent oxidation of methylglyoxal to pyruvate.[36] However, in mammalian systems this enzyme activity appears to be highly species specific as well as organ specific (rat liver; sheep kidney, lung, and adrenal). The enzyme is present in some bacterial systems along with glyoxalase I.[23,24]

B. Glyoxalase I

This zinc metalloenzyme is the best studied member of the glyoxalase system. Highly purified preparations of the enzyme have been obtained from yeast (MW ~32,000–35,000, monomer)[37,38] and from mammalian tissues (MW ~43,000–46,000, dimer)[38] such as pig[39] and human[40] erythrocytes as well as rat[41] and mouse[42] liver. The kinetic properties and substrate specificity of the enzyme seem to be well suited to its particular function in the cell. The catalytic mechanism of the enzyme reflects, in part, the special ability of divalent sulfur to stabilize α-carbanion intermediates.

a. Substrate Specificity

Glyoxalase I has the capacity to use as substrates a range of structurally different α-ketoaldehydes, RC(O)CHO. These include aliphatic α-ketoaldehydes [eg, R = CH_3—, $HOCH_2$—, $(CH_3)_3C$—] as well as the aromatic α-ketoaldehyde phenylglyoxal (R = C_6H_5) and several of its para-substituted derivatives.[43] Hydrophobic interactions between enzyme and substrate may be an important component of binding, on the basis of the observation that as the hydrophobicity of *S*-alkyl and para-substituted *S*-(arylalkyl)glutathione derivatives increase, their inhibition constants with yeast glyoxalase I decrease.[44,45]

b. Kinetic Mechanism

Different kinetic mechanisms can be envisioned for glyoxalase I, given the different equilibrium forms of methylglyoxal in aqueous solutions containing glutathione (reaction 9-3). That the thiohemiacetals (H_R and/or H_S) serve directly as substrates for the enzyme is supported by the observation that enzyme-catalyzed product formation is limited by the rate of thiohemiacetal formation, even under conditions where dehydration of the α-ketoaldehyde is not entirely rate determining.[43] Subsequent experiments using carbonic anhydrase to speed up the hydration equilibrium failed to detect conditions that would allow product formation to proceed more rapidly than thiohemiacetal formation.[46] Thus, a two-substrate kinetic mechanism in which

α-ketoaldehyde and glutathione separately serve as substrates for the enzyme appears to be of minor importance in comparison to a one-substrate mechanism in which the thiohemiacetals serve as substrates. That both H_R and H_S are used directly by glyoxalase I from yeast and from porcine erythrocytes was demonstrated by the ability of high concentrations of glyoxalase I to convert both [³H]H_R and [³H]H_S (derived from phenylglyoxal and [³H]glutathione) to ³H-labeled product before exchange with unlabeled glutathione.[45] The absence of substrate stereospecificity in this enzyme may provide a mechanism whereby the decomposition rates (k_1 and k_4) do not limit the metabolic conversion of methylglyoxal to D-lactate in cells, as discussed in section 5.

c. Kinetic Properties

Glyoxalase I is a highly efficient catalyst on the basis of the studies of Vander Jagt and co-workers in which the kinetic properties of the enzyme could be analyzed in terms of a simple Michaelis–Menten scheme (reaction 9-7).[37]

$$E + S \underset{k_2}{\overset{k_1}{\rightleftarrows}} E{\cdot}S \xrightarrow{k_3} E + P \qquad (9\text{-}7)$$

The enzyme from yeast obeys Briggs–Haldane kinetics ($k_3 \gg k_2$) since the isotope effect on k_{cat} ($k_3{}^H/k_3{}^D = 3.3$) is approximately equal to the isotope effect on K_m ($K_m{}^H/K_m{}^D = (k_2 + k_3{}^H)/(k_2 + k_3{}^D) \simeq k_3{}^H/k_3{}^D$), using phenylglyoxal vs α-deuteriophenylglyoxal as substrates. This is one criterion for an optimally efficient enzyme.[47] Using methylglyoxal vs perdeuteriomethylglyoxal (²H₃CC(O)C²HO) as substrates, the isotope effects on k_{cat} (2.9 ± 0.2) and K_m (1.7 ± 0.2) were also similar, although not identical. Moreover, at neutral pH the magnitudes of k_{cat}/K_m for the yeast enzyme {k_{cat}/K_m = (7 × 10^2/s)/(2 × 10^{-4} M) = 3.5×10^6/M/s for phenylglyoxal–thiohemiacetal}[37] and for the human erythrocyte enzyme {k_{cat}/K_m = (5.9 × 10^2/s per subunit)/(5.7 × 10^{-5} M) = 1×10^7/M/s for methylglyoxal–thiohemiacetal}[48] can be compared to that for the diffusion-controlled triose-phosphate isomerase reaction ($k_{cat}/K_m = 2.4 \times 10^8$/M/s for unhydrated glyceraldehyde-3-phosphate).[49]

d. Catalytic Mechanism

Early investigators concluded that the catalytic mechanism of glyoxalase I most likely involved an intramolecular hydride transfer on the basis of minimal ($\leq 4\%$) incorporation of solvent tritium or deuterium into S-D-lactoylglutathione when the glyoxalase I reaction was carried out in isotopically labeled water (reaction 9-8).[50,51]

$$H_3C-\overset{\overset{O}{\|}}{C}-\overset{\overset{OH}{|}}{\underset{\underset{H}{|}}{C}}-SG \longrightarrow \underset{H_3C}{\overset{HO}{\diagdown}}\overset{\overset{*H}{|}}{\underset{\underset{O}{\|}}{C}}\diagdown \overset{}{\underset{}{C}} \diagdown SG \tag{9-8}$$

This mechanism is analogous to the intramolecular Cannizzaro rearrangement of the hydrate of phenylglyoxal to form mandelic acid that takes place in aqueous base without exchange with solvent protons (reaction 9-9).[52]

$$C_6H_5\overset{\overset{O}{\|}}{C}-\overset{\overset{O^-}{|}}{\underset{\underset{*H}{|}}{C}}-OH \rightleftharpoons C_6H_5\overset{\overset{OH}{|}}{\underset{\underset{*H}{|}}{C}}-\overset{\overset{O}{\|}}{C}-O^- \tag{9-9}$$

However, several recent experimental observations strongly support an enediol-proton-transfer mechanism presumably involving an active site base (reaction 9-10).

$$H_3C-\overset{\overset{O}{\|}}{C}-\overset{\overset{OH}{|}}{\underset{\underset{H}{|}}{C}}-SG \rightleftharpoons H_3C-\overset{\overset{O^-}{|}}{C}=\overset{\overset{OH}{|}}{C}-SG \rightleftharpoons H_3C-\overset{\overset{OH}{|}}{\underset{\underset{H}{|}}{C}}-\overset{\overset{O}{\|}}{C}-SG \tag{9-10}$$

$$\underset{B}{\cdot\cdot} \qquad \underset{B}{\overset{H}{\underset{\cdot\cdot}{}}} \qquad \underset{B}{\cdot\cdot}$$

First, reexamination of the early isotope exchange studies in D_2O solvent, using methylglyoxal in the presence of glyoxalases I and II, demonstrated low but significant incorporation of deuterium into D-lactate (detected on the basis of NMR measurements).[53] The fact that deuterium incorporation increased with temperature (15% at 25°C, 22% at 35°C) indicated that proton transfer takes place within a highly protected active site only partially accessible to solvent. Second, glyoxalase I catalyzes the partitioning of fluoromethylglyoxal and glutathione between S-fluorolactoylglutathione and S-pyruvylglutathione; this is most easily explained in terms of a common enediol intermediate (reaction 9-11).[54] Third, glyoxalase I reportedly catalyzes the reduction of the flavin 10-ethylisoalloxazine in the presence of phenylglyoxal and glutathione.[55,56] This can be explained by the selective capacity of flavins to oxidize an enediol intermediate, in this case bound to an active site.[57–59] Thus, the preponderance of the evidence now available favors an enediol intermediate, analogous to that proposed for the sugar isomerases.[60]

The ability of glyoxalase I to use both H_R and H_S as substrates is the least understood aspect of the mechanism. One possibility is that either one of the two stereochemical pathways depicted in Figure 9-1 is operative, depending upon which diastereomer binds to the active site. This assumes that a single

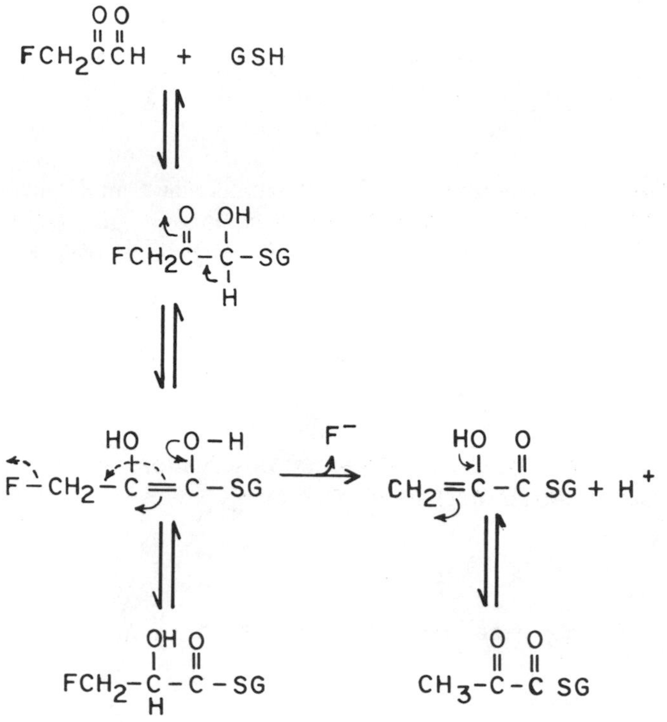

$$FCH_2\overset{\overset{O}{\|}}{C}\overset{\overset{O}{\|}}{C}H \quad + \quad GSH$$

$$FCH_2\overset{\overset{O}{\|}}{C}-\overset{\overset{OH}{|}}{\underset{\underset{H}{|}}{C}}-SG$$

$$F^-CH_2-\overset{+}{C}=\overset{\overset{O-H}{|}}{C}-SG \quad \overset{F^-}{\longrightarrow} \quad CH_2=\overset{\overset{HO}{|}}{C}-\overset{\overset{O}{\|}}{C} SG + H^+$$

$$FCH_2-\overset{\overset{OH}{|}}{\underset{\underset{H}{|}}{C}}-\overset{\overset{O}{\|}}{C}-SG \qquad\qquad CH_3-\overset{\overset{O}{\|}}{C}-\overset{\overset{O}{\|}}{C} SG$$

(9-11)

active site base is responsible for proton transfer, in analogy to the mechanism envisioned for the sugar isomerases. In this case, the active site must be capable of accommodating both a *cis*- and a *trans*-enediol intermediate. Recently, a *cis*-enediol intermediate has been indirectly implicated as the kinetically competent intermediate on the basis of stereospecific processing of "mirror image" substrates by glyoxalase I.[61] If this is correct, the enzyme may have the ability to catalyze the interconversion of bound H_R and bound H_S by a mechanism wherein only bound H_S is on a direct path to bound product via a *cis*-enediol intermediate. Phosphoglucose isomerase represents an analogous situation in that both the α and β anomers of glucose-6-phosphate are used as substrates.[60] This enzyme appears to catalyze the interconversion of the bound anomers, and a *cis*-enediol is the exclusive intermediate along the reaction pathway. Alternatively, a two-base mechanism can be envisioned for glyoxalase I that would allow direct processing of both H_R and H_S through a *cis*-enediol intermediate. This mechanism requires that the abstracted proton be transferable between the active site bases so as to allow stereospecific protonation at C-2 of the enediol intermediate. The special capacity of the active site to bind and use both H_R and H_S may be related to the ability of the enzyme to bind with nearly equal affinity both the cis- and the trans-isomers of para-substituted S-(phenylethenyl)glutathione

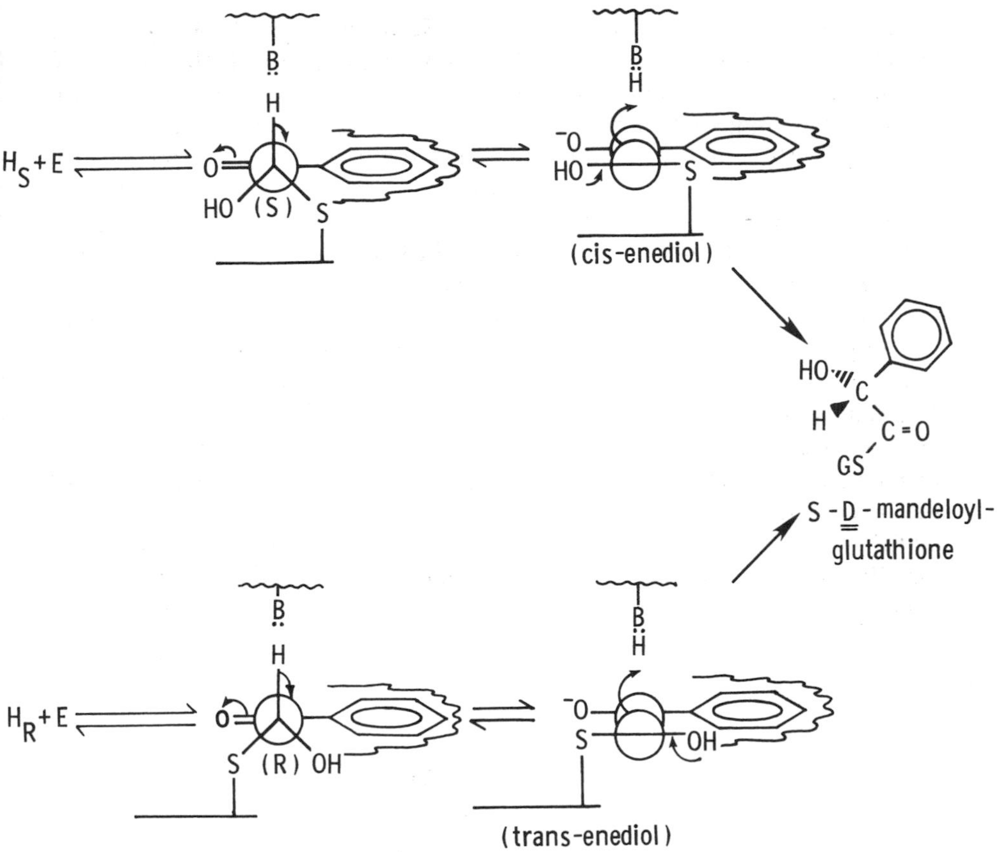

Figure 9-1. Stereochemistry of the glyoxalase I reaction. Two possible stereochemical pathways are shown to S-D-mandeloylglutathione by an enediol-proton-transfer mechanism, depending on whether H_S or H_R is the true substrate for the enzyme. The bound diastereotopic thiohemiacetals, due to phenylglyoxal and GSH, are shown as viewed along the C-1 and C-2 bond axis. In both pathways, a single active site base is envisioned to catalyze intramolecular proton transfer between C-1 and C-2. The transferred proton must add to the si face of the carbonyl of bound substrate in order to generate the correct stereochemistry of product, thus establishing a direct relationship between the stereochemistry of the substrate and that of the enediol intermediate.

derivatives, originally designed as apolar analogs of the isomeric enediol intermediate along the reaction pathway.[45,62]

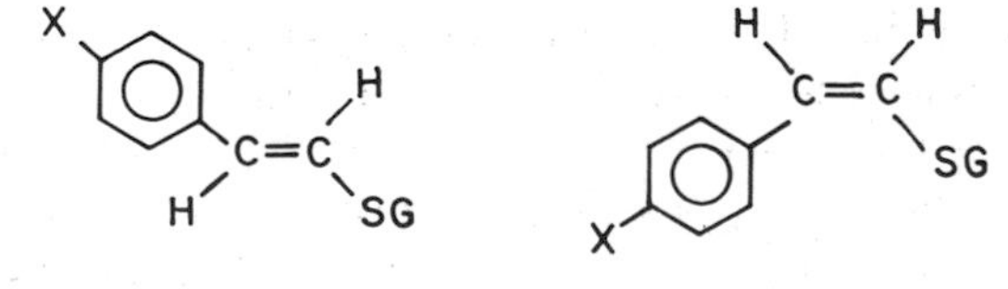

Where X = H⁻ , H₃C⁻, Cl⁻ , C₆H₅⁻

The possible involvement in the catalytic mechanism of enzyme-bound metal ion has received considerable recent attention. Both the yeast and mammalian enzymes contain one zinc ion per subunit that is required for activity.[63] For the enzyme from erythrocytes, the zinc ions can be replaced with Mg^{2+}, Mn^{2+}, and Co^{2+} to give specific activities of 101, 67, and 66% of that for the native enzyme.[64] A distorted octahedral geometry for the enzyme-bound metal ion is suggested on the basis of optical and electron paramagnetic resonance (EPR) studies of the Co^{2+}-substituted enzyme[65] and extended X-ray absorption fine structure (EXAFS) studies of the Zn^{2+}-substituted enzyme.[66] That enzyme-bound S-D-lactoylglutathione is in the outer coordination sphere of the Mn^{2+}-substituted enzyme is consistent with distance measurements obtained from nuclear magnetic relaxation studies.[67] Subsequent distance measurements, using the Mn^{2+}-substituted enzyme with S-D-lactoylglutathione, enriched in ^{13}C at the lactoyl carbonyl carbon, provided enough information to orient the carbonyl oxygen to point toward the metal ion, but also indicated that the oxygen is too distant for direct coordination to the metal ion.[68] A catalytically important water molecule may intervene between the carbonyl oxygen and the metal ion and facilitate catalysis by a general acid–base mechanism.[67]

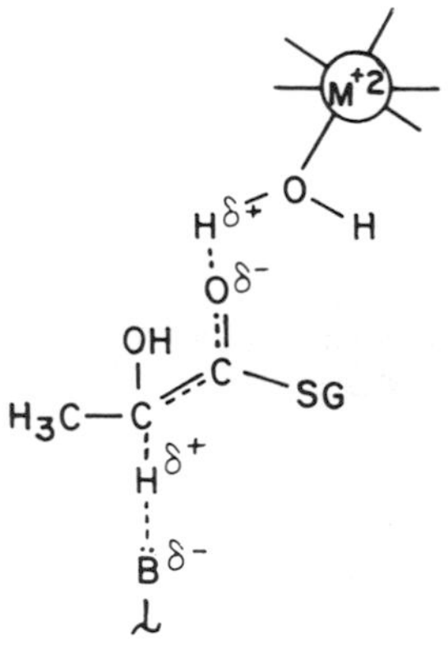

e. Model Reactions

A general-base-mediated proton transfer mechanism for glyoxalase I involving an enediol intermediate is reasonable on the basis of chemical model studies. Both imidazole and $H_2PO_4^-$ catalyze the conversion of glutathione and methyl- or phenylglyoxal to the corresponding α-hydroxy thioesters.[69] These reactions proceed by mechanisms involving complete exchange with solvent protons. In addition, diazabicyclo[2.2.2]octane catalyzes a similar dismutation of the thiohemiacetals formed from 2-mercaptoethanol and various para-substituted phenylglyoxals.[70] The near-maximal primary kinetic deuterium isotope effect on these reactions ($k_H/k_D = 5.9 \pm 0.9$) and a $\rho = +0.9$ obtained from a Hammett-δ correlation suggest a delocalized, carbanionlike transition state during rate-determining proton transfer to the catalytic base. By comparison, the intramolecular Cannizzaro rearrangement of para-substituted phenylglyoxals in strong aqueous base gives a $\rho = +2.0$, indicating substantially greater negative charge development at the β carbon in the transition state.[71]

A model reactionmost analogous to that postulated for glyoxalase I is the the intramolecular general-base-catalyzed rearrangement of the thiohemiacetal formed from phenylglyoxal and the bifunctional reagent N-(2-mercaptoethyl)morpholine (reaction 9-12).[72]

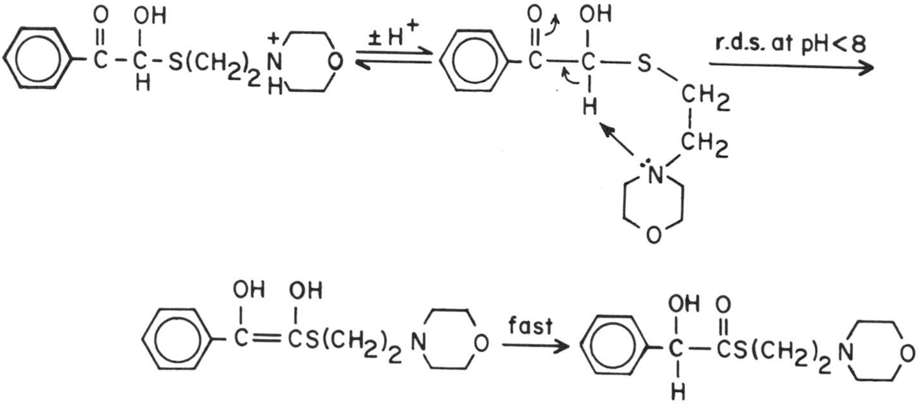

(9-12)

Similarly, Franzen reported much earlier that N,N-dialkylaminoethanethiols in the presence of [1-^{2}H]phenyl- or [1-^{2}H]methylglyoxal in H_2O solvent form the corresponding [2-^{2}H]α-hydroxythioesters, an observation in support of an intramolecular Cannizzaro rearrangement.[73,74] Infrared spectroscopy was used to monitor the deuterium content of the product. However, reexamination of this reaction using more sensitive nuclear magnetic resonance (NMR) methods to measure the isotopic content of reactant and product species demonstrated complete isotopic exchange with solvent.[69] In addition, an enediol intermediate is implicated in the reaction mechanism on the basis of flavin-trapping experiments.[75,76] Thus, model studies indicate that the preferred reaction pathway for the base-catalyzed dismutation of α-ketothiohemiacetals, near neutral pH, involves an enediol intermediate.

f. α-Carbanion Stabilization by Divalent Sulfur

The thiohemiacetal function seems to be well suited to an enediol-proton-transfer mechanism for glyoxalase I and its corresponding model reactions.

TABLE 9-2. Comparative Carbon Acid Acidities of Ethers and Thioethers Measured in Dimethyl Sulfoxide Solvent[80]

R	$RXCH_2SO_2Ph^a$		$RXCH_2COPh^b$		$RXCH_2CN^c$	
	X = S	X = O	X = S	X = O	X = S	X = O
Ph	20.5(26.8)^d	27.9(25.7)	17.1(20.8)	21.1(19.9)	20.8(28.1)	28.1(27.3)
Me	23.4(27.8)	30.7(27.2)	—	22.9(21.2)	—	—

^{a} For RX = CH$_3$—, pK_a = 31.0.
^{b} For RX = CH$_3$—, pK_a = 24.4.
^{c} For RX = CH$_3$—, pK_a = 32.5.
^{d} The values in parenthesis are the calculated pK_a's obtained from $\sigma_I\rho_I$ correlations that would apply if acidity was controlled exclusively by differences in the polar inductive properties (σ_I) of RS— vs RO—.

Divalent sulfur increases the kinetic and thermodynamic acidity of an adjacent C—H more than other second- and third-row heteroatoms.[77] For example, this is evidenced by the relative magnitudes of the rate constants for the potassium amide-catalyzed dedeuteration of the following homologous series in liquid ammonia solvent:[78] $C_6H_5N(C^2H_3)_2$, $k = 1$ (relative); $C_6H_5OC^2H_3$, $k = 40$; $C_6H_5P(C^2H_3)_2$, $k = 3 \times 10^4$; $C_6H_5SC^2H_3$, $k = 2 \times 10^8$. In addition, the tert-butoxide-catalyzed detritiation of the ethylmercaptan benzaldehyde thioacetal, $C_6H_5C^3H(SEt)_2$, proceeds approximately 10^5 times faster than the detritiation of the corresponding oxygen acetal.[79] Moreover, the special capacity of sulfur to stabilize α-carbanions is reflected in the lower pK_a's (4–7 pK_a units) of various thioethers in comparison to the corresponding oxygen ethers (Table 9-2). Thus, for glyoxalase I, the thiohemiacetal function of substrate may promote stabilization of the incipient carbanion in the transition state leading to the bound enediol intermediate.

The nature of the orbital/electronic interactions that account for carbanion stabilization by sulfur in model compounds suggests the potential for stereoelectronic control of the glyoxalase I reaction. Since carbon acid pK_a's of thioethers are significantly lower than those expected on the basis of polar inductive effects alone, some additional effect must be operative (Table 9-2). In the past, $(2p—3d)_\pi$ bonding has often been invoked as an explanation for the unusual stabilizing effect of sulfur, a bonding interaction not available to the first-row heteroatoms nitrogen and oxygen.[79,81,82] This can be expressed by the following resonance forms, symbolic of back donation of electrons into vacant 3d orbitals:

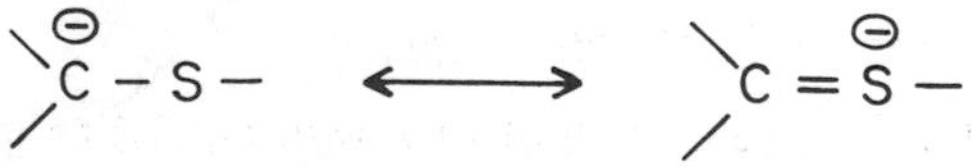

In orbital terms, two of the five vacant 3d orbitals of sulfur that are in orthogonal planes are available for overlap with the occupied 2p orbital of the carbanion carbon:

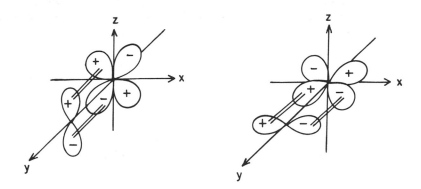

Thus, $(p\text{—}d)_\pi$ bonding will be relatively insensitive to the angle of rotation between the bonding orbitals in comparison to $(p\text{—}p)_\pi$ bonding between carbon atoms.

However, the concept of $(p\rightarrow d)_\pi$ bonding has never been uniformly accepted, either on the basis that the 3d orbitals of sulfur are too diffuse to function effectively in chemical bonding or that the basic concept is simply unnecessary in order to explain the properties of sulfur compounds.[83–90] The results of recent experimental and computational investigations are in accordance with this viewpoint. Structure–activity studies of the "contact ion pair" acidities of substituted dithianes in cyclohexylamine solvent failed to reveal any evidence for delocalization of charge.[90] An ab initio self-consistent field (SCF) study of $HSCH_3$ and $HSCH_2^-$ shows that 3d-type functions result in an equivalent lowering of the total energy of both species and, therefore, d orbitals have essentially no effect on the intrinsic basicity of the anion.[91] Moreover, the calculated electron density differences between these species suggests that polarization of the SH group may be an important contributing factor to carbanion stability.

$$\overset{\ominus}{\underset{/}{\overset{\backslash}{C}}}\text{—}\overset{\delta^+ \rightarrow \delta^-}{S}\text{—}$$

Similar conclusions were obtained from a comparative quantum chemical study of the proton affinities of carbanions adjacent to oxygen and sulfur.[92] Thus, in qualitative terms, the greater carbon acidities of thioethers vs oxygen ethers may reflect, in part, the greater atomic polarizability of sulfur vs oxygen.[93] To emphasize the potential conformational dependence of the stabilizing effect of sulfur, a molecular orbital model has been proposed in which the greater stability of >CSR$^-$ in comparison to >COR$^-$ is due to a hyperconjugative interaction between the lone pair on carbon and the low-lying δ^*_{SR} orbital.[94] This kind of interaction is optimized in the conformationally constrained bicyclic trithiane **1** and may explain why the rate constant for dedeuteration of **1** exceeds that of **2** by a factor of approximately 10^3.[95]

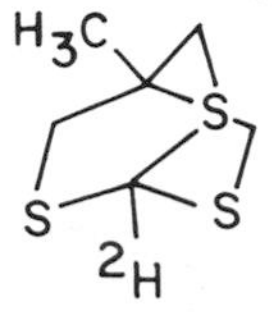

1 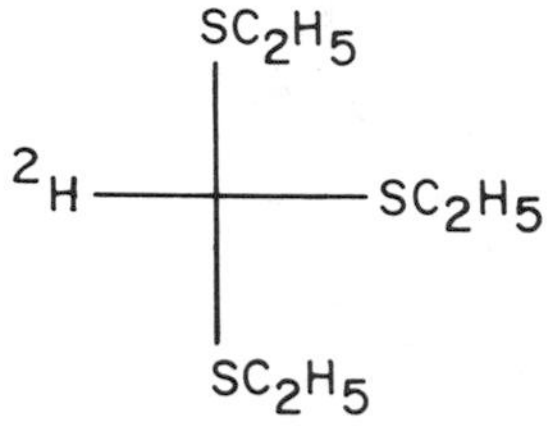   2

If indeed d orbital involvement in carbanion stabilization has been overestimated in the past and conformationally dependent factors predominate, the capacity of glyoxalase I to stabilize a conformation of bound thiohemiacetal analogous to **1** may contribute significantly to the catalytic rate enhancement.

C. Glyoxalase II

This enzyme is the second member of the glyoxalase pathway (reaction 9-1). With the exception of some neoplastic tissues,[35] the enzyme is as ubiquitous as glyoxalase I. Glyoxalase II has been purified to homogeneity or near homogeneity from rat erythrocytes (MW 21,900),[96] human liver (MW 22,900),[97] and mouse liver (MW 29,500).[98]

a. Substrate Specificity and Kinetic Properties

Glyoxalase II has a broad substrate specificity for various thioesters of glutathione. Consistent with its proposed role in the glyoxalase pathway, the enzyme (human liver) is most active with α-hydroxythioesters of glutathione:[97] S-Lactoylglutathione ($V_{max} = 1.0$ (relative), $K_m = 0.19$ mM), S-mandeloylglutathione ($V_{max} = 0.05$, $K_m = 0.016$ mM), S-glyceroylglutathione ($V_{max} = 0.62$, $K_m = 0.109$ mM), and S-glycolylglutathione ($V_{max} = 0.39$, $K_m = 0.07$ mM). The enzyme from human liver[97] and from rat erythrocytes[99] has a broad pH-activity optimum centered around neutral pH. Like glyoxalase I, glyoxalase II is a highly active catalyst near neutral pH; for the enzyme from rat erythrocytes, $k_{cat}/K_m = 1.6 \times 10^6/M/s$ with S-D-lactoylglutathione.[99]

b. Catalytic Mechanism

The mechanism of action of glyoxalase II is still in the formative stages of investigation. Some preliminary studies have been carried out on the enzyme from rat erythrocytes.[99] Attempts at chemical modification using phenylmethane sulfonic acid, N-ethylmaleimide, and 5,5'-dithiobis(2-nitrobenzoate) ion failed to reveal any evidence for the involvement of a serine or a cysteine residue in the catalytic mechanism. This raises the interesting possibility that the mechanism of action is different than that for the serine and sulfhydryl proteases. An active site imidazole has been implicated in the catalytic mechanism on the basis of inactivation studies using diethyl pyrocarbonate and on the basis of photoinactivation experiments using methylene blue. The possible involvement of an arginine residue in either substrate binding or catalysis is suggested by inactivation experiments using high concentrations of phenylglyoxal (20 mM, pH 8).

4. KINETIC MODEL FOR THE CONVERSION OF METHYLGLYOXAL TO D-LACTATE IN ERYTHROCYTES

A quantitative kinetic model for the title conversion in mammalian erythrocytes has been formulated (Figure 9-2).[100] The model is based upon (1) the measured or calculated rate constants associated with the equilibria of reaction 9-3, (2) the measured intracellular concentrations of glutathione in mammalian erythrocytes (Table 9-1), and (3) the kinetic properties and measured intracellular concentrations of glyoxalases I and II in erythrocytes (Table 9-3). The model provides important insights into the relative significance of the enzymatic and nonenzymatic kinetic barriers to the conversion of methylglyoxal to D-lactate within a relatively simple, uncompartmentalized cell. The methods and reasoning used to evaluate the microscopic rate constants given in the legend of Figure 9-2 are summarized in the next subsection.

A. Thiohemiacetal and Hydration Equilibria

The average observed rate constant for the decomposition of H_R and H_S [$(k_1 + k_4)/2$], due to the diastereotopic thiohemiacetals formed from phenylglyoxal and glutathione, has been estimated to be in the range 10–40/s (pH 7) on the basis of the results of isotope-trapping experiments.[45] Subsequent stopped-flow measurements demonstrated that (1) the observed decomposition rate constants for H_R and H_S are approximately equal to one another ($k_1 \simeq k_4$) and (2) the decomposition of the diastereomers is primarily catalyzed by hydroxide ion according to the rate law $k_{obsd} = k_{OH}[OH^-]$, where $k_{OH} = 1.1 \times 10^8/M/s$.[100] A similar rate law applies to the decomposition of thiohemiacetals formed from acetaldehyde and ethanethiol, $k_{OH} = 0.8 \times$

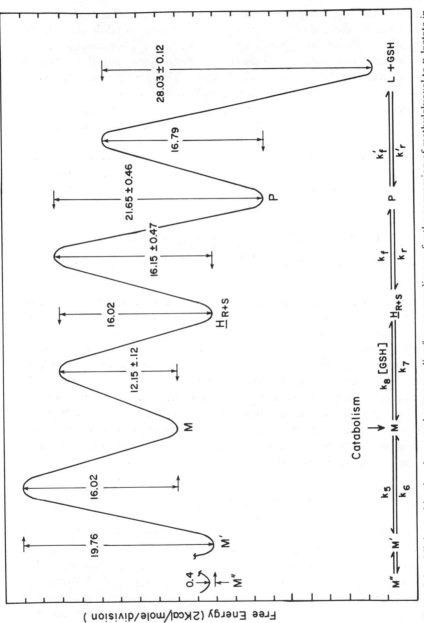

Figure 9-2. Minimum kinetic scheme and corresponding free energy diagram for the conversion of methylglyoxal to D-lactate in mammalian erythrocytes where M is methylglyoxal, M' and M'' are mono- and dihydrates of methylglyoxal (respectively). H_{R+S} is $H_R + H_S$, P is S-D-lactoylglutathione and L is D-lactate. Kinetic barriers ($\Delta G^{\ddagger}$) were calculated using $k = [k' \times (T/h)] \exp[-\Delta G^{\ddagger}/(RT)]$ where k = rate constant, k' = Boltzmann's constant, T = 298 K, h = Planck's constant, and R = gas constant. The magnitudes of the rate constants were deduced on the basis of the rationale given in the text: $k_7 \simeq 11/s$, $k_8 \simeq 3.1 \times 10^6/M/s$, $k_8[\text{GSH}] \simeq (6.2$–$9.3) \times 10^3/s$, $k_5 \simeq 11/s$, $k_6 \simeq 0.02/s$, $k_f \simeq 4$–$20/s$, $k_r \simeq (3.6$–$18) \times 10^{-4}/s$, $k_f' \simeq 3/s$, $k_r' \simeq 6.8 \times 10^{-6}/M/s$, and $k_r'[\text{GSH}]$

TABLE 9-3. Activity of Glyoxalases I and II in Erythrocytes Freshly Obtained from Different Mammalian Sources

Source	Assay conditions	Activity[a] (units/mL)	K_m (mM)	$(k_{cat} \times E_t/K_m)^b$ (per s)
Glx I				
Pig[c]	Imidazole/HCl (25 mM), pH 7 (25°C)	63 ± 9	0.14[d]	7.5 ± 1.1
Pig[e]	Imidazole/HCl (25 mM), pH 7 (30°C)	54	0.14[d]	6.4
Rat[f]	Phosphate (44 mM), KCl (0.1M), MgCl₂ (10 mM); pH 7 (25°C)	114	0.09	21.1
Human[g]	Imidazole/HCl (25 mM), pH 7 (30°C)	58	0.13	7.4
Human[h]	Phosphate (44 mM), KCl (0.1M); pH 7 (25°C)	26	0.12	3.6
Glx II				
Rat[i]	Tris (50 mM), pH 7.4 (25°C)	~34	0.18[j]	~3.1

[a] Defined as micromoles of *S*-D-lactoylglutathione formed/m/mL packed erythrocytes under maximum velocity conditions.

[b] Calculated as being equal to (activity)/($K_m \times 60$ s/m).

[c] Taken from Creighton.[100] The reported activity is the average value and range of values obtained from three separate determinations on the same sample of whole pig blood.

[d] Taken from Griffis et al.[45]

[e] Based upon the work of Aronsson and Mannervik.[39] Activity under maximum velocity conditions was calculated from their reported activity (34 units/mL) obtained under assay conditions in which the glutathione (*R*,*S*)methylglyoxal–thiohemiacetal concentration would have been ~0.24 mM on the basis of [GSH]ₜ = 0.66 mM, [methylglyoxal]ₜ = 2 mM and an association constant for the thiohemiacetal of 333/*M*.[101]

[f] Based upon the work of Han et al.[102] The activity per milliliter of packed erythrocytes was calculated on the basis of their reported units of activity per milliliter of whole blood (49 units/mL) and a hematocrit of 0.43.[103]

[g] Based upon the work of Aronsson et al.[40] Activity under maximum velocity conditions was calculated on the basis of the correction described in footnote *e*.

[h] Based upon the work of Schimandle and Vander Jagt on the Glo 2-1 allozyme.[104] The activity per milliliter of packed erythrocytes was calculated on the basis of their reported units of activity per milliliter of whole blood (11.8 units/mL) and a hematocrit of 0.45.[103]

[i] Taken from Ball and Vander Jagt.[96]

[j] Taken from Ball and Vander Jagt.[99]

$10^8/M/s$.[105] Given the insensitivity of the rate law to aldehyde structure, the decomposition rate constants for the methylglyoxal– and phenylglyoxal–thiohemiacetals are assumed to be approximately equal. Therefore, the rate constant k_7 (= $k_1 \simeq k_4$) was calculated for pH 7 from the rate law $k_7 = 1.1 \times 10^8/M/s$ [OH⁻].

The rate constant k_8 must be equal to $k_2 + k_3$ (reaction 9-3). The latter two rate constants are, in turn, approximately equal to one another, since $k_1 \simeq k_4$ and since $[H_R] \simeq [H_S]$ at equilibrium in D₂O solvent on the basis of NMR measurements.[100] The magnitude of k_8 is ultimately obtained from the definition

$$K_S = [H_{R+S}]/([GSH][M]) = k_8/k_7 \tag{9-13}$$

The magnitude of K_S is related to experimentally obtainable parameters by

$$K_S = K_S{}^{obsd}(1 + K_h) \qquad (9\text{-}14)$$

where $K_S{}^{obsd} = [H_{R+S}]/\{[GSH]([M] + [M'] + [M''])\} = 333 \pm 70/M$ [phosphate buffer (67 mM, pH 7), $\Gamma/2 = 0.2\ M]^{101}$ and $K_h = ([M'] + [M''])/[M]$. The hydration constant, K_h, is a large number since aldehydic proton resonances are not apparent in D_2O solutions of methylglyoxal. Nevertheless, the magnitude of K_h can be obtained from the equality

$$K_h = K_h'(1 + K_h'') \qquad (9\text{-}15)$$

where $K_h' = [M']/[M]$ and $K_h'' = [M'']/[M']$. K_h' was estimated from the Hammett–Taft correlation $\log K_h^\circ = 1.68\ \sigma^* - 0.02$ derived from the hydration constants for seven different aliphatic aldehydes spanning the range $K_h^\circ = 1.4$ to 2.78×10^4 (Figure 9-3).[106] Therefore, $K_h^\circ = K_h' = 5625$ when

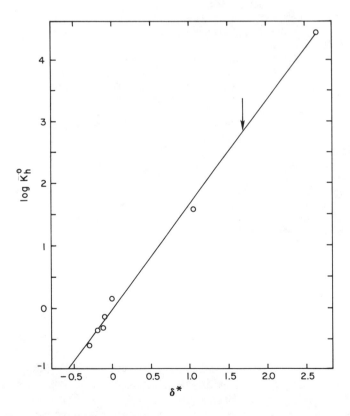

Figure 9-3. Hammett–Taft correlation for the hydration constants (K_h°) of aliphatic aldehydes in aqueous solutions.[106] The magnitude of $K_h^\circ\ (= K_h')$ for methylglyoxal was obtained by interpolation at $\sigma^* = 1.65$ (arrow) for the $H_3CC{=}O$ substituent.[107]

$\sigma^* = 1.65$ for the substituent $H_3CC{=}O$.[108] Moreover, $K_h'' \simeq 0.5$ on the basis of the relative equilibrium distribution of M' and M'' in D_2O solvent determined by NMR.[100,108] Therefore, $K_h \simeq 8438$ (Equation 9-15), $K_S \simeq 2.8 \times 10^5/M$ (Equation 9-14), and $k_8 \simeq 3.1 \times 10^6/M/s$ (Equation 9-13). The range for the maximum value of $k_8[GSH]$ was calculated on the basis of intracellular $[GSH] = 2$–3 mM determined for human, rat, and sheep erythrocytes (Table 9-1).

The magnitude of k_6 is obtained from the reported initial velocity (v_i) of thiohemiacetal formation (monitored spectrophotometrically) from methylglyoxal (5 mM) and GSH under conditions where v_i is independent of the concentration of GSH (0.5–5.0 mM), presumably due to rate-determining dehydration of methylglyoxal: $v_i = (1.6 \pm 0.2) \times 10^{-5}$ M/s, phosphate buffer (44 mM, pH 7), $\Gamma/2 = 0.2$ M, 5°C.[43] The fact that the same initial velocity was observed by including high concentrations of glyoxalase I in this system and following thioester product formation suggests that the dehydration of M', and not M'', is reflected in v_i (reaction 9-16).

$$
\underset{(M')}{\overset{\displaystyle \underset{H}{\overset{O \quad OH}{\underset{|}{\overset{\|\quad|}{CH_3C{-}C{-}OH}}}}}{} \xrightarrow{k_6} H_2OCH_3\overset{\displaystyle \overset{O\quad O}{\overset{\|\quad\|}{C{-}CH}}}{} \xrightarrow[\text{(Glx I)}]{GSH} P \qquad (9\text{-}16)
$$

Therefore, $v_i = k_6[M']$. Assuming that v_i increases approximately twofold per 10°C rise in temperature, $k_6(25°C) \simeq [4(1.6 \times 10^{-5}\ M/s)]/[0.67(5 \times 10^{-3}\ M)] \simeq .02/s$. Since $K_h' = k_5/k_6$, $k_5 \simeq 11/s$.

B. Enzymatic Reactions

For the glyoxalase I reaction, a range for $k_f[= (k_{cat}/K_m) \times E_t]$ was calculated from the V_{max} units of enzyme per milliliter of centrifugally packed pig, rat, and human erythrocytes (Table 9-3). Human erythrocytes reportedly contain three isozyme forms of glyoxalase I.[40] However, the isozymes are kinetically indistinguishable.[104] The range for k_r was calculated using the Haldane relationship: $K_{eq} = [P]/[H_{R+S}] = k_f/k_r = 1.1 \times 10^4$ (sodium phosphate, 100 mM, pH 7).[48]

For the glyoxalase II reaction, $k_f'[= (k_{cat}/K_m) \times E_t]$ was calculated using the single available literature value for the activity of the enzyme in rat erythrocytes (Table 9-3). An *approximate* value for k_r' was obtained from the approximation $K_{eq} = ([L][GSH])/[P] = k_f'/k_r' \simeq 4.4 \times 10^5\ M$. The magnitude of the equilibrium constant was, in turn, obtained from the free energy of hydrolysis in dilute aqueous solutions (pH 7) of thioesters of acetic acid, $\Delta G = 7.7$ kcal/mol.[109]

5. OPTIMIZATION OF EFFICIENCY IN THE GLYOXALASE PATHWAY

In several respects the efficiency of the glyoxalase pathway at removing methylglyoxal from erythrocytes is governed by the same fundamental kinetic and thermodynamic factors that govern the efficiencies of individual enzyme molecules. The efficiency of the pathway is limited by different factors depending upon the steady-state concentrations of intermediates in the pathway: $[M]_{ss}$, $[GSH]_{ss}$, $[H_{R+S}]_{ss}$, and $[P]_{ss}$. Since these concentrations are not known experimentally, two general cases are now considered.

A. Case I: $[H_{R+S}]_{ss}$, $[P]_{ss} \ll K_m$'s

Under conditions where the steady-state concentrations of metabolites are small in comparison to the K_m values for glyoxalases I ($K_m \simeq 0.1$ mM) and II ($K_m \simeq 0.2$ mM), the apparent rate constants for the enzyme-catalyzed steps reduce to $(k_{cat}/K_m) \times E_t$ and $[GSH]_t \simeq [GSH]_{ss}$. There is mounting evidence that many intracellular enzymes that do not occupy control points in metabolism operate under conditions where [substrate] $< K_m$, as discussed by Albery and Knowles.[110] For example, this is true for several of the glycolytic enzymes.[111]

The free energy diagram of Figure 9-2 illustrates the relative kinetic importance of the enzymatic and nonenzymatic barriers to the conversion of methylglyoxal to D-lactate under these conditions. The hydration of methylglyoxal is kinetically unimportant, since the apparent rate constant for formation of H_{R+S} ($k_8[GSH]_t$) is approximately 500 to 10^3-fold larger than the rate constant for hydration (k_5); the overall barrier to the formation of D-lactate from methylglyoxal is small in comparison to the kinetic barrier to hydration. Therefore, the steady-state rate of appearance of D-lactate (v) will conform to equation 9-17, derived on the basis of the assumption that $k_r' \simeq 0$.

$$v = \frac{k_{catp}[GSH]_t[M]_{ss}}{K_{mp} + [M]_{ss}} \tag{9-17}$$

where

$$k_{catp} = \frac{k_f'k_f}{k_f' + k_r + k_f} \simeq 1.7 \text{ to } 2.6/s$$

$$K_{mp} = \frac{(k_7 k_r) + [k_f'(k_7 + k_f)]}{k_8(k_f' + k_r + k_f)} \simeq 1.3 \text{ to } 2.1 \ \mu M$$

The form of Equation 9-17 is identical to that of the Michaelis–Menten equation. However, under the conditions of case I, $[M]_{ss} \ll K_{mp}$ since $[GSH]_t \simeq [GSH]_{ss}$ and $K_{mp}/[M]_{ss} = [GSH]_{ss}/([P]_{ss} + [H_{R+S}]_{ss})$. The latter equality is analogous to that for an enzymatic reaction in which $K_m/[$sub-strate$] = [$enzyme$]/\Sigma[$enzyme-bound substrate forms$]$.[47] Thus, Equation 9-17 reduces to Equation 9-18.

$$v = \frac{k_{catp}}{K_{mp}} [GSH]_t[M]_{ss} \qquad (9\text{-}18)$$

The efficiency of the glyoxalase pathway may be defined in terms of an efficiency function (E_f) where

$$E_f = v/v_m \qquad (9\text{-}19)$$

and $v_m = k_8([GSH]_t[M]_{ss})$. This function reflects how closely the velocity of the pathway approaches the limiting bimolecular rate of addition of GSH to methylglyoxal (v_m). Hence, maximum efficiency $(E_f = 1)$ is achieved when the enzyme-catalyzed steps are so rapid that the rate of formation of H_{R+S} is the rate-determining step in the pathway; ie, $v = v_m$. On this basis, the actual efficiency of the pathway is 50% of maximal, using the numerical values of k_{catp} and K_{mp} in Equation 9-20, obtained by combining Equations 9-18 and 9-19.

$$E_f = \frac{K_{catp}}{K_{mp}k_8} \simeq 0.5 \qquad (9\text{-}20)$$

In this sense, the efficiency of the glyoxalase pathway parallels that of the well-studied triose-phosphate isomerase reaction on the basis of an analogous efficiency function $(E_f \simeq 0.6)$ that reflects the degree to which the diffusional step is rate determining when $[$substrate$] \ll K_m$.[110]

However, there is an additional component of efficiency in a metabolic pathway that is not readily apparent from the magnitude of E_f. From the standpoint of cellular energetics, the benefit to the cell of increasing E_f must be balanced against the energy expended in order to synthesize enzyme protein. The concentration of glyoxalase I in erythrocytes seems to reflect such a balance. The apparent rate constant for enzyme-catalyzed conversion of H_{R+S} to $P(k_f \simeq 4\text{-}20/s)$ is approximately equal to the decomposition rate constant for $H_{R+S}(k_7 = 11/s$, pH 7). Therefore, the concentration of glyoxalase I is just that necessary in order to lower the top of the free energy barrier for catalysis to that for the interconversion of H_{R+S} and GSH plus methylglyoxal. The synthesis of more enzyme in order to further reduce the catalytic barrier would result in only small increases (no more than twofold) in the apparent rate constant for conversion of methylglyoxal plus GSH to P.

This phenomenon is graphically illustrated by the log plot of Figure 9-4 showing the variation of E_f vs k_f according to Equation 9-21, obtained by combining Equation 9-20 with the kinetic definitions of k_{catp} and K_{mp}.

$$E_f = \frac{k'_f}{k_7/K_{eq} + [(k'_f k_7)/k_f] + k'_f} \tag{9-21}$$

where $K_{eq} = k_f/k_r = 1.1 \times 10^4$. All of the experimental values for k_f (Table 9-3) fall near the "break point" in efficiency centered about k_7.

A similar analysis for glyoxalase II reveals that k'_f is approximately 10^4 times larger than the break point in efficiency near k_r (Figure 9-4). This means that the cell contains approximately 10^4 times more glyoxalase II than is necessary for maximal efficiency of the pathway when $[M]_{ss} \ll K_{mp}$. However, there is a reasonable explanation for the apparent excess of glyoxalase II in the cell. If indeed the concentration of this enzyme was such that $k'_f \simeq k_r$, K_{mp} would decrease by roughly a factor of 10^3. This, in turn, would increase the chances that $[M]_{ss} > K_{mp}$. Under these conditions, most of the GSH in the cell is converted to P_{ss}, and the velocity of the pathway is now limited by the small number of units of glyoxalase II (see the following subsection). Indeed, the amount of glyoxalase II in the cell is roughly that

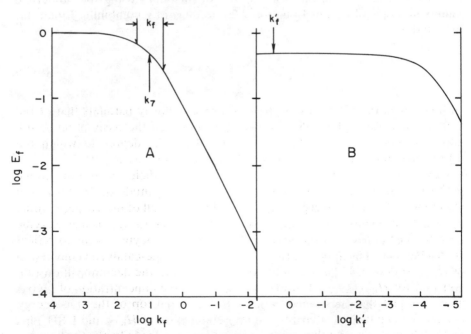

Figure 9-4. Log plot of the variation of E_f vs k_f (A) and vs k'_f (B) according to Equation 9-21. The solid lines were calculated by varying either k_f or k'_f and keeping the other constants fixed at the values shown in the legend to Figure 9-2. For the variation of E_f vs k'_f, a value of 10/s was used for k_f.

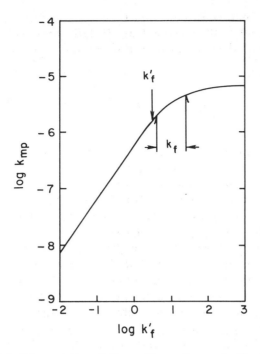

Figure 9-5. Log plot of the variation of K_{mp} vs k_f' according to the kinetic definition of K_{mp} (Equation 9-17). The solid line was calculated by varying k_f' and keeping the other constants fixed at the values shown in the legend to Figure 9-2; a value of 10/s was used for k_f.

necessary ($k_f' \simeq k_f$) in order to *increase* K_{mp} to within a factor of two ($K_{mp} \simeq (k_7 + k_f)/2 k_8$) of the theoretical maximum [if $k_f' \gg k_f$, $K_{mp} \simeq (k_7 + k_f)/k_8$] for the amount of glyoxalase I in the cell (Figure 9-5).

B. Case II: $[H_{R+S}]_{ss}$, $[P]_{ss} \gg K_m$'s

Clearly, the rate of formation of D-lactate cannot exceed the V_{max} units ($k_{cat} \times E_t$) of glyoxalase II in the cell. This upper limit would be reached under conditions where the flux through the pathway is sufficiently high ($[M]_{ss} > K_{mp}$) so that the concentrations of intermediary metabolites are large in comparison to the K_m's of the glyoxalase enzymes. In this concentration range, v is bounded by $k_{cat} \times E_t$ for glyoxalase II, and E_f becomes much less than 1 (Equation 9-19).

C. Substrate Specificity of Glyoxalase I

The capacity of glyoxalase I to use both diastereotopic thiohemiacetals (H_{R+S}) with nearly equal k_{cat}/K_m efficiencies enhances the overall efficiency of the glyoxalase pathway under case I conditions. If only one of the diaste-reomers were used by the enzyme (say, H_R), the apparent rate constant for conversion of both diastereomers to P must contain k_7, since $k_{app} = 0.5 \, k_7/[1 + (k_7/k_f)]$:

$$H_S \underset{0.5 \, k_7}{\overset{0.5 \, k_7}{\rightleftharpoons}} H_R \xrightarrow{k_f} P \qquad (9\text{-}22)$$

In the nonstereospecific case, $k_{app} = k_f$. Therefore, the kinetic advantage of nonstereospecific substrate usage by the enzyme is $2[(k_7 + k_f)/k_7] \simeq 3$ to 6 on the basis of the range of k_f values in Table 9-3.

Finally, it is instructive to consider possible explanations for the apparent inability of glyoxalase I to use GSH and methylglyoxal as substrates within the context of the efficiency of the glyoxalase pathway. Under case I condi-tions, the efficiency of the pathway is primarily limited by the rate of addi-tion of GSH to methylglyoxal ($k_8[\text{GSH}]_t$) (Figure 9-2). In principle, this limitation could be overcome if the enzyme had the ability to use not only H_{R+S} but also GSH and methylglyoxal directly. However, a change in sub-strate specificity alone would not enhance the efficiency of the pathway since $k_f \ll k_8[\text{GSH}]_t$. Either one or both of two strategies would have to be employed in order for $k_f > k_8[\text{GSH}]_t$: Either the cell must synthesize more enzyme and/or the catalytic efficiency (k_{cat}/K_m) of the enzyme must increase so that k_f increases by more than 500 to 10^3-fold. The former strategy in-volves the disadvantageous expenditure of more ATP by the cell. With respect to the latter strategy, there *may* not be room for orders of magnitude improvement in k_{cat}/K_m since glyoxalase I is already an efficient catalyst for H_{R+S} as previously discussed. Therefore, the substrate specificity of glyox-alase I may very well be optimal when both the kinetic and bioenergetic aspects of the glyoxalase pathway are considered.

In summary, the above analysis emphasizes the apparent high degree of adaptation between the kinetic and molecular properties of the glyoxalase system and its probable metabolic role in cells.

6. THE FORMALDEHYDE DEHYDROGENASE/S-FORMYLGLUTATHIONE HYDROLASE ENZYME SYSTEM

This apparent detoxification pathway (reaction 9-2) is not as well understood as the glyoxalase system. Nevertheless, the results of recent studies suggest that the two enzyme systems have features in common.

Formaldehyde dehydrogenase activity was first identified in and purified from beef liver in 1955.[112] The enzyme was originally thought to produce formate as a product.[112–114] However, the production of formate in those early experiments is now recognized to have been due to the presence of a contaminating S-formylglutathione hydrolase in crude preparations of the enzyme.[115–119] Studies by Uotila and Koivusalo, using homogeneous enzyme from human liver, clearly demonstrated that S-formylglutathione is the true product of the formaldehyde dehydrogenase reaction.[115] Thus, the NAD(P)⁺/glutathione-dependent conversion of formaldehyde to formate in tissues is due to the action of two enzymes.

A. Physiological Function

Both formaldehyde dehydrogenase and S-formylglutathione hydrolase are widely distributed in living cells.[115,120] Formaldehyde dehydrogenase activity is abundant in various animal tissues[115,120,123] and has been identified in plant, yeast, and bacterial extracts.[115,124–128]

Several different metabolic pathways are potentially available for removing cytotoxic formaldehyde from cells other than that involving formaldehyde dehydrogenase. These include the NAD⁺-dependent conversion of formaldehyde to formate catalyzed by unspecific aldehyde dehydrogenase,[129,131] the hydrogen peroxide-dependent oxidation of formaldehyde to formate catalyzed by catalase,[132] and the reaction of formaldehyde with tetrahydrofolate to form N^5,N^{10}-methylenetetrahydrofolate.[133] Consistent with the idea that the formaldehyde dehydrogenase pathway plays an important role in the chemical removal of formaldehyde from cells is the finding that the depletion of tissue glutathione in rats by exposure to methylene chloride results in a significant increase (30–40%) in the concentration of tissue formaldehyde.[134,135] Nevertheless, the relative physiological importance of the different formaldehyde-utilizing pathways is not clearly understood.

Assignment of a clear physiological role to the formaldehyde dehydrogenase pathway has been further complicated by past reports that methylglyoxal also serves as a substrate for the dehydrogenase isolated from bakers yeast,[113] *Candida biodinii*,[116] *Pichia* sp. NRRL-Y-11328,[118] pea seeds,[125] *E. coli*,[121] rat liver,[121] and human liver.[115] The enzyme preparations from the latter three sources reportedly have maximum velocities with methylglyoxal that are 85–115% of those with formaldehyde. On this basis, formaldehyde dehydrogenase could function in the detoxification of methylglyoxal in competition with the glyoxalase pathway.[121] However, subsequent kinetic and product-analysis studies on the reactions catalyzed by the enzyme from bovine liver and from human liver failed to reveal any evidence that methylglyoxal is a substrate for the dehydrogenase.[136] Some, if not all, past reports that methylglyoxal serves as a substrate for the dehydrogenase *may* be explained by the demonstrated presence of contaminating formaldehyde in some commercial preparations of methylglyoxal. Therefore, the substrate specificity of the dehydrogenase is fully consistent with a role in formaldehyde metabolism; there is no clear evidence for a role in methylglyoxal metabolism.

B. Formaldehyde Dehydrogenase

The dehydrogenase has been purified to homogeneity or near homogeneity from several different biological sources (MW $\approx 80,000$).[115,116,118,124,125] The enzyme is a dimer composed of similar subunits. The amino acid composition of the enzyme from yeast (*Pichia pastoris*) was recently determined.[124]

a. Substrate Specificity

The enzyme appears to be highly specific for formaldehyde. Simple aliphatic and aromatic aldehydes do not serve as substrates.[115] A report that the human liver enzyme can use hydroxypyruvaldehyde, kethoxal [$CH_3CH(OC_2H_5)C(O)CHO$], and glyoxal ($OHCCHO$) has not been reproduced.[136] In addition, no evidence has been found that the bovine liver enzyme can use these compounds as substrates.[136] Besides NAD^+, $NADP^+$ also serves as a coenzyme for the human liver enzyme.[115] However, the enzymes from several strains of yeast[113,116,118] and from rat liver[121] are specific for NAD^+.

b. Kinetic Mechanism and Kinetic Properties

Like glyoxalase I, formaldehyde dehydrogenase potentially operates on different equilibrium forms of formaldehyde in aqueous solutions containing glutathione (reaction 9-4). Therefore, a three-substrate kinetic mechanism could be operative in which formaldehyde, glutathione, and NAD^+ serve as substrates for the enzyme. Alternatively, a two-substrate mechanism could apply in which S-hydroxymethylglutathione and NAD^+ are substrates for the enzyme. That S-hydroxymethylglutathione can serve directly as a substrate for the bovine liver enzyme is indicated by the observation that, under certain conditions, the dehydrogenase reaction can be made to proceed more rapidly than the rate of decomposition of S-hydroxymethylglutathione to give formaldehyde and glutathione.[136] Moreover, the results of a steady-state kinetic analysis of the human liver enzyme are most easily rationalized on the basis of random addition of NAD^+ and S-hydroxymethylglutathione to the enzyme followed by random product release.[137,138] While these observations support a two-substrate mechanism for the dehydrogenase, they do not rigorously exclude a three-substrate mechanism under all conditions.

In analogy to the substrate specificity of the glyoxalase system, the ability of the dehydrogenase to use S-hydroxymethylglutathione directly may be an adaptation to potentially large amounts of this substrate form in cells. Of the total formaldehyde in cells, more than 50% is probably as the thiohemiacetal on the basis of the dissociation constant for the thiohemiacetal ($K_d = 1.5$

mM, pH 8)[115] and the estimated concentrations of glutathione in cells (Table 9-1). Not enough information is available to provide a rigorous kinetic analysis of the dehydrogenase pathway like that given for the glyoxalase system (Figure 9-2).

The pH dependence of the activity of the human liver enzyme has been studied.[115] In the forward direction, the pH optimum of activity in a standard enzyme assay is near pH 8; in the reverse direction the pH optimum is near pH 6. However, it is not clear whether the pH dependence of the activity reflects variations in k_{cat} and/or K_m for substrates. The catalytic efficiency of the bovine liver enzyme with S-hydroxymethylglutathione (saturating NAD$^+$, pH 8) is significantly less than that of glyoxalase I: $k_{cat}/K_m \simeq$ $(1.4/s)(8 \times 10^{-6}\ M) \simeq 1.8 \times 10^5/M/s/subunit$.[136]

c. Catalytic Mechanism

Formaldehyde dehydrogenase is an A-side-specific dehydrogenase, transferring the pro-R proton at C-4 of the nicotinamide ring of NADH to bound S-formylglutathione.[121] With respect to the potential involvement of metal ions in catalysis, the enzyme has been reported to be completely inhibited by 1,10-phenanthroline.[120] However, this observation has not been reproduced, nor were several other chelating agents found to inhibit the enzyme.[121] With regard to the possible involvement of sulfhydryl groups in the active site, saturating levels of NAD$^+$ or NADH provide either complete or partial protection from inactivation of the enzyme by different sulfhydryl group reagents.[121]

C. S-Formylglutathione Hydrolase

This enzyme is also widely distributed in the cytosol of cells. The hydrolase has been purified to homogeneity from human liver (MW 52,500, apparent homodimer)[139] and partially purified from the liver and kidney tissues of other animals,[140] as well as from human erythrocytes,[141] $E.\ coli$,[140] and pea seeds.[125]

a. Substrate Specificity/Kinetic Properties

The properties of the human liver enzyme have been most carefully studied.[139] The enzyme has a pH activity optimum near pH 7 and is highly specific for S-formylglutathione among several different S-acylglutathione derivatives tested as substrates; the enzyme does not use S-D-lactoylglutathione. The catalytic efficiency of the enzyme is comparable to that of glyoxalase II: $k_{cat}/K_m = (2.05 \times 10^3/s/subunit)/2.9 \times 10^{-4}\ M = 7 \times 10^6/M/s$. The wide occurrence and substrate specificity of the enzyme are fully consistent with the idea that formaldehyde dehydrogenase and the hydrolase are part of the same metabolic pathway.

b. Catalytic Mechanism

The detailed features of the catalytic mechanism of the hydrolase are unknown. The enzyme from human liver is partially inhibited by amino and sulfhydryl group reagents. However, it is unclear whether these functionalities play essential roles in catalysis.[139] As in the case of glyoxalase II, there is no evidence for a catalytically essential serine residue, on the basis that the activity of the enzyme is insensitive to the presence of organophosphates (diisopropylphosphorofluoridate, diethyl p-nitrophenyl phosphate) that are known to inhibit serine proteases.

7. FUTURE DIRECTIONS

Several questions remain to be answered with respect to the glutathione-dependent aldehyde oxidation reactions. Foremost among them are the following: (1) By what mechanism is glyoxalase I able to use both H_R and H_S as substrates and what is the nature of the active site base(s) involved in the enediol-proton-transfer mechanism? (2) How does the efficiency and/or K_{mp} of the glyoxalase pathway vary among different cell types, particularly in neoplastic tissues that contain substantial amounts of glyoxalase I but reduced or undetectable levels of glyoxalase II?[35] (3) Quantitatively, what are the most important metabolic sources of methylglyoxal in different cell types? (4) What are the detailed mechanistic features of the thioester hydrolases, glyoxalase II, and S-formylglutathione hydrolase? (5) Finally, what is the quantitative significance of the formaldehyde dehydrogenase/S-formylglutathione hydrolase system relative to other formaldehyde-utilizing pathways?

ACKNOWLEDGMENTS

We wish to thank Professors Ralph M. Pollack and John W. Kozarich for helpful discussions. Research from the author's laboratory was supported by grants from the National Institutes of Health and the American Cancer Society.

REFERENCES

1. Hopkins, F. G. *Biochem. J.* **1921**, *15*, 286.
2. Hopkins, F. G. *J. Biol. Chem.* **1929**, *84*, 269.
3. Kendall, E. C.; Mckenzie, B. F.; Mason, H. L. *J. Biol. Chem.* **1929**, *84*, 657.
4. Pirie, N. W.; Pinhey, K. G. *J. Biol. Chem.* **1929**, *84*, 321.
5. Harington, C. R.; Mead, T. H. *Biochem. J.* **1935**, *29*, 1602.
6. Kosower, N. S.; Kosower, E. M. *Int. Rev. Cytol.* **1978**, *54*, 109.
7. Larsson, A.; Orrenius, S.; Holmgren, A.; Mannervik, B. (Eds.) "Functions of Glutathione," Raven Press: New York, 1983.

8. Harden, A. "Alcoholic Fermentation," 4th ed.; Longmans, Green: London, 1932, pp. 109–146.
9. Knox, W. E. In "The Enzymes," Vol. 2; Boyer, P. D.; Lardy, H.; and Myrback, K., Eds.; Academic Press: New York, 1960, pp. 253–294.
10. Needham, J.; Lehmann, H. *Biochem. J.* **1937,** *31,* 1913.
11. Riddle, V. M.; Lorenz, F. W. *J. Biol. Chem.* **1968,** *243,* 2718.
12. Richard, J. P. *J. Am. Chem. Soc.* **1984,** *106,* 4926.
13. Riddle, V. M.; Lorenz, F. W. *Biochem. Biophys. Res. Commun.* **1973,** *50,* 27.
14. Penninckx, M. J.; Jaspers, C. J.; Legrain, M. J. *J. Biol. Chem.* **1983,** *258,* 6030.
15. Cooper, R. A.; Anderson, A. *FEBS Lett.* **1970,** *11,* 273.
16. Hopper, D. J., and Cooper, R. A. *FEBS Lett.* **1971,** *13,* 213.
17. Freedberg, W. B.; Kistler, W. S.; Lin, E. C. C. *J. Bacteriol.* **1971,** *108,* 137.
18. Cooper, R. A. *Eur. J. Biochem.* **1974,** *44,* 81.
19. Sato, J.; Wang, Y.; van Eys, J. *J. Biol. Chem.* **1980,** *255,* 2046.
20. Browne, C. A.; Campbell, I. D.; Kiener, P. A.; Phillips, D. C.; Waley, S. G.; and Wilson, I. A. *J. Mol. Biol.* **1976,** *100,* 319.
21. Campbell, I. D.; Jones, R. B.; Kiener, P. A.; Waley, S. G. *Biochem. J.* **1979,** *179,* 607.
22. Rose, I. A. *Biochemistry* **1984,** *23,* 5893.
23. Higgins, I. J.; Turner, J. M. *Biochim. Biophys. Acta* **1969,** *184,* 464.
24. Willetts, A. J.; Turner, J. M. *Biochim. Biophys. Acta* **1970,** *222,* 668.
25. Gibson, J. E. "Formaldehyde Toxicity"; Hemisphere Publishing Corp.: Washington, New York, London, 1983.
26. Gillette, J. R. *Adv. Pharmacol.* **1966,** *4,* 219.
27. Koivusalo, M. In "International Encyclopedia of Pharmacology and Therapeutics," Vol. 20; Tremoliers, J., Ed.; Pergamon Press: Oxford and New York, 1970, pp. 465–505.
28. Cooney, C. L.; and Levine, D. W. *Adv. Appl. Microbiol.* **1972,** *15,* 337.
29. Sellin, S.; Aronsson, A.-C., Eriksson, L. E. G.; Larson, K.; Tibbelin, G.; Mannervik, B. In "Functions of Glutathione: Biochemical, Physiological, Toxicological and Clinical Aspects," Larsson, A.; Orrenius, S.; Holmgren, A.; and Mannervik, B., Eds.; Raven Press: New York, 1983, pp. 187–197.
30. Jordan, F.; Cohen, J. F.; Wang, C.-T.; Wilmott, J. M.; Hall, S. S. *Drug Metab. Rev.* **1983,** *14,* 723.
31. Douglas, K. T.; Shinkai, S. *Angew. Chem. Intl. Ed.* **1985,** *24,* 31.
32. Egyud, L. G.; Szent-Gyorgyi, A. *Proc. Natl. Acad. Sci. U.S.A.* **1966,** *56,* 203.
33. Mannervik, B. In "Enzymatic Basis of Detoxication," Vol. 2; Jakoby, W. B., Ed.; Academic Press: New York, 1980, pp. 263–273.
34. Larsen, K.; Aronsson, A.-C.; Marmstal, E.; Mannervik, B. *Comp. Biochem. Physiol. B* **1985,** *82,* 625.
35. Jerzykowski, T.; Winter, R.; Matuszewski, W.; Piskorska, D. *Int. J. Biochem.* **1978,** *9,* 853.
36. Monder, C. *J. Biol. Chem.* **1967,** *242,* 4603.
37. Vander Jagt, D. L.; Han, L.-P. B. *Biochemistry* **1973,** *12,* 5161.
38. Marmstal, E.; Aronsson, A.-C.; Mannervik, B. *Biochem. J.* **1979,** *183,* 23.
39. Aronsson, A.-C.; Mannervik, B. *Biochem. J.* **1977,** *165,* 503.
40. Aronsson, A.-C.; Tibbelin, G.; Mannervik, B. *Anal. Biochem.* **1979,** *92,* 390.
41. Marmstal, E.; Mannervik, B. *Biochim. Biophys. Acta* **1979,** *566,* 362.
42. Mannervik, B.; Aronsson, A.-C.; Marmstal, E.; Tibbelin, G. *Methods Enzymol.* **1981,** *77,* 297.
43. Vander Jagt, D. L.; Daub, E.; Krohn, J. A.; Han, L.-P. B. *Biochemistry* **1975,** *14,* 3669.
44. Vince, R.; Wadd, W. B. *Biochem. Biophys. Res. Commun.* **1969,** *35,* 593.
45. Griffis, C. E. F.; Ong, L. H.; Buettner, L.; Creighton, D. J. *Biochemistry* **1983,** *22,* 2945.
46. Marmstal, E., and Mannervik, B. *FEBS Lett.* **1981,** *131,* 301.
47. Fersht, A. "Enzyme Structure and Mechanism," Freeman: San Francisco, **1977.**
48. Sellin, S.; Mannervik, B. *J. Biol. Chem.* **1983,** *258,* 8872.

49. Albery, W. J.; Knowles, J. R. *Biochemistry* **1976**, *15*, 5627.
50. Franzen, V. *Chem. Ber.* **1956**, *89*, 1020.
51. Rose, I. A. *Biochim. Biophys. Acta* **1957**, *25*, 214.
52. Hine, J.; Koser, G. F. *J. Org. Chem.* **1971**, *36*, 3591.
53. Hall, S. S.; Doweyko, A. M.; and Jordan, F. *J. Am. Chem. Soc.* **1976**, *98*, 7460.
54. Kozarich, J. W.; Chari, R. V. J.; Wu, J. C.; Lawrence, T. L. *J. Am. Chem. Soc.* **1981**, *103*, 4593.
55. Ueda, K.; Shinkai, S.; Douglas, K. T. *J. Am. Chem. Soc., Chem. Commun.* **1984**, 371.

56. Douglas, K. T.; Quilter, A. J.; Shinkai, S.; Ueda, K. *Biochim. Biophys. Acta* **1985**, *829*, 119.
57. Shinkai, S.; Kunitake, T.; Bruice, T. C. *J. Am. Chem. Soc.* **1974**, *96*, 7140.
58. Shinkai, S.; Yamashita, T.; Manabe, O. *J. Chem. Soc., Chem. Commun.* **1979**, 301.
59. Hemmerich, P.; Massey, V.; Fenner, H. *FEBS Lett.* **1977**, *84*, 415.
60. Rose, I. A. *Adv. Enzymol. Relat. Areas Mol. Biol.* **1975**, *43*, 491.
61. Chari, R. V. J.; Kozarich, J. W. *J. Am. Chem. Soc.* **1983**, *105*, 7169.
62. Creighton, D. J.; Weiner, A.; and Buettner, L. *Biophys. Chem.* **1980**, *11*, 265.
63. Aronsson, A.-C.; Marmstal, E.; Mannervik, B. *Biochem. Biophys. Res. Commun.* **1978**, *81*, 1235.
64. Sellin, S.; Aronsson, A.-C.; Mannervik, B. *Acta Chem. Scand. Ser. B* **1980**, *34*, 541.
65. Sellin, S.; Eriksson, L. E. G.; Aronsson, A.-C.; Mannervik, B. *J. Biol. Chem.* **1983**, *258*, 2091.
66. Garcia-Iniquez, L.; Powers, L.; Chance, B.; Sellin, S.; Mannervik, B.; Mildvan, A. S. *Biochemistry* **1984**, *23*, 685.
67. Sellin, S.; Rosevear, P. R.; Mannervik, B.; and Mildvan, A. S. *J. Biol. Chem.* **1982**, *257*, 10023.
68. Rosevear, P. R.; Chari, R. V. J.; Kozarich, J. W.; Sellin, S.; Mannervik, B.; Mildvan, A. S. *J. Biol. Chem.* **1983**, *258*, 6823.
69. Hall, S. S.; Doweyko, A. M.; Jordan, F. *J. Am. Chem. Soc.* **1978**, *100*, 5934.
70. Douglas, K. T.; Demircioglu, H. *J. Chem. Soc., Perkin Trans. 2* **1985**, 1951.
71. Vander Jagt, D. L.; Han, L.-P. B.; Lehman, C. H. *J. Org. Chem.* **1972**, *37*, 4100.
72. Okuyama, T.; Komoguchi, S.; Fueno, T. *J. Am. Chem. Soc.* **1982**, *104*, 2582.
73. Franzen, V. *Chem. Ber.* **1955**, *88*, 1361.
74. Franzen, V. *Chem. Ber.* **1957**, *90*, 623.
75. Shinkai, S.; Yamashita, T.; Kusano, Y.; Manabe, O. *Chem. Lett.* **1979**, 1323.
76. Shinkai, S.; Yamashita, T.; Kusano, Y.; Manabe, O. *J. Am. Chem. Soc.* **1981**, *103*, 2070.
77. Tagaki, W. In "Organic Chemistry of Sulfur," Oae, S., ed.; Plenum Press: New York, London, 1977.
78. Shatenshtein, A. I.; Gvozdeva, H. A. *Tetrahedron* **1969**, *25*, 2749.
79. Oae, S.; Tagaki, W.; Ohno, A. *Tetrahedron* **1964**, *20*, 417.
80. Bordwell, F. G.; Van Der Puy, M.; Vanier, N. R. *J. Org. Chem.* **1976**, *41*, 1885.
81. Price, C. C.; Oae, S. "Sulfur Bonding"; Ronald Press: New York, **1962**, pp. 55–60.
82. Doering, W. von E.; Hoffmann, A. K. *J. Am. Chem. Soc.* **1955**, *77*, 521.
83. Craig, D. P.; Maccoll, A.; Nyholm, R. S.; Orgel, L. E.; Sutton, L. E. *J. Chem. Soc. (London)*, **1954**, 332.
84. Rundle, R. E. *Survey Prog. Chem.* **1963**, *1*, 81.
85. Havinga, E. E.; Wiebenga, E. H. *Recueil* **1959**, *78*, 724.
86. Pimentel, G. C. *J. Chem. Phys.* **1951**, *19*, 446.
87. Coulson, C. A. *Nature (London)* **1969**, *221*, 1106.
88. Musher, J. I. *Angew. Chem.* **1969**, *8*, 54.
89. Musher, J. I. *J. Am. Chem. Soc.* **1972**, *95*, 1320.
90. Streitwieser, A.; Ewing, S. P. *J. Am. Chem. Soc.* **1975**, *97*, 190.
91. Streitwieser, A.; Williams, J. E. *J. Am. Chem. Soc.* **1975**, *97*, 191.
92. Bernardi, F.; Csizmadia, I. G.; Mangini, A.; Bernhard, S.; Whangbo, M.-H.; Wolfe, S. *J. Am. Chem. Soc.* **1975**, *97*, 2209.

93. Thorhallsson, J.; Fisk, C.; Fraga, S. *Theor. Chim. Acta (Berlin)* **1968**, *10*, 388.
94. Epiotis, N. D.; Yates, R. L.; Bernardi, F.; Wolfe, S. *J. Am. Chem. Soc.* **1976**, *98*, 5435.
95. Oae, S.; Tagaki, W.; Ohno, A. *J. Am. Chem. Soc.* **1961**, *83*, 5036.
96. Ball, J. C.; Vander Jagt, D. L. *Anal. Biochem.* **1979**, *98*, 472.
97. Uotila, L. *Biochemistry* **1973**, *12*, 3944.
98. Oray, B.; Norton, S. J. *Biochim. Biophys. Acta* **1980**, *611*, 168.
99. Ball, J. C.; Vander Jagt, D. L. *Biochemistry* **1981**, *20*, 899.
100. Creighton, D. J.; Migliorini, M.; Paurmotabbed, T.; Guha, M. K. *Biochemistry*, **1988**, *27*, 7376.
101. Vander Jagt, D. L.; Han, L.-P. B.; Lehman, C. H. *Biochemistry* **1972**, *11*, 3735.
102. Han, L.-P. B.; Davison, L. M.; Vander Jagt, D. L. *Biochim. Biophys. Acta* **1976**, *445*, 486.
103. Sanderson, J. H.; Phillips, C. E. "An Atlas of Laboratory Animal Haematology"; Oxford University Press: Oxford, 1981.
104. Schimandle, C. M.; Vander Jagt, D. L. *Arch. Biochem. Biophys.* **1979**, *195*, 261.
105. Barnett, R. E.; Jencks, W. P. *J. Am. Chem. Soc.* **1969**, *91*, 6758.
106. Greenzaid, P.; Luz, Z.; Samuel, D. *J. Am. Chem. Soc.* **1967**, *89*, 749.
107. Taft, R. W. In "Steric Effects in Organic Chemistry"; Newman, M. S., Ed.; Wiley: New York, 1956.
108. Kanchuger, M. S.; Byers, L. D. *J. Am. Chem. Soc.* **1979**, *101*, 3005.
109. Jencks, W. P.; Cordes, S.; Carriuole, J. *J. Biol. Chem.* **1960**, *235*, 3608.
110. Albery, W. J.; Knowles, J. R. *Biochemistry* **1976**, *15*, 5631.
111. Hess, B. *Symp. Soc. Exp. Biol.* **1973**, *27*, 105.
112. Strittmatter, P.; Ball, E. G. *J. Biol. Chem.* **1955**, *213*, 445.
113. Rose, Z. B.; Racker, E. *J. Biol. Chem.* **1962**, *237*, 3279.
114. Kato, N.; Tamaoki, T.; Tani, Y.; Ogata, K. *Agric. Biol. Chem.* **1972**, *36*, 2411.
115. Uotila, L.; Koivusalo, M. *J. Biol. Chem.* **1974**, *249*, 7653.
116. Schutte, H.; Flossdorf, J.; Sahm, H.; Kula, M. R. *Eur. J. Biochem.* **1976**, *62*, 151.
117. Van Dijken, P.; Oostra-Demkes, G. J.; Otto, R.; Harder, W. *Arch. Microbiol.* **1976**, *111*, 77.
118. Patel, R. N.; Hou, C. T.; Derelanko, P. *Arch. Biochem. Biophys.* **1983**, *221*, 135.
119. Uotila, L. *Biochemistry* **1973**, *12*, 3938.
120. Goodman, J. T.; Tephly, T. R. *Biochem. Biophys. Acta* **1971**, *252*, 489.
121. Uotila, L.; Koivusalo, M. In "Functions of Glutathione: Biochemical, Biophysical, Toxicological, and Clinical Aspects"; Larsson, A.; Orrenius, S.; Holmgren, A.; and Mannervik, B., Eds.; p. 175, Raven Press: New York, 1983.
122. Castle, S. L.; Board, P. G. *Human Hered.* **1982**, *32*, 222.
123. Khan, P. M.; Wijnin, L. M. M.; Hagemeijer, A.; Pearson, P. L. *Cytogenet. Cell Genet.* **1982**, *38*, 112.
124. Allias, J. J.; Louktibi, A.; Baratti, J. *Agric. Biol. Chem.* **1983**, *47*, 1509.
125. Uotila, L.; Koivusalo, M. *Arch. Biochem. Biophys.* **1979**, *196*, 33.
126. Ben-Bassat, A.; Goldberg, I. *Biochim. Biophys. Acta* **1977**, *497*, 586.
127. Boulton, C. A.; Large, P. J. *J. Gen. Microbiol.* **1977**, *101*, 151.
128. Cox, R. B.; Quayle, J. R. *Biochem. J.* **1975**, *150*, 569.
129. Koivusalo, M.; Koivula, T.; Uotila, L. In "Enzymology of Carbonyl Metabolism: Aldehyde Dehydrogenase and Aldo/Keto Reductase"; Weiner, H.; and Wermuth, B., Eds.; Allan R. Liss: New York, 1982, p. 155.
130. Cinti, D. L.; Keyes, S. R.; Lemelin, M. A.; Denk, H.; Schenkman, J. B. *J. Biol. Chem.* **1976**, *251*, 1571.
131. Denk, H.; Moldeus, P. W.; Schulz, R. A.; Schenkman, J. B.; Keyes, S. R.; Cinti, D. L. *J. Cell Biol.* **1976**, *69*, 589.
132. Waydhas, C.; Weigl, K.; Lies, H. *Eur. J. Biochem.* **1978**, *89*, 143.
133. Kallen, R. G.; Jencks, W. P. *J. Biol. Chem.* **1966**, *241*, 5851.
134. Dodd, D. E.; Bus, J. S.; Barrow, C. S. *Toxicol. Appl. Pharmacol.* **1982**, *62*, 228.
135. Heck, H. d'A.; White, E. L.; Casanova-Schmitz, M. *Biomed. Mass Spectrom.* **1982**, *9*, 347.

136. Pourmotabbed, T.; Creighton, D. J. *J. Biol. Chem.* **1986,** *261,* 14240.
137. Uotila, L.; Mannervik, B. *Biochem. J.* **1979,** *177,* 869.
138. Uotila, L.; Mannervik, B. *Biochim. Biophys. Acta* **1980,** *616,* 153.
139. Uotila, L.; Koivusalo, M. *J. Biol. Chem.* **1974,** *249,* 7664.
140. Uotila, L.; Koivusalo,M. *Methods Enzymol.* **1981,** *77,* 314.
141. Uotila, L. *Biochim. Biophys. Acta* **1979,** *580,* 277.

Addendum

CHAPTER 1

The binding of stable phosphorous-based transition state analogues to thermolysin (TLN) and carboxypeptidase A (CPA) has received recent crystallographic attention. The binding of intact phosphonamidates, phosphoramidates, and phosphonates to TLN has been described,[1-3] and another set of TLN-phosphonamidate examples have been recently reported.[4,5] It is found that TLN inhibitors which possess P_1 glycine residues or analogues thereof display anomalous binding characteristics relative to those which bear P_1 phenylalanine side chains. Similarly, an intact CPA phosphonamidate inhibitor bearing a P_1 glycine residue binds anomalously relative to those inhibitors bearing P_1 phenylalanine side chains.[6,7] Moreover, this particular inhibitor, the phosphonamidate analogue (see Figure 1-10, Chapter 1) of the peptide substrate Cbz-Gly-Phe, binds in intact form at pH 8.5 and as hydrolysis products at pH 7.5.[7,8] The implication that the enzyme may host the phosphonamidate hydrolysis, as well as the anomalous binding mode of the intact inhibitor (the Cbz-Gly moiety resides in a hydrophobic cleft normally occupied by P_1 side chains), are topics which will receive attention in future chemical and structural studies.

The recent observation that CPA binds the hydrate form of an unactivated ketonic substrate analogue supports the hypothesis that the enzyme in fact may participate in carbonyl hydration reactions at its active site.[9] Because the chemical reaction of hydration parallels the presumed first step of an actual proteolytic reaction (i.e., attack of promoted water at a scissile peptide carbonyl), we now classify the hydrated aldehyde and ketone inhibitors as *reaction coordinate analogues*.[7] A reaction coordinate analogue is a chemically reactive substrate analogue. It can undergo a reversible chemical reaction identical to the first elementary step(s) of catalysis, yet it cannot complete the chemistry of the entire catalytic cycle due to a subsequently insurmountable barrier (e.g., expulsion of hydride or carbanion). Reaction coordinate analogues differ from stable tetrahedral transition state analogues in that the latter type of enzyme inhibitor is chemically inert, whereas the former is chemically reactive. Moreover, since the chemical reactivity of the reaction coordinate analogue parallels the reversible reactivity of substrates, it cannot be classified as an irreversible mechanism-based or suicide enzyme inhibitor. Reaction coordinate analogues are perhaps the best candidates for X-ray study because their binding involves chemistry which can be detected over the time course of typical X-ray crystallographic data collection. Compounds now designated as reaction coordinate analogues serve as inhibitors of many other proteolytic enzymes in addition to the zince proteases (e.g., serine, aspartyl, and sulfhydryl proteases), and typical inhibitor functional groups include aldehydes, ketones, boronic acids, and alkylboronic acids.

REFERENCES

1. Bartlett, P. A.; Marlowe, C. K. *Science* **1987**, *235*, 569–571.
2. Tronrud, D. E.; Holden, H. M.; Matthews, B. W. *Science* **1987**, *235*, 571–574.
3. Tronrud, D. E.; Monzingo, A. F.; Matthews, B. W. *Eur. J. Biochem.* **1986**, *157*, 261–268.
4. Bartlett, P. A.; Marlowe, C. K. *Biochemistry* **1987**, *26*, 8553–8561.
5. Holden, H. M.; Tronrud, D. E.; Monzingo, A. F.; Weaver, L. H.; Matthews, B. W. *Biochemistry* **1987**, *26*, 8542–8553.
6. Jacobsen, N. E.; Bartlett, P. A. *J. Am. Chem. Soc.* **1981**, *103*, 654–657.
7. Christianson, D. W.; Lipscomb, W. N. *J. Am. Chem. Soc.*, in press.
8. Christianson, D. W.; Lipscomb, W. N. *J. Am. Chem. Soc.* **1986**, *108*, 545–546.
9. Shoham, G.; Christianson, D. W.; Oren, D. A. *Proc. Natl. Acad. Sci. USA* **1988**, *85*, 684–688.

General Index

A

Ab initio approaches, to charge-relay catalysis, 154–155
Acetoacetate decarboxylase, 43–45
Aconitase, 48
  addition/elimination reaction, 48
Activation energy, spatiotemporal hypothesis and, 84–85
Acylation, in protolytic catalysis by enzymes, 120–124
N-acylimidazoles, hydrolysis of, 3
Addition/elimination reactions, 47–48
Adenylosuccinase, 48
  addition/elimination reaction, 48
D-Alanyl-D-alanyl-carboxypeptidase, 2
Aldopyranose dehydrogenases, 66–67
  stereoelectronics, 66–70
Alkenes,
  epoxidation mechanism, 262–272
  epoxidation of, 262–272
Alkyl hydroperoxides,
  catalase reaction with iron proto-porphyrin-IX group species, 245–248
  manganese (III) porphyrin reactions, 248–249
  mechanism of oxygen transfer, 236–245
  rebound mechanism for oxygen insertion in C—H bonds, 249–253
Amino acid decarboxylases, 46
Amino acids,
  amino acid residues, 38–40
  See also Enzymes; Proteins
6-Amino-5-oxo-3H,5H-uracil intermediates, synthesis of, 321–326
Angiotensin converting enzyme, 21
Anomeric effect, 32
  and 2,3-dichloro-1,4-dioxacyclohexane, 33–34

explanations of, 32–33
generalized anomeric effect, 35–37
kinetic anomeric effect, 36–37
reverse anomeric effect, 34–36
Anomers, 65
Antibody, 110
Antibonding orbitals, 29
Arabinose dehydrogenase, 67
Arg-127 in CPA, 18–21
Argand diagram, 9
Arginosuccinase, 48
  addition/elimination reaction, 48
Aromatic rings, hydroxylation, 342–349
Asn-144 in CPA, 19–21
Asp-102 in α-chymotrypsin, charge-relay system, 130–133
Aspartate aminotransferase, 43, 47
Aspartate ammonia lyase, 48
Aspartic acid residue, chymotrypsin catalysis and, 86–87
Asp-52 in lysozyme, electrostatic stabilization and, 108–109
Axial group, stereoelectronic position regarding, 45
Axial ligands, 201–203
  Desulfovibrio cytochromes, 202
Azide ion, difference Fourier method in study of, 6
Azotobacter cytochromes, 216

B

Bachovchin's study, of charge-relay chain, 137–140
Benzamide, 88
5-Benzamido-2-benzyl-4-oxopentanoic acid (BOP), 12–20
2-Benzyl-3-formylpropanoic acid (BFP), 12–20
2-Benzyl-3-p-methoxybenzoylpropanoic acid (BMP), 12–20

2-Benzyl-4-oxo-5,5,5-trifluoropenta-
noic acid (TFP), 12–20
N-[[[(Benzyloxycarbonyl)amino]-
methyl]hydroxyphosphinyl]-L-
phenylalanine (ZGP), 16–18
BFP, *See* 2-Benzyl-3-formylpropanoic
acid (BFP)
Binding,
binding energy, 76
definition of, 95
ketones and, 12–17
lone pair exchange repulsion
(LPER), 109
site-directed mutagenesis (SDM)
and, 21–23
of zinc proteases, 1–2
Biological porphyrin metalation, *See*
Metals; Porphyrins
Bipyridyl, 111
BMP,
CPA–BMP reactions, 18–20
See 2-Benzyl-3-*p*-methoxybenzoyl-
propanoic acid (BMP)
Boat conformers, 55–56
decarboxylation and, 44–46
nicotinamide ring syn, 57–58
Bonding,
of N-acylimidazoles, 3
and anomeric effect, 32–35
classical versus stereoelectronic
theory, 29–31
esterolytic reactions and zinc-coordinated water, 3–5
orbital steering, 82–84
proteolytic reactions and zinc-
coordinated water, 3–5
PT (proton transfer) in protolytic
catalysis and, 124–133
stereoelectronics and, 27–71
transition states and enolization,
41–42
zinc-coordinated water, 3–5
BOP (5-Benzamido-2-benzyl-4-oxo-
pentanoic acid), 12–20
Bredt's rule, 29
and anti-Bredt compounds, 41–
42
Bridging, protonic in enzyme catalysis, 119–123

C

C-C bonds, stereoelectronics and, 29–
32
C-H bonds, rebound mechanism for
oxygen insertion in C-H bonds,
249–253
Carbonyl functions,
hydroxylation of aromatic rings,
342–349
nucleophilic addition reactions,
339–342
zinc interactions, 3–5
zinc-carbonyl interactions, 3–5
Carbonyl group, stereoelectronics,
61–63
Carbonyl moiety, binding and, 12–13
Carboxyl group,
glyoxalase I, 360–364
stereoelectronic analysis, 43–46
Carboxylates, stereoelectronic rules
and, 49–51
Carboxylic acid, 80–84
Carboxymuconic acid, 49, 50
cyclization of, 49
β-Carboxymuconolactone, 49–50
Carboxypeptidase A, 2, 4, 11, addendum
Carboxypeptidase A, 2, 4, 11–21
See also Zinc exoprotease carboxypeptidase A (CPA)
Carboxypeptidase B, 2
Catalase,
iron protoporphyrin-IX group
chemistry, 228–230
See also Iron protoporphyrin-IX
group species
Catalysis,
ab initio approaches, 154–155
Bachovchin's study, 137–140
charge-relay system, 130–136
chemical models of charge-relay
catalysis, 141–146
of chymotrypsin, 86–88
coupling modes, 127–130
crystallographic studies, 136–138,
140–141
of enzymes, 76–112
EVB approach, 146–152

formaldehyde dehydrogenase, 381
S-Formylglutathione hydrolase
    enzyme system, 382
glyoxalase I, 360–364
glyoxalase II, 369
HAR (heavy atom reorganization)
    in protolytic, 124–133
heavy-atom reorganization and,
    126–133
and intramolecular reactions be-
    tween reagents, 79–82
ISCRF approach, 153–154
lysozyme and rate accelerations,
    105–109
MAR diagrams and, 126–131
nitrophenylacetate, 101–103
NMR spectroscopy, 138–141
nonprotein systems and strain-
    induced accelerations, 111–112
PDLD method, 150, 152
proteins that catalyze conversions,
    109–111
protolytic, 120–164
proton transfer, 126–133
    and protonic binding, 119–123
    proton-inventory studies, 156–163
    PT approach in protolytic, 124–
    133
in reactant and transition states,
    133–136
Rogers and Bruice model, 142–143,
    145–146
SCSSD method, 150–152
7 + 10 model, 143–146
6 + 8 model, 142–143, 145–146
transition-state stabilization and,
    130–133
via reversible covalent bond forma-
    tion, 93
zinc catalysts, 3–5
See also Enzymes; Protolytic catal-
    ysis
Catalytic tetrad, 4, 11, 20
    of zinc proteases, 4–5
Catalytic triad, definition of, 122
Chair conformer, decarboxylation
    and, 44–46
Charge-relay catalysis, 130–136
    ab initio approaches, 154–155

Bachovchin's study, 137–140
chemical models for, 141–146
crystallographic studies, 136–138,
    140–141
EVB approach, 146–152
ISCRF approach, 153–154
NMR spectroscopy, 138–141
PDLD method, 150, 152
proton-inventory studies, 156–163
Rogers and Bruice model, 142–143,
    145–146
SCSSD method, 150–152
7 + 10 model, 143–146
6 + 8 model, 142–143, 145–146
theoretical models, 146–156
transition-state stabilization and,
    130–133
Charge-relay system,
    definition of, 122
    methods used for studying, 123–124
    protolytic catalysis and, 120–124
Chemical models,
    of charge-relay catalysis, 141–146
    of lactonization, 82–89
    in study of charge-relay system,
        123–124
    theoretical models, 142–147
2-chloro-1,3-dioxane, 31
2-chlorooxacyclohexane, anomeric
    effect, 32–33
Chlorophylls,
    ferrochelatase inhibition of and
        chlorophyll biosynthesis, 307–
        309
    metalation of, 286–287
Cholesterol, conversion to lanosterol,
    44–45
Chorismate, 53–54
Chymotrypsin, 85–95
    alternative enzyme mechanisms, 85
    imidazoylphenyl acetates, 90
    mechanism of, 87
    mechanisms of catalysis, 86–88
    models for, 88–93
    protolytic catalysis and, 120–123
Circular dichroism, Desulfovibrio
    cytochromes and, 202
Cis-trans isomerism, 29
Clostridium acetobutylicum, 43–44

Co(II)/(III) redox reaction, 111
Cobalamins, metalation of, 287–288
Collagenase, 21
Collision theory, definition of, 76
Concerted reactions, 49
Conformational behavior,
  of dimethoxymethane, 35–37
  in dioxacyclohexane system, 31–32
  of imidazoyl group, 34–35
  of keto steroids, 36–39
Conformational changes,
  axial ligands, 201–203
  large conformational changes, 199–
    200
  in proteins, 199–204
  small conformational changes, 200–
    201
Conformationally immobile systems,
  79
Corey-Sneen rule, 36, 41–42
Coupling modes, of PT and HAR,
  127–130
Covalent bonds, orbital steering, 82–
  84
CPA, See Zinc exoprotease carboxypeptidase A (CPA)
Crown ether, enzyme mimic, 98–101
Crystallographic studies, of catalysis,
  1–23, 136–138, 140–141
Crystallography, in study of charge-
  relay system, 123–124
Cyclodextrin, 98–100
β-Cyclodextrin, 98–99
1,4-Cyclohexadiene systems, 55–56
Cyclohexane, ground state energy, 78
Cyclohexanes, substituent effects, 31
Cyclohexyl halides, rate accelerations
  and strain, 102–103
Cyclopropane, ground state energy,
  78
Cytochrome P-450,
  N,N-dimethylaniline N-oxides reac-
    tions and, 255–262
  epoxidation of alkenes, 262–272
  iron protoporphyrin-IX group
    chemistry, 228–230
Cytochromes,
  conformational changes in proteins
    and, 199–204

Cytochrome P-450, 228–230
Desulfovibrio cytochromes, 202
and electron transfer, 185–192
electron transfer and, 170–217
electron transfer in protein-organo-
    metallic complexes, 198–199
electron transfer in protein-protein
    complexes, 196–198
intermolecular electron transfer,
    204–214
intramolecular electron transfer,
    196–204
molecular geometry of, 208–211
multiheme, 215–216
protein structure, 171–174
rebound mechanism for oxygen
    insertion in C-H bonds, 249–
    253
redox potential and electron trans-
    fer, 186–187
See also Iron protoporphyrin-IX
    group species

D

Deacylation, in protolytic catalysis by
    enzymes, 120–124
Dealkylation reactions, mechanisms
    of, 253–262
Deazaflavins, 58
Decarboxylation, 43
  boat conformer, 44–46
  chair conformer, 44–46
  of β-keto acids, 43–46
Dehydrogenases, 56
  stereoelectronic arguments regard-
    ing, 56–58
  stereoelectronics of hydrolyses, 66–
    70
Dehydroquinate,
  catalyzed reactions of synthase, 49
  conversion to dehydroshikimate,
    51–52
  dehydratase, 49
  synthase, 49, 51
Dehydroshikimate,
  conversion to protocatechuate, 49
  from dehydroquinate, 51–52

Desulfovibrio cytochromes, 202, 215–216

Deuterium oxide, 120
  proton-inventory studies, 156–163
Dicarboxylic semiesters, anhydride formation, 80–82
2,3-Dichloro-1,4-dioxacyclohexane, and anomeric effect, 33–34
Diels-Alder adduct formation, 95
Diels-Alder reactions, 103
Difference electron density map, 6, 7
Difference Fourier method,
  Argand diagrams, 9
  and study of enzyme-ligand interaction, 5–11
Differential complementarity, of enzymes, 134
4a,5-Dihydro-4a-hydroperoxy-5-alkyl-lumiflavins, 331–338
4a,5-Dihydro-4a-hydroperoxyflavin, synthesis of, 321–326
Dihydrophenylalanine, 52–53
Diisopropyl fluorophosphate (DFP) derivatives, 138–140
Dimethoxymethane, conformational behavior, 35–37
N,N-dimethylaniline N-oxides, reactions with metal (III) porphyrins, 255–262
Dioxacyclohexane system, substituent effects, 31–32
DNA, 96, 100
  enzyme mimic, 100–102
Donor/acceptor orientation, and electron transfer, 192–193
Driving force, and electron transfer, 184–188, 190–192

E

"Effect", definition of, 31
"Effective molarity", definition of, 81
EL complex, *See* Enzyme-ligand interaction
EL model, *See* Enzyme-ligand interaction
Electron density,
  Argand diagrams, 9

difference Fourier method in study of, 6–11
Electron transfer, 169–217
  calculations from crystal structures, 189
  complex formation, 211–215
  conformational changes in proteins and, 199–204
  controlling factors, 183–196
  cytochrome b_5, 187–188
  cytochrome c redox potential, 186–187
  distance and, 184
  donor/acceptor orientation, 192–193
  driving force, 184–188
  electrostatic control of, 204–206
  Franck-Condon principle, 174–177
  at heme edge, 206–211
  HOMO-LUMO interactions, 177–180
  intermolecular, 204–214
  intervening material effects, 194–195
  intramolecular, 196–204
  Marcus inverted region, 169
  measurements of, 177–183
  model systems, 188–189
  molecular geometry of in cytochromes, 208–211
  in protein-organometallic complexes, 198–199
  protein-protein complexes, 196–198
  rate as function of driving force, 190–192
  rate as function of temperature, 189–190
  redox potential control, 184–188
  reorganization energy, 188–192
  site-directed mutagenesis, 170
  spin state and, 195
  theory of, 174–177
Electrostatic control, of electron transfer, 204–206
Electrostatic generation, of higher valent iron-oxo species, 233–234
Electrostatic stabilization,
  Asp-52 in lysozyme and, 108–109
  spatiotemporal hypothesis, 85

Empirical valence bond (EVB)
   method, 146–152
Enkephalinase, 21
Enolase, 48
   addition/elimination reaction, 48
Enolizations, 41–43
   and decarboxylations of beta-keto
      acids, 43–46
Enolpyruvylshikimate-3-phosphate,
   53
Enoyl-CoA hydratase, catalyzed reactions, 49
Enthalpy of formation, 78
Enzyme catalysis, See Catalysis;
   Enzymes
Enzyme inhibitors, 2, 21
   aldehyde, 12
   hydroxamate, 4
   ketone, 12
   phosphonamidate, 2, 15, 16, adden-
      dum
   phosphonate, 2, 16
   reaction coordinate analogue, ad-
      dendum
   transition-state analogue, 2, 12,
      addendum
   α-trifluoroketone, 12
   unblocked dipeptide, 4
Enzyme mimics, 96–103
   definition of, 77
Enzyme model,
   for chymotrypsin, 86–93
   definition of, 77
Enzyme-ligand interaction,
   Argand diagrams, 9
   difference Fourier method, 5–11
Enzymes, 76–112
   addition/elimination reactions, 48
   alternative enzyme mechanisms and
      chymotrypsin, 85
   amino acid residues, 38–40
   6-amino-5-oxo-3H,5H-uracil inter-
      mediates, 321–326
   anti-elimination modes, 48
   Argand diagrams, 9
   Bachovchin's study of charge-relay
      chain, 137–140
   5-Benzamido-2-benzyl-4-oxopen-
      tonic acid (BOP), 12–17

2-Benzyl-3-formylpropanoic acid
   (BFP), 12–17
2-Benzyl-3-p-methoxybenzoylpropa-
   noic acid (BMP), 12–16
2-Benzyl-4-oxo-5,5,5-trifluoropenta-
   noic acid (TFP), 12–17
carbonyl functions and nucleophilic
   additions, 339–342
chemical models of charge-relay
   catalysis, 141–146
chymotrypsin, 85–95
crystallographic studies, 136–138,
   140–141
decarboxylations of β-keto acids,
   43–46
difference Fourier method, 5–11
differential complementarity, 134
4a,5-dihydro-4a-hydroperoxy-5-
   alkyllumiflavins derived, 331–
   338
4a,5-dihydro-4a-hydroperoxyflavin,
   321–326
dihydrophenylalanine reactions, 52–
   53
diisopropyl fluorophosphate (DFP)
   derivatives, 138–140
enzyme mimics, 96–103
5-ethyl-4a,5-dihydro-4a-hydro-
   peroxylumiflavin, 338–339
and evolutionary biology, 39–40
flavin reduction, 58
flavoenzyme chemical studies, 316–
   318
formaldehyde, 353–382
formaldehyde dehydrogenase(s)
   and, 378–382
S-formylglutathione hydrolase enzyme system, 379–380
glutathione-dependent aldehyde
   oxidation reactions, 353–382
glyoxalase enzyme system, 358–369
glyoxalase I and II reactions, 373
ground state destabilization and, 79
HAR (heavy atom reorganization)
   in protolytic catalysis, 124–133
heavy-atom reorganization and
   catalysis, 126–133
hydroxylation of aromatic rings,
   342–349

imidazoylphenyl acetates and, 90
iron protoporphyrin-IX group, 228–230
kinetic studies, 124
large stereoelectronic effects in enzyme reactions, 41–43
lysozyme, 105–109
metalation reactions, 287–300
methods used for studying, 123–124
methylglyoxal, 353–382
monoisopropylphosphoryl (MIP) group, 137
NMR spectroscopy, 138–141
nonprotein systems and strain-induced accelerations, 111–112
orbital steering, 82–84
proteins and catalysis, 109–111
protolytic catalysis, 120–164
proton transfer in catalysis, 119–123
proton transfer processes, 126–133
proton-inventory studies, 156–163
PT (proton transfer) in protolytic catalysis, 124–133
pyridoxal, 43, 46–47
pyridoxal enzymes, 46–47
rate accelerations and proximity, 95–103
rate specificities of enzymes and mimics, 96–99
site-directed mutagenesis (SDM) and, 21–23
stereoelectronic analysis of enzyme reactions, 27, 38–70
transition-state stabilization and charge-relay system, 130–136
trypsin and pancreatic trypsin inhibitor (PTI), 137
ZGP and, 16–18
zinc proteases, 1–23
See also Iron protoporphyrin-IX group species; Stereoelectronics; Zinc proteases
Epoxidation, of alkenes, 262–272
Epoxides, rate accelerations and strain, 102–103
Equatorial position, stereoelectronic position regarding, 45
Erythrose 4-phosphate, 54

Esterolytic reactions, zinc-coordinated water and, 3–5
5-Ethyl-4a,5-dihydro-4a-hydroperoxy-lumiflavin, 338–339
EVB approach, charge-relay catalysis, 146–152
Evolution,
   of enzymes, 39
   of proteins, 39–40
"Exo anomeric effect", 56–59

F

Fatty acid synthetase, 49
FBE (Formal bonding enthalpy), 78
Ferrochelatase, 292–296
   inhibition of and chlorophyll biosynthesis, 307–309
   lead poisoning, 310
   pathological conditions, 309–310
   structural studies of, 297–307
   zinc protoporphyrin IX formation, 310
Flavin mixed-function oxidases,
   6-amino-5-oxo-$3H,5H$-uracil intermediates, 321–326
   chemical studies of, 316–318
   4a,5-dihydro-4a-hydroperoxy-5-alkyllumiflavins, 331–338
   1,5-dihydroflavins and, 326–331
   4a,5-dihydro-4a-hydroperoxyflavin synthesis, 321–326
   5-ethyl-4a,5-dihydro-4a-hydroperoxylumiflavin, 338–339
   historical background, 319–320
   synthesis of, 321–326
Flavins,
   6-amino-5-oxo-$3H,5H$-uracil intermediates, 321–326
   carbonyl functions and nucleophilic additions, 339–342
   chemical studies of, 316–318
   4a,5-dihydro-4a-hydroperoxy-5-alkyllumiflavins, 331–338
   1,5-dihydroflavins and, 326–331
   4a,5-dihydro-4a-hydroperoxyflavin, 321–326
   5-ethyl-4a,5-dihydro-4a-hydroperoxylumiflavin, 338–339

Flavins (*cont.*)
  hydroxylation of aromatic rings,
    342–349
  reduction of, 58
  See also Flavin mixed-function
    oxidases
Flavoenzymes,
  5-ethyl-4a,5-dihydro-4a-hydro-
    peroxylumiflavin, 338–339
  hydroxylation of aromatic rings,
    342–349
FME (Formal medium enthalpy), 78
Formal bonding enthalpy (FBE), 78
Formal medium enthalpy (FME), 78
Formal pair enthalpy (FPE), 78
Formal steric enthalpy (FSE), 78
Formaldehyde, 353–382
  dehydrogenase activity, 378–382
  distribution of in cells, 358
  glyoxalase enzyme system and,
    358–369
Formaldehyde dehydrogenase, 380–
  381
  catalytic mechanisms, 381
  kinetic mechanism and kinetic
    properties, 380–381
  substrate specificity, 380
S-Formylglutathione hydrolase en-
  zyme system, 378–382
  catalytic mechanisms, 382
  physiological function, 379–380
  substrate specificity, 381
Fourier methods, *See* Difference
  Fourier method
FPE (Formal polar enthalpy), 78
Franck-Condon principle, 174–177
FSE (Formal steric enthalpy), 78
Fucose dehydrogenase, 67, 70
Fumarase, 48

G

Galactose dehydrogenase, 67
Gem-diol, 14, 15
Generalized anomeric effect,
  definition of, 35
  See also Anomeric effect
Glu-270 in CPA, 18–21
Glu-72 in CPA, 19–21

Glucose dehydrogenase, 67
Glucose-6-phosphate dehydrogenase,
  67, 68
Glutathione, 353–382
  distribution of in cells, 356
  glyoxalase enzyme system and,
    358–369
Glutathione reductase, 58, 59
Glycosides, stereoelectronics of hydrolyses, 66–70
Glycosidic bond, 56, 63
Glycyl-L-tyrosine, 3–5
Glyoxalase enzyme system and, 358–
  369
  glyoxalase I, 359–368
  glyoxalase II, 368–369
  physiological function, 358–359
  See also Glyoxalase I; Glyoxalase
    II
Glyoxalase I, 359–368
  α-Carbanion stabilization by diva-
    lent sulfur, 365–368
  catalytic mechanism, 360–364
  kinetic mechanism, 359–360
  kinetic properties, 360
  model reactions, 364–365
  substrate specificity, 359, 377–378
Glyoxalase II, 368–369
  catalytic mechanisms, 369
  substrate specificity and kinetic
    properties, 368
Glyoxalase pathway, optimization of
  efficiency in, 373
Ground state destabilization, 79
Group transfer reactions, 59

H

Heavy-atom reorganization,
  coupling modes, 127–130
  MAR diagrams and, 126–133
  in protolytic catalysis, 124–133
Heme formation, 289–292
Heteroatom oxygenation, 5-ethyl-
  4a,5-dihydro-4a-hydroperoxy-
  lumiflavin, 338–339
Higher valent iron-oxo species, 230–
  234

Higher valent manganese-oxo species, 234–236
His-69 in CPA, 19–21
Histidine ammonia-lyase, addition/ elimination reaction, 48
Histidine residues, chymotrypsin catalysis and, 86–87
HOMO-LUMO interactions, of electron transfer, 177–180
pro-R Hydrogen, 57–58
pro-S Hydrogen, 57–58
Hydrolysis,
    of N-acylimidazoles, 3
    of chymotrypsin and other enzymes, 92–95
    Rogers and Bruice model, 142–143, 145–146
    Zinc exoprotease carboxypeptidase A (CPA) and, 18–19
Hydroperoxides,
    catalase reaction, 245–248
    manganese (III) porphyrin reactions, 248–249
    metal (III) porphyrins and, 236–245
Hydroxamic acid inhibitors, 3
Hydroxy acid sets, and Menger's criteria, 82–84
Hydroxydecanoylthioester dehydratase, 49
β-Hydroxyisobutyrate, 49
Hydroxylation, of aromatic rings, 342–349
2-hydroxymethylbenzamide, lactonization of, 88–89
Hydroxyphenylpyruvate, 54, 55, 70
    biosynthesis of, 54

I

Imidazole,
    catalysis processes, 88–89
    protonated versus unprotonated, 34–36
Imidazoles, and alternative enzyme mechanisms, 85
Imidazoyls, conformational behaviors, 34–35
Inhomogeneous self-consistent reac-

tion field (ISCRF) approach, 153–154
Intermolecular electron transfer, 204–214
Intramolecular anhydride formation, 80
Intramolecular electron transfer, 196–204
Intramolecular lactonization, 83
Intramolecular reactions, 75–111
    between reagents, 79–82, 95–96
    and catalysis of enzymes, 76–112
    of chymotrypsin and other enzymes, 92–95
    lone pair exchange repulsion (LPER), 109
    lysozyme catalysis and strain, 105–109
    Menger's criteria and, 82–84
    nonprotein systems and strain-induced accelerations, 111–112
    proteins and catalysis, 109–111
    rate accelerations and proximity, 95–103
    rate accelerations and strain, 102–112
    rate specificities of enzymes and mimics, 96–99
    spatiotemporal hypothesis, 84–85
    strain and, 102–110
Iron, See Metals; Porphyrins
Iron protoporphyrin-IX group species,
    catalase reaction, 245–248
    chemistry of, 228–230
    dealkylation reactions, 253–262
    N,N-dimethylaniline N-oxides reactions, 255–262
    enzymes of, 228–230
    epoxidation of alkenes, 262–272
    higher valent iron-oxo chemical preparation, 230–233
    higher valent iron-oxo electrostatic generation, 233–234
    higher valent manganese-oxo species, 234–236
    manganese (III) porphyrin reactions, 248–249
    mechanism of oxygen transfer, 236–245

Iron protoporphyrin-IX group species
    (*cont.*)
    rebound mechanism for oxygen
        insertion in C-H bonds, 249–
        253
ISCRF approach, 153–154

K

β-keto acids, decarboxylations of, 43–
    46
Keto acids, enzyme mimic, 98–101
Keto steroids, stereoelectronics of,
    36–39
Ketones, 12–17
    keto/hydrate binding, 15
    stereoelectronic rules and, 49–51
    See also specific names
Kinetic anomeric effect, 36–38
    definition of, 36
    See also Anomeric effect

L

Lactamase, 62–63
D-Lactate, methylglyoxal conversion,
    369–373
Lactonization,
    of 2-hydroxymethylbenzamide, 88–
        89
    Menger's criteria and, 82–84
    rate accelerations and strain, 103–
        108
Lanosterol, conversion to cholesterol,
    44–45
Lanosterol oxidative decarboxylation,
    44
Lead poisoning, 310
Ligands,
    Desulfovibrio cytochromes, 202
    difference Fourier method in study
        of, 5–11
    See also Enzyme-ligand interaction
Lone pair exchange repulsion, 108–
    109
Lone pair exchange repulsion
    (LPER), and electrostatic strain,
    109
Low temperature crystallography, 22

LPER (Lone pair exchange repul-
    sion), 108–109
Lysozyme, 105–109
    rate accelerations and strain, 105–
        109

M

Malease, 48
    addition/elimination reaction, 48
Manganese,
    higher valent manganese-oxo spe-
        cies, 234–236
    manganese (III) porphyrin reac-
        tions, 248–249
    rebound mechanism for oxygen
        insertion in C-H bonds, 249–
        253
    See Metals; Porphyrins
Map of alternate route, *See* MAR
    diagrams
MAR diagrams, 126–133
Marcus inverted region, 169
Menger's criteria, 82–84
Mesaconate, 51
Metalation,
    chlorophylls and, 286–287
    cobalamins and, 287–288
    enzymatic, 287–300
    ferrochelatase, 292–307
    ferrochelatase inhibition of and
        chlorophyll biosynthesis, 307–
        309
    heme formation, 289–292
    metal-ligand dissociation, 283–284
    outer-sphere complexation, 282–283
    porphyrin reactions, 279–311
    porphyrin ring distortion, 281–282
    in vitro, 280–286
    in vivo, 286–307
Metals,
    catalase reaction, 245–248
    dealkylation reactions, 253–262
    N,N-dimethylaniline N-oxides reac-
        tions, 255–262
    electron transfer in protein-organo-
        metallic complexes, 198–199
    epoxidation of alkenes, 262–272

ferrochelatase and zinc protoporphyrin IX formation, 310
ferrochelatase inhibition of and
chlorophyll biosynthesis, 307–
309
formation of complexes in vivo,
286–307
manganese (III) porphyrin reactions, 248–249
mechanism of oxygen transfer, 236–
245
metal (III) porphyrins, 236–245
metal-ligand dissociation, 283–284
pathological conditions associated
with ferrochelatase, 309–310
porphyrin metalation reactions in
biochemistry, 279–311
porphyrin ring distortion, 281–282
rebound mechanism for oxygen
insertion in C-H bonds, 249–
253
in vitro porphyrin reactions, 280–
286
See also Iron protoporphyrin-IX
group species
Methacrylate, 49
conversion to β-hydroxyisobutyrate, 49
Methyl aspartase, 51
L-erythro-β-methylaspartate, 51
Methyl glutaconyl CoA hydratase, 49
Methyl groups, removal from C-4 of
steroids, 44–45
Methyl transfer, 60
Methylaspartate, 51
Methylglyoxal, 353–382
conversion to d-lactate, 369–373
distribution of in cells, 357–358
glyoxalase enzyme system and,
358–369
Methylglyoxal conversion to d-lactate, 369–373
3-Methyl-5-hydroxypent-2-enoic acid,
52
Mevalonic acid, 52
conversion to 3-methyl-5-hydroxy-
pent-2-enoic acid, 52
Micelle, 101–102
Michaelis complex, 14

Missense mutations, of S. cerevisiae,
216–217
Mixed-function oxidases,
catalase reaction, 245–248
dealkylation reactions, 253–262
higher valent iron-oxo chemical
preparation, 230–233
higher valent iron-oxo electrostatic
generation, 233–234
higher valent manganese-oxo species, 234–236
manganese (III) porphyrin reactions, 248–249
mechanism of oxygen transfer, 236–
245
See also Iron protoporphyrin-IX
group species
Modeling, in study of charge-relay
system, 123–124
of charge-relay catalysis, 141–146
Molecular dynamics, 22
site-directed mutagenesis (SDM)
and, 21–23
Molecular geometry,
of electron transfer in cytochromes,
208–211
lone pair exchange repulsion
(LPER), 109
Molecular orbital calculations,
HOMO-LUMO interactions, 177–
180
nicotinamide cofactors, 56–59
spin state and electron transfer, 195
stereoelectronics and, 27–71
See also Electron transfer
Molecular orbital theory,
"exo anomeric effect", 56–59
transition states and enolization,
41–42
Molecular oxygen, 1,5-dihydroflavins
and, 326–331
Monoisopropylphosphoryl (MIP)
group, 137
Monooxygen transfer, from 4a,5-
dihydro-4a-hydroperoxy-5-alkyl-
lumiflavins, 331–338
Muconate cycloisomerase, 49, 50
cis-cis-muconate cycloisomerase,
catalyzed reactions, 49–50

Multiheme cytochromes, 215–216
Myoglobin, 6
    difference Fourier method in study
        of, 6

N

Natural selection, 39
Nicotinamide adenine dinucleotide, 56
Nicotinamide cofactors, 56–59
Nitrophenylacetate, catalysis, 101–103
NMR spectroscopy,
    of catalysis, 138–141
    of diisopropyl fluorophosphate
        (DFP) derivatives, 138–140
    in study of charge-relay system,
        123–124
NMR studies,
    of conformational changes in pro-
        teins, 199–204
    of electron transfer, 177–185
    of multiheme cytochromes, 216–217
    of protein conformational change,
        170
Noncovalently enforced proximity,
    rate accelerations and, 95–103
Noncovalently enforced strain, rate
    acceleration and, 104–112
Nonoptimal orbital overlap, cyclopro-
    pane, 78
Nuclear magnetic resonance, *See*
    NMR spectroscopy
Nucleophic displacement, 59
Nucleophilic additions, carbonyl func-
    tions and, 339–342

O

Olefins,
    biosynthesis of, 47–53
    twisting of, 40
Oleic acid,
    addition/elimination reaction, 48
    dehydratase, 48
Orbital overlap effects, 33–37, 41
Orbital steering, 82–84

Organic chemistry,
    stereoelectronics and, 29–40
    traditional versus stereoelectronic
        theory, 29–32
Oxacyclohexanes, substituent effects,
    31
Oxidoreductases, 56–59
Oxygen,
    catalase reaction with iron proto-
        porphyrin-IX group species,
        245–248
    4a,5-dihydro-4a-hydroperoxy-5-
        alkyllumiflavins, 331–338
    1,5-dihydroflavins and, 326–331
    N,N-dimethylaniline N-oxides reac-
        tions, 255–262
    epoxidation of alkenes, 262–272
    5-ethyl-4a,5-dihydro-4a-hydro-
        peroxylumiflavin, 338–339
    mechanism of oxygen transfer, 236–
        245
    rebound mechanism for oxygen
        insertion in C-H bonds, 249–
        253

P

p orbitals, stereoelectronics and, 29–
    32
P-450, *See* Cytochrome P-450
Pancreatic trypsin inhibitor (PTI), 137
PDLD method, charge-relay catalysis,
    152
Peptidase G, 2
Peptide bonds, CPA interactions, 18–
    19
Percarboxylic acids, mechanism of
    oxygen transfer, 236–245
Peroxidase,
    iron protoporphyrin-IX group
        chemistry, 228–230
    See also Iron protoporphyrin-IX
        group species
Phase angle, 8, 9
Phenylalanine ammonia lyase, 48, 52
Phenylalanine ammonia-lyase, addi-
    tion/elimination reaction, 48
Phosphonamidate inhibitors, 16–18

Phosphoramides, 2–3
Phosphoryl transfer, 60
Pi orbital, and decarboxylations of
    beta-keto acids, 44–45
Pi system,
  and carboxyl group, 44–45
  stereoelectronics and, 29–32
  triose-phosphate isomerase cata-
      lyzed reactions, 41–43
PMSF, NMR spectroscopy of, 138–
    139
Porphyrins,
  catalase reaction, 245–248
  enzyme reactions, 228–230
  ferrochelatase inhibition of and
      chlorophyll biosynthesis, 307–
      309
  higher valent iron-oxo chemical
      preparation, 230–233
  higher valent iron-oxo electrostatic
      generation, 233–234
  higher valent manganese-oxo spe-
      cies, 234–236
  manganese (III) porphyrin reac-
      tions, 248–249
  mechanism of oxygen transfer, 236–
      245
  metalation reactions in biochemis-
      try, 279–311
  metal-ligand dissociation, 283–284
  outer-sphere complexation, 282–283
  proton loss and nitrogen atoms, 284
  ring distortion, 281–282
  in vitro metalation, 280–286
  in vivo metalation, 286–307
  See also Flavin mixed-function
      oxidases; Iron protoporphyrin-
      IX group species
Proteases, 62
Proteins,
  amino acid residues, 38–40
  Asn-144 in CPA, 19–21
  catalysis and, 109–111
  conformational change, 170, 199–
      204
  electron transfer in protein-organo-
      metallic complexes, 198–199
  electron transfer in protein-protein
      complexes, 196–198

evolutionary implications, 39–40
Glu-72 in CPA, 19–21
His-69 in CPA, 19–21
protein dipoles-Langevin dipoles
    (PDLD) approach, 150, 152
protein-ligand interaction and differ-
    ence Fourier method, 5–11
Ser-197 in CPA, 19–21
structure of cytochromes, 171–174
Tyr-248 in CPA, 19–22
Proteolytic reaction, N-[[[(Benzyloxy-
    carbonyl)amino]methyl]hydroxy-
    phosphinyl]-L-phenylalanine, 16–
    18
Proteolytic reactions,
  of zinc proteases, 4–5
  zinc-coordinated water and, 3–5
Protocatechuate, 49
Protolytic catalysis, 120–164
  basic issues in, 124–130
  charge-relay system, 130–136
  charge-relay system and, 120–124
  coupling modes, 127–130
  crystallographic studies, 136–138,
      140–141
  HAR (heavy atom reorganization),
      124–133
  MAR diagrams and, 126–133
  NMR spectroscopy, 138–141
  PT (proton transfer) in, 124–133
  transition-state stabilization and,
      130–133
Protolytic hydrolysis, 7 + 10 model,
    143–146
Proton transfer,
  coupling modes, 127–130
  in enzyme catalysis, 119–123
  MAR diagrams and, 126–133
  in protolytic catalysis, 124–133
Proton-inventory studies, 156–163
  in study of charge-relay system, 124
Proximity, 95
  definition of, 79
  rate accelerations and, 95–103
Pseudomonas, 48, 67
Pseudomonas cytochromes, 215–216
Pyridoxal enzymes, 43, 46–47
Pyridoxamine, 98–100
5-pyrophosphomevalonate, 48

R

Rate accelerations,
  lysozyme and, 105–109
  nonprotein systems, 111–112
  proteins and catalysis, 109–111
  proximity and, 95–103
  strain and, 102–112
Reagents,
  catalysis and intramolecular reactions, 79–82
  rate accelerations and proximity, 95–103
Rebound mechanism, for oxygen insertion in C-H bonds, 249–253
Redox potential, and electron transfer, 184–188
Reverse anomeric effect, 34–36
  definition of, 34
Ribonuclease, 61
tRNA synthetase, 109
Rogers and Bruice model, of charge-relay catalysis, 142–143, 145–146

S

S. cerevisiae, 216–217
  axial ligands and, 203
S. griseus, 138
Schiff's base, and decarboxylations of beta-keto acids, 43–44
SCSSD method, charge-relay catalysis, 150–152
SDM, *See* site-directed mutagenesis (SDM)
Ser-197, 19–21
Serine proteases, charge-relay system and, 123–124
Serine residues, chymotrypsin catalysis and, 86–87
7 + 10 model of charge-relay catalysis, 143–146
Site-directed mutagenesis, 21, 110, 170
Site-directed mutagenesis (SDM), 21–23
6 + 8 model of charge-relay catalysis, 142–143, 145–146
Solvent effect, 34

Spatiotemporal hypothesis, 84–85
Spectroscopy,
  structural studies of ferrochelatase, 297–307
  studies of catalysis, 138–141
  in study of charge-relay system, 123–124
  See also NMR spectroscopy
Spin state, and electron transfer, 195
Stereoelectronics, 27–71
  aldopyranose dehydrogenases, 66–70
  anomeric effect, 31–33
  anti and syn eliminations, 47–48
  carbonyl group reactions, 61–63
  definition of, 27–29
  dihydrophenylalanine and, 52–53
  and enolizations, 41–43
  of enzyme reactions, 37–55
  and evolutionary biology, 39–40
  first rule, 42
  glycosidic centers in reactions, 63–66
  group transfer reactions, 59–61
  large effects in enzyme reactions, 41–43
  magnitude of effects, 40–41
  medium-sized stereoelectronic effects, 47–56
  and nicotinamide cofactors, 56–59
  in organic chemistry, 28–37
  oxidoreductases, 56–59
  principles of, 28–41
  of pyridoxal enzymes, 46–47
  second rule, 44
  small stereoelectronic effects, 56–70
  theory of, 27–28
  triose-phosphate isomerase catalyzed reactions, 41–43
  vinylogous dehydrations, 53–56
  See also Anomeric effect
Steroid biosynthesis, 45–46
Strain, 78, 102
  definition of, 77–79
  ground state destabilization and, 79
  lactonization effects, 103
  lone pair exchange repulsion (LPER), 109

lysozyme and rate accelerations,
105–109
nonprotein systems and strain-
induced accelerations, 111–112
rate accelerations and, 102–112
Stress, definition of, 104
Structural anomeric effect,
definition of, 31
See Anomeric effect
Structural theory, definition of, 28–29
Substituent effects, 31
Sugars, substituent effects, 31–32
Syn elimination, 49–52

T

Temperature, and electron transfer,
189–190
Tetrahedral intermediate, 15, 17, 18
TFP, *See* 2-Benzyl-4-oxo-5,5,5-tri-
fluoropentanoic acid (TFP)
Thermolysin, 2, 4, addendum
Thioesters, stereoelectronic rules and,
49–51
L-*threo*-β-methylaspartate, 51
TLN, *See* Zinc endoprotease thermolysin (TLN)
Transamination, 98–100
Transition state stabilization, 1, 3, 5
in charge-relay catalysis, 130–133
and HAR (heavy atom reorganization) in protolytic catalysis,
124–133
spatiotemporal hypothesis and, 85
strain and, 79
Transition states,
analogues of, 109
anti and syn, 47–48
and enolization, 41–42
MAR diagrams and, 126–133
stabilization, 1, 3, 5, 79, 85, 124–
133
zinc ion analogues, 16–18
Triacetoxyxylopyranose, and imidazoyl group, 34–35
Triose-phosphate isomerase, catalyzed reactions, 41–43
Trypsin, NMR evidence, 137

Twisted molecular bonds, stereoelectronics and, 29–30
Tyr-248 in CPA, 19–21

V

Vinylogous dehydrations, 53–56

W

Water, zinc ion and water molecules,
3–5

X

X-ray crystallographic studies, 1–13,
136–138, 140–141
and anomeric effect, 33–34
of CPA, 4–5
Glycyl-L-tyrosine, 3–5
of ketones, 12–17
protolytic catalysis by enzymes,
120–124
site-directed mutagenesis (SDM)
and, 21–23
of TLN, 4–5
of zinc proteases, 1–2

Z

Zero coupling, HAR and PT in protolytic catalysis and, 124–133
ZGP (N-[[[(Benzyloxycarbonyl)-
amino]methyl]hydroxyphos-
phinyl]-L-phenylalanine), 16–18
Zinc, 1–23
catalytic role of, 2–5
ferrochelatase and zinc protoporphyrin IX formation, 310
zinc-carbonyl interactions, 3–5
zinc-coordinated water, 3–5
Zinc endoprotease thermolysin
(TLN), 2
X-ray crystallographic studies, 4–5
See also Zinc exoprotease carboxypeptidase A; Zinc proteases

Zinc exoprotease carboxypeptidase A
(CPA), 11–21
Arg-127 and, 18–19
Asn-144 and, 19–21
BFP and, 12–17
BMP and, 12–17
BOP and, 12–17
CPA-BMP complex, 18–20
Glu-270 and, 18–19
Glu-72 and, 19–21
His-69, 19–21
hydrolytic reactions, 18–19
inhibitor interactions, 12–14
Ser-197, 19–21
site-directed mutagenesis (SDM)
and, 21–23
TFP and, 12–17
Tyr-248, 19–21

X-ray crystallographic studies, 2,
4–5
ZGP and, 16–18
Zinc ions,
Glycyl-L-tyrosine and, 3–5
ion-dipole interactions, 3–4
proteolytic reactions, 4–5
transition state analogues, 16–18
ZGP and, 16–18
Zinc proteases, 1–23
D-alanyl-D-carboxypeptidase, 2
binding of, 1–2
definition of, 1
role of zinc ion, 2–5
site-directed mutagenesis (SDM),
21–23
zinc exoprotease carboxypeptidase
A (CPA), 2